ELSEVIER'S DICTIONARY OF REPTILES

ELSEVIER'S DICTIONARY OF REPTILES

in

Latin, English, German, French and Italian

compiled by

M. WROBEL
London, United Kingdom

2004

ELSEVIER

Amsterdam – Boston – Heidelberg – London – New York – Oxford
Paris – San Diego – San Francisco – Singapore – Sydney – Tokyo

ELSEVIER B.V.
Sara Burgerhartstraat 25
P.O. Box 211
1000 AE Amsterdam
The Netherlands

ELSEVIER Inc.
525 B Street, Suite 1900
San Diego, CA 92101-4495
USA

ELSEVIER Ltd
The Boulevard, Langford Lane
Kidlington, Oxford OX5 1GB
UK

ELSEVIER Ltd
84 Theobalds Road
London WC1X 8RR
UK

First edition 2004

Library of Congress Cataloging in Publication Data
A catalog record is available from the Library of Congress.

British Library Cataloguing in Publication Data
A catalogue record is available from the British Library.

ISBN: 0-444-51499-6

∞ The paper used in this publication meets the requirements of ANSI/NISO Z39.48-1992 (Permanence of Paper).
Printed in The Netherlands.

Preface

This dictionary has been compiled with the aim of giving an overview of the English, German, French and Italian names of reptiles. The Basic Table contains, in alphabetical order, the scientific names of families, genera, species and some sub-species with, again in alphabetical order, their identified names in English, French, German, and Italian, which are given in the singular for species and sub-species and in the plural for other terms. The relevant order and family are shown for each term and the taxonomy is that of the sources consulted.

The criteria to determine which names are hyphenated or shown as two or more words are based on the frequency they are used by, and the relative importance I have attached to, the sources consulted.

My special thanks go to Katayoun Asadollahzadeh-Sarab who prepared this work for publication and to my wife Myra for her support and encouragement.

Murray Wrobel

Contents

Abbreviations

The following abbreviations denote a general, but not necessarily exclusive, use in the various areas.

(Ants)	-	One or more of the French-speaking islands of the West Indies
(ANZ)	-	Australia and New Zealand
(CSA)	-	Central and Southern Africa
(ISC)	-	Indian Subcontinent
(NA)	-	North America
(Qué)	-	Quebec
(SEA)	-	South East Asia
(VIP)	-	Virgin Islands and Puerto Rico
(WI)	-	One or more of the English-speaking islands of the West Indies

Basic Table

A

1 ***Ablepharus***
Squamata - Scincidae
e Ocellated Skinks; Lidless Skinks; Snake-eyed Skinks
d Natteraugenskinke; Natteraugen
f Abléphares

2 ***Ablepharus bivitattus***
Squamata - Scincidae
e Twin-striped Skink; Two-streaked Snake-eyed Skink
d Gestreiftes Natterauge
f Abléphare à deux bandes

3 ***Ablepharus deserti***
Squamata - Scincidae
e Desert Lidless Skink
i Ablefaro bivittato

4 ***Ablepharus grayanus***
Squamata - Scincidae
e Minor Snake-eyed Skink
d Kleines Natterauge

5 ***Ablepharus kitaibelii***
Squamata - Scincidae
e Juniper Skink; Balkan Skink; Snake-eyed Skink; European Copper Skink
d Johannisechse
f Abléphare de Kitaibel; Abléphare sans paupières; Scinque sans paupières; Scinque abléphare
i Scinco occhi di serpente

6 ***Ablepharus pannonicus***
Squamata - Scincidae
e Asian Snake-eyed Skink
d Asiatisches Natterauge
i Ablefaro de Kitaibel

7 ***Abronia***
Squamata - Anguidae
e Arboreal Alligator Lizards; Abronias; Gutezpaslias (NA)
d Baumschleichen; Bromelienschleichen
f Abronies

8 ***Abronia anzuetoi***
Squamata - Anguidae
e Anzuetoi Arboreal Alligator Lizard

9 ***Abronia aurita***
Squamata - Anguidae
e Cope's Arboreal Alligator Lizard

10 ***Abronia bogerti***
Squamata - Anguidae
e Bogert's Arboreal Alligator Lizard; Bogert's Abronia

11 ***Abronia campbelli***
Squamata - Anguidae
e Campbell's Arboreal Alligator Lizard

12 ***Abronia chiszari***
Squamata - Anguidae
e Chiszar's Arboreal Alligator Lizard

13 ***Abronia deppei***
Squamata - Anguidae
e Deppe's Arboreal Alligator Lizard; Gutezpaslia (NA)
f Abronie de Deppe

14 ***Abronia fuscolabialis***
Squamata - Anguidae
e Deppe's Arboreal Alligator Lizard

15 ***Abronia gaiophantasma***
Squamata - Anguidae
e Brilliant Arboreal Alligator Lizard

16 ***Abronia graminea***
Squamata - Anguidae
e Terrestrial Arboreal Alligator Lizard

17 ***Abronia leurolepis***
Squamata - Anguidae
e White-scaled Arboreal Alligator Lizard

18 ***Abronia lythrochila***
Squamata - Anguidae
e Red-lipped Arboreal Alligator Lizard

19 ***Abronia matudai***
Squamata - Anguidae
e Matuda's Arboreal Alligator Lizard

20 ***Abronia mitchelli***
Squamata - Anguidae
e Mitchell's Arboreal Alligator Lizard

21 ***Abronia mixteca***
Squamata - Anguidae
e Mixtecan Arboreal Alligator Lizard

22 ***Abronia montecristoi***
Squamata - Anguidae
e Montecristo Arboreal Alligator Lizard

23 ***Abronia oaxacae***
Squamata - Anguidae
e Oaxacan Arboreal Alligator Lizard

24 ***Abronia ochoterenai***
Squamata - Anguidae
e Northern Chioapas Arboreal Alligator Lizard

25 ***Abronia ornatus***
Squamata - Anguidae
e Ornate Arboreal Alligator Lizard

26 ***Abronia ornelasi***
Squamata - Anguidae
e Cerro Baul Arboreal Alligator Lizard

27 ***Abronia ramirezi***
Squamata - Anguidae
e Ramirez's Arboreal Alligator Lizard

28 ***Abronia reidi***
Squamata - Anguidae
e Reid's Arboreal Alligator Lizard

29 ***Abronia salvadorensis***
Squamata - Anguidae
e Salvador Arboreal Alligator Lizard

30 ***Abronia smithi***
Squamata - Anguidae
e Smith's Arboreal Alligator Lizard

31 ***Abronia taeniata***
Squamata - Anguidae
e Bromeliad Arboreal Alligator Lizard; Streaked Gutezpalia; Bromeliad Alligator Lizard
d Gestreifte Baumschleiche
f Abronie à bandes

32 ***Abronia vasconcelosi***
Squamata - Anguidae
e Bocourt's Arboreal Alligator Lizard

33 ***Acalyptophis***
Squamata - Elapidae
e Perron's Sea Snakes

34 ***Acalyptophis peroni***
Squamata - Elapidae
e Perron's Sea Snake

35 ***Acanthochelys***
Testudines - Chelidae
e South American Sideneck Turtles
d Plattschildkröten
f Platémydes

36 ***Acanthochelys macrocephala***
Testudines - Chelidae
e Pantanal Swamp Turtle; Big-headed Pantanal Swamp Turtle; Pantanal Big-headed Turtle
d Großkopfplattschildkröte
f Platémyde à grosse tête

37 ***Acanthochelys pallidipectoris***
Testudines - Chelidae
e Chaco Sideneck Turtle; Chaco Side-necked Turtle
d Spornplattschildkröte; Sporenplattschildkröte
f Platémyde à éperons

38 ***Acanthochelys radiolata***
Testudines - Chelidae
e Brazilian Radiolated Swamp Turtle
d Strahlenplattschildkröte
f Platémyde radiolée

39 ***Acanthochelys spixii***
Testudines - Chelidae
e Spinyneck Swamp Turtle; Black Spine-necked Swamp Turtle; Spix's Snake-necked Turtle; Black Spiny-necked Swamp Turtle; Spiny-necked Swamp Turtle; Spiny-necked Turtle
d Stachelhalsplattschildkröte
f Platémyde de Spix

40 ***Acanthodactylus***
Squamata - Lacertidae
e Fringe-fingered Lizards; Sand Spine-toed Lizards; Fringe-toed Lizards
d Fransenfinger; Fransenfingereidechsen
f Acanthodactyles

41 ***Acanthodactylus arabicus***
Squamata - Lacertidae
e Arabian Fringe-fingered Lizard

42 ***Acanthodactylus aureus***
Squamata - Lacertidae
e Golden Fringe-fingered Lizard; Golden Fringe-toed Lizard
f Acanthodactyle doré

43 ***Acanthodactylus bedriaga***
Squamata - Lacertidae
e Bédriaga's Fringe-fingered Lizard

44 ***Acanthodactylus blanci***
Squamata - Lacertidae
e White Fringe-fingered Lizard

45 ***Acanthodactylus blanfordii***
Squamata - Lacertidae
e Blanford's Fringe-fingered Lizard

46 ***Acanthodactylus boskianus***
Squamata - Lacertidae
e Bosc's Fringe-fingered Lizard; Bosc's Lizard; Bosc's Spinefoot; Bosc's Fringe-toed Lizard; Bosc's Sandracer
d Bosc-Fransenfinger
f Acanthodactyle de Bosc; Acanthodactyle rugueux
i Acantodattilo di Bosk

47 ***Acanthodactylus boueti***
Squamata - Lacertidae
e Chabanaud's Fringe-fingered Lizard

48 ***Acanthodactylus busacki***
Squamata - Lacertidae
e Busack's Fringe-fingered Lizard; Busack's Leopard Fringe-toed Lizard
f Acanthodactyle de Busack

49 ***Acanthodactylus cantoris***
Squamata - Lacertidae
e Indian Fringe-fingered Lizard; Indian Fringe-toed Lizard (ISC); Indian Fringe-toed Sand Lizard
d Dünenfransenfinger; Pakistan-Fransenfinger

50 ***Acanthodactylus dumerili***
Squamata - Lacertidae

e Duméril's Fringe-fingered Lizard; Duméril's Fringe-toed Lizard
f Acanthodactyle de Duméril

51 ***Acanthodactylus erythrurus***
Squamata - Lacertidae
e Fringe-fingered Lizard; Fringe-toed Lizard; Common Fringe-toed Lizard; Spiny-footed Lizard; European Fringefinger
d Europäischer Fransenfinger; Gewöhnlicher Fransenfinger; Sägefinger
f Acanthodactyle commun
i Acantodattilo eritruro; Acantodatillo coda-rossa

52 ***Acanthodactylus felicis***
Squamata - Lacertidae
e Cat Fringe-fingered Lizard

53 ***Acanthodactylus gongrorhynchatus***
Squamata - Lacertidae
e Saudi Fringe-fingered Lizard

54 ***Acanthodactylus grandis***
Squamata - Lacertidae
e Giant Fringe-fingered Lizard; Giant Fringe-toed Lizard

55 ***Acanthodactylus guineensis***
Squamata - Lacertidae
e Guinea Fringe-fingered Lizard

56 ***Acanthodactylus haasi***
Squamata - Lacertidae
e Haas's Fringe-fingered Lizard

57 ***Acanthodactylus longipes***
Squamata - Lacertidae
e Long Fringe-fingered Lizard; Long-footed Fringe-toed Lizard
f Acanthodactyle à longs pieds

58 ***Acanthodactylus maculatus***
Squamata - Lacertidae
e Spotted Fringe-fingered Lizard; Spotted Leopard Fringe-toed Lizard
d Fransenfinger
f Acanthodactyle tacheté

59 ***Acanthodactylus masirae***
Squamata - Lacertidae
e Masira Fringe-fingered Lizard

60 ***Acanthodactylus micropholis***
Squamata - Lacertidae
e Yellowtail Fringe-fingered Lizard; Yellow-tailed Lizard; Persian Fringe-toed Lizard

61 ***Acanthodactylus opheodurus***
Squamata - Lacertidae
e Arnold's Fringe-fingered Lizard

62 ***Acanthodactylus pardalis***
Squamata - Lacertidae
e Egyptian Fringe-fingered Lizard; Leopard Fringe-toed Lizard; Egyptian Leopard Lizard
d Gefleckter Fransenfinger; Pantherfransenfinger; Gefleckter Sandläufer

63 ***Acanthodactylus robustus***
Squamata - Lacertidae
e Robust Fringe-fingered Lizard

64 ***Acanthodactylus savignyi***
Squamata - Lacertidae
e Savigny's Fringe-fingered Lizard; Savigny's Spinefoot; Savigny's Fringe-toed Lizard
f Acanthodactyle de Savigny

65 ***Acanthodactylus schmidti***
Squamata - Lacertidae
e Schmidt's Fringe-fingered Lizard; Schmidt's Fringe-toed Lizard

66 ***Acanthodactylus schreiberi***
Squamata - Lacertidae

e Schreiber's Fringe-fingered Lizard

67 ***Acanthodactylus scutellatus***
Squamata - Lacertidae
e Nidua Fringe-fingered Lizard; Nidua Lizard; Shielded Spinefoot; Nidua Fringe-toed Lizard; Fringe-toed Lizard
f Acanthodactyle pommelé

68 ***Acanthodactylus spinicauda***
Squamata - Lacertidae
e Domergue's Fringe-fingered Lizard

69 ***Acanthodactylus tilburyi***
Squamata - Lacertidae
e Tilbury's Fringe-fingered Lizard

70 ***Acanthodactylus tristrami***
Squamata - Lacertidae
e Lebanon Fringe-fingered Lizard

71 ***Acanthodactylus yemenicus***
Squamata - Lacertidae
e Yemen Fringe-fingered Lizard

72 ***Acanthophis***
Squamata - Elapidae
e Death Adders; Deaf Adders (ANZ)
d Todesottern
i Vipere della morte

73 ***Acanthophis antarcticus***
Squamata - Elapidae
e Common Death Adder; Death Adder; Southern Death Adder; Deaf Adder (ANZ)
d Todesotter; Gewöhnliche Todesotter
i Vipera della morte

74 ***Acanthophis antarcticus antarcticus***
Squamata - Elapidae
d Australische Todesotter

75 ***Acanthophis antarcticus laevis***
Squamata - Elapidae
e New Guinea Death Adder
d Neuguinea Todesotter

76 ***Acanthophis praelongus***
Squamata - Elapidae
e Northern Death Adder
d Nôrdliche Todesotter
i Vipera della morte del Nord

77 ***Acanthophis pyrrhus***
Squamata - Elapidae
e Desert Death Adder
d Wüstentodesotter
i Vipera della morte del deserto

78 ***Acanthosaura***
Squamata - Agamidae
e Pricklenapes; Angle-headed Lizards
d Nackenstachler; Winkelkopfagamen

79 ***Acanthosaura armata***
Squamata - Agamidae
e Armoured Pricklenape; Horned Anglehead; Mountain Horned Dragon
d Nackenstachler; Winkelkopfagame; Malaysischer Nackenstachler

80 ***Acanthosaura capra***
Squamata - Agamidae
e Green Pricklenape; Mountain Horned Dragon; Mountain Horned Lizard
d Grüner Nackenstachler; Vietnam-Nackenstachler
f Lézard cornu des montagnes; Dragon cornu des montagnes; Dragon des montagnes

81 ***Acanthosaura crucigera***
Squamata - Agamidae
e Pseudemoia; Horned Mountain Dragon; Mountain Horned Dragon
d Nackenstachler; Winkelkopfagame
f Dragon des montagnes

82 ***Acanthosaura lepidogaster***
Squamata - Agamidae
e Brown Pricklenape
d Brauner Nackenstachler; Nackenstachler

83 ***Achalinus***
Squamata - Colubridae
e Odd-scaled Snakes

84 ***Achalinus ater***
Squamata - Colubridae
e Bourret's Odd-scaled Snake

85 ***Achalinus formosanus***
Squamata - Colubridae
e Formosa Odd-scaled Snake

86 ***Achalinus formosanus chigirai***
Squamata - Colubridae
e Yaeyama Odd-scaled Snake

87 ***Achalinus hainanus***
Squamata - Colubridae
e Hainan Odd-scaled Snake

88 ***Achalinus jinggangensis***
Squamata - Colubridae
e Zong's Odd-scaled Snake

89 ***Achalinus meiguensis***
Squamata - Colubridae
e Szechwan Odd-scaled Snake

90 ***Achalinus niger***
Squamata - Colubridae
e Black Odd-scaled Snake

91 ***Achalinus rufescens***
Squamata - Colubridae
e Boulenger's Odd-scaled Snake

92 ***Achalinus spinalis***
Squamata - Colubridae
e Peters's Odd-scaled Snake; Japanese Odd-scaled Snake

93 ***Aclys***
Squamata - Gekkonidae
e Elegant Legless Lizards

94 ***Aclys concinna***
Squamata - Gekkonidae
e Elegant Legless Lizard

95 ***Acontias***
Squamata - Scincidae
e Lance Skinks; Greater Legless Skinks (CSA); Dart Skinks
d Lanzenskinke
f Acontias

96 ***Acontias breviceps***
Squamata - Scincidae
e Shorthead Lance Skink; Short-headed Legless Skink (CSA)

97 ***Acontias gracilicauda***
Squamata - Scincidae
e Slendertail Lance Skink; Thin-tailed Legless Skink (CSA); Slender-tailed Legless Skink (CSA)

98 ***Acontias gracilicauda gracilicauda***
Squamata - Scincidae
e Slender-tailed Legless Skink (CSA)

99 ***Acontias lineatus***
Squamata - Scincidae
e Lined Lance Skink; Striped Legless Skink (CSA); Woodbush Legless Skink (CSA)

100 ***Acontias litoralis***
Squamata - Scincidae
e Broadley's Lance Skink; Coastal Legless Skink (CSA)

101 ***Acontias meleagris***
Squamata - Scincidae

e Linnaeus's Lance Skink; Cape Legless Skink (CSA)
d Dunkler Lanzenskink; Jap-Lanzenskink
f Acontias pintade

102 ***Acontias meleagris meleagris***
Squamata - Scincidae
e Cape Legless Skink

103 ***Acontias percivali***
Squamata - Scincidae
e Percival's Lance Skink; Percival's Limbless Skink (CSA); Teita Mountains Limbless Skink
d Westlicher Lanzenskink

104 ***Acontias percivali occidentalis***
Squamata - Scincidae
e Percival's Legless Skink

105 ***Acontias plumbeus***
Squamata - Scincidae
e Giant Lance Skink; Giant Legless Skink (CSA); Giant Limbless Skink (CSA)
d Riesenlanzenskink

106 ***Acontophiops***
Squamata - Scincidae
e Legless Skinks; Woodbush Legless Skinks

107 ***Acontophiops lineatus***
Squamata - Scincidae
e Woodbush Legless Skink

108 **Acrochordidae**
Squamata
e Wart Snakes; File Snakes; Stilleto Snakes; Acrochordids
d Warzenschlangen
f Acrochordidés

109 ***Acrochordus***
Squamata - Acrochordidae
e Java Wart Snakes; Wart Snakes; Indian Wart Snakes; Javan Wart Snakes; File Snakes; Elephant's Trunk Snake
d Warzenschlangen
f Acrochordes

110 ***Acrochordus arafurae***
Squamata - Acrochordidae
e Arafura File Snake (ANZ); McDowell's Wart Snake
d Arafura-Warzenschlange

111 ***Acrochordus granulatus***
Squamata - Acrochordidae
e Little File Snake (ANZ); Marine File Snake; Indian Wart Snake; Granular-scaled Wart Snake; Rasp-skinned Water Snake; Asiatic File Snake (ISC); Wart Snake (ISC); Granulated File Snake
d Südostasiatische Warzenschlange; Indische Warzenschlange; Javanische Warzenschlangen; Zwergwarzenschlange
f Chersydre à bandes

112 ***Acrochordus javanicus***
Squamata - Acrochordidae
e Java File Snake; Elephant's-trunk Snake; Javan File Snake; Javan Wart Snake; File Snake
d Javanische Warzenschlange; Elefantenrüsselschlange
f Acrochorde de Java

113 ***Adelophis***
Squamata - Colubridae
e Meadow Snakes

114 ***Adelophis copei***
Squamata - Colubridae
e Cope's Mountain Meadow Snake

115 ***Adelophis foxi***
Squamata - Colubridae
e Fox's Mountain Meadow Snake

116 ***Adelphicos***
Squamata - Colubridae
e American Burrowing Snakes; Highland Earth Snakes

117 ***Adelphicos daryi***
Squamata - Colubridae
e Dary's Burrowing Snake; Highland Earth Snake

118 ***Adelphicos quadrivirgatus***
Squamata - Colubridae
e Middle American Burrowing Snake; Four-lined Earth Snake

119 ***Adelphicos veraepacis***
Squamata - Colubridae
e Stuart's Burrowing Snake; Highland Earth Snake

120 ***Adenorhinos***
Squamata - Viperidae
e Mountain Bush Vipers

121 ***Adenorhinos barbouri***
Squamata - Viperidae
e Uruzungwe Mountain Bush Viper

122 ***Adolfus***
Squamata - Lacertidae
e Forest Lizards

123 ***Adolfus africanus***
Squamata - Lacertidae
e Multi-scaled Forest Lizard

124 ***Adolfus alleni***
Squamata - Lacertidae
e Alpine Meadow Lizard

125 ***Adolfus jacksoni***
Squamata - Lacertidae
e Jackson's Lizard; Jackson's Forest Lizard
d Tanzania-Eidechse

Squamata - Lacertidae
e Sparse-scaled Forest Lizard

127 ***Aeluroscalabotes***
Squamata - Eublepharidae
e Cat Geckos
d Lidgeckos; Katzengeckos
f Geckos-chat

128 ***Aeluroscalabotes felinus***
Squamata - Eublepharidae
e Cat Gecko
d Fuchsgesichtlidgecko; Katzengecko
f Gecko-chat

129 ***Afroablepharus***
Squamata - Scincidae
e Dwarf Skinks

130 ***Afroablepharus duruarus***
Squamata - Scincidae
e Monard's Dwarf Skink

131 ***Afroablepharus megalurus***
Squamata - Scincidae
e Nieden's Dwarf Skink

132 ***Afroablepharus seydeli***
Squamata - Scincidae
e Seydel's Dwarf Skink

133 ***Afroablepharus tancredi***
Squamata - Scincidae
e Boulenger's Dwarf Skink; Abyssinian Lidless Skink

134 ***Afroablepharus wahlbergi***
Squamata - Scincidae
e Werner's Dwarf Skink; Wahlberg's Dwarf Skink; Snake-eyed Skink (CSA)
d Wahlbergs Schlangenauge

135 ***Afroablepharus wilsoni***
Squamata - Scincidae
e Wilson's Dwarf Skink

136 ***Afroedura***
Squamata - Gekkonidae
e Rock Geckos; Flat Geckos (CSA); Thick-tailed Geckos
d Felsengeckos

137 ***Afroedura africana***
Squamata - Gekkonidae
e African Rock Gecko; African Flat Gecko (CSA)

138 ***Afroedura amatolica***
Squamata - Gekkonidae
e Amatola Rock Gecko; Amatola Flat Gecko (CSA)

139 ***Afroedura bogerti***
Squamata - Gekkonidae
e Bogert's Rock Gecko

140 ***Afroedura hawequensis***
Squamata - Gekkonidae
e Cape Rock Gecko; Hawequa Flat Gecko (CSA)

141 ***Afroedura karroica***
Squamata - Gekkonidae
e Inland Rock Gecko; Karoo Flat Gecko (CSA)

142 ***Afroedura nivaria***
Squamata - Gekkonidae
e Drakensberg Rock Gecko; Mountain Flat Gecko (CSA)
d Schneefelsengecko

143 ***Afroedura pondolia***
Squamata - Gekkonidae
e Pondo Rock Gecko; Pondo Flat Gecko (CSA)

144 ***Afroedura pondolia multiporis***
Squamata - Gekkonidae
e Pondo Thick-tailed Gecko (CSA)

145 ***Afroedura tembulica***
Squamata - Gekkonidae
e Queenstown Rock Gecko; Tembo Flat Gecko (CSA)

146 ***Afroedura transvaalica***
Squamata - Gekkonidae
e Transvaal Rock Gecko; Transvaal Flat Gecko (CSA); African Flat Gecko; Flat-tailed Gecko
d Transvaal-Felsengecko

147 ***Afroedura transvaalica transvaalica***
Squamata - Gekkonidae
e Transvaal Thick-tailed Gecko (CSA)

148 ***Afronatrix***
Squamata - Colubridae
e African Brown Water Snakes; Brown Water Snakes
d Afrikanische Schwimmnattern

149 ***Afronatrix anoscopus***
Squamata - Colubridae
e African Brown Water Snake; Brown Water Snake
d Afrikanische Schwimmnatter

150 ***Agama***
Squamata - Agamidae
e Typical Agamas; Spiny Agamas; Agamas; Kenya Rock Lizards
d Agamen; Echte Agamen
f Agames
i Agame

151 ***Agama aculeata***
Squamata - Agamidae
e Ground Agama
d Rote Stachelagame
f Agame aiguilloné

152 ***Agama agama***
Squamata - Agamidae
e Common Agama; Common Rock Agama; Rock Agama; Rainbow Lizard; Red-headed Rock Agama; African Redhead Agama; African Rainbow Lizard
d Siedleragame
f Agame des colons; Margouillat; Petit margouillat
i Agama

153 ***Agama agama lionotus***
Squamata - Agamidae
e Common Agama; Blue Agama
d Blaue Siedlergame; Anchietas Agame

154 ***Agama anchietae***
Squamata - Agamidae
e Anchieta's Agama; Desert Agama
d Wüstenagame

155 ***Agama atra***
Squamata - Agamidae
e Southern Rock Agama; South African Rock Agama; South African Mountain Agama; Black Agama; Rock Agama (CSA); Black-necked Tree Agama (CSA)
d Schwarze Agame; Felsenagame; Bergagame
f Agame sombre; Agame commun; Agame des colons

156 ***Agama bocourti***
Squamata - Agamidae
e Bocourt's Agama

157 ***Agama bottegi***
Squamata - Agamidae
e Somali Agama

158 ***Agama boueti***
Squamata - Agamidae
e Mali Agama
d Mali-Dornschwanz

159 ***Agama boulengeri***
Squamata - Agamidae
e Boulenger's Agama
d Boulengers Agame
f Agame de Boulenger

160 ***Agama caudispinosa***
Squamata - Agamidae
e Elmenteita Rock Agama

161 ***Agama chernovi***
Squamata - Agamidae
e Chernov's Agama

162 ***Agama cornii***
Squamata - Agamidae
e Scortecci's Agama

163 ***Agama cristata***
Squamata - Agamidae
e Mocquard's Agama

164 ***Agama doriae***
Squamata - Agamidae
e Nigeria Agama

165 ***Agama etoshae***
Squamata - Agamidae
e Etosha Agama

166 ***Agama gracilimembris***
Squamata - Agamidae
e Benin Agama

167 ***Agama hartmanni***
Squamata - Agamidae
e Hartmann's Agama

168 ***Agama hispida***
Squamata - Agamidae
e Common Spiny Agama; Spiny Agama (CSA)

d Stachelagame
f Agame épineux; Agame terrestre

169 ***Agama impalearis***
Squamata - Agamidae
e Bibron's Agama; Atlas Agama
d Atlas-Agame
f Agame de Bibron

170 ***Agama insularis***
Squamata - Agamidae
e Insular Agama

171 ***Agama kirkii***
Squamata - Agamidae
e Kirk's Rock Agama; Kirk's Agama
d Kirks Agame
f Agame de Kirk

172 ***Agama makarikarica***
Squamata - Agamidae
e De Waal's Agama

173 ***Agama mehelyi***
Squamata - Agamidae
e Méhely's Agama
d Méhelys Agame
f Agame de Méhely

174 ***Agama microtympanum***
Squamata - Agamidae
e Small-eared Agama

175 ***Agama mossambica***
Squamata - Agamidae
e Mozambique Agama; Moçambique Agama
d Mozambique-Agame; Moçambique-Agame; Mozambik-Agame
f Agame de Mozambique

176 ***Agama mwanzae***
Squamata - Agamidae
e Mwanza Rock Agama
d Mwanza-Agame
f Agame de Mwanza

177 ***Agama neumanni***
Squamata - Agamidae
e Neumann's Agama

178 ***Agama pakistanica***
Squamata - Agamidae
e Pakistan Agama

179 ***Agama paragama***
Squamata - Agamidae
e False Agama
d Falsche Siedleragame

180 ***Agama persimilis***
Squamata - Agamidae
e Similar Agama

181 ***Agama planiceps***
Squamata - Agamidae
e Namib Rock Agama; Namibian Rock Agama (CSA); Rock Agama
d Felsenagame; Stachelagame; Heldenagame

182 ***Agama robecchii***
Squamata - Agamidae
e Robecchi's Agama

183 ***Agama rueppelli***
Squamata - Agamidae
e Arboreal Agama

184 ***Agama sankaranica***
Squamata - Agamidae
e Senegal Agama

185 ***Agama sennariensis***
Squamata - Agamidae
e Werner's Agama

186 ***Agama spinosa***
Squamata - Agamidae
e Lanza's Spiny Agama

187 ***Agama sylvanus***
Squamata - Agamidae
e Forest Agama

188 ***Agama weidholzi***
Squamata - Agamidae
e Gambia Agama

189 **Agamidae**
Squamata
e Agamas; Agamids; Dragon Lizards
d Agamen
f Agames; Agamidés

190 ***Agamura***
Squamata - Gekkonidae
e Spider Geckos
d Spinnengeckos

191 ***Agamura femoralis***
Squamata - Gekkonidae
e Pakistan Gecko; Sharp-tailed Spider Gecko

192 ***Agamura gastropholis***
Squamata - Gekkonidae
e Werner's Spider Gecko

193 ***Agamura missonei***
Squamata - Gekkonidae
e De Witte's Gecko

194 ***Agamura persica***
Squamata - Gekkonidae
e Spider Gecko; Blunt-tailed Spider Gecko
d Spinnengecko
f Gecko araignée

195 ***Agkistrodon***
Squamata - Viperidae
e Copperheads; Cottonmouths and Mocassins; Cottonmouths; Cottonmouths and Copperheads; American Moccassins
d Mokassinschlangen; Dreieckskföpfe; Dreieckskopfottern; Mokassinottern
f Mocassins; Agkistrodons; Mocassins d'eau
i Serpenti testa di rame

196 ***Agkistrodon bilineatus***
Squamata - Viperidae
e Cantil; Mexican Moccasin
d Mokassinschlange; Mexikanische Nasenotter; Mexikanische Mokassinotter; Mexikanische Mokassinschlange
f Mocassin des tropiques

197 ***Agkistrodon bilineatus bilieanatus***
Squamata - Viperidae
e Mexican Cantil
d Mexikanische Mokassinotter

198 ***Agkistrodon bilineatus howardgloydi***
Squamata - Viperidae
e Castellana

199 ***Agkistrodon bilineatus russeolus***
Squamata - Viperidae
e Yucatan Cantil

200 ***Agkistrodon bilineatus taylori***
Squamata - Viperidae
e Ornate Cantil; Taylor's Mocassin
d Tamaulipas Mokassinotter

201 ***Agkistrodon blomhoffi***
Squamata - Viperidae
e Japanese Mamushi; Mamushi; Short-tailed Viper
d Mamushi

202 ***Agkistrodon caliginosus***
Squamata - Viperidae
e Dusky Mamushi; Caliginous Mamushi

203 ***Agkistrodon contortrix***
Squamata - Viperidae
e Copperhead (NA); American Copperhead; Copperhead Moccasin (NA); Highland Moccasin (NA)
d Kupferkopf
f Mocassin à tête cuivrée; Tête-cuivrée
i Testa di rame

204 ***Agkistrodon contortrix contortrix***
Squamata - Viperidae
e Southern Copperhead
d Südlicher Kupferkopf

205 ***Agkistrodon contortrix laticinctus***
Squamata - Viperidae
e Broad-banded Copperhead
d Breitbändiger Kupferkopf; Breitbandkupferkopf

206 ***Agkistrodon contortrix mokasen***
Squamata - Viperidae
e Northern Copperhead; Upland Moccasin; Chunkhead (NA); Deaf Adder (NA); Pilot Snake (NA)
d Nördlicher Kupferkopf

207 ***Agkistrodon contortrix phaeogaster***
Squamata - Viperidae
e Osage Copperhead
d Osagen-Kupferkopf

208 ***Agkistrodon contortrix pictigaster***
Squamata - Viperidae
e Trans-Pecos Copperhead
d Transpecos Kupferkopf

209 ***Agkistrodon halys***
Squamata - Viperidae
e Siberian Pit Viper; Haly's Pit Viper; Haly's Viper; Central Asian Viper; Shchitomordnik
d Halys Otter; Halys Schlange; Haly-Schlange; Halys Grubenotter
f Agkistrodon des steppes de la Russie; Moccassin d'Europe; Vipère d'Haly
i Crotalo caspico

210 ***Agkistrodon himalayanus***
Squamata - Viperidae
e Himalaya Pit Viper; Himalayan Pit Viper (ISC); Himalayan Mamushi; Himalayan Palm Viper
d Himalaja-Grubenotter
f Agkistrodon de l'Himalaya

211 ***Agkistrodon intermedius***
Squamata - Viperidae
e Central Asian Pit Viper; Asian Pit Viper
d Zentralasiatische Grubenotter
f Mocassin

212 ***Agkistrodon intermedius saxatilis***
Squamata - Viperidae
e Korean Rock Viper

213 ***Agkistrodon monticola***
Squamata - Viperidae
e Likiang Pit Viper; Large-spotted Viper; Mountain Pit Viper
d Likiang-Grubenotter

214 ***Agkistrodon piscivorus***
Squamata - Viperidae
e Water Mocassin; Cottonmouth (NA); Cottonmouth Moccasin; Cottonmouth (NA)
d Wassermokassinotter; Wassermokassinschlange; Baumwollmaul
f Mocassin d'eau

215 ***Agkistrodon piscivorus conanti***
Squamata - Viperidae
e Florida Cottonmouth

d Florida-Wassermokassinschlange

216 ***Agkistrodon piscivorus leucostoma***
Squamata - Viperidae
e Western Cottonmouth; Water Moccasin; Cottonmouth Moccasin (NA); Moccasin
d Westliche Wassermokassinschlange

217 ***Agkistrodon piscivorus piscivorus***
Squamata - Viperidae
e Eastern Cottonmouth
d Östliche Wassermokassinschlange

218 ***Agkistrodon saxatilis***
Squamata - Viperidae
e Brown Mamushi

219 ***Agkistrodon shedaoensis***
Squamata - Viperidae
e Zhao Pit Viper

220 ***Agkistrodon strauchi***
Squamata - Viperidae
e Strauch's Pit Viper; Strauch's Mamushi

221 ***Ahaetulla***
Squamata - Colubridae
e Asian Longnose Whip Snakes; Vine Snakes; Whip Snakes; Bronze Snakes; Long-nosed Tree Snakes
d Bronzennattern; Peitschennattern; Baumschnüffler
i Serpenti delle vigne asiatici

222 ***Ahaetulla dispar***
Squamata - Colubridae
e Indian Bronzeback

223 ***Ahaetulla fasciolata***
Squamata - Colubridae
e Speckle-headed Whip Snake

224 ***Ahaetulla fronticincta***
Squamata - Colubridae
e Günther's Whip Snake

225 ***Ahaetulla mycterizans***
Squamata - Colubridae
e Malayan Green Whip Snake; Common Green Whip Snake
d Baumschnüffler

226 ***Ahaetulla nasuta***
Squamata - Colubridae
e Longnose Whip Snake; Long-nosed Tree Snake; Long-nosed Whip Snake; Long-nosed Vine Snake; Vine Snake (ISC); Common Green Whip Snake (ISC); Indian Green Vine Snake
d Baumschnüffler; Lianennatter; Peitschennatter
f Nasique

227 ***Ahaetulla perroteti***
Squamata - Colubridae
e Western Ghats Bronzeback

228 ***Ahaetulla prasina***
Squamata - Colubridae
e Oriental Whip Snake; Green Vine Snake; Malayan Whip Snake; Boie's Whip Snake; Grass-green Whip Snake; Himalaya-Malayan Green Whip Snake; Common Green Vine Snake
d Grüner Baumschnüffler

229 ***Ahaetulla pulverulenta***
Squamata - Colubridae
e Brown-speckled Whip Snake

230 ***Ailuronyx***
Squamata - Gekkonidae
e Duméril's Geckos

231 ***Ailuronyx seychellensis***
Squamata - Gekkonidae
e Bronze Gecko; Seychelles Skin-

sloughing Gecko
d Seychellen-Bronzefingergecko; Bronzegecko; Seychellen-Hausgecko
f Gecko de bronze; Gecko qui pèle des Seychelles

232 ***Ailuronyx trachygaster***
Squamata - Gekkonidae
e Yellow Gecko

233 ***Aipysurus***
Squamata - Elapidae
e Sea Snakes; Olive Sea Snakes
d Braune Seeschlangen

234 ***Aipysurus apraefrontalis***
Squamata - Elapidae
e White-spotted Sea Snake; Sahul Reef Snake
d Riff Seeschlange; Sahul-Riffschlange

235 ***Aipysurus duboisii***
Squamata - Elapidae
e Dubois's Sea Snake

236 ***Aipysurus eydouxii***
Squamata - Elapidae
e Northern Sea Snake

237 ***Aipysurus foliosquamata***
Squamata - Elapidae
e Red-brown Sea Snake

238 ***Aipysurus fuscus***
Squamata - Elapidae
e Brown Sea Snake; Timor Reef Snake
d Riff Seeschlange; Timor-Riffschlange

239 ***Aipysurus laevis***
Squamata - Elapidae
e Olive Sea Snake; Olive-brown Sea Snake

240 ***Aipysurus tenuis***
Squamata - Elapidae
e Arafura Sea Snake

241 ***Algyroides***
Squamata - Lacertidae
e Algyroideses; Keeled Lizards
d Kieleidechsen; Kielechsen
f Algyroïdes
i Algiroidi

242 ***Algyroides fitzingeri***
Squamata - Lacertidae
e Fitzinger's Algyroides; Pygmy Algyroides
d Zwergkieleidechse; Tyrrhenische Kieleidechse; Sardische Kieleidechse; Sardinische Kieleidechse; Tyrrhenische Kielechse
f Algyroïde tyrrhénien; Algyroïde de Fitzinger
i Algiroide nano; Algiroide di Fitzinger

243 ***Algyroides marchi***
Squamata - Lacertidae
e Spanish Keeled Lizard; Spanish Algyroides; Spanish Lizard
d Peloponnesische Kieleidechse; Spanische Kieleidechse; Spanische Kielechse
f Algyroïde d'Espagne; Algyroïde de Valverde
i Algiroide di March

244 ***Algyroides moreoticus***
Squamata - Lacertidae
e Greek Keeled Lizard; Greek Algyroides
d Ionische Kieleidechse; Peloponnesische Kieleidechse; Peloponnesische Kielechse
f Algyroïde du Péloponnèse
i Algiroide greco

245 ***Algyroides nigropunctatus***
Squamata - Lacertidae

e Blue-throated Keeled Lizard; Dalmatian Algyroides; Keeled Lizard
d Griechische Kieleidechse; Prachtkieleidechse; Blaukehlige Kieleidechse; Schwarzpunktierte Kieleidechse; Blaukehlige Kielechse
f Algyroïde à gorge bleue; Algyroïde à points noirs
i Algiroide magnifico; Algiroide punteggiato

246 ***Alligator***
Crocodylia - Alligatoridae
e Alligators
d Alligatoren
f Alligators
i Alligatori

247 ***Alligator mississippiensis***
Crocodylia - Alligatoridae
e American Alligator; Mississippi Alligator; Louisiana Alligator; Pike-headed Alligator
d Mississippi-Alligator; Hechtalligator
f Alligator américain; Alligator du Mississippi

248 ***Alligator sinensis***
Crocodylia - Alligatoridae
e Chinese Alligator; Janzi Alligator; Yangtse Alligator
d China-Alligator; Yangtse-Alligator
f Alligator chinois; Alligator du Chine

249 **Alligatoridae**
Squamata
e Alligators; Alligators and Caimans
d Alligatoren
f Alligatoridés

250 ***Alopoglossus***
Squamata - Teiidae
e Northern Teiids; Shade Lizards
d Kleintejus

251 ***Alopoglossus andreanus***
Squamata - Teiidae
e Andes Teiid

252 ***Alopoglossus angulatus***
Squamata - Teiidae
e Northern Teiid; Large-scaled Shade Lizard

253 ***Alopoglossus atriventrus***
Squamata - Teiidae
e Duellman's Teiid; Keel-bellied Shade Lizard

254 ***Alopoglossus buckleyi***
Squamata - Teiidae
e Buckley's Teiid; Smooth-bellied Shade Lizard

255 ***Alopoglossus festae***
Squamata - Teiidae
e Peracca's Teiid

256 ***Alpetonotus***
Squamata - Lacertidae
e Longtail Lizards

257 ***Alpetonotus dorsalis***
Squamata - Lacertidae
e Stejneger's Longtail Lizard

258 ***Alsophis***
Squamata - Colubridae
e Racers

259 ***Alsophis anomalus***
Squamata - Colubridae
e Hispaniola Racer; Hispaniolan Brown Racer

260 ***Alsophis antillensis***
Squamata - Colubridae
e Antilles Racer; Leeward Racer (WI)
d Dominikanische Schlanknatter

261 ***Alsophis ater***
Squamata - Colubridae
e Jamaica Racer; Jamaican Racer (WI)

262 ***Alsophis biserialis***
Squamata - Colubridae
e Galapagos Racer

263 ***Alsophis cantherigerus***
Squamata - Colubridae
e Cuban Racer; Cuban Ground Snake
d Kuba-Schlanknatter

264 ***Alsophis melanichinus***
Squamata - Colubridae
e La Vega Racer; Hispaniolan Green Racer

265 ***Alsophis occidentalis***
Squamata - Colubridae
e Western Racer

266 ***Alsophis portoricensis***
Squamata - Colubridae
e Puerto Rican Racer; Ground Snake

267 ***Alsophis rijersmai***
Squamata - Colubridae
e Anguilla Racer; Anguilla Bank Racer (WI)

268 ***Alsophis rufiventris***
Squamata - Colubridae
e Saba Racer; West Indian Racer; Orange-bellied Racer (WI)

269 ***Alsophis sanctaecrucis***
Squamata - Colubridae
e St. Croix Racer

270 ***Alsophis vudii***
Squamata - Colubridae
e Bimini Racer; Bahamian Racer (WI)
d Bahama-Natter; Bahamas-Peitschennatter

271 ***Alsophylax***
Squamata - Gekkonidae
e Even-fingered Geckos
d Geradfingergeckos; Geradfinger

272 ***Alsophylax laevis***
Squamata - Gekkonidae
e Southern Even-fingered Gecko
d Südlicher Geradfingergecko

273 ***Alsophylax loricatus***
Squamata - Gekkonidae
e Strauch's Even-fingered Gecko

274 ***Alsophylax pipiens***
Squamata - Gekkonidae
e Caspian Even-fingered Gecko; Caspian Straight-fingered Gecko
d Kaspischer Geradfinger; Kaspischer Geradfingergecko
f Alsophyle de la Caspienne
i Alsofilace caspico

275 ***Alsophylax przewalskii***
Squamata - Gekkonidae
e Xinjiang Even-fingered Gecko

276 ***Alsophylax tadjikiensis***
Squamata - Gekkonidae
e Tadjikistan Even-fingered Gecko

277 ***Alsophylax tokabejevi***
Squamata - Gekkonidae
e Kirghizia Even-fingered Gecko

278 ***Amapasaurus***
Squamata - Teiidae
e Four-toed Amapasauruses

279 ***Amapasaurus tetradactylus***
Squamata - Teiidae
e Four-toed Amapasaurus

280 ***Amastridium***
Squamata - Colubridae
e Rustyhead Snakes; Rusty-headed Snakes

281 ***Amastridium sapperi***
Squamata - Colubridae
e Rustyhead Snake; Northern Rusty-headed Snake

282 ***Amastridium veliferum***
Squamata - Colubridae
e Rustyhead Snake; Southern Rusty-headed Snake; Rusty-headed Snake

283 ***Amblyodipsas***
Squamata - Colubridae
e Purple-glossed Snakes; White-lipped Snakes

284 ***Amblyodipsas concolor***
Squamata - Colubridae
e Natal Purple-glossed Snake; Purple-glossed Snake (CSA)

285 ***Amblyodipsas dimidiata***
Squamata - Colubridae
e Mpwapwa Purple-glossed Snake

286 ***Amblyodipsas katengensis***
Squamata - Colubridae
e Katanga Purple-glossed Snake; Tanganyika White-lipped Snake

287 ***Amblyodipsas micropthalma***
Squamata - Colubridae
e Eyeless Purple-glossed Snake; Eastern Purple-glossed Snake (CSA); Eastern White-lipped Snake (CSA); Purple Glossed Snake
d Purpurne Glanzschlange

288 ***Amblyodipsas polylepis***
Squamata - Colubridae
e Common Purple-glossed Snake; Purple-glossed Snake (CSA)

289 ***Amblyodipsas rodhaini***
Squamata - Colubridae
e Rodhain's Purple-glossed Snake

290 ***Amblyodipsas teitana***
Squamata - Colubridae
e Teitana Purple-glossed Snake

291 ***Amblyodipsas unicolor***
Squamata - Colubridae
e Dull Purple-glossed Snake

292 ***Amblyodipsas ventrimaculata***
Squamata - Colubridae
e Kalahari Purple-glossed Snake; Barotse Burrowing Snake (CSA)

293 ***Amblyrhinchus***
Squamata - Iguanaidae
e Marine Iguanas
d Meerechsen; Galapagos-Meerechsen
f Amblyrhynques

294 ***Amblyrhynchus cristatus***
Squamata - Iguanaidae
e Marine Iguana; Galapagos Marine Iguana
d Meerechse; Galapagos-Meerechse
f Amblyrhynque à crête; Iguane marin des Galapagos; Iguane marin

295 ***Amblyrhynchus cristatus venustissimus***
Squamata - Iguanaidae
e Hood Island Marine Iguana

296 ***Ameiva***
Squamata - Teiidae
e Ameivas; Runners; Racerunners
d Ameiven
f Améives; Améivas; Téjus des Caraïbes

297 ***Ameiva ameiva***
Squamata - Teiidae

e Giant Ameiva; Comon Ameiva; Jungle Runner; Ground Lizard; South American Ground Lizard; Amazon Racerunner; Neotropical Ameiva (WI)
d Grüne Ameive; Ameiven-Teju; Ameive
f Améive géant; Améiva géante

298 ***Ameiva anomala***
Squamata - Teiidae
e Echternacht's Ameiva

299 ***Ameiva auberi***
Squamata - Teiidae
e Auber's Ameiva; Cuban Ameiva (WI)

300 ***Ameiva bifrontata***
Squamata - Teiidae
e Cope's Ameiva

301 ***Ameiva bridgesi***
Squamata - Teiidae
e Bridges's Ameiva

302 ***Ameiva chaitzami***
Squamata - Teiidae
e Chaitzam's Ameiva
d Tigerameive

303 ***Ameiva chrysolaema***
Squamata - Teiidae
e Common Ameiva; Common Teiid; Hispaniolan Giant Ameiva
d Haiti-Ameive
f Améiva vulgaire; Améive vulgaire

304 ***Ameiva cineracea***
Squamata - Teiidae
e Guadeloupe Ameiva

305 ***Ameiva corax***
Squamata - Teiidae
e Censky's Ameiva

306 ***Ameiva corvina***
Squamata - Teiidae
e Sombrero Ameiva

307 ***Ameiva dorsalis***
Squamata - Teiidae
e Jamaica Ameiva

308 ***Ameiva edracantha***
Squamata - Teiidae
e Bocourt's Ameiva

309 ***Ameiva erythrocephala***
Squamata - Teiidae
e St. Christopher Ameiva; Orange-throated Ameiva (WI)

310 ***Ameiva exsul***
Squamata - Teiidae
e Puerto Rican Ameiva; Common Ground Lizard ; Common Puerto Rican Ameiva; Puerto Rican Giant Ameiva (W)

311 ***Ameiva festiva***
Squamata - Teiidae
e Middle American Ameiva; Central American Ameiva; Central American Whiptail Lizard

312 ***Ameiva fuscata***
Squamata - Teiidae
e Dominican Ameiva; Dominica Ameiva (WI)

313 ***Ameiva griswoldi***
Squamata - Teiidae
e Griswold's Ameiva; Antigua Bank Ameiva (WI)

314 ***Ameiva leberi***
Squamata - Teiidae
e Hispaniola Ameiva

315 ***Ameiva leptophrys***
Squamata - Teiidae

e Delicate Ameiva

316 ***Ameiva lineolata***
Squamata - Teiidae
e Dwarf Teiid; Pygmy Blue-tailed Ameiva
d Zwergameive
f Améive linéolée; Améiva linéolée

317 ***Ameiva major***
Squamata - Teiidae
e Martinique Ameiva

318 ***Ameiva maynardi***
Squamata - Teiidae
e Great Inagua Ameiva; Bahamian Ameiva (WI)

319 ***Ameiva nicefori***
Squamata - Teiidae
e Dunn's Ameiva

320 ***Ameiva orcesi***
Squamata - Teiidae
e Peters's Ameiva

321 ***Ameiva pleei***
Squamata - Teiidae
e Caribbean Ameiva; Anguilla Bank Ameiva (WI)

322 ***Ameiva pluvianotata***
Squamata - Teiidae
e St. Croix Ameiva; Montserrat Ameiva (WI)

323 ***Ameiva polops***
Squamata - Teiidae
e Montserrat Ameiva; St. Croix Ground Lizard ; St. Croix Ameiva
d St.-Croix-Ameive

324 ***Ameiva quadrilineata***
Squamata - Teiidae
e Four-lined Ameiva

325 ***Ameiva septemlineata***
Squamata - Teiidae
e Seven-lined Ameiva

326 ***Ameiva taeniura***
Squamata - Teiidae
e Haitian Ameiva; Hispaniolan Blue-tailed Ameiva
d Bänderschwanzameive

327 ***Ameiva undulata***
Squamata - Teiidae
e Rainbow Ameiva; Metallic Ameiva

328 ***Ameiva wetmorei***
Squamata - Teiidae
e Wetmore's Ameiva; Blue-tailed Ground Lizard ; Puerto Rican Blue-tailed Ameiva

329 ***Amphibolurus***
Squamata - Agamidae
e Australian Dragon Lizards; Lashtails (ANZ)
d Australische Bodenagamen; Bartagamen
f Amphibolures
i Anfiboluri

330 ***Amphibolurus centralis***
Squamata - Agamidae
e Centralian Lashtail

331 ***Amphibolurus muricatus***
Squamata - Agamidae
e Jacky Lizard; Australian Bloodsucker; Jacky Lashtail (ANZ); Jacky Dragon; Dragon Lizard (ANZ); Tree Dragon (ANZ)
d Australischer Blutsauger; Felsenechse

332 ***Amphibolurus nobbi***
Squamata - Agamidae
e Nobbi; Nobbi Lashtail (ANZ); Nobbi Dragon (ANZ)

333 ***Amphibolurus norrisi***
Squamata - Agamidae
e Norris's Dragon Lizard; Mallee Heath Lashtail (ANZ); Norris's Dragon (ANZ); Dragon Lizard (ANZ)

334 ***Amphibolurus parviceps***
Squamata - Agamidae
e Coastal Earless Lizard

335 ***Amphiesma***
Squamata - Colubridae
e Keelbacks

336 ***Amphiesma atemporalis***
Squamata - Colubridae
e Tonkin Keelback

337 ***Amphiesma beddomei***
Squamata - Colubridae
e Nilgiri Keelback; Beddome's Keelback

338 ***Amphiesma bitaeniata***
Squamata - Colubridae
e Kutkai Keelback

339 ***Amphiesma boulengeri***
Squamata - Colubridae
e Tai-yong Keelback

340 ***Amphiesma conspicillata***
Squamata - Colubridae
e Redbelly Keelback; Red-bellied Keelback

341 ***Amphiesma craspedogaster***
Squamata - Colubridae
e Kuatun Keelback

342 ***Amphiesma deschauenseei***
Squamata - Colubridae
e Northern Keelback

343 ***Amphiesma flavifrons***
Squamata - Colubridae
e Sabah Keelback

344 ***Amphiesma groundwateri***
Squamata - Colubridae
e Groundwater's Keelback

345 ***Amphiesma inas***
Squamata - Colubridae
e Malayan Mountain Keelback

346 ***Amphiesma johannis***
Squamata - Colubridae
e Johann's Keelback

347 ***Amphiesma khasiensis***
Squamata - Colubridae
e Khasi Hills Keelback; Boulenger's Keelback

348 ***Amphiesma metusia***
Squamata - Colubridae
e Wa Shan Keelback

349 ***Amphiesma miyajimae***
Squamata - Colubridae
e Maki's Keelback

350 ***Amphiesma modesta***
Squamata - Colubridae
e Modest Keelback

351 ***Amphiesma monticola***
Squamata - Colubridae
e Wynad Keelback

352 ***Amphiesma nicobariensis***
Squamata - Colubridae
e Nicobar Island Keelback

353 ***Amphiesma octolineata***
Squamata - Colubridae
e Eight-lined Keelback

354 ***Amphiesma optata***
Squamata - Colubridae
e Mount Omei Keelback

355 ***Amphiesma parallela***
Squamata - Colubridae
e Yunnan Keelback; Anderson's Keelback

356 ***Amphiesma pealii***
Squamata - Colubridae
e Assam Keelback

357 ***Amphiesma petersi***
Squamata - Colubridae
e Peters's Keelback

358 ***Amphiesma platyceps***
Squamata - Colubridae
e Himalayan Keelback; Blyth's Keelback

359 ***Amphiesma popei***
Squamata - Colubridae
e Pope's Keelback

360 ***Amphiesma pryeri***
Squamata - Colubridae
e Pryer's Keelback

361 ***Amphiesma pryeri concelarum***
Squamata - Colubridae
e Myiako Keelback

362 ***Amphiesma pryeri ishigakiense***
Squamata - Colubridae
e Yaeyama Keelback

363 ***Amphiesma pryeri pryeri***
Squamata - Colubridae
e Pryer's Keelback Snake

364 ***Amphiesma sanguinae***
Squamata - Colubridae
e Cameroon Keelback; Smedley's Keelback

365 ***Amphiesma saravacensis***
Squamata - Colubridae
e Sarawak Keelback; Sarawak Mountain Keelback; Speckle-bellied Keelback

366 ***Amphiesma sauteri***
Squamata - Colubridae
e Kosempo Keelback

367 ***Amphiesma sieboldii***
Squamata - Colubridae
e Sikkim Keelback

368 ***Amphiesma stolata***
Squamata - Colubridae
e Buff-striped Keelback (ISC); Striped Keelback
d Asiatische Streifennatter; Gelbhalswassernatter; Gelbbandwassernatter

369 ***Amphiesma venningi***
Squamata - Colubridae
e Chin Hills Keelback

370 ***Amphiesma vibakari***
Squamata - Colubridae
e Japanese Keelback; Vibrakari Keelback

371 ***Amphiesma vibakari danzyoensis***
Squamata - Colubridae
e Danzyo Keelback

372 ***Amphiesma xenura***
Squamata - Colubridae
e Wall's Keelback

373 ***Amphiesmoides***
Squamata - Colubridae
e Ornate Snakes

374 ***Amphiesmoides ornaticeps***
Squamata - Colubridae
e Werner's Ornate Snake

375 ***Amphiglossus***
Squamata - Scincidae
e Common Skinks; Madagascar Skinks
f Amphiglosses

376 ***Amphiglossus andranovahensis***
Squamata - Scincidae
e Andranovaho Skink

377 ***Amphiglossus ankodabensis***
Squamata - Scincidae
e Ankodabe Skink

378 ***Amphiglossus ardouini***
Squamata - Scincidae
e Yellow Skink

379 ***Amphiglossus astrolabi***
Squamata - Scincidae
e Diving Skink
f Amphiglosse d'Astrolabe

380 ***Amphiglossus decaryi***
Squamata - Scincidae
e Rock Skink

381 ***Amphiglossus frontoparietalis***
Squamata - Scincidae
e Boulenger's Skink

382 ***Amphiglossus gastrostictus***
Squamata - Scincidae
e O'Shaughnessy's Madagascar Skink

383 ***Amphiglossus igneocaidtus***
Squamata - Scincidae
e Redtail Skink

384 ***Amphiglossus intermedius***
Squamata - Scincidae
e Stripeneck Skink

385 ***Amphiglossus johannae***
Squamata - Scincidae
e Johanna's Skink

386 ***Amphiglossus macrocercus***
Squamata - Scincidae
e Black-striped Skink

387 ***Amphiglossus macrolepis***
Squamata - Scincidae
e Rusty Skink

388 ***Amphiglossus melanopleura***
Squamata - Scincidae
e Common Madagascar Skink; Madagascar Skink

389 ***Amphiglossus melanurus***
Squamata - Scincidae
e Spotted Skink

390 ***Amphiglossus mouroundavae***
Squamata - Scincidae
e Morondava Skink

391 ***Amphiglossus ornaticeps***
Squamata - Scincidae
e Grey Skink

392 ***Amphiglossus poecilops***
Squamata - Scincidae
e Toamasina Skink

393 ***Amphiglossus polleni***
Squamata - Scincidae
e Madagascar Coastal Skink

394 ***Amphiglossus reticulatus***
Squamata - Scincidae
e Reticulate Skink

395 ***Amphiglossus splendidus***
Squamata - Scincidae
e Splendid Skink

396 ***Amphiglossus stumpffi***
Squamata - Scincidae
e Stumpff's Skink

397 ***Amphiglossus tsaratananensis***
Squamata - Scincidae
e Tsaratanan Skink

398 ***Amphiglossus valhalle***
Squamata - Scincidae
e Valhalle Skink

399 ***Amphiglossus waterloti***
Squamata - Scincidae
e Water Skink

400 ***Amphisbaena***
Squamata - Amphisbaenidae
e Worm Lizards; Ringed Lizards; Amphisbaenas
d Doppelschleichen; Eigentliche Doppelschleichen
f Amphisbènes

401 ***Amphisbaena alba***
Squamata - Amphisbaenidae
e Red Worm Lizard; Pale Worm Lizard; White-bellied Worm Lizard; Giant Amphisbaena
d Rote Doppelschleiche
f Amphisbène blanc

402 ***Amphisbaena angustifrons***
Squamata - Amphisbaenidae
e South American Worm Lizard

403 ***Amphisbaena bahiani***
Squamata - Amphisbaenidae
e Bahama Worm Lizard; Puerto Rican Grey Amphisbaena

404 ***Amphisbaena bakeri***
Squamata - Amphisbaenidae
e Baker's Worm Lizard; La Cadena Blind Snake

405 ***Amphisbaena caeca***
Squamata - Amphisbaenidae
e Blind Worm Lizard; Common Blind Snake

406 ***Amphisbaena camura***
Squamata - Amphisbaenidae
e Crooked Worm Lizard; Puerto Rican Grey Amphisbaena

407 ***Amphisbaena carlgansi***
Squamata - Amphisbaenidae
e Cuban Pink Amphisbaena

408 ***Amphisbaena carvalhoi***
Squamata - Amphisbaenidae
e Carvalho's Worm Lizard

409 ***Amphisbaena cubana***
Squamata - Amphisbaenidae
e Cuban Worm Lizard; Cuban Brown Amphisbaena

410 ***Amphisbaena darwini***
Squamata - Amphisbaenidae
e Darwin's Worm Lizard; Darwin's Ringed Lizard
d Darwins Doppelschleiche
f Amphisbène de Darwin

411 ***Amphisbaena dubia***
Squamata - Amphisbaenidae
e Uncertain Worm Lizard

412 ***Amphisbaena fenestrata***
Squamata - Amphisbaenidae
e Cope's Worm Lizard; Virgin Islands Blind Snake ; Virgin Islands Amphisbaena (NA)

413 ***Amphisbaena fuliginosa***
Squamata - Amphisbaenidae
e Speckled Worm Lizard; Black-and-white Worm Lizard; Spotted Worm Lizard
d Gefleckte Doppelschleiche
f Amphisbène enfumé

414 ***Amphisbaena fuliginosa bassleri***
Squamata - Amphisbaenidae
e Mottled Amphisbaena

415 ***Amphisbaena gonavensis***
Squamata - Amphisbaenidae
e Gonave Worm Lizard; Gonave Amphisbaena

416 ***Amphisbaena gracilis***
Squamata - Amphisbaenidae
e Slender Worm Lizard

417 ***Amphisbaena heathi***
Squamata - Amphisbaenidae
e Heath's Worm Lizard

418 ***Amphisbaena hogei***
Squamata - Amphisbaenidae
e Hoge's Worm Lizard

419 ***Amphisbaena innocens***
Squamata - Amphisbaenidae
e Innocent Worm Lizard; Tiburon Amphisbaena
d Schuldlose Doppelschleiche

420 ***Amphisbaena leeseri***
Squamata - Amphisbaenidae
e Urucum Worm Lizard

421 ***Amphisbaena leucocephala***
Squamata - Amphisbaenidae
e White-headed Worm Lizard

422 ***Amphisbaena manni***
Squamata - Amphisbaenidae
e Mann's Worm Lizard; Hispaniolan Dwarf Amphisbaena

423 ***Amphisbaena mertensi***
Squamata - Amphisbaenidae
e Mertens's Worm Lizard

424 ***Amphisbaena mitchelsii***
Squamata - Amphisbaenidae
e Mitchell's Worm Lizard

425 ***Amphisbaena munoai***
Squamata - Amphisbaenidae
e Munoa Worm Lizard

426 ***Amphisbaena neglecta***
Squamata - Amphisbaenidae
e Neglected Worm Lizard

427 ***Amphisbaena nigricauda***
Squamata - Amphisbaenidae
e Black-headed Worm Lizard

428 ***Amphisbaena occidentalis***
Squamata - Amphisbaenidae
e Western Worm Lizard

429 ***Amphisbaena pericensis***
Squamata - Amphisbaenidae
e Perico Worm Lizard

430 ***Amphisbaena petrei***
Squamata - Amphisbaenidae
e Rio Grande Worm Lizard

431 ***Amphisbaena plumbea***
Squamata - Amphisbaenidae
e Lead Worm Lizard

432 ***Amphisbaena polygrammica***
Squamata - Amphisbaenidae
e Werner's Worm Lizard

433 ***Amphisbaena prunicolor***
Squamata - Amphisbaenidae
e Plum-coloured Worm Lizard

434 ***Amphisbaena ridleyi***
Squamata - Amphisbaenidae
e Ridley's Worm Lizard

435 ***Amphisbaena roberti***
Squamata - Amphisbaenidae
e Robert's Worm Lizard

436 ***Amphisbaena rozei***
Squamata - Amphisbaenidae
e Roze's Worm Lizard

437 ***Amphisbaena schmidti***
Squamata - Amphisbaenidae
e Schmidt's Worm Lizard; Triangle-headed Blind Snake ; Puerto Rican Dusky Amphisbaena

438 ***Amphisbaena silvestrii***
Squamata - Amphisbaenidae
e Silvestri's Worm Lizard

439 ***Amphisbaena slatery***
Squamata - Amphisbaenidae
e Slater's Worm Lizard

440 ***Amphisbaena slevini***
Squamata - Amphisbaenidae
e Slevin's Worm Lizard

441 ***Amphisbaena spurelli***
Squamata - Amphisbaenidae
e Spurell's Worm Lizard

442 ***Amphisbaena steindachneri***
Squamata - Amphisbaenidae
e Steindachner's Worm Lizard

443 ***Amphisbaena stejnegeri***
Squamata - Amphisbaenidae
e Stejneger's Worm Lizard

444 ***Amphisbaena vanzolini***
Squamata - Amphisbaenidae
e Vanzolini's Worm Lizard

445 ***Amphisbaena vermicularis***
Squamata - Amphisbaenidae
e Wagler's Worm Lizard

446 ***Amphisbaena violacea***
Squamata - Amphisbaenidae
e Violet Worm Lizard; Violet Round-snouted Amphisbaenian (CSA)

447 ***Amphisbaena xera***
Squamata - Amphisbaenidae
e Dry Worm Lizard; Dryland Blind Snake ; Puerto Rican Xeric Amphisbaena

448 **Amphisbaenidae**
Squamata
e Worm Lizards; Tropical Worm Lizards; Ringed Lizards
d Doppelschleichen; Eigentliche Doppelschleichen
f Amphisbénidés

449 ***Amplorhinus***
Squamata - Colubridae
e Many-spotted Snakes; Cape Many-spotted Snakes

450 ***Amplorhinus multimaculatus***
Squamata - Colubridae
e Many-spotted Snake; Cape Reed Snake; Cape Many-spotted Snake (CSA); Reed Snake (CSA)

451 ***Amyda***
Testudines - Testudinidae
e Asiatic Softshells; Asiatic Softshell Turtles
d Weichschildkröten

452 ***Amyda***
Testudines - Trionychidae
e Asiatic Softshells

453 ***Amyda cartilaginea***
Testudines - Testudinidae
e Asiatic Softshell; Asiatic Softshell Turtle; Asian Softshell Turtle; Black-rayed Softshell Turtle
d Knorpelweichschildkröte
f Trionyx cartilagineuse
i Tartaruga dal guscio molle asiatica

454 ***Amyda cartilaginea***
Testudines - Trionychidae
e Asiatic Softshell

455 ***Anadia***
Squamata - Teiidae
e Anadias

456 ***Anadia altliserrania***
Squamata - Teiidae
e Harris's Anadia

457 ***Anadia bitaeniata***
Squamata - Teiidae
e Two-banded Anadia

458 ***Anadia blakei***
Squamata - Teiidae
e Blake's Anadia

459 ***Anadia bogotensis***
Squamata - Teiidae
e Bogota Anadia

460 ***Anadia brevifrontalis***
Squamata - Teiidae
e Shorthead Anadia

461 ***Anadia hobarti***
Squamata - Teiidae
e Hobart's Anadia

462 ***Anadia marmorata***
Squamata - Teiidae
e Spotted Anadia

463 ***Anadia metallica***
Squamata - Teiidae
e Metallic Anadia

464 ***Anadia ocellata***
Squamata - Teiidae
e Ocellated Anadia

465 ***Anadia pamplonensis***
Squamata - Teiidae
e Pamplona Anadia

466 ***Anadia petersi***
Squamata - Teiidae
e Peters's Anadia

467 ***Anadia pulchella***
Squamata - Teiidae
e Ruthven's Anadia

468 ***Anadia rhombifera***
Squamata - Teiidae
e Rhombifer Anadia

469 ***Anadia steyeri***
Squamata - Teiidae
e Steyer's Anadia

470 ***Anadia vittata***
Squamata - Teiidae
e Boulenger's Anadia

471 ***Anatosaura***
Squamata - Teiidae
e Anatosauras

472 ***Anatosaura brachylepis***
Squamata - Teiidae
e Dixon's Anatosaura

473 ***Anatosaura collaris***
Squamata - Teiidae
e Collared Anatosaura

474 ***Anatosaura vanzolinia***
Squamata - Teiidae
e Vanzolini's Anatosaura

475 ***Ancylocranium***
Squamata - Amphisbaenidae
e Sharp-snouted Worm Lizards

476 ***Ancylocranium barkeri***
Squamata - Amphisbaenidae
e Lindi Sharp-snouted Worm Lizard

477 ***Ancylocranium ionidesi***
Squamata - Amphisbaenidae
e Kilwa Sharp-snouted Worm Lizard

478 ***Ancylocranium somalicum***
Squamata - Amphisbaenidae
e Somali Sharp-snouted Worm Lizard

479 ***Androngo***
Squamata - Scincidae
e Greater Burrowing Skinks

480 ***Androngo alluaudi***
Squamata - Scincidae
e Brygoo's Burrowing Skink

481 ***Androngo crenni***
Squamata - Scincidae
e Banded Burrowing Skink

482 ***Androngo elongata***
Squamata - Scincidae
e Elongate Burrowing Skink

483 ***Anelytropsis***
Testudines - Dibamidae
e Mexican Blind Lizards
d Mexikanische Schlangenechsen
f Lézards-fouisseur du Mexique

484 ***Anelytropsis papillosus***
Testudines - Dibamidae
e Mexican Blind Lizard
d Mexikanische Schlangenechse
f Lézard-fouisseur du Mexique

485 ***Angolosaurus***
Squamata - Cordylidae
e Dune Plated Lizards; Desert Plated Lizard (CSA); Sand Dune Lizard
d Sandschildechsen
f Angolosaures

486 ***Angolosaurus skoogi***
Squamata - Cordylidae
e Sand Dune Plated Lizard; Desert Plated Lizard (CSA)
d Sandschildechse

487 **Anguidae**
Crocodylia
e Anguid Lizards; Lateral-fold Lizards; Alligator Lizards (NA)
d Schleichen; Wirtelechsen
f Lézards terrestres; Anguidés; Orvets

488 ***Anguis***
Squamata - Anguidae
e Slow Worms; Anguine Lizards
d Blindschleichen; Eigentliche Schleichen
f Orvets; Anguines

489 ***Anguis fragilis***
Squamata - Anguidae
e Slow Worm; Blind Worm
d Blindschleiche; Haselwurm; Glasschlange; Bruchschleiche
f Orvet; Orvet fragile; Orvet commun; Serpent de verre; Serpent aveugle
i Orbettino; Angue fragile; Sordone; Giazòl

490 **Aniliidae**
Squamata
e Coral Pipe Snakes; Cylinder Snakes; Pipe Snakes; South American Pipe Snakes
d Rollschlangen; Wühlschlangen; Eigentliche Rollschlangen
f Aniliidés

491 ***Anilius***
Squamata - Aniliidae
e Coral Pipe Snakes; False Coral Snakes; Pipe Snakes
d Korallenrollschlangen
f Serpents-tube d'Amérique du Sud

492 ***Anilius scytale***
Squamata - Aniliidae
e Coral Pipe Snake; Coral Cylinder Snake; Scarlet Pipe Snake; South American Pipe Snake; False Coral Snake; Red Pipe Snake
d Korallenrollschlange
f Serpent-tube d'Amérique du Sud; Faux serpent-corail; Serpent rouleau; Serpent à deux têtes; Faux corail

493 ***Anisolepis***
Squamata - Iguanaidae
e Tree Lizards

494 ***Anisolepis grilli***
Squamata - Iguanaidae
e Boulenger's Tree Lizard

495 ***Anisolepis undulatus***
Squamata - Iguanaidae
e Wiegmann's Tree Lizard

496 ***Anniella***
Squamata - Aniliidae
e Legless Lizards; American Legless Lizards; North American Legless Lizard
d Ringelschleichen
f Lézards apodes

497 ***Anniella geronimensis***
Squamata - Aniliidae
e Baja California Legless Lizard
d Geronimo-Ringelschleiche
f Lézard apode de Geronimo

498 ***Anniella pulchra***
Squamata - Aniliidae
e California Legless Lizard
d Kalifornische Ringelschleiche
f Lézard apode de Californie

499 ***Anniella pulchra nigra***
Squamata - Aniliidae
e Black Legless Lizard
d Schwarze Ringelschleiche; Dunkle Ringelschleiche

500 ***Anniella pulchra pulchra***
Squamata - Aniliidae
e Silvery Legless Lizard

501 **Anniellidae**
Squamata
e Shovel-snouted Legless Lizards; California Legless Lizards (NA)
d Ringelschleichen
f Anniellidés; Lézards apodes

502 ***Anolis***
Squamata - Iguanaidae
e Anoles; American Chameleons
d Saumfinger; Anolis; Saumfingerechsen
f Anolis
i Anolidi

503 ***Anolis altitudinalis***
Squamata - Iguanaidae
e Turquino Green-mottled Anole

504 ***Anolis achilles***
Squamata - Iguanaidae
e Achilles's Anole

505 ***Anolis acutus***
Squamata - Iguanaidae
e Sharp Anole; St. Croix Tree Lizard ; St. Croix Anole (NA)

506 ***Anolis aeneus***
Squamata – Iguanaidae
e Bronze Anole; Grenada Bush Anole (WI)

507 ***Anolis aequatorialis***
Squamata - Iguanaidae
e Equatorial Anole

508 ***Anolis agassizi***
Squamata - Iguanaidae
e Agassiz's Anole

509 ***Anolis agueroi***
Squamata - Iguanaidae
e Cabo Cruz Bearded Anole

510 ***Anolis ahli***
Squamata - Iguanaidae
e Ahl's Anole; Escambray Blue-eyed Anole

511 ***Anolis alayoni***
Squamata - Iguanaidae
e Guantanamo Twig Anole

512 ***Anolis albi***
Squamata - Iguanaidae
e Light Anole

513 ***Anolis albimaculatus***
Squamata - Iguanaidae
e White-spotted Anole

514 ***Anolis alfaroi***
Squamata - Iguanaidae
e Small-fanned Bush Anole

515 ***Anolis aliniger***
Squamata - Iguanaidae
e La Vega Anole; Hispaniolan Orange-armpit Anole

516 ***Anolis allisoni***
Squamata - Iguanaidae
e Allison's Anole; Cuban Blue Anole

517 ***Anolis allogus***
Squamata - Iguanaidae
e Bueycito Anole; Spanish Flag Anole

518 ***Anolis altae***
Squamata - Iguanaidae
e High Anole

519 ***Anolis altavelensis***
Squamata - Iguanaidae
e Noble's Anole; Alto Velo Gracile Anole

520 ***Anolis altivelensis***
Squamata - Iguanaidae
e Hassler's Anole

521 ***Anolis alumina***
Squamata - Iguanaidae
e Shiny Anole; Barahona Grass Anole

522 ***Anolis alutaceus***
Squamata - Iguanaidae
e Monte Verde Anole; Blue-eyed Twig Anole

523 ***Anolis andianus***
Squamata - Iguanaidae
e Andes Anole

524 ***Anolis anfiloquioi***
Squamata - Iguanaidae
e Anfiloqui Anole; Brown-eyed Bush Anole

525 ***Anolis angusticeps***
Squamata - Iguanaidae

e Cienfuegos Anole; Cuban Twig Anole (WI)

526 ***Anolis anisolepis***
Squamata - Iguanaidae
e Chiappas Ornate Anole

527 ***Anolis annectens***
Squamata - Iguanaidae
e Annexe Anole

528 ***Anolis antioquiae***
Squamata - Iguanaidae
e Antioquia Anole

529 ***Anolis antonii***
Squamata - Iguanaidae
e Anton's Anole

530 ***Anolis apollinaris***
Squamata - Iguanaidae
e Boulenger's Anole

531 ***Anolis aquaticus***
Squamata - Iguanaidae
e Water Anole

532 ***Anolis aregenteolus***
Squamata - Iguanaidae
e Guantanamo Anole; Cuban Trunk Anole

533 ***Anolis argillaceus***
Squamata - Iguanaidae
e Bay Anole; Cuban Dark-bark Anole

534 ***Anolis armouri***
Squamata - Iguanaidae
e Armoured Anole; Black-throated Stout Anole

535 ***Anolis attenuatus***
Squamata - Iguanaidae
e Slender Anole

536 ***Anolis auratus***
Squamata - Iguanaidae
e Grass Anole
d Grasanolis; Kolumbianischer Baumleguan

537 ***Anolis baccatus***
Squamata - Iguanaidae
e Bocourt's Anole

538 ***Anolis bahorucaensis***
Squamata - Iguanaidae
e Baoruca Anole; Baoruco Long-snouted Anole

539 ***Anolis baleatus***
Squamata - Iguanaidae
e Puerto Plata Anole; Dominica Giant Anole

540 ***Anolis baracoae***
Squamata - Iguanaidae
e Baracoa Anole; Baracoa Giant Anole

541 ***Anolis barahonae***
Squamata - Iguanaidae
e Barahona Anole; Barahona Giant Anole

542 ***Anolis barbatus***
Squamata - Iguanaidae
e Western Bearded Anole

543 ***Anolis barkeri***
Squamata - Iguanaidae
e Barker's Anole

544 ***Anolis bartschi***
Squamata - Iguanaidae
e West Cuban Anole; Pinar del Rio Cliff Anole
d Höhlenanolis

545 ***Anolis bateatus***
Squamata - Iguanaidae

e River Anole

546 ***Anolis bimaculatus***
Squamata - Iguanaidae
e Panther Anole; Statia Bank Tree Anole (WI)

547 ***Anolis binotatus***
Squamata - Iguanaidae
e Two-marked Anole

548 ***Anolis biporcatus***
Squamata - Iguanaidae
e Neotropical Green Anole; Central American Green Anole
d Großer Honduras-Anolis

549 ***Anolis birama***
Squamata - Iguanaidae
e Branch Anole; Cuban Big-eared Anole

550 ***Anolis biscutiger***
Squamata - Iguanaidae
e Twinshield Anole

551 ***Anolis bitectus***
Squamata - Iguanaidae
e Roof Anole

552 ***Anolis blanquillanus***
Squamata - Iguanaidae
e Hummelinck's Anole

553 ***Anolis boettgeri***
Squamata - Iguanaidae
e Boettger's Anole

554 ***Anolis bombiceps***
Squamata - Iguanaidae
e Surprise Anole; Blue-lipped Anole

555 ***Anolis bonairensis***
Squamata - Iguanaidae
e Ruthven's Anole

556 ***Anolis bourgeoei***
Squamata - Iguanaidae
e Bourgeoe's Anole

557 ***Anolis bransfordii***
Squamata - Iguanaidae
e Bransford's Anole

558 ***Anolis breedlovei***
Squamata - Iguanaidae
e Breedlove's Anole

559 ***Anolis bremeri***
Squamata - Iguanaidae
e Herradura Anole; Cuban Variegated Anole

560 ***Anolis brevirostris***
Squamata - Iguanaidae
e Shortnose Anole; Gracile Desert Anole

561 ***Anolis bruneti***
Squamata - Iguanaidae
e Brunet's Anole

562 ***Anolis brunneus***
Squamata - Iguanaidae
e Crooked Island Anole; Crooked-Acklins Green Anole (WI)

563 ***Anolis calimae***
Squamata - Iguanaidae
e Ayala's Anole

564 ***Anolis capito***
Squamata - Iguanaidae
e Bighead Anole; Big-headed Anole

565 ***Anolis caquetae***
Squamata - Iguanaidae
e Caqueta Anole

566 ***Anolis carolinensis***
Squamata - Iguanaidae
e Green Anole; Alligator Lizard (NA); Fence Lizard (NA); Green Lizard (NA); Common Green Anole; Carolina Anole (WI); American Chameleon; American Anole
d Rotkehlanolis; Amerikanisches Chamäleon
f Anolis vert; Anolis de la Caroline

567 ***Anolis carolinensis carolinensis***
Squamata - Iguanaidae
e Northern Green Anole; Green Lizard; Carolina Anole; Green Anole

568 ***Anolis carolinensis seminolus***
Squamata - Iguanaidae
e Southern Green Anole

569 ***Anolis carpenteri***
Squamata - Iguanaidae
e Carpenter Anole; Carpenter's Anole

570 ***Anolis casildae***
Squamata - Iguanaidae
e Casilda's Anole; Giant Anole

571 ***Anolis caudalis***
Squamata - Iguanaidae
e Cochran's Gianthead Anole; Gonave Gracile Anole

572 ***Anolis centralis***
Squamata - Iguanaidae
e Central Anole; Central Pallid Anole

573 ***Anolis chloris***
Squamata - Iguanaidae
e Boulenger's Green Anole; Green Anole

574 ***Anolis chlorocyanus***
Squamata - Iguanaidae
e Hispaniola Green Anole; Northern Green Anole ; Blue-green Anole
d Haiti-Anole; Laubanole
f Anolis vert-bleu

575 ***Anolis chocorum***
Squamata - Iguanaidae
e Chocolate Anole; Choco Green Anole

576 ***Anolis christophei***
Squamata - Iguanaidae
e King Christophe Anole; Big-fanned Trunk Anole

577 ***Anolis chrysolepis***
Squamata - Iguanaidae
e Goldenscale Anole
d Goldstreifenanolis
f Anolis chrysolépide

578 ***Anolis chrysolepis scypheus***
Squamata - Iguanaidae
e Yellow-tongued Anole

579 ***Anolis chrysolepis tandae***
Squamata - Iguanaidae
e Blue-throated Anole

580 ***Anolis clivicola***
Squamata - Iguanaidae
e Mountain Anole; Turquino Fern Anole

581 ***Anolis cobanensis***
Squamata - Iguanaidae
e Stuart's Anole

582 ***Anolis coelestinus***
Squamata - Iguanaidae
e Jeremie Anole; Southern Green Anole

583 ***Anolis compressicauda***
Squamata - Iguanaidae
e Malposo Scaly Anole

584 ***Anolis concolor***
Squamata - Iguanaidae
e Isla San Andres Anole

585 ***Anolis confusus***
Squamata - Iguanaidae
e Cabo Cruz Trunk Anole

586 ***Anolis conspersus***
Squamata - Iguanaidae
e Grand Cayman Anole; Caymans Blue-fanned Anole (WI)

587 ***Anolis cooki***
Squamata - Iguanaidae
e Cook's Anole; Dryland Anole ; Guanica Pallid Anole

588 ***Anolis crassulus***
Squamata - Iguanaidae
e Ornate Anole

589 ***Anolis cristatellus***
Squamata - Iguanaidae
e Crested Anole; Man Lizard ; Puerto Rican Crested Anole
d Kammanolis
f Petit anolis à crête

590 ***Anolis cristatellus cristatellus***
Squamata - Iguanaidae
e Puerto Rican Crested Anole

591 ***Anolis cristifer***
Squamata - Iguanaidae
e Cristifer Anole

592 ***Anolis cumingi***
Squamata - Iguanaidae
e Balsas Anole

593 ***Anolis cupeyalensis***
Squamata - Iguanaidae
e Cupeyal Anole; Stripe-bellied Grass Anole

594 ***Anolis cupreus***
Squamata - Iguanaidae
e Copper Anole; Dry Forest Anole

595 ***Anolis cuprinus***
Squamata - Iguanaidae
e Chiapas Anole

596 ***Anolis curtus***
Squamata - Iguanaidae
e Short Anole

597 ***Anolis cuvieri***
Squamata - Iguanaidae
e Cuvier's Anole; Puerto Rican Giant Anole
d Riesenanolis

598 ***Anolis cyanopleurus***
Squamata - Iguanaidae
e Yateras Anole; Green Fern Anole

599 ***Anolis cybotes***
Squamata - Iguanaidae
e Largehead Anole; Large-headed Anole; Hispaniolan Stout Anole
d Dickkopfanolis
f Anolis à grosse tête

600 ***Anolis cybotes cybotes***
Squamata - Iguanaidae
e Common Large-headed Anole

601 ***Anolis cymbops***
Squamata - Iguanaidae
e Cope's Veracruz Anole

602 ***Anolis damulus***
Squamata - Iguanaidae
e Cope's Smooth Anole

603 ***Anolis darlingtoni***
Squamata - Iguanaidae
e Darlington's Anole;

La Hotte Twig Anole

604 ***Anolis delafuente***
Squamata - Iguanaidae
e Guamuhaya Anole; Sierra de Trinidad Crested Anole

605 ***Anolis deltae***
Squamata - Iguanaidae
e Delta Anole

606 ***Anolis desechensis***
Squamata - Iguanaidae
e Heatwole's Anole; Desecheo Anole

607 ***Anolis dissimilis***
Squamata - Iguanaidae
e Odd Anole

608 ***Anolis distichoides***
Squamata - Iguanaidae
e Rosen's Anole

609 ***Anolis distichus***
Squamata - Iguanaidae
e Bark Anole; Yellow-throated Anole; Bahaman Bark Anole; Hispaniolan Gracile Anole (WI)
d Zaunanolis; Rindenanolis
f Anolis écorce

610 ***Anolis distichus dominicensis***
Squamata - Iguanaidae
e Green Bark Anole

611 ***Anolis distichus floridanus***
Squamata - Iguanaidae
e Florida Bark Anole

612 ***Anolis dolfusianus***
Squamata - Iguanaidae
e Coffee Anole

613 ***Anolis dolichocephalus***
Squamata - Iguanaidae
e Place Negre Anole; La Hotte Long-snouted Anole

614 ***Anolis duellmani***
Squamata - Iguanaidae
e Duellman's Pygmy Anole

615 ***Anolis dunni***
Squamata - Iguanaidae
e Dunn's Anole

616 ***Anolis equestris***
Squamata - Iguanaidae
e Knight Anole; Giant Chameleon; Cuban Giant Anole
d Ritteranolis
f Anolis chevalier; Anolis à écharpe
i Anolide di Cuba

617 ***Anolis equestris equestris***
Squamata - Iguanaidae
e Western Knight Anole

618 ***Anolis ernestwilliamsi***
Squamata - Iguanaidae
e Ernest's Anole; Carrot Rock Anole

619 ***Anolis etheridgei***
Squamata - Iguanaidae
e Etheridge's Anole; Montane Bush Anole

620 ***Anolis eugenegrahami***
Squamata - Iguanaidae
e Eugene's Anole; Black Stream Anole

621 ***Anolis eulaemus***
Squamata - Iguanaidae
e Good Anole

622 ***Anolis evermanni***
Squamata - Iguanaidae
e Eversmann's Anole; Emerald Anole ; Puerto Rican Emerald Anole

623 ***Anolis extrema***
Squamata - Iguanaidae
e Extreme Anole

624 ***Anolis fasciata***
Squamata - Iguanaidae
e Banded Anole; Barbados Anole (WI)

625 ***Anolis ferreus***
Squamata - Iguanaidae
e Morne Constant Anole; Marie-Galante Anole (WI)
d Großer Kammanolis

626 ***Anolis festae***
Squamata - Iguanaidae
e Veronica's Anole

627 ***Anolis fitchi***
Squamata - Iguanaidae
e Fitch's Anole

628 ***Anolis forbesi***
Squamata - Iguanaidae
e Forbes's Anole

629 ***Anolis foresti***
Squamata - Iguanaidae
e Forest Anole

630 ***Anolis fowleri***
Squamata - Iguanaidae
e Fowler's Anole; Green-banded Anole

631 ***Anolis fraseri***
Squamata - Iguanaidae
e Fraser's Anole

632 ***Anolis frenatus***
Squamata - Iguanaidae
e Bridled Anole

633 ***Anolis fugitivus***
Squamata - Iguanaidae
e Moa Anole; Green-headed Grass Anole

634 ***Anolis fungosus***
Squamata - Iguanaidae
e Myers's Anole

635 ***Anolis fuscoauratus***
Squamata - Iguanaidae
e Brown-eared Anole
f Anolis brun doré

636 ***Anolis fuscoauratus fuscoauratus***
Squamata - Iguanaidae
e Slender Anole

637 ***Anolis gadovii***
Squamata - Iguanaidae
e Gadow's Anole

638 ***Anolis garmani***
Squamata - Iguanaidae
e Jamaica Giant Anole; Jamaican Giant Anole; Jamaican Anole
d Grüner Jamaika-Saumfinger; Jamaika-Anolis

639 ***Anolis garridoi***
Squamata - Iguanaidae
e Escambray Twig Anole

640 ***Anolis gemnosus***
Squamata - Iguanaidae
e O'Shaughnessy's Anole

641 ***Anolis gibbiceps***
Squamata - Iguanaidae
e Hook Anole

642 ***Anolis gingivinus***
Squamata - Iguanaidae
e Anguilla Anole; Anguilla Bank Anole (WI)

643 ***Anolis godmani***
Squamata - Iguanaidae
e Godman's Anole

644 ***Anolis graciliceps***
Squamata - Iguanaidae
e Charm Anole

645 ***Anolis grahami***
Squamata - Iguanaidae
e Graham's Anole; Jamaican Turquoise Anole (WI)

646 ***Anolis granuliceps***
Squamata - Iguanaidae
e Granular Anole

647 ***Anolis grayi***
Squamata - Iguanaidae
e Gray's Anole

648 ***Anolis grisea***
Squamata - Iguanaidae
e Grey Anole; St. Vincent Tree Anole (WI)

649 ***Anolis guafe***
Squamata - Iguanaidae
e Cabo Cruz Banded Anole

650 ***Anolis guamuhaya***
Squamata - Iguanaidae
e Escambray Bearded Anole

651 ***Anolis guazima***
Squamata - Iguanaidae
e Sierra Anole; Turquino Twig Anole

652 ***Anolis gundlachi***
Squamata - Iguanaidae
e Gundlach's Anole; Yellow-bearded Anole ; Yellow-chinned Anole

653 ***Anolis haetianus***
Squamata - Iguanaidae
e Tiburon Anole; La Hotte Stout Anole

654 ***Anolis haguei***
Squamata - Iguanaidae
e Hague's Anole

655 ***Anolis hendersoni***
Squamata - Iguanaidae
e Henderson's Anole; La Selle Long-snouted Anole

656 ***Anolis hernandezi***
Squamata - Iguanaidae
e Hernandez's Anole

657 ***Anolis homolechis***
Squamata - Iguanaidae
e Habana Anole; Cuban White-fanned Anole

658 ***Anolis huilae***
Squamata - Iguanaidae
e Huila Anole

659 ***Anolis humilis***
Squamata - Iguanaidae
e Humble Anole; Ground Anole

660 ***Anolis ibague***
Squamata - Iguanaidae
e Ibague Anole

661 ***Anolis imias***
Squamata - Iguanaidae
e Imias Anole; Imias Rock Anole

662 ***Anolis inexpectata***
Squamata - Iguanaidae
e Estrada's Anole

663 ***Anolis insignis***
Squamata - Iguanaidae
e Decorated Anole; Short-legged Giant Anole

664 ***Anolis insolitus***
Squamata - Iguanaidae
e La Palma Anole; Cordillera Central Twig Anole

665 ***Anolis intermedius***
Squamata - Iguanaidae
e Intermediate Anole; Gray Lichen Anole (NA); Grey Lichen Anole

666 ***Anolis isolepis***
Squamata - Iguanaidae
e Jatibonico Anole

667 ***Anolis isthmicus***
Squamata - Iguanaidae
e Tehuantepec Anole

668 ***Anolis jacare***
Squamata - Iguanaidae
e Jacare Anole

669 ***Anolis johnmeyeri***
Squamata - Iguanaidae
e Meyer's Anole; John Meyer's Anole

670 ***Anolis juangundlachi***
Squamata - Iguanaidae
e Finca Ceres Anole

671 ***Anolis jubar***
Squamata - Iguanaidae
e Cubitas Anole

672 ***Anolis kemptoni***
Squamata - Iguanaidae
e Kempton's Anole

673 ***Anolis koopmani***
Squamata - Iguanaidae
e Koopman's Anole; Brown Red-bellied Anole

674 ***Anolis krugi***
Squamata - Iguanaidae
e Krug's Anole; Upland Grass Anole ; Olive Bush Anole

675 ***Anolis laevis***
Squamata - Iguanaidae
e Smooth Anole

676 ***Anolis laeviventris***
Squamata - Iguanaidae
e White Anole

677 ***Anolis leachii***
Squamata - Iguanaidae
e Leach's Anole; Barbuda Bank Tree Anole (WI)

678 ***Anolis lemurinus***
Squamata - Iguanaidae
e Ghost Anole

679 ***Anolis lerneri***
Squamata - Iguanaidae
e Lerner's Anole

680 ***Anolis limifrons***
Squamata - Iguanaidae
e Border Anole; Slender Anole

681 ***Anolis lineatopus***
Squamata - Iguanaidae
e Stripefoot Anole; Jamaican Grey Anole (WI)
d Strichfußanolis

682 ***Anolis lineatus***
Squamata - Iguanaidae
e Striped Anole

683 ***Anolis liogaster***
Squamata - Iguanaidae
e Guerreran Anole

684 ***Anolis lionotus***
Squamata - Iguanaidae
e Lion Anole

685 ***Anolis lividus***
Squamata - Iguanaidae
e Plymouth Anole; Montserrat Anole (WI)

686 ***Anolis longiceps***
Squamata - Iguanaidae
e Navassa Anole

687 ***Anolis longitibialis***
Squamata - Iguanaidae
e Isla Beata Anole; Barahona Stout Anole

688 ***Anolis loveridgei***
Squamata - Iguanaidae
e Loveridge's Anole

689 ***Anolis loysianus***
Squamata - Iguanaidae
e Peach Anole

690 ***Anolis luciae***
Squamata - Iguanaidae
e St. Lucia Anole

691 ***Anolis lucius***
Squamata - Iguanaidae
e Cave Anole
d Höhlenanolis; Geckoanolis
f Anolis à museau de brochet

692 ***Anolis luteogularis***
Squamata - Iguanaidae
e White-throated Anole

693 ***Anolis luteosignifer***
Squamata - Iguanaidae
e Cayman Brac Anole

694 ***Anolis lynchi***
Squamata - Iguanaidae
e Lynch's Anole

695 ***Anolis macrini***
Squamata - Iguanaidae
e Macrinius's Anole

696 ***Anolis macrolepis***
Squamata - Iguanaidae
e Big-scaled Anole

697 ***Anolis maculigua***
Squamata - Iguanaidae
e Marked Anole

698 ***Anolis maculiventris***
Squamata - Iguanaidae
e Blotchbelly Anole

699 ***Anolis marcanoi***
Squamata - Iguanaidae
e La Horma Anole; Red-fanned Stout Anole

700 ***Anolis mariarum***
Squamata - Iguanaidae
e Blemished Anole

701 ***Anolis marmoratus***
Squamata - Iguanaidae
e Leopard Anole; Guadeloupe Anole (WI)
f Anolis à tête marbrée

702 ***Anolis marron***
Squamata - Iguanaidae
e Jacmel Anole; Jacmel Gracile Anole

703 ***Anolis matudai***
Squamata - Iguanaidae
e Matuda's Anole

704 ***Anolis maynardi***
Squamata - Iguanaidae
e Maynard's Anole; Caymans Green Anole (WI)

705 ***Anolis megalopithecus***
Squamata - Iguanaidae
e Ruida's Anole

706 ***Anolis megapholidotus***
Squamata - Iguanaidae
e Large-scaled Anole

707 ***Anolis menta***
Squamata - Iguanaidae
e Mixed Anole

708 ***Anolis meridionalis***
Squamata - Iguanaidae
e Meridian Anole

709 ***Anolis mestrei***
Squamata - Iguanaidae
e Pinar del Rio Anole

710 ***Anolis microlepidotus***
Squamata - Iguanaidae
e Guerreran Oak Anole

711 ***Anolis microlepis***
Squamata - Iguanaidae
e Chiapas Smooth Anole

712 ***Anolis microtus***
Squamata - Iguanaidae
e Tiny Anole

713 ***Anolis milleri***
Squamata - Iguanaidae
e Miller's Anole

714 ***Anolis mimus***
Squamata - Iguanaidae
e Santiago Anole

715 ***Anolis monensis***
Squamata - Iguanaidae
e Mona Anole

716 ***Anolis monticola***
Squamata - Iguanaidae
e Foothill Anole; La Hotte Bush Anole

717 ***Anolis nannodes***
Squamata - Iguanaidae
e Little Anole

718 ***Anolis nasofrontalis***
Squamata - Iguanaidae
e Nose Anole

719 ***Anolis naufragus***
Squamata - Iguanaidae
e Hidalgo Anole

720 ***Anolis nebuloides***
Squamata - Iguanaidae
e False Clouded Anole

721 ***Anolis nebulosus***
Squamata - Iguanaidae
e Clouded Anole

722 ***Anolis nelsoni***
Squamata - Iguanaidae
e Nelson's Anole

723 ***Anolis nigropunctatus***
Squamata - Iguanaidae
e Black-spotted Anole

724 ***Anolis noblei***
Squamata - Iguanaidae
e Holguin Anole

725 ***Anolis notopholis***
Squamata - Iguanaidae
e Scalyback Anole

726 ***Anolis nubilus***
Squamata - Iguanaidae
e Redonda Anole

727 ***Anolis occultus***
Squamata - Iguanaidae
e Limestone Anole; Pygmy Anole

728 ***Anolis oculatus***
Squamata - Iguanaidae
e Eyed Anole; Puerto Rican Twig Anole ; Dominica Anole (WI)

729 ***Anolis oligaspis***
Squamata - Iguanaidae
e West Indies Anole

730 ***Anolis olssoni***
Squamata - Iguanaidae
e Monte Cristi Anole; Desert Grass Anole
d Grasanolis

731 ***Anolis omiltemanus***
Squamata - Iguanaidae
e Omilteme Anole

732 ***Anolis onca***
Squamata - Iguanaidae
e Bulky Anole

733 ***Anolis opalinus***
Squamata - Iguanaidae
e Bluefield's Anole; Opal-bellied Anole (WI)

734 ***Anolis ophiolepis***
Squamata - Iguanaidae
e Snakescale Anole
d Grasanolis

735 ***Anolis ortonii***
Squamata - Iguanaidae
e Orton's Anole; Bark Anole

736 ***Anolis oxycephalus***
Squamata - Iguanaidae
e Red-headed Anole

737 ***Anolis oxylophus***
Squamata - Iguanaidae
e Sharp-crested Anole; Stream Anole
d Wasseranolis

738 ***Anolis pachypus***
Squamata - Iguanaidae
e Thick Anole

739 ***Anolis parilis***
Squamata - Iguanaidae
e Ecuador Anole

740 ***Anolis parvauritus***
Squamata - Iguanaidae
e Little-eared Anole

741 ***Anolis parvicirculata***
Squamata - Iguanaidae
e Berriozabal Anole

742 ***Anolis paternus***
Squamata - Iguanaidae
e Nueva Gerona Anole

743 ***Anolis pentaprion***
Squamata - Iguanaidae
e Lichen Anole

744 ***Anolis perraccae***
Squamata - Iguanaidae
e Peracca's Anole

745 ***Anolis petersii***
Squamata - Iguanaidae
e Peters's Anole

746 ***Anolis phyllorhinus***
Squamata - Iguanaidae
e Bat Anole

747 ***Anolis pigmaequestris***
Squamata - Iguanaidae
e Cayo Frances Anole

748 ***Anolis pinchoti***
Squamata - Iguanaidae
e Crab Cay Anole

749 ***Anolis placidus***
Squamata - Iguanaidae
e Placid Anole; Neiba Twig Anole

750 ***Anolis poecilopus***
Squamata - Iguanaidae
e Dappled Anole

751 ***Anolis pogus***
Squamata - Iguanaidae
e Bearded Anole; Anguilla Bank Bush Anole (WI)

752 ***Anolis polylepis***
Squamata - Iguanaidae
e Many-scaled Anole

753 ***Anolis polyrhachis***
Squamata - Iguanaidae
e Sierra Juarez Anole

754 ***Anolis poncensis***
Squamata - Iguanaidae
e Ponce Anole; Dryland Grass Anole ; Ponce Small-fanned Anole

755 ***Anolis porcatus***
Squamata - Iguanaidae
e Cuban Green Anole
d Baumstammanole; Großer Honduras-Anolis

756 ***Anolis princeps***
Squamata - Iguanaidae
e First Anole

757 ***Anolis proboscis***
Squamata - Iguanaidae
e Horned Anole

758 ***Anolis procellaris***
Squamata - Iguanaidae
e Stormy Anole

759 ***Anolis pseudotigrinus***
Squamata - Iguanaidae
e False Tiger Anole

760 ***Anolis pulchellus***
Squamata - Iguanaidae
e Puerto Rican Anole; Sharp-mouthed Lizard ; Puerto Rican Bush Anole
d Puerto-Rico-Grasanolis
f Anolis gentil

761 ***Anolis pumilis***
Squamata - Iguanaidae
e Garrido's Anole

762 ***Anolis punctatus***
Squamata - Iguanaidae
e Spotted Anole
f Anolis ponctué

763 ***Anolis purpurescens***
Squamata - Iguanaidae
e Purple Anole

764 ***Anolis pygmaeus***
Squamata - Iguanaidae
e Chiapas Pygmy Anole

765 ***Anolis quadriocellifer***
Squamata - Iguanaidae
e Peninsula Anole

766 ***Anolis quaggulus***
Squamata - Iguanaidae
e Flabby Anole

767 ***Anolis quercorum***
Squamata - Iguanaidae
e Oaxacan Oak Anole

768 ***Anolis reconditus***
Squamata - Iguanaidae
e Blue Mountains Anole

769 ***Anolis rhombifer***
Squamata - Iguanaidae
e Rhombifer Anole

770 ***Anolis richardi***
Squamata - Iguanaidae
e Richard's Anole; Grenada Tree Anole (WI)
f Anolis de Richard

771 ***Anolis ricordi***
Squamata - Iguanaidae
e Haitian Green Anole; Haitian Giant Anole
d Riesenanolis; Haiti-Risenanolis
f Anolis de Ricord

772 ***Anolis rimarum***
Squamata - Iguanaidae
e Marmelade Anole; Artibonite Bush Anole

773 ***Anolis riparius***
Squamata - Iguanaidae
e Riverbank Anole

774 ***Anolis rivalis***
Squamata - Iguanaidae
e Neighbour

775 ***Anolis robinsoni***
Squamata - Iguanaidae
e Robinson's Anole

776 ***Anolis rodriguezi***
Squamata - Iguanaidae
e Rodriguez's Anole; Border Anole; Smooth Anole

777 ***Anolis roosevelti***
Squamata - Iguanaidae
e Culebra Giant Anole; Virgin Islands Giant Anole
d Roosevelts Anolis

778 ***Anolis roquet***
Squamata - Iguanaidae
e Savanna Anole; Savannah Anole; Martinique Anole (WI)

779 ***Anolis roquet aeneus***
Squamata - Iguanaidae
d Blauer Karibik-Anolis

780 ***Anolis roquet guyananensis***
Squamata - Iguanaidae
d Blauer Anolis

781 ***Anolis roquet roquet***
Squamata - Iguanaidae
d Martinique-Anolis

782 ***Anolis roquet sumnus***
Squamata - Iguanaidae
d Gefleckter Anolis

783 ***Anolis rubribarbus***
Squamata - Iguanaidae
e Sagua de Tanamo Anole

784 ***Anolis rupinae***
Squamata - Iguanaidae
e Castillon Anole; Banded Red-bellied Anole

785 ***Anolis sabanus***
Squamata - Iguanaidae
e Saban Anole; Saba Anole (WI)
d Saba-Anolis; Pantheranolis

786 ***Anolis sagrei***
Squamata - Iguanaidae

e Brown Anole; West Indian Chameleon; Cuban Brown Anole (WI)
d Braune Anolis; Bahama-Anolis; Haiti-Anolis; Kuba-Anolis; Sagrei-Anolis
f Anolis brun; Anolis de Sagra

787 ***Anolis sagrei ordinatus***
Squamata - Iguanaidae
e Bahaman Brown Anole

788 ***Anolis sagrei sagrei***
Squamata - Iguanaidae
e Cuban Brown Anole
d Kuba-Anolis

789 ***Anolis sagrei stejnegeri***
Squamata - Iguanaidae
e Key West Anole

790 ***Anolis sallaei***
Squamata - Iguanaidae
e Sallae's Anole

791 ***Anolis salvini***
Squamata - Iguanaidae
e Salvin's Anole

792 ***Anolis santamartae***
Squamata - Iguanaidae
e Santa Marta Anole

793 ***Anolis schiedii***
Squamata - Iguanaidae
e Schied's Anole

794 ***Anolis schmidti***
Squamata - Iguanaidae
e Schmidt's Anole

795 ***Anolis schwartzi***
Squamata - Iguanaidae
e Schwartz's Anole; Statia Bank Bush Anole (WI)

796 ***Anolis scriptus***
Squamata - Iguanaidae
e Silver Key Anole; Southern Bahamas Anole (WI)

797 ***Anolis semilineatus***
Squamata - Iguanaidae
e Santo Domingo Anole; Hispaniolan Grass Anole
d Grasanolis

798 ***Anolis sericeus***
Squamata - Iguanaidae
e Silky Anole

799 ***Anolis sheplani***
Squamata - Iguanaidae
e Cabral Anole; Baoruco Twig Anole

800 ***Anolis shrevei***
Squamata - Iguanaidae
e Shreve's Anole; Cordillera Central Stout Anole

801 ***Anolis simmonsi***
Squamata - Iguanaidae
e Simmons's Anole

802 ***Anolis singularis***
Squamata - Iguanaidae
e Porcupine Anole; Hispaniolan Green Twig Anole

803 ***Anolis smallwoodi***
Squamata - Iguanaidae
e Smallwood's Anole

804 ***Anolis smaragdinus***
Squamata - Iguanaidae
e South Point Anole; Bahamian Green Anole (WI)

805 ***Anolis sminthus***
Squamata - Iguanaidae
e Mouse Anole

806 ***Anolis solitarius***
Squamata - Iguanaidae
e Solitare Anole

807 ***Anolis spectrum***
Squamata - Iguanaidae
e Mantanza's Anole

808 ***Anolis squamulatus***
Squamata - Iguanaidae
e Small-scaled Anole

809 ***Anolis strahmi***
Squamata - Iguanaidae
e Strahl's Anole; Baoruca Stout Anole

810 ***Anolis stratulus***
Squamata - Iguanaidae
e St. Thomas Anole; Salmon Lizard ; Barred Anole; Puerto Rican Spotted Anole

811 ***Anolis subocularis***
Squamata - Iguanaidae
e Pacific Anole

812 ***Anolis sulcifrons***
Squamata - Iguanaidae
e Grooved Anole

813 ***Anolis taylori***
Squamata - Iguanaidae
e Taylor's Anole

814 ***Anolis tigrinus***
Squamata - Iguanaidae
e Tiger Anole

815 ***Anolis townsendi***
Squamata - Iguanaidae
e Townsend's Anole

816 ***Anolis trachyderma***
Squamata - Iguanaidae
e Roughskin Anole; Common Forest Anole

817 ***Anolis transversalis***
Squamata - Iguanaidae
e Transverse Anole; Banded Tree Anole

818 ***Anolis trinitatis***
Squamata - Iguanaidae
e Trinidad Anole; St. Vincent Bush Anole (WI)

819 ***Anolis tropidogaster***
Squamata - Iguanaidae
e Tropical Anole

820 ***Anolis tropidolepis***
Squamata - Iguanaidae
e Swift Anole; Cloud Forests Anole

821 ***Anolis tropidonotus***
Squamata - Iguanaidae
e Greater Scaly Anole

822 ***Anolis uniformis***
Squamata - Iguanaidae
e Lesser Scaly Anole

823 ***Anolis utowanae***
Squamata - Iguanaidae
e Utowana Anole

824 ***Anolis valencienni***
Squamata - Iguanaidae
e Short-tail Anole; Sword-tailed Iguanid; Jamaican Twig Anole (WI)
d Schwertschwanzanolis; Schwertschwanzleguan
f Anolis de Valenciennes

825 ***Anolis vanidicus***
Squamata - Iguanaidae
e Vanidicus Anole

826 ***Anolis vaupesianus***
Squamata - Iguanaidae
e Williams's Anole

827 ***Anolis ventrimaculatus***
Squamata - Iguanaidae
e Speckled Anole

828 ***Anolis vermiculatus***
Squamata - Iguanaidae
e Vinales Anole
d Wasseranolis
f Anolis vermiculé

829 ***Anolis villai***
Squamata - Iguanaidae
e Country Anole

830 ***Anolis vittigerus***
Squamata - Iguanaidae
e Garland Anole

831 ***Anolis vociferans***
Squamata - Iguanaidae
e Roaring Anole

832 ***Anolis wattsi***
Squamata - Iguanaidae
e Watts's Anole; Barbuda Bank Bush Anole (WI)

833 ***Anolis websteri***
Squamata - Iguanaidae
e Webster's Anole; Yellow-bellied Desert Anole

834 ***Anolis whitemani***
Squamata - Iguanaidae
e Whiteman's Anole; Pallid Stout Anole

835 ***Anolis woodi***
Squamata - Iguanaidae
e Wood's Anole
d Chamäleone

836 **Anomalepididae**
Squamata
e Blind Snakes; American Blind Snakes
d Amerikanische Blindschlangen; Blindschlangen
f Anomalepididés
i Anomalepidi

837 ***Anomalepis***
Squamata - Anomalepididae
e South American Blind Snakes; South American Worm Snakes
d Amerikanische Blindschlangen; Südamerikanische Blindschlangen

838 ***Anomalepis aspinosus***
Squamata - Anomalepididae
e Taylor's Peru Blind Snake; Taylor's Peru Worm Snake

839 ***Anomalepis colombius***
Squamata - Anomalepididae
e Colombian Blind Snake; Colombian Worm Snake

840 ***Anomalepis flavapices***
Squamata - Anomalepididae
e Ecuador Blind Snake; Ecuador Worm Snake

841 ***Anomalepis mexicanus***
Squamata - Anomalepididae
e Mexican Blind Snake; Mexican Worm Snake; Anomalous Blind Snake

842 ***Anomalopus***
Squamata - Scincidae
e Lesser Burrowing Skinks

843 ***Anomalopus brevicollis***
Squamata - Scincidae
e Shortneck Burrowing Skink

844 ***Anomalopus gowi***
Squamata - Scincidae
e Gow's Burrowing Skink

845 ***Anomalopus leuckartii***
Squamata - Scincidae
e Leuckart's Burrowing Skink

846 ***Anomalopus mackayi***
Squamata - Scincidae
e Mackay's Burrowing Skink

847 ***Anomalopus pluto***
Squamata - Scincidae
e Brown Burrowing Skink

848 ***Anomalopus swansoni***
Squamata - Scincidae
e Swanson's Burrowing Skink

849 ***Anomalopus verreauxi***
Squamata - Scincidae
e Verreaux's Burrowing Skink; Verreaux's Skink; Three-clawed Worm Skink

850 ***Anomochilus***
Squamata - Uropeltidae
e Common Pipe Snakes
d Sumatra-Rollschlangen
i Serpenti cilindrici dell'Asia sudorientale

851 ***Anomochilus leonardi***
Squamata - Uropeltidae
e Leonard's Pipe Snake

852 ***Anomochilus weberi***
Squamata - Uropeltidae
e Pipe Snake; Weber's Cylinder Snake
d Sumatra-Rollschlange

853 ***Anops***
Squamata - Amphisbaenidae
e King's Worm Lizards
d Kings Doppelschleichen
f Amphisbènes de King

854 ***Anops kingii***
Squamata - Amphisbaenidae
e King's Worm Lizard
d Kings Doppelschleiche
f Amphisbène de King

855 ***Antaresia***
Squamata - Boidae
e Antares Pythons

856 ***Antaresia childreni***
Squamata - Boidae
e Children's Python; North Australian Python
d Gefleckter Python; Childrens Python
f Liasis de Children
i Pitone di Children

857 ***Antaresia maculosa***
Squamata - Boidae
e Spotted Python
i Pitone maculato

858 ***Antaresia perthensis***
Squamata - Boidae
e Perth Pygmy Python; Pymy Python; Western Water Python; Anthill Python
d Australischer Zwergpython

859 ***Antaresia stimsoni***
Squamata - Boidae
e Large-blotched Antares Python; Stimson's Python; Large-blotched Python

860 ***Antillophis***
Squamata - Colubridae
e Antillean Snakes
d Antillen-Nattern

861 ***Antillophis andreai***
Squamata - Colubridae
e Common Antillean Snake

862 ***Antillophis parvifrons***
Squamata - Colubridae
e Cope's Antillean Snake; Hispaniolan Lesser Racer (WI)

863 ***Apalone***
Testudines - Trionychidae
e North American Softshells; North american Softshells
d Weichschildkröten
f Tortues molles

864 ***Apalone ater***
Testudines - Trionychidae
e Black Spiny Softshell Turtle; Mexico Softshell Turtle; Black Soft-shelled Turtle; Cuatro Cienegas Soft-shelled Turtle; Cuatro Cienegas Black Turtle
d Schwarze Weichschildkröte
f Trionyx épineux noir; Tortue noire
i Tartaruga a guscio molle della Florida

865 ***Apalone ferox***
Testudines - Trionychidae
e Florida Softshell; Southern Soft-shelled Turtle (NA); Florida Softshell Turtle
d Florida-Weichschildkröte; Wilde Dreiklaue; Beißschildkröte
f Tortue molle de Floride; Trionyx de Floride

866 ***Apalone mutica***
Testudines - Trionychidae
e Smooth Softshell; Softshell; Brown Soft-shelled Turtle (NA); Spineless Soft-shelled Turtle (NA); Smooth Soft-shelled Turtle; Smooth Softshell Turtle
d Glattrandweichschildkröte
f Tortue molle lisse; Trionyx mutique

867 ***Apalone mutica calvata***
Testudines - Trionychidae
e Gulf Coast Smooth Softshell; Gulf Coast Smooth Softshell Turtle
f Trionyx mutique du Golfe

868 ***Apalone mutica mutica***
Testudines - Trionychidae
e Midland Smooth Softshell; Smooth Softshell; Midland Smooth Softshell Turtle
f Ytionyx mutique commune
i Tartaruga a guscio molle liscia

869 ***Apalone spinifera***
Testudines - Trionychidae
e Spiny Softshell; Spiny Soft-shelled Turtle (NA); Spiny Softshell Turtle; Eastern Spiny Softshell
d Dornrandweichschildkröte
f Tortue molle à épines
i Tartaruga a guscio molle spinosa

870 ***Apalone spinifera aspera***
Testudines - Trionychidae
e Gulf Coast Spiny Softshell; Gulf Coast Soft-shelled Turtle; Gulf Coast Spiny Softshell Turtle
f Trionyx épineux du Golfe

871 ***Apalone spinifera emoryi***
Testudines - Trionychidae
e Texas Spiny Softshell; Texas Spiny Soft-shelled Turtle; Texas Softshell; Texas Spiny Softshell Turtle; Rio Grande Spiny Softshell Turtle
f Trionyx épineux du Sud

872 ***Apalone spinifera guadalupensis***
Testudines - Trionychidae
e Guadalupe Spiny Softshell; Guadalupe Spiny Soft-shelled Turtle
f Trionyx épineux de Guadalupe

873 ***Apalone spinifera hartwegi***
Testudines - Trionychidae

e Western Spiny Softshell; Western Spiny Soft-shelled Turtle; Western Spiny Softshell Turtle
f Trionyx épineux de l'Ouest

874 ***Apalone spinifera pallida***
Testudines - Trionychidae
e Pallid Spiny Softshell; Pallid Spiny Soft-shelled Turtle; Pallid Spiny Softshell Turtle
f Trionyx épineux pâle

875 ***Apalone spinifera spinifera***
Testudines - Trionychidae
e Eastern Spiny Softshell; Baby Spiny Soft-shelled Turtle (NA); Eastern Spiny Soft-shelled Turtle; Eastern Spiny Softshell Turtle; Spiny Softshell
f Tortue molle à épines

876 ***Aparallactus***
Squamata - Colubridae
e Centipede-eaters; Centipede-eater Snakes
d Tausendfüßerfresser

877 ***Aparallactus capensis***
Squamata - Colubridae
e Cape Centipede-eater; Cape Centipede-eater Snake (CSA); Cape Black-headed Snake; Centipede-eater
d Tausendfüßerfresser

878 ***Aparallactus capensis capensis***
Squamata - Colubridae
e Black-headed Centipede-eater

879 ***Aparallactus guentheri***
Squamata - Colubridae
e Black Centipede-eater; Zanzibar Centipede-eter; Günther's Snake; Günther's Centipede-eater (CSA)

880 ***Aparallactus jacksoni***
Squamata - Colubridae
e Kilimanjaro Centipede-eater

881 ***Aparallactus lineatus***
Squamata - Colubridae
e Lined Centipede-eater

882 ***Aparallactus lunulatus***
Squamata - Colubridae
e Reticulate Centipede-eater; Banded Centipede-eater; Plumbeous Centipede-eater; Reticulated Centipede-eater (CSA); Blotched-back Centipede-eater (CSA)

883 ***Aparallactus lunulatus lunulatus***
Squamata - Colubridae
e Barred Centipede-eater

884 ***Aparallactus modestus***
Squamata - Colubridae
e Western Forest Centipede-eater

885 ***Aparallactus moeruensis***
Squamata - Colubridae
e Zaire Centipede-eater

886 ***Aparallactus nigriceps***
Squamata - Colubridae
e Mozambique Centipede-eater

887 ***Aparallactus turneri***
Squamata - Colubridae
e Malindi Centipede-eater

888 ***Aparallactus werneri***
Squamata - Colubridae
e Usambara Centipede-eater

889 ***Aperopristis***
Squamata - Iguanaidae
e Peracca's Iguanas

890 ***Aperopristis paronae***
Squamata - Iguanaidae
e Peracca's Iguana

891 ***Aphaniotis***
Squamata - Agamidae
e Earless Agamas; Blue-mouthed Agamas
d Blaumaul-Agamen
Squamata - Agamidae

892 *Aphaniotis acutirostris*
e Indonesia Earless Agama; Earless Agama
d Blaumaul-Agame

893 ***Aphaniotis fusca***
Squamata - Agamidae
e Dusky Earless Agama

894 ***Aphaniotis nasuta***
Squamata - Agamidae
e Sabah Earless Agama

895 ***Aphaniotis ornata***
Squamata - Agamidae
e Ornate Earless Agama

896 ***Aplopeltura***
Squamata - Colubridae
e Blunthead Snakes; Blunthead Slug Snakes; Blunt-headed Tree Snakes; Indian Snail-eaters

897 ***Aplopeltura boa***
Squamata - Colubridae
e Blunthead Snake; Blunthead Slug Snake; Blunt-headed Tree Snake; Indian Snail-eater

898 ***Apodora***
Squamata - Boidae
e Papua Pythons

899 ***Apodora papuanus***
Squamata - Boidae
e Papua Python; Papuan Olive Python; Papuan Water Python

900 ***Aporosaura***
Squamata - Lacertidae
e Sanddivers; Shovel-snouted Lizards (CSA); Sand Lizards; Dune Lizards
d Sandechsen

901 ***Aporosaura anchieta***
Squamata - Lacertidae
e Namib Sanddiver; Shovel-snouted Lizard (CSA); Anchieta's Dune Lizard
d Sandtaucher; Sandechse

902 ***Apostolepis***
Squamata - Colubridae
e Burrowing Snakes; Worm Snakes

903 ***Apostolepis ambiniger***
Squamata - Colubridae
e South American Burrowing Snake

904 ***Apostolepis assimilis***
Squamata - Colubridae
e Reinhardt's Burrowing Snake

905 ***Apostolepis barrioi***
Squamata - Colubridae
e Barrio's Burrowing Snake

906 ***Apostolepis caerensis***
Squamata - Colubridae
e Gomes's Burrowing Snake

907 ***Apostolepis coronata***
Squamata - Colubridae
e Crowned Burrowing Snake

908 ***Apostolepis dorbignyo***
Squamata - Colubridae
e Bolivian Burrowing Snake

909 ***Apostolepis erythronota***
Squamata - Colubridae
e Peters's Burrowing Snake

910 ***Apostolepis flavotorquata***
Squamata - Colubridae
e Central Burrowing Snake

911 ***Apostolepis goiasensis***
Squamata - Colubridae
e Goias Burrowing Snake

912 ***Apostolepis intermedia***
Squamata - Colubridae
e Matto Grosso Burrowing Snake

913 ***Apostolepis longicaudata***
Squamata - Colubridae
e Longhead Burrowing Snake

914 ***Apostolepis niceforoi***
Squamata - Colubridae
e Amazon Burrowing Snake; Amazon Worm Snake

915 ***Apostolepis nigroterminata***
Squamata - Colubridae
e Peru Burrowing Snake; Peruvian Worm Snake

916 ***Apostolepis quinquelineata***
Squamata - Colubridae
e Guyana Burrowing Snake

917 ***Apostolepis rondoni***
Squamata - Colubridae
e Amaral's Burrowng Snake

918 ***Apostolepis tenuis***
Squamata - Colubridae
e Ruthven's Burrowing Snake

919 ***Apostolepis ventrimaculata***
Squamata - Colubridae
e Spotted Burrowing Snake

920 ***Apostolepis villaricae***
Squamata - Colubridae
e Villarica Burrowing Snake

921 ***Apostolepis vittatus***
Squamata - Colubridae
e Cope's Burrowing Snake

922 ***Aprasia***
Squamata - Gekkonidae
e Blunt-tail Legless Geckos; Blunt-tail Legless Lizards; Blunt-tailed Scaly-foots
d Schmuckflossenfüße

923 ***Aprasia aurita***
Squamata - Gekkonidae
e Eared Legless Gecko; Eared Legless Lizard; Mallee Worm Lizard (ANZ)

924 ***Aprasia fusca***
Squamata - Gekkonidae
e Brown Legless Gecko; Brown Legless Lizard

925 ***Aprasia haroldi***
Squamata - Gekkonidae
e Harold's Legless Gecko; Harold's Legless Lizard

926 ***Aprasia inaurita***
Squamata - Gekkonidae
e Earless Legless Gecko; Earless Legless Lizard; Pink-nosed Worm Lizard (ANZ)

927 ***Aprasia parapulchella***
Squamata - Gekkonidae
e Mongolo Legless Gecko; Mongolo Legless Lizard; Pink-tailed Worm Lizard (ANZ)
d Schmuckflossenfuß

928 ***Aprasia pseudopulchella***
Squamata - Gekkonidae
e Flinders Legless Gecko; Flinders Legless Lizard

929 ***Aprasia pulchella***
Squamata - Gekkonidae
e Coastal Legless Gecko; Coastal Legless Lizard
d Schmuckflossenfuß

930 ***Aprasia repens***
Squamata - Gekkonidae
e Hasty Legless Gecko; Hasty Legless Lizard

931 ***Aprasia rostrata***
Squamata - Gekkonidae
e Beaked Legless Gecko; Beaked Legless Lizard

932 ***Aprasia smithi***
Squamata - Gekkonidae
e Smith's Legless Gecko; Smith's Legless Lizard
d Wurmflossenfuß

933 ***Aprasia striolata***
Squamata - Gekkonidae
e Striped Legless Gecko; Striped Legless Lizard; Striped Legless Skink (ANZ)

934 ***Apterygodon***
Squamata - Scincidae
e Borneo Skinks

935 ***Apterygodon vittatum***
Squamata - Scincidae
e Borneo Skink

936 ***Aptycholaemos***
Squamata - Iguanaidae
e Longtail Iguanas

937 ***Aptycholaemos longicauda***
Squamata - Iguanaidae
e Argentine Longtail Iguana; Long-tailed Anole

938 ***Aristelliger***
Squamata - Gekkonidae
e Caribbean Geckos

939 ***Aristelliger barbouri***
Squamata - Gekkonidae
e Striped Caribbean Gecko; Inagua Gecko (WI)

940 ***Aristelliger cochranae***
Squamata - Gekkonidae
e Cochran's Caribbean Gecko; Navassa Gecko

941 ***Aristelliger georgeensis***
Squamata - Gekkonidae
e St. George Island Gecko

942 ***Aristelliger hechti***
Squamata - Gekkonidae
e Hecht's Caribbean Gecko; Caicos Gecko (WI)

943 ***Aristelliger lar***
Squamata - Gekkonidae
e Giant Caribbean Gecko; Hispaniolan Giant Gecko

944 ***Aristelliger praesignis***
Squamata - Gekkonidae
e Spotted Caribbean Gecko; Jamaican Croaking Gecko (WI)

945 ***Arizona***
Squamata - Colubridae
e Glossy Snakes; Faded Snakes
d Arizona-Nattern; Erznattern
f Couleuvres lustrées

946 ***Arizona elegans***
Squamata - Colubridae
e Glossy Snake; Faded Snake
d Erznatter; Arizona-Natter
f Couleuvre élégante

947 ***Arizona elegans arenicola***
Squamata - Colubridae
e Texas Glossy Snake

948 ***Arizona elegans candida***
Squamata - Colubridae
e Mojave Glossy Snake

949 ***Arizona elegans eburnata***
Squamata - Colubridae
e Desert Glossy Snake

950 ***Arizona elegans elegans***
Squamata - Colubridae
e Kansas Glossy Snake; Smooth-scaled Rat Snake (NA); Glossy Snake (NA); Faded Snake (NA)

951 ***Arizona elegans noctivaga***
Squamata - Colubridae
e Arizona Glossy Snake
d Arizona-Natter

952 ***Arizona elegans occidentalis***
Squamata - Colubridae
e California Glossy Snake; Western Glossy Snake
d Westliche Arizona-Natter

953 ***Arizona elegans philipi***
Squamata - Colubridae
e Painted Desert Glossy Snake

954 ***Arizona pacata***
Squamata - Colubridae
e Peninsular Glossy Snake

955 ***Arrhyton***
Squamata - Colubridae
e Island Racers

956 ***Arrhyton ainictum***
Squamata - Colubridae
e Cuban Island Racer

957 ***Arrhyton dolichura***
Squamata - Colubridae
e Habana Island Racer

958 ***Arrhyton landoi***
Squamata - Colubridae
e Schwartz's Island Racer

959 ***Arrhyton taeniatum***
Squamata - Colubridae
e Guenther's Island Racer

960 ***Arrhyton tanyplectum***
Squamata - Colubridae
e San Vicente Island Racer

961 ***Arrhyton vittatum***
Squamata - Colubridae
e Common Island Racer; Island Racer

962 ***Arthrosaura***
Squamata - Gymnophthalmidae
e Arthrosauras; Creek Lizards

963 ***Arthrosaura kocki***
Squamata - Gymnophthalmidae
e Orangebelly Arthrosaura; Eastern Creek Lizard

964 ***Arthrosaura reticulata***
Squamata - Gymnophthalmidae
e Yellowbelly Arthrosaura

965 ***Arthrosaura reticulata reticulata***
Squamata - Gymnophthalmidae
e Western Creek Lizard

966 ***Arthrosaura synaptolepis***
Squamata - Gymnophthalmidae
e Donnelly's Arthrosaura

967 ***Arthroseps***
Squamata - Teiidae
e Arthrosepses

968 ***Arthroseps fluminensis***
Squamata - Teiidae
e Forest Arthroseps

969 ***Arthroseps werneri***
Squamata - Teiidae
e Sao Paolo Arthroseps

970 ***Asaccus***
Squamata - Gekkonidae
e Middle Eastern Geckos

971 ***Asaccus elisae***
Squamata - Gekkonidae
e Werner's Gecko
f Gecko d'Elisa
i Fillodattilo di Elisa

972 ***Asaccus gallagheri***
Squamata - Gekkonidae
e Gallagher's Gecko

973 ***Asaccus griseonotus***
Squamata - Gekkonidae
e Grey-marked Gecko

974 ***Asiocolotes***
Squamata - Gekkonidae
e Golubev's Geckos

975 ***Asiocolotes depressus***
Squamata - Gekkonidae
e Low-lying Gecko; Mountain Dwarf Gecko

976 ***Asiocolotes levitoni***
Squamata - Gekkonidae
e Leviton's Gecko

977 ***Aspidelaps***
Squamata - Elapidae
e Shieldnose Cobras; Shield-nosed Cobras; Shield-nosed Snakes; African Coral Snakes; Shield Snakes; Shieldnose Snakes
d Schildkobras; Nasenkobras; Schildnasenkobras

978 ***Aspidelaps lubricus***
Squamata - Elapidae
e South African Coral Snake; Cape Coral Snake; Coral Snake (CSA)
d Kap-Zwergschildkobra; Südafrikanische Korallenkobra; Südafrikanische Korallenschlange

979 ***Aspidelaps lubricus infuscatus***
Squamata - Elapidae
d Westafrikanische Korallenschlange

980 ***Aspidelaps lubricus lubricus***
Squamata - Elapidae
d Südafrikanische Korallenschlange

981 ***Aspidelaps scutatus***
Squamata - Elapidae
e Shieldnose Cobra; Shield-nosed Cobra; Shieldnose Snake; Shield Snake (CSA); Shield-nosed Snake
d Schildkobra; Schildnasenkobra; Schildschlange

982 ***Aspidelaps scutatus intermedius***
Squamata - Elapidae
d Schildschlange

983 ***Aspidelaps scutatus scutatus***
Squamata - Elapidae
d Schildschlange

984 ***Aspideretes***
Testudines - Trionychidae
e Indian Softshells
d Indische Weichschildkröten

985 ***Aspideretes gangeticus***
Testudines - Trionychidae
e Indian Softshell; Ganges Soft-shelled Turtle; Ganges Softshell (ISC); Indian Softshell Turtle (ISC); Indian

Soft-shelled Turtle (ISC); Ganga Soft-shelled Turtle (ISC); Ganges Softshell Turtle
d Ganges-Weichschildkröte
f Trionyx du Gange; Tortue du Gange
i Tartaruga dal guscio molle del Gange

986 ***Aspideretes hurum***
Testudines - Trionychidae
e Indian Peacock Softshell; Peacock Soft-shelled Turtle; Peacock Softshell (ISC); Peacock-marked Soft-shelled Turtle (ISC); Indian Peacock Softshell Turtle; Peacock Softshell Turtle
d Pfauenaugeweichschildkröte; Pfauenaugenweichschildkröte
f Trionyx à ocelles; Trionyx paon

987 ***Aspideretes leithii***
Testudines - Trionychidae
e Leith's Softshell; Nagpur Soft-shelled Turtle; Deccan Softshell (ISC); Peninsular Soft-shelled Turtle (ISC); Leith's Softshell Turtle; Nagpur Softshell Turtle
f Trionyx de Leith

988 ***Aspideretes nigricans***
Testudines - Trionychidae
e Black Softshell; Black Soft-shelled Turtle; Chittagong Softshell (ISC); Black Softshell Turtle; Bostami Softshell Turtle; Dark-coloured Soft-shelled Turtle
d Dunkle Weichschildkröte
f Trionyx noirâtre; Trionyx sombre
i Tartaruga dal guscio molle scura; Tartaruga a guscio molle scura

989 ***Aspidites***
Squamata - Boidae
e Womas; Black-headed Pythons
d Schwarzkopfpythone; Womas
f Pythons à tête noire

990 ***Aspidites melanocephalus***
Squamata - Boidae
e Black-headed Python
d Schwartzkopfpython
f Python à tête noire d'Australie

991 ***Aspidites ramsayi***
Squamata - Boidae
e Woma; Ramsay's Python; Woma Python
d Woma; Ramsays Python

992 ***Aspidomorphus***
Squamata - Elapidae
e Collared Adders

993 ***Aspidomorphus lineaticollis***
Squamata - Elapidae
e Collared Adder

994 ***Aspidomorphus muelleri***
Squamata - Elapidae
e Müller's Adder

995 ***Aspidomorphus schlegelii***
Squamata - Elapidae
e Schlegel's Adder

996 ***Aspidura***
Squamata - Colubridae
e Rough-sided Snakes; Aspiduras

997 ***Aspidura brachyorrhus***
Squamata - Colubridae
e Boie's Rough-sided Snake; Boie's Aspidura

998 ***Aspidura copii***
Squamata - Colubridae
e Cope's Rough-sided Snake; Cope's Aspidura

999 ***Aspidura deraniyagalae***
Squamata - Colubridae
e Sri Lanka Rough-sided Snake

1000 ***Aspidura drummondhayi***
Squamata - Colubridae
e Drummond-Hay's Rough-sided Snake; Drummond-Hay's Roughside

1001 ***Aspidura guentheri***
Squamata - Colubridae
e Günther's Rough-sided Snake; Günther's Aspidura

1002 ***Aspidura trachyprocta***
Squamata - Colubridae
e Common Rough-sided Snake; Rough-sided Snake; Common Roughside

1003 ***Astrotia***
Squamata - Elapidae
e Stokes's Sea Snakes; Malayan Sea Snakes
d Stokes-Schlangen

1004 ***Astrotia stokesii***
Squamata - Elapidae
e Stokes's Sea Snake; Malayan Sea Snake
d Stokes-Schlange

1005 ***Ateuchosaurus***
Squamata - Scincidae
e Oriental Ateuchosauruses; Unarmoured Skinks; Ateuchosauruses

1006 ***Ateuchosaurus chinensis***
Squamata - Scincidae
e Chinese Ateuchosaurus

1007 ***Ateuchosaurus pellopleurum***
Squamata - Scincidae
e Japanese Ateuchosaurus

1008 ***Atheris***
Squamata - Viperidae
e Bush Vipers; African Bush Vipers; Tree Vipers; Hairy Bush Vipers
d Buschvipern; Baumvipern
f Vipères d'arbres; Vipères arbricoles

1009 ***Atheris ceratophorus***
Squamata - Viperidae
e Usambara Mountain Bush Viper
d Usambara Buschviper

1010 ***Atheris chloroechis***
Squamata - Viperidae
e Green Bush Viper; Green Tree Viper
d Grüne Buschviper; Grüner Baumviper
f Vipère d'arbres

1011 ***Atheris desaixi***
Squamata - Viperidae
e Ashe's Bush Viper

1012 ***Atheris hindii***
Squamata - Viperidae
e Kinangop Mountain Bush Viper; Hind's Viper; Hind's Montane Viper
d Ostafrikanische Bergotter; Hind-Viper
f Vipère de Hind

1013 ***Atheris hispidus***
Squamata - Viperidae
e Laurent's Mounatin Bush Viper; Hairy Bush Viper
d Rauschuppenbuschviper; Kielschuppige Buschviper

1014 ***Atheris katangensis***
Squamata - Viperidae
e Katanga Mountain Bush Viper
d Katanga-Buschviper

1015 ***Atheris nitschei***
Squamata - Viperidae
e Great Lakes Bush Viper
d Schwarzgrüne Buschviper; Nitschesche Baumviper

1016 ***Atheris squamiger***
Squamata - Viperidae
e Rough-scaled Bush Viper; Green Bush Viper
d Rauhschuppige Buschviper;

Blattgrüne Buschviper; Rankschuppige Buschviper

1017 ***Atheris superciliaris***
Squamata - Viperidae
e Lowland Swamp Viper; Lowland Viper
d Blattgrüne Buschviper; Sumpfviper; Sumpfotter

1018 **Atractaspididae**
Squamata
e Mole Vipers
d Erdvipern; Maulwurfsnattern; Erdottern
f Atractaspididés

1019 ***Atractaspis***
Squamata - Atractaspididae
e Mole Vipers; Burrowing Asps (CSA); Burrowing Vipers; Burrowing Adders (CSA); Stilleto Snakes (CSA); Side-stabbing Snakes (CSA); Assassin Snakes
d Erdvipern; Erdottern
f Atractaspides
i Atrattaspidi

1020 ***Atractaspis alterima***
Squamata - Atractaspididae
e Western Forest Mole Viper; Western Forest Burrowing Adder; Mole Viper
d Erdviper

1021 ***Atractaspis battersbyi***
Squamata - Atractaspididae
e Battersby's Mole Viper; Mole Viper; Mole Viper
d Erdviper

1022 ***Atractaspis bibronii***
Squamata - Atractaspididae
e Bibron's Mole Viper; Bibron's Burrowing Asp; Southern Burrowing Asp (CSA); Bibron's Burrowing Viper; Mole Viper; Oviparous Adder; Bibron's Stilleto Snake (CSA); Mole Adder (CSA)
d Südliche Erdotter; Erdviper; Bibrons Erdviper
f Atractaspide de Bibron

1023 ***Atractaspis boulengeri***
Squamata - Atractaspididae
e Boulenger's Mole Viper; Mole Viper
d Erdviper

1024 ***Atractaspis coalescens***
Squamata - Atractaspididae
e Perret's Mole Viper; Mole Viper
d Erdviper

1025 ***Atractaspis congica***
Squamata - Atractaspididae
e Eastern Congo Mole Viper; Eastern Congo Burrowing Asp (CSA); Mole Viper
d Erdviper

1026 ***Atractaspis corpulenta***
Squamata - Atractaspididae
e Fat Mole Viper; Mole Viper
d Erdviper

1027 ***Atractaspis dahomeyensis***
Squamata - Atractaspididae
e Bocage's Mole Viper; Mole Viper
d Erdviper

1028 ***Atractaspis engdahli***
Squamata - Atractaspididae
e Engdahl's Mole Viper; Mole Viper
d Erdviper

1029 ***Atractaspis fallax***
Squamata - Atractaspididae

e False Mole Viper; Mole Viper
d Erdviper

1030 ***Atractaspis irregularis***
Squamata - Atractaspididae
e Variable Mole Viper; Variable Burrowing Adder; Mole Viper; Common Burrowing Viper
d Gewöhnliche Erdotter; Erdviper
f Atractaspide vulgaire

1031 ***Atractaspis leucomelas***
Squamata - Atractaspididae
e White Mole Viper; Mole Viper
d Erdviper

1032 ***Atractaspis microlepidota***
Squamata - Atractaspididae
e Small-scaled Mole Viper; Small-scaled Burrowing Asp; Small-scaled Burrowing Adder; South Arabian Mole Viper
d Erdviper

1033 ***Atractaspis micropholis***
Squamata - Atractaspididae
e Common Mole Viper; Mole Viper
d Erdviper

1034 ***Atractaspis phillipsi***
Squamata - Atractaspididae
e Phillips's Mole Viper

1035 ***Atractaspis reticulata***
Squamata - Atractaspididae
e Reticulate Mole Viper; Mole Viper
d Erdviper

1036 ***Atractaspis scortecci***
Squamata - Atractaspididae
e Scortecci's Mole Viper; Mole Viper
d Erdviper

1037 ***Atractus***
Squamata - Colubridae
e Ground Snakes; Burrowing Spindle Snakes; Earth Snakes
d Spindelnattern

1038 ***Atractus albuquerqei***
Squamata - Colubridae
e Albuquerque Ground Snake

1039 ***Atractus alphonsehogei***
Squamata - Colubridae
e Alphonse's Ground Snake; Striped Earth Snake

1040 ***Atractus andinus***
Squamata - Colubridae
e Andean Ground Snake

1041 ***Atractus arrangoi***
Squamata - Colubridae
e Colombian Ground Snake

1042 ***Atractus badius***
Squamata - Colubridae
e Boie's Ground Snake
f Faux corail (Ants)

1043 ***Atractus balzani***
Squamata - Colubridae
e Bolivian Ground Snake

1044 ***Atractus biseriatus***
Squamata - Colubridae
e Two-lined Ground Snake

1045 ***Atractus bocki***
Squamata - Colubridae
e Bock's Ground Snake

1046 ***Atractus bocki***
Squamata - Colubridae
e Bocourt's Ground Snake

1047 ***Atractus boettgeri***
Squamata - Colubridae

e Boettger's Ground Snake

1048 ***Atractus boulengeri***
Squamata - Colubridae
e Boulenger's Ground Snake

1049 ***Atractus canedi***
Squamata - Colubridae
e Scroccho's Ground Snake

1050 ***Atractus carrioni***
Squamata - Colubridae
e Parker's Ground Snake

1051 ***Atractus clarki***
Squamata - Colubridae
e Clark's Ground Snake

1052 ***Atractus collaris***
Squamata - Colubridae
e Collared Ground Snake; Ring-necked Earth Snake

1053 ***Atractus crassicaudatus***
Squamata - Colubridae
e Thickhead Ground Snake; Thicktail Ground Snake

1054 ***Atractus duidensis***
Squamata - Colubridae
e Venezuela Ground Snake

1055 ***Atractus dunni***
Squamata - Colubridae
e Dunn's Ground Snake

1056 ***Atractus ecuadorensis***
Squamata - Colubridae
e Ecuadorean Ground Snake

1057 ***Atractus elaps***
Squamata - Colubridae
e Black Ground Snake; Ornate Earth Snake

1058 ***Atractus emigdioi***
Squamata - Colubridae
e Emigdio's Ground Snake

1059 ***Atractus emmeli***
Squamata - Colubridae
e Emmel's Ground Snake

1060 ***Atractus erythromelas***
Squamata - Colubridae
e Red-black Ground Snake

1061 ***Atractus favae***
Squamata - Colubridae
e Filippi's Ground Snake

1062 ***Atractus flammigerus***
Squamata - Colubridae
e Flaming Ground Snake; Blotched Earth Snake

1063 ***Atractus fuliginosus***
Squamata - Colubridae
e Hallowell's Ground Snake

1064 ***Atractus gaigae***
Squamata - Colubridae
e Gaige's Ground Snake

1065 ***Atractus guentheri***
Squamata - Colubridae
e Günther's Ground Snake

1066 ***Atractus indistinctus***
Squamata - Colubridae
e Indistinct Ground Snake

1067 ***Atractus insipidus***
Squamata - Colubridae
e Border Ground Snake

1068 ***Atractus iridescens***
Squamata - Colubridae
e Iridescent Ground Snake

1069 ***Atractus lancinii***
Squamata - Colubridae
e Lancini's Ground Snake

1070 ***Atractus lasallei***
Squamata - Colubridae
e Lasalle's Ground Snake

1071 ***Atractus latifrons***
Squamata - Colubridae
e Broadhead Ground Snake; Wedge-tailed Earth Snake

1072 ***Atractus lehmanni***
Squamata - Colubridae
e Lehmann's Ground Snake

1073 ***Atractus limitaneus***
Squamata - Colubridae
e Limitan Ground Snake

1074 ***Atractus loveridgei***
Squamata - Colubridae
e Loveridge's Ground Snake

1075 ***Atractus maculatus***
Squamata - Colubridae
e Spotted Ground Snake

1076 ***Atractus major***
Squamata - Colubridae
e Brown Ground Snake; Giant Earth Snake

1077 ***Atractus manizalensis***
Squamata - Colubridae
e Prado's Ground Snake

1078 ***Atractus mariselae***
Squamata - Colubridae
e Marisela's Ground Snake

1079 ***Atractus melanogaster***
Squamata - Colubridae
e Blackbelly Ground Snake

1080 ***Atractus melas***
Squamata - Colubridae
e Dark Ground Snake

1081 ***Atractus micheli***
Squamata - Colubridae
e French Guyana Ground Snake

1082 ***Atractus modestus***
Squamata - Colubridae
e Modest Ground Snake

1083 ***Atractus multicinctus***
Squamata - Colubridae
e Banded Ground Snake

1084 ***Atractus nicefori***
Squamata - Colubridae
e Northern Ground Snake

1085 ***Atractus nigricauda***
Squamata - Colubridae
e Black-headed Ground Snake

1086 ***Atractus nigriventris***
Squamata - Colubridae
e Black-ventered Ground Snake

1087 ***Atractus obesus***
Squamata - Colubridae
e Fat Ground Snake

1088 ***Atractus obtusirostris***
Squamata - Colubridae
e Bignose Ground Snake

1089 ***Atractus occidentalis***
Squamata - Colubridae
e Western Ground Snake

1090 ***Atractus occipitoalbus***
Squamata - Colubridae

e Grey Ground Snake

1091 ***Atractus oculotemporalis***
Squamata - Colubridae
e Hispanic Ground Snake

1092 ***Atractus pamplonensis***
Squamata - Colubridae
e Pamplona Ground Snake

1093 ***Atractus paucidens***
Squamata - Colubridae
e Despax's Ground Snake

1094 ***Atractus pauciscutatus***
Squamata - Colubridae
e Little-scaled Ground Snake

1095 ***Atractus peruvianus***
Squamata - Colubridae
e Peru Ground Snake

1096 ***Atractus poeppigi***
Squamata - Colubridae
e Basin Ground Snake; Black-backed Earth Snake

1097 ***Atractus punctiventris***
Squamata - Colubridae
e Pointed Ground Snake

1098 ***Atractus resplendens***
Squamata - Colubridae
e Resplendent Ground Snake

1099 ***Atractus reticulatus***
Squamata - Colubridae
e Reticulate Ground Snake

1100 ***Atractus riveroi***
Squamata - Colubridae
e Rivero's Ground Snake

1101 ***Atractus roulei***
Squamata - Colubridae
e Roule's Ground Snake

1102 ***Atractus sanctamartae***
Squamata - Colubridae
e St, Marta's Ground Snake

1103 ***Atractus sanguineus***
Squamata - Colubridae
e Bloody Ground Snake

1104 ***Atractus schach***
Squamata - Colubridae
e Schach's Ground Snake

1105 ***Atractus serranus***
Squamata - Colubridae
e Sao Paolo Ground Snake

1106 ***Atractus steyermarki***
Squamata - Colubridae
e Steyermark's Ground Snake

1107 ***Atractus torquatus***
Squamata - Colubridae
e Neckband Ground Snake; Rusty Earth Snake

1108 ***Atractus trihedrurus***
Squamata - Colubridae
e Southern Ground Snake

1109 ***Atractus trilineatus***
Squamata - Colubridae
e Three-lined Ground Snake

1110 ***Atractus trivittatus***
Squamata - Colubridae
e Three-banded Ground Snake

1111 ***Atractus univittatus***
Squamata - Colubridae
e One-banded Ground Snake

1112 ***Atractus variegatus***
Squamata - Colubridae
e Variegated Ground Snake

1113 ***Atractus ventrimaculatus***
Squamata - Colubridae
e Speckled Ground Snake

1114 ***Atractus vertebralis***
Squamata - Colubridae
e Vertebral Ground Snake

1115 ***Atractus vertebrolineatus***
Squamata - Colubridae
e Striped Ground Snake

1116 ***Atractus wagleri***
Squamata - Colubridae
e Wagler's Ground Snake

1117 ***Atractus werneri***
Squamata - Colubridae
e Werner's Ground Snake

1118 ***Atractus zidocki***
Squamata - Colubridae
e Zidock's Ground Snake

1119 ***Atretium***
Squamata - Colubridae
e Keelbacks; Olivaceous Keelbacks

1120 ***Atretium schistosum***
Squamata - Colubridae
e Split Keelback; Olivaceous Keelback (ISC); Plumbaeous Keelback Water Snake; Olivaceous Keelback Water Snake

1121 ***Atretium yunnanensis***
Squamata - Colubridae
e Yunnan Olive Keelback

1122 ***Atropoides***
Squamata - Viperidae
e Jumping Pit Vipers
d Springende Lanzenottern

1123 ***Atropoides nummifer***
Squamata - Viperidae
e Jumping Pit Viper; Jumping Viper
d Springende Lanzenotter

1124 ***Atropoides olmec***
Squamata - Viperidae
e Olmecan Pit Viper

1125 ***Atropoides picadoi***
Squamata - Viperidae
e Picado's Pit Viper
d Picados Springende Lanzenotter

1126 ***Aulura***
Squamata - Amphisbaenidae
e Barbour's Worm Lizards

1127 ***Aulura anomola***
Squamata - Amphisbaenidae
e Barbour's Worm Lizard

1128 ***Austrelaps***
Squamata - Elapidae
e Australian Copperheads; Copperhead Snakes (ANZ); Copperheads (ANZ)
d Australische Kupferköpfe

1129 ***Austrelaps labialis***
Squamata - Elapidae
e Kangaroo Island Copperhead; Pygmy Copperhead

1130 ***Austrelaps ramsayi***
Squamata - Elapidae
e Highland Copperhead

1131 ***Austrelaps superbus***
Squamata - Elapidae
e Australian Copperhead; Australian

Lowlands Copperhead; Copperhead (ANZ)
d Australischer Kupferkopf

1132 ***Azemiops***
Squamata - Viperidae
e Fea's Vipers; Fea Vipers
d Fea-Vipern

1133 ***Azemiops feae***
Squamata - Viperidae
e Fea's Viper; Fea Viper; White-headed Viper
d Fea-Viper

B

1134 ***Bachia***
Squamata - Gymnophthalmidae
e Bachias; Worm Lizards
d Wüstentejus; Wühletejus

1135 ***Bachia abendrothii***
Squamata - Gymnophthalmidae
e Abendroth's Bachia
d Abendroths Schlangenteju

1136 ***Bachia alleni***
Squamata - Gymnophthalmidae
e Allen's Bachia

1137 ***Bachia barbouri***
Squamata - Gymnophthalmidae
e Barbour's Bachia

1138 ***Bachia bicolor***
Squamata - Gymnophthalmidae
e Two-coloured Bachia

1139 ***Bachia blairi***
Squamata - Gymnophthalmidae
e Blair's Bachia

1140 ***Bachia bresslaui***
Squamata - Gymnophthalmidae
e Bresslau's Bachia

1141 ***Bachia cuvieri***
Squamata - Gymnophthalmidae
e Cuvier's Bachia

1142 ***Bachia dorbignyi***
Squamata - Gymnophthalmidae
e D'Orbigny's Bachia

1143 ***Bachia guianensis***
Squamata - Gymnophthalmidae
e Guyana Bachia

1144 ***Bachia heteropa***
Squamata - Gymnophthalmidae
e La Guiara Bachia; Earless Worm Lizard

1145 ***Bachia huallagana***
Squamata - Gymnophthalmidae
e Dixon's Bachia

1146 ***Bachia intermedia***
Squamata - Gymnophthalmidae
e Noble's Bachia

1147 ***Bachia pallidiceps***
Squamata - Gymnophthalmidae
e Cope's Bachia

1148 ***Bachia panopila***
Squamata - Gymnophthalmidae
e Thomas's Bachia

1149 ***Bachia peruana***
Squamata - Gymnophthalmidae
e Peru Bachia; Peruvian Worm Lizard

1150 ***Bachia scolecoides***
Squamata - Gymnophthalmidae
e Vanzolini's Bachia

1151 ***Bachia talpa***
Squamata - Gymnophthalmidae
e Ruthven's Bachia

1152 ***Bachia trisanale***
Squamata - Gymnophthalmidae

1153 ***Bachia trisanale trisanale***
Squamata - Gymnophthalmidae

e Striped Worm Lizard

1154 ***Bachia vermiformis***
Squamata - Gymnophthalmidae
e Small-legged Bachia; Brown Worm Lizard

1155 ***Baikia***
Squamata - Amphisbaenidae
e West African Worm Lizards

1156 ***Baikia africana***
Squamata - Amphisbaenidae
e West African Worm Lizard

1157 ***Barisia***
Squamata - Anguidae
e Alligator Lizards
d Krokodilschleichen
f Lézards-alligator

1158 ***Barisia imbricata***
Squamata - Anguidae
e Imbricate Alligator Lizard
d Kiel-Krokodilschleiche

1159 ***Barisia levicollis***
Squamata - Anguidae
e Chihuahuan Alligator Lizard; Chihuaha Alligator Lizard

1160 ***Barisia rudicollis***
Squamata - Anguidae
e Roughneck Alligator Lizard
f Gerrhonote à cou rude

1161 ***Barkudia***
Squamata - Scincidae
e Madras Spotted Skinks

1162 ***Barkudia insularis***
Squamata - Scincidae
e Madras Spotted Skink

1163 ***Basiliscus***
Squamata - Iguanaidae
e Basilisks
d Basilisken
f Basilisques; Basilics

1164 ***Basiliscus basiliscus***
Squamata - Iguanaidae
e Common Basilisk; Longtail Iguana; Helmeted Basilisk; Brown Basilisk; American Basilisk; Central American Basilisk
d Helmbasilisk
f Basilisque à bandes

1165 ***Basiliscus galeritus***
Squamata - Iguanaidae
e Western Basilisk

1166 ***Basiliscus plumifrons***
Squamata - Iguanaidae
e Green Basilisk; Plumed Basilisk; Double-crested Basilisk; Jesus Christ Lizard
d Stirnlappenbasilisk; Federbuschbasilisk
f Lézard peureux
i Basilisco piumato

1167 ***Basiliscus vittatus***
Squamata - Iguanaidae
e Brown Basilisk; Banded Basilisk; Striped Basilisk
d Streifenbasislisk
f Basilisque à bandes; Basilisque vert; Basilic vert

1168 ***Batagur***
Testudines - Emydidae
e Batagurine Turtles; Bastagurs; River Terrapins; Tuntongs
d Batagurschildkröten
i Tartarughe fluviale indiane

1169 ***Batagur basca***
Testudines - Emydidae

e Batagurine Turtle; Batagur; Common Batagur; River Terrapin (ISC); Tuntong; Batagur Terrapin (ISC); Batagur River Terrapin (ISC)
d Batagurschildkröte
f Batagur malais; Tortue fluviale indienne
i Tartaruga fluviale indiana

1170 ***Bavayia***
Squamata - Gekkonidae
e New Caledonian Geckos

1171 ***Bavayia crassicollis***
Squamata - Gekkonidae
e Roux's Caledonian Gecko

1172 ***Bavayia cyclura***
Squamata - Gekkonidae
e Günther's New Caledonian Gecko

1173 ***Bavayia montana***
Squamata - Gekkonidae
e Mountain New Caledonian Gecko

1174 ***Bavayia ornata***
Squamata - Gekkonidae
e Ornate New Caledonian Gecko

1175 ***Bavayia savageii***
Squamata - Gekkonidae
e Savage's New Caledonian Gecko

1176 ***Bavayia septuiclavis***
Squamata - Gekkonidae
e Sadlier's New Caledonian Gecko

1177 ***Bavayia validiclavis***
Squamata - Gekkonidae
e Key New Caledonian Gecko

1178 ***Bedriagaia***
Squamata - Lacertidae
e African Tree Lizards

1179 ***Bedriagaia tropidopholis***
Squamata - Lacertidae
e African Tree Lizard

1180 **Bipedidae**
Squamata
e Two-legged Worm Lizards
d Zweifußdoppelschleichen
f Bipédidés

1181 ***Bipes***
Squamata - Bipedidae
e Two-legged Worm Lizards
d Handwühlen
f Bipes

1182 ***Bipes biporus***
Squamata - Bipedidae
e Five-toed Worm Lizard; Two-footed Worm Lizard; Two-legged Worm Lizard; Ajolote; Mole Lizard (NA); Common Two-legged Worm Lizard
d Handwühle
f Bipes à deux pores

1183 ***Bipes canaliculatus***
Squamata - Bipedidae
e Four-toed Worm Lizard; Four-toed Mole Lizard

1184 ***Bipes tridactylus***
Squamata - Bipedidae
e Three-toed Worm Lizard; Three-toed Mole Lizard

1185 ***Bitis***
Squamata - Viperidae
e Puff Adders; African Adders (CSA)
d Puffottern

1186 ***Bitis arietans***
Squamata - Viperidae
e Puff Adder; Common Puff Adder
d Gemeine Puffotter; Puffotter; Gewöhnliche Puffotter

f Vipère heurtante; Trigonocéphale; Vipère du Cap
i Vipera soffiante

1187 ***Bitis atropos***
Squamata - Viperidae
e Berg Adder (CSA); Mountain Adder
d Bergotter; Bergpuffotter; Atroposviper
f Vipère montagnarde

1188 ***Bitis atropos unicolor***
Squamata - Viperidae
e Mountain Adder

1189 ***Bitis caudalis***
Squamata - Viperidae
e Horned Puff Adder; Cape Horned Adder (CSA); Horned Adder (CSA); Sidewinder Adder; Horned Viper
d Gehörnte Puffotter; Gehörnte Zwergpuffotter
i Bitis cornuta

1190 ***Bitis cornuta***
Squamata - Viperidae
e Many-horned Adder
d Büschelbrauenotter; Südafrikanische Hornviper
f Vipère cornue

1191 ***Bitis cornuta cornuta***
Squamata - Viperidae
e Western Many-horned Adder

1192 ***Bitis gabonica***
Squamata - Viperidae
e Gaboon Viper; Gaboon Adder (CSA)
d Gabun-Viper; Gabun-Otter
f Vipère du Gabon
i Vipera del Gabon

1193 ***Bitis gabonica gabonica***
Squamata - Viperidae
e East African Gabun Viper
d Ostafrikanische Gabun-Viper

1194 ***Bitis gabonica rhinocerus***
Squamata - Viperidae
e West African Gabun Viper
d Westafrikanische Gabun-Viper

1195 ***Bitis heraldica***
Squamata - Viperidae
e Bocage's Horned Adder

1196 ***Bitis inornata***
Squamata - Viperidae
e Hornless Adder; Plains Mountain Adder; Plain Mountain Adder (CSA); Cape Puff Adder
d Hornlose Gabunviper

1197 ***Bitis nasicornis***
Squamata - Viperidae
e Rhinoceros Viper
d Nashornviper
f Vipère à cornes; Vipère rhinocéros

1198 ***Bitis olduvaiensis***
Squamata - Viperidae
e Olduvai Viper

1199 ***Bitis parvicula***
Squamata - Viperidae
e Ethiopia Viper

1200 ***Bitis peringueyi***
Squamata - Viperidae
e Péringuey's Adder; Dwarf Puff Adder; Péringuey's Viper; Desert Adder
d Namib-Viper; Zwergpuffotter; Peringuey-Wüstenotter; Péringueys Wüstenotter
f Vipère du désert de Namibie; Vipère du désert de Namib

1201 ***Bitis schneideri***
Squamata - Viperidae

e Namaqua Dwarf Adder; Namaqua Puff Adder
d Namaqua-Zwergpuffotter; Namaqua-Zwergotter

1202 ***Bitis worthingtoni***
Squamata - Viperidae
e Kenya Horned Viper

1203 ***Bitis xeropaga***
Squamata - Viperidae
e Desert Mountain Adder

1204 ***Blanus***
Squamata - Amphisbaenidae
e Mediterranean Worm Lizards
d Netzwühlen

1205 ***Blanus cinereus***
Squamata - Amphisbaenidae
e Mediterranean Worm Lizard; Iberian Worm Lizard; Grey Burrowing Lizard
d Maurische Netzwühle; Ringelschleiche; Maurische Ringelwühle
f Amphisbène cendré
i Blano cinereo; Anfisbena cinerea; Blano cinerino

1206 ***Blanus strauchi***
Squamata - Amphisbaenidae
e Turkish Worm Lizard; Ringed Amphisbaenian
d Kleinasiatische Netzwühle
f Amphisbène de Strauch
i Blano di Strauch; Anfisbena turca

1207 ***Blythia***
Squamata - Colubridae
e Reticulate Snakes

1208 ***Blythia reticulata***
Squamata - Colubridae
e Blyth's Reticulate Snake

1209 ***Boa***
Squamata - Boidae
e Boa Constrictors; Boas; Common Boas
d Boas; Abgottschlangen
f Boas

1210 ***Boa constrictor***
Squamata - Boidae
e Boa Constrictor; Neotropical Boa (WI)
d Abgottschlange; Königsschlange; Königsboa
f Couleuvre de terre (Ants); Boa constrictor; Boa constricteur
i Boa costrittore

1211 ***Boa constrictor amarali***
Squamata - Boidae
e Amaral's Boa
d Amarals Boa

1212 ***Boa constrictor constrictor***
Squamata - Boidae
e Red-tailed Python; True Redtail Boa; Red-tailed Boa

1213 ***Boa constrictor imperator***
Squamata - Boidae
e Hogg's Island Boa; Common Boa Constrictor
d Abgottschlange; Kaiserboa
f Boa de l'île de Hogg
i Boa costrittore; Boa coda rossa

1214 ***Boa constrictor occidentalis***
Squamata - Boidae
e Argentine Boa
d Südboa; Argentinische Rotschwanzboa; Argentinische Königsschlange
f Boa constricteur de l'Argentine

1215 ***Boa constrictor ortoni***
Squamata - Boidae
d Boa Constrictor

1216 ***Boa dumerili***
Squamata - Boidae
e Duméril's Boa; Duméril's Ground Boa; Duméril's Madagascar Boa
d Dumérils Madagaskar-Boa; Duméril-Boa; Südliche Madagaskar-Boa; Madagaskar-Boa
f Boa de Duméril
i Boa di Duméril

1217 ***Boa madagascariensis***
Squamata - Boidae
e Madagascar Boa; Madagascan Boa; Madagascar Boa Constrictor; Madagascan Ground Boa; Madagascar Tree Boa; Malagasy Tree Boa; Madagascar Ground Boa
d Madagaskar-Boa; Gewöhnliche Madagaskar-Boa; Nördliche Madagaskar-Boa; Malagassischer Hundskopfschlinger; Madagaskar-Hundskopfboa; Madagaskar-Hundskopfschlinger; Madagassische Erdboa
f Boa de Madagascar

1218 ***Boa manditra***
Squamata - Boidae
e Madagascar Tree Boa; Madagascan Tree Boa
f Boa arboricole de Madagascar; Boa canin de Madagascar

1219 ***Bogertia***
Squamata - Gekkonidae
e Bogert's Geckos

1220 ***Bogertia lutzae***
Squamata - Gekkonidae
e Bogert's Gecko

1221 ***Bogertophis***
Squamata - Colubridae
e Rat Snakes; Desert Rat Snakes
d Rattennattern

1222 ***Bogertophis rosaliae***
Squamata - Colubridae
e Baja California Rat Snake; Santa Rosa Rat Snake

1223 ***Bogertophis subocularis***
Squamata - Colubridae
e Trans-Pecos Rat Snake; Davis Mountain Rat Snake; Davis Mountain Pilot Snake
d Transpecos-Kletternatter; Transpecos-Rattennatter
f Élaphe Trans-Pecos
i Serpente dei ratti Trans-Pecos

1224 ***Bogertophis subocularis subocularis***
Squamata - Colubridae
e Trans-Pecos Rat Snake
f Couleuvre de Mont Davis

1225 **Boidae**
Squamata
e Boas; Boas and Pythons; Giant Snakes; Boids
d Riesenschlangen; Boaschlangen; Pythonschlangen
f Boïdés; Boas vrais
i Boidi

1226 ***Boiga***
Squamata - Colubridae
e Mangrove Snakes; Tree Snakes; Tree Cat Snakes
d Nachtbaumnattern; Trugnattern
f Serpents d'arbre
i Serpenti delle mangrovie; Serpenti gatto

1227 ***Boiga angulata***
Squamata - Colubridae
e Leyte Cat Snake

1228 ***Boiga barnesi***
Squamata - Colubridae
e Barnes's Cat Snake

1229 ***Boiga blandingii***
Squamata - Colubridae
e Blanding's Cat Snake; Blanding's Tree Snake
d Blandings Nachtbaumnatter

1230 ***Boiga ceylonensis***
Squamata - Colubridae
e Sri Lanka Cat Snake; Günther's Cat Snake; Ceylon Cat Snake (ISC)

1231 ***Boiga cyanea***
Squamata - Colubridae
e Green Cat Snake; Green Cat-eyed Snake; Green Tree Snake
d Grüne Nachtbaumnatter; Grüne Katzenaugennatter
i Occhi di gatto verde

1232 ***Boiga cynodon***
Squamata - Colubridae
e Dog-tooth Cat Snake; Dog-toothed Cat Snake; Large-toothed Tree Snake; Boie's Tree Snake; Black-barred Tree Snake; Dog-toothed Mangrove Snake
d Katzenaugennachtbaumnatter

1233 ***Boiga dendrophila***
Squamata - Colubridae
e Mangrove Snake; Yellow-ringed Cat Snake; Striking Mangrove Cat Snake
d Mangrovennachtbaumnatter; Ulaborong
f Serpent d'arbre; Serpent des paletuviers
i Serpente delle mangrovie; Serpe delle mangrovie

1234 ***Boiga dightoni***
Squamata - Colubridae
e Pirmad Cat Snake

1235 ***Boiga drapiezii***
Squamata - Colubridae
e White-spotted Cat Snake

1236 ***Boiga flavescens***
Squamata - Colubridae
e Yellow Snake

1237 ***Boiga forsteni***
Squamata - Colubridae
e Forsten's Snake; Forsten's Cat Snake (ISC)

1238 ***Boiga gokool***
Squamata - Colubridae
e Arrowback Tree Snake; Arrow-backed Tree Snake; Tree Snake; Eastern Gamma (ISC)

1239 ***Boiga hexagonata***
Squamata - Colubridae
e Grey Tree Snake; Stoliczka's Tree Snake

1240 ***Boiga irregularis***
Squamata - Colubridae
e Brown Tree Snake; Brown Cat Snake; Eastern Brown Tree Snake
d Braune Nachtbaumnatter

1241 ***Boiga jaspidea***
Squamata - Colubridae
e Jasper Cat Snake

1242 ***Boiga kraepelini***
Squamata - Colubridae
e Kelung Cat Snake

1243 ***Boiga multifasciata***
Squamata - Colubridae
e Many-banded Tree Snake; Himalayan Tree Snake

1244 ***Boiga multimaculata***
Squamata - Colubridae
e Marble Cat Snake; Many-spotted Cat Snake; Spotted Cat Snake; Large-spotted Cat Snake
d Vielfleckennachtbaumnatter

1245 ***Boiga nigriceps***
Squamata - Colubridae
e Red Cat Snake; Dark-headed Cat Snake

1246 ***Boiga ocellata***
Squamata - Colubridae
e Grey Cat Snake

1247 ***Boiga philippina***
Squamata - Colubridae
e Philippine Cat

1248 ***Boiga pulverulenta***
Squamata - Colubridae
e Fischer's Cat Snake

1249 ***Boiga saengsomi***
Squamata - Colubridae
e Banded Green Cat Snake

1250 ***Boiga trigonata***
Squamata - Colubridae
e Indian Gamma Snake; Rear-fanged Snake; Common Tree Snake; Indian Gamma (ISC); Indian Cat Snake (ISC); Common Indian Cat Snake
d Dreiecknachtbaumnatter

1251 ***Bolyeria***
Squamata - Bolyeriidae
e Round Island Burrowing Boas; Round Island Boas
d Mauritius-Boas; Bolyer-Schlangen
f Bolyerides

1252 ***Bolyeria multocarinata***
Squamata - Bolyeriidae
e Round Island Burrowing Boa; Round Island Boa
d Mauritius-Boa
f Bolyeride
i Boa dell'isola di Round

1253 **Bolyeriidae**
Squamata
e Round Island Boas
d Bolyers-Schlangen; Mauritius-Boas
f Bolyeriidés

1254 ***Bothriechis***
Squamata - Viperidae
e Palm Pit Vipers
d Palmenlanzenottern; Palmenvipern

1255 ***Bothriechis aurifer***
Squamata - Viperidae
e Yellow-blotched Palm Pit Viper
d Gelbgefleckte Palmenlanzenotter

1256 ***Bothriechis bicolor***
Squamata - Viperidae
e Guatemala Palm Pit Viper; Guatemalan Palm Pit Viper

1257 ***Bothriechis lateralis***
Squamata - Viperidae
e Side-striped Palm Pit Viper; Coffee Palm Viper
d Grüngelbe Palmenlanzenotter

1258 ***Bothriechis marchi***
Squamata - Viperidae
e March's Palm Pit Viper; Green Palm Pit Viper
d Marchs Palmenlanzenotter

1259 ***Bothriechis nigroviridis***
Squamata - Viperidae
e Black-speckled Palm Pit Viper

1260 ***Bothriechis rowleyi***
Squamata - Viperidae
e Rowley's Palm Pit Viper
d Rowleys Palmenlanzenotter

1261 ***Bothriechis schlegeli***
Squamata - Viperidae
e Eyelash Palm Pit Viper; Eyelash Viper; Eyelash Pit Viper; Common

Eyelash Pitviper; Eyelash Lancehead; Schlegel's Palm Viper
d Schlegelsche Greifschwanzlanzenotter; Greifschwanzlanzenotter; Schlegelsche Lanzenotter; Greifschwanzpalmenlanzenotter
f Vipère de Schlegel

1262 ***Bothriechis supraciliaris***
Squamata - Viperidae
e Taylor's Eyelash Pit Viper

1263 ***Bothriopsis***
Squamata - Viperidae
e Forest Pit Vipers

1264 ***Bothriopsis alticola***
Squamata - Viperidae
e Andean Forest Pit Viper

1265 ***Bothriopsis bilineata***
Squamata - Viperidae
e Two-striped Forest Pit Viper; Green Jararaca; Two-lined Forest Pit Viper
d Grüne Jaracara
f Vipère arboricole; Vrai Jacquot

1266 ***Bothriopsis bilineata smaragdina***
Squamata - Viperidae
e Western Striped Forest Pit Viper

1267 ***Bothriopsis medusa***
Squamata - Viperidae
e Venezuela Forest Pit Viper

1268 ***Bothriopsis oligolepis***
Squamata - Viperidae
e Inca Forest Pit Viper

1269 ***Bothriopsis peruviana***
Squamata - Viperidae
e Peru Forest Pit Viper

1270 ***Bothriopsis punctata***
Squamata - Viperidae
e Chocoan Forest Pit Viper

1271 ***Bothriopsis taeniata***
Squamata - Viperidae
e Speckled Forest Pit Viper
d Castelnauds Lanzenotter
f Vipère arboricole

1272 ***Bothrochilus***
Squamata - Boidae
e Ringed Pythons; New Ireland Ringed Pythons; New Guinea Dwarf Pythons
d Zwergpythone

1273 ***Bothrochilus boa***
Squamata - Boidae
e Ringed Python; Bismarck Ringed Python; New Ireland Ringed Python
d Neuguinea-Zwergpython; Zwergpython; Bismarcks Python

1274 ***Bothrolycus***
Squamata - Colubridae
e Günther's Black Snakes

1275 ***Bothrolycus ater***
Squamata - Colubridae
e Günther's Black Snake

1276 ***Bothrophthalmus***
Squamata - Colubridae
e Red-black Striped Snakes; Striped Snakes
d Augengrubennattern

1277 ***Bothrophthalmus lineatus***
Squamata - Colubridae
e Red-black Striped Snake; Red-and-black Striped Snake
d Gestreifte Augengrubennatter

1278 ***Bothrops***
Squamata - Viperidae
e Lanceheads; American Lanceheads; Lancehead Snakes; Tree Pit Vipers

d Lanzenottern; Gewöhnliche Lanzenottern; Amerikanische Lanzenottern; Amerikanische Lanzenschlangen
f Fer-de-lances; Vipères fer-de-lance
i Jaracara; Yaracara; Yarar

1279 ***Bothrops alternatus***
Squamata - Viperidae
e Urutu Lancehead; Urutu; Jararaca de Rabo (NA); Halfmoon Lancehead; Urutu Pit Viper
d Uruturu; Halbmondlanzenotter; Urutu
f Urutu

1280 ***Bothrops ammodytoides***
Squamata - Viperidae
e Patagonia Lancehead; Yararanata; Patagonian Pit Viper
d Yararanata

1281 ***Bothrops andianus***
Squamata - Viperidae
e Andean Lancehead

1282 ***Bothrops asper***
Squamata - Viperidae
e Terciopelo; Central American Lancehead
d Fer-de-Lance
i Terciopelo

1283 ***Bothrops atrox***
Squamata - Viperidae
e Fer-de-lance; Common Lancehead; Lancehead; Barba Amarilla (NA); Caicaca Lancehead; South American Lancehead
d Caicaca; Gewöhnliche Lanzenotter; Lanzenotter; Lanzenschlange; Gewöhnliche Lanzenschlange
f Vipère fer-de-lance; Fer-de-lance; Grage (Ants)
i Ferro di lancia

1284 ***Bothrops barnetti***
Squamata - Viperidae
e Barnett's Lancehead

1285 ***Bothrops brazili***
Squamata - Viperidae
e Brazil's Lancehead; Brazil's Pit Viper; Velvety Lancehead
d Brazils Lanzenotter
f Grage petit carreaux (Ants)

1286 ***Bothrops caribbaeus***
Squamata - Viperidae
e St. Lucia Lancehad; St. Lucia Serpent; St. Lucia Pit Viper; St. Lucia Viper (WI)
d St.Lucia-Lanzenotter

1287 ***Bothrops cotiara***
Squamata - Viperidae
e Cotiara

1288 ***Bothrops erythromelas***
Squamata - Viperidae
e Caatinga Lancehead
d Rotschwarze Lanzenotter

1289 ***Bothrops fonsecai***
Squamata - Viperidae
e Fonseca's Lancehead; Fonseca's Pit Viper

1290 ***Bothrops iglesiasi***
Squamata - Viperidae
e Sertao Lancehead; Iglesias's Pit Viper

1291 ***Bothrops insularis***
Squamata - Viperidae
e Golden Lancehead; Island Lancehead; Island Viper; Jaracara Ilhoa (NA)
d Insel-Lanzenotter; Queimada-Grande-Insellanzenotter
f Vipère insulaire

1292 ***Bothrops itapetiningae***
Squamata - Viperidae
e Sao Paolo Lancehead; Cotiarinha (NA)
d Cotiarinha

1293 ***Bothrops jararaca***
Squamata - Viperidae
e Jararaca
d Jararaca
f Jararaca

1294 ***Bothrops jararacussu***
Squamata - Viperidae
e Jararacussu (NA); Jararacucu (NA)
d Jaracussu
f Jaracussu

1295 ***Bothrops jonathoni***
Squamata - Viperidae
e Jonathan's Lancehead

1296 ***Bothrops lanceolatus***
Squamata - Viperidae
e Martinique Lancehead; Martinican Pit Viper; Martinique Viper (WI)
d Martinique-Lanzenotter
f Fer-de-lance; Vipère jaune

1297 ***Bothrops leucurus***
Squamata - Viperidae
e Whitetail Lancehead

1298 ***Bothrops lojanus***
Squamata - Viperidae
e Lojan Lancehead; Lojan Pit Viper

1299 ***Bothrops marajoensis***
Squamata - Viperidae
e Marajo Lancehead; Ilha Marajo Lancehead; Marajoran Pit Viper

1300 ***Bothrops micropthalmus***
Squamata - Viperidae
e Small-eyed Lancehead; Small-eyed Pit Viper; Cope's Pit Viper

1301 ***Bothrops moojeni***
Squamata - Viperidae
e Brazilian Lancehead; Caissaca
d Caissaca; Brasilianische Lanzenotter

1302 ***Bothrops neuwiedi***
Squamata - Viperidae
e Neuwied's Lancehead; Jaracara Pintada (NA); Neuwied's Pit Viper; Prince Neuwied's Pit Viper
d Neuwied-Lanzenotter; Newuwieds Lanzenotter; Lanzenotter

1303 ***Bothrops pictus***
Squamata - Viperidae
e Desert Lancehead

1304 ***Bothrops pirajai***
Squamata - Viperidae
e Piraja's Lancehead

1305 ***Bothrops pradoi***
Squamata - Viperidae
e Prado's Lancehead; Prado's Pit Viper

1306 ***Bothrops pulcher***
Squamata - Viperidae
e Dusky Lancehead

1307 ***Bothrops sanctaecrucis***
Squamata - Viperidae
e Bolivian Lancehead

1308 ***Bothrops venezuelensis***
Squamata - Viperidae
e Venezuela Lancehead; Venezuelan Pit Viper
d Lanzenotter

1309 ***Boulengerina***
Squamata - Elapidae
e Water Cobras; Water Snakes

d Wasserkobras
f Couleuvres d'eau

1310 ***Boulengerina annulata***
Squamata - Elapidae
e Ringed Water Cobra; Banded Water Cobra; Ringed Water Snake; Water Cobra
d Geringelte Wasserkobra; Wasserkobra
f Couleuvre anulaire

1311 ***Boulengerina annulata stormsi***
Squamata - Elapidae
e Storm's Water Cobra
d Storms Wasserkobra

1312 ***Boulengerina christyi***
Squamata - Elapidae
e Christy's Water Cobra

1313 ***Brachylophus***
Squamata - Iguanaidae
e Fiji Iguanas; Fijian Iguanas; Banded Iguanas; Fijian Crested Iguanas
d Fidschi-Leguane; Kurzkammleguane; Fidji-Leguane
f Brachylophes

1314 ***Brachylophus fasciatus***
Squamata - Iguanaidae
e Fiji Banded Iguana ; Banded Iguana
d Fidschi-Leguan; Kurzkammleguan; Fidji-Leguan
f Brachylophe à bandes

1315 ***Brachylophus vitiensis***
Squamata - Iguanaidae
e Fiji Crested Iguana
d Kammleguan

1316 ***Brachymeles***
Squamata - Scincidae
e Short-legged Skinks; Brachymeleses
f Brachymèles

1317 ***Brachymeles apus***
Squamata - Scincidae
e Hikida's Short-legged Skink

1318 ***Brachymeles bicolor***
Squamata - Scincidae
e Two-coloured Short-legged Skink

1319 ***Brachymeles bonitae***
Squamata - Scincidae
e Pretty Short-legged Skink
f Brachymèle de Bonite

1320 ***Brachymeles boulengeri***
Squamata - Scincidae
e Boulenger's Short-legged Skink

1321 ***Brachymeles cebuensis***
Squamata - Scincidae
e Rabor's Short-legged Skink

1322 ***Brachymeles elerae***
Squamata - Scincidae
e Common Short-legged Skink

1323 ***Brachymeles gracilis***
Squamata - Scincidae
e Pathfinder Short-legged Skink

1324 ***Brachymeles pathfinderi***
Squamata - Scincidae
e Brown's Short-legged Skink

1325 ***Brachymeles samarensis***
Squamata - Scincidae
e Mindanao Short-legged Skink

1326 ***Brachymeles schadenbergi***
Squamata - Scincidae
e Mindanao Short-legged Skink

1327 ***Brachymeles talinis***
Squamata - Scincidae
e Duméril's Short-legged Skink

1328 ***Brachymeles tridactylus***
Squamata - Scincidae
e Three-fingered Short-legged Skink

1329 ***Brachymeles vermis***
Squamata - Scincidae
e Taylor's Short-legged Skink

1330 ***Brachymeles wrighti***
Squamata - Scincidae
e Wright's Short-legged Skink

1331 ***Brachyophidium***
Squamata - Uropeltidae
e Mountain Burrowing Snakes

1332 ***Brachyophidium rhodogaster***
Squamata - Uropeltidae
e Palni Mountain Burrowing Snakes

1333 ***Brachyophis***
Squamata - Colubridae
e Revoil's Short Snakes

1334 ***Brachyophis revoili***
Squamata - Colubridae
e Revoil's Short Snake

1335 ***Brachyorrhus***
Squamata - Colubridae
e White Racers

1336 ***Brachyorrhus albus***
Squamata - Colubridae
e White Racer

1337 ***Brachysaura***
Squamata - Agamidae
e Hardwick's Bloodsuckers; Night Agamas

1338 ***Brachysaura minor***
Squamata - Agamidae
e Hardwick's Bloodsucker

1339 ***Bradypodion***
Squamata - Chamaeleonidae
e Dwarf Chameleons
d Zwergchamäleone; Bunte Zwergchamäleone
f Caméléons nains

1340 ***Bradypodion adolfifriderici***
Squamata - Chamaeleonidae
e Ituri Forest Chameleon

1341 ***Bradypodion caffrum***
Squamata - Chamaeleonidae
e Transkei Dwarf Chameleon

1342 ***Bradypodion damaranum***
Squamata - Chamaeleonidae
e Kenya Dwarf Chameleon; Knysa Dwarf Chameleon (CSA)
d Knysna-Zwergchamäleon; Buntes Zwergchamäleon

1343 ***Bradypodion dracomontanum***
Squamata - Chamaeleonidae
e Drakensberg Dwarf Chameleon
d Drakensberg-Zwerchamäleon

1344 ***Bradypodion gutturale***
Squamata - Chamaeleonidae
e Robertson's Dwarf Chameleon

1345 ***Bradypodion karroicum***
Squamata - Chamaeleonidae
e Karoo Dwarf Chameleon
d Karoo-Zwergchamäleon

1346 ***Bradypodion melanocephalum***
Squamata - Chamaeleonidae
e Black-headed Dwarf Chameleon
d Zwergchmäleon; Schwarzkopfzwergchamäleon

1347 ***Bradypodion nemorale***
Squamata - Chamaeleonidae
e Zululand Dwarf Chameleon

1348 ***Bradypodion oxyrhinum***
Squamata - Chamaeleonidae
e Rednose Dwarf Chameleon

1349 ***Bradypodion pumilum***
Squamata - Chamaeleonidae
e Cape Dwarf Chameleon; Dwarf Chameleon (CSA)
d Buntes Zwergchamäleon
f Caméléon nain

1350 ***Bradypodion setaroi***
Squamata - Chamaeleonidae
e Setaro's Dwarf Chameleon

1351 ***Bradypodion setcersi***
Squamata - Chamaeleonidae
e Raw's Dwarf Chameleon

1352 ***Bradypodion taeniabronchum***
Squamata - Chamaeleonidae
e Smith's Dwarf Chameleon

1353 ***Bradypodion tenue***
Squamata - Chamaeleonidae
e Matschie's Dwarf Chameleon

1354 ***Bradypodion thamnobates***
Squamata - Chamaeleonidae
e Natal Midlands Dwarf Chameleon

1355 ***Bradypodion transvaalense***
Squamata - Chamaeleonidae
e Transvaal Dwarf Chameleon

1356 ***Briba***
Squamata - Gekkonidae
e Brazilian Geckos

1357 ***Briba brasiliana***
Squamata - Gekkonidae
e Amaral's Brazilian Gecko

1358 ***Bronchocela***
Squamata - Agamidae
e Common Bloodsuckers; Slender Agamas
d Schönechsen
f Bronchocèles

1359 ***Bronchocela celebensis***
Squamata - Agamidae
e Sulawesi Bloodsucker

1360 ***Bronchocela cristatella***
Squamata - Agamidae
e Borneo Bloodsucker; Bornean Bloodsucker; Green Crested Lizard
d Borneo-Schönechse; Blutsaugeragame; Langschwanzagame
f Bronchocèle cristatelle

1361 ***Bronchocela danieli***
Squamata - Agamidae
e Daniel's Bloodsucker

1362 ***Bronchocela hayeki***
Squamata - Agamidae
e Sumatra Bloodsucker

1363 ***Bronchocela jubata***
Squamata - Agamidae
e Nicobar Island Bloodsucker
d Mähnenschönechse; Grüne Schönechse
f Bronchocèle à crinière

1364 ***Bronchocela marmorata***
Squamata - Agamidae
e Marbled Bloodsucker
d Marmorschönechse
f Bronchocèle marbré; Galéote marbré

1365 ***Bronchocela smaragdina***
Squamata - Agamidae
e Günther's Bloodsucker; Emerald Bloodsucker
d Smaragdschönechse
f Bronchocèle d'émeraude

1366 ***Bronia***
Squamata - Amphisbaenidae
e Brazilian Worm Lizards

1367 ***Bronia brasiliana***
Squamata - Amphisbaenidae
e Brazilian Worm Lizard

1368 ***Brookesia***
Squamata - Chamaeleonidae
e Leaf Chameleons; Stump-tailed Chameleons; Stub-tailed Chameleons
d Erdchmäleone; Stummelschwanzchamäleone
f Brookesies

1369 ***Brookesia antoetrae***
Squamata - Chamaeleonidae
e Brygoo's Leaf Chameleon

1370 ***Brookesia betschii***
Squamata - Chamaeleonidae
e Blanc's Leaf Chameleon

1371 ***Brookesia decaryi***
Squamata - Chamaeleonidae
e Spiny Leaf Chameleon
d Madagaskar-Erdchamäleon

1372 ***Brookesia dentata***
Squamata - Chamaeleonidae
e Toothed Leaf Chameleon

1373 ***Brookesia ebenaui***
Squamata - Chamaeleonidae
e Common Leaf Chameleon

1374 ***Brookesia griveaudi***
Squamata - Chamaeleonidae
e Marojejy Leaf Chameleon

1375 ***Brookesia karachi***
Squamata - Chamaeleonidae
e Naturelle Leaf Chameleon

1376 ***Brookesia lambertoni***
Squamata - Chamaeleonidae
e Fito Leaf Chameleon

1377 ***Brookesia legenderi***
Squamata - Chamaeleonidae
e Nosy Be Leaf Chameleon

1378 ***Brookesia minima***
Squamata - Chamaeleonidae
e Minute Leaf Chameleon; Tiny Ground Chameleon
d Nossi-Bé-Zwergchamäleon; Nosy-Be Stummelschwanzchamäleon; Nosy-Be; Stummelschwanzchamäleon
f Brookesie de Nossi-Bé

1379 ***Brookesia nasus***
Squamata - Chamaeleonidae
e Elongate Leaf Chameleon

1380 ***Brookesia peramata***
Squamata - Chamaeleonidae
e Antsingy Leaf Chameleon; Armoured Chameleon; Armoured Stub-tailed Chamaeleon; Armoured Leaf Chamaeleon
d Panzerchamäleon

1381 ***Brookesia peyrieresi***
Squamata - Chamaeleonidae
e Antongil Leaf Chameleon

1382 ***Brookesia ramanantsoai***
Squamata - Chamaeleonidae
e Ramanantsoa's Leaf Chameleon

1383 ***Brookesia stumpffi***
Squamata - Chamaeleonidae
e Plated Leaf Chameleon; Spiny Chameleon
d Stachelchamäleon

1384 ***Brookesia superciliaris***
Squamata - Chamaeleonidae

e Brown Leaf Chameleon
d Augenzipfel-Stummelschwanzchamäleon

1385 ***Brookesia therezieni***
Squamata - Chamaeleonidae
e Perinet Leaf Chameleon

1386 ***Brookesia thieli***
Squamata - Chamaeleonidae
e Domergue's Leaf Chameleon

1387 ***Brookesia tuberculata***
Squamata - Chamaeleonidae
e Mount d'Ambre Leaf Chameleon
d Madagaskar-Zwergchamäleon

1388 ***Brookesia vadoni***
Squamata - Chamaeleonidae
e Iraka River Leaf Chameleon

1389 ***Brookesia valeriae***
Squamata - Chamaeleonidae
e Rekworthy's Leaf Chameleon

1390 ***Bungarus***
Squamata - Elapidae
e Indian Kraits; Kraits; Bungars
d Kraite; Bungaren
f Bongares

1391 ***Bungarus andamanensis***
Squamata - Elapidae
e South Andaman Krait

1392 ***Bungarus bungaroides***
Squamata - Elapidae
e Northeastern Krait; Cantor's Krait; North Eastern Hill Krait

1393 ***Bungarus caeruleus***
Squamata - Elapidae
e Indian Krait; Common Indian Krait (ISC); Blue Krait; Common Krait (ISC); Banded Krait
d Indischer Krait; Gewöhnlicher Krait; Gewöhnlicher Bungar; Blauer Krait
f Bongare commun

1394 ***Bungarus candidus***
Squamata - Elapidae
e Malayan Krait; Common Krait
d Malaiischer Krait; Blauer Krait

1395 ***Bungarus ceylonicus***
Squamata - Elapidae
e Ceylon Krait; Karawala; Sri Lankan Krait
d Karawala

1396 ***Bungarus fasciatus***
Squamata - Elapidae
e Banded Krait
d Gebänderter Krait; Gelber Bungar; Gelber Krait; Bänderkrait
f Bongare rayé; Bongare annelé

1397 ***Bungarus flaviceps***
Squamata - Elapidae
e Red-headed Krait; Yellow-headed Krait
d Rotkopfkrait
f Bongare à tète jaune

1398 ***Bungarus javanicus***
Squamata - Elapidae
e Java Krait
d Javanischer Bungar; Javanischer Krait

1399 ***Bungarus lividus***
Squamata - Elapidae
e Lesser Krait; Lesser Black Krait

1400 ***Bungarus magnimaculatus***
Squamata - Elapidae
e Burmese Krait

1401 ***Bungarus multicinctus***
Squamata - Elapidae

e Many-banded Krait
d Vielbindenbungar; Südchinesischer Vielbandenbungar; Vielbindenkrait

1402 ***Bungarus niger***
Squamata - Elapidae
e Greater Krait; Black Krait; Greater Black Krait
d Schwarzer Bungar

1403 ***Bungarus sindanus***
Squamata - Elapidae
e Sind Krait

1404 ***Bunopus***
Squamata - Gekkonidae
e Arabian Geckos

1405 ***Bunopus crassicauda***
Squamata - Gekkonidae
e Thickhead Rock Gecko

1406 ***Bunopus spatalurus***
Squamata - Gekkonidae
e Spacious Rock Gecko; Arabian Rock Gecko

1407 ***Bunopus tuberculatus***
Squamata - Gekkonidae
e Baluch Rock Gecko
d Kleinere Warzengecko

C

1408 ***Cabrita***
Squamata - Lacertidae
e Cabritas; Snake-eyes

1409 ***Cabrita jerdonii***
Squamata - Lacertidae
e Jerdon's Cabrita; Jerdon's Snake-eye (ISC)

1410 ***Cabrita leschenaulti***
Squamata - Lacertidae
e Leschenault's Cabrita; Leschenault's Snake-eye (ISC); Snake-eyed Lacerta

1411 ***Cacophis***
Squamata - Elapidae
e Dwarf Crowned Snakes; Crowned Snakes

1412 ***Cacophis harriettae***
Squamata - Elapidae
e White Crowned Snake

1413 ***Cacophis krefftii***
Squamata - Elapidae
e Dwarf Crowned Snake (ANZ); Red-bellied Black Snake

1414 ***Cacophis squamulosus***
Squamata - Elapidae
e Golden Crowned Snake (ANZ); Red-bellied Snake; Golden-crowned Snake

1415 ***Cadae***
Squamata - Amphisbaenidae
e Dickerson's Worm Lizards

1416 ***Cadae blanoides***
Squamata - Amphisbaenidae
e Isla de Pinos Worm Lizard; Cuban Spotted Amphisbaena

1417 ***Cadae palirostrata***
Squamata - Amphisbaenidae
e Dickerson's Worm Lizard; Cuban Sharp-nosed Amphisbaena

1418 ***Caiman***
Crocodylia - Alligatoridae
e Speckled Caimans; Caimans
d Kaimane; Brillenkaimane
f Caïmans
i Caimani

1419 ***Caiman crocodilus***
Crocodylia - Alligatoridae
e Spectacled Caiman; Narrow-snouted Spectacled Caiman; Babilla; Common Caiman
d Krokodilkaiman; Brillenkaiman
f Caïman à lunettes

1420 ***Caiman crocodilus apaporiensis***
Crocodylia - Alligatoridae
e Rio Apaporis Spectacled Caiman; Rio Apaporis Caiman
d Rio-Apaporis-Brillenkaiman
f Caïman du Rio Apaporis

1421 ***Caiman crocodilus fuscus***
Crocodylia - Alligatoridae
e Brown Caiman

1422 ***Caiman crocodilus yacara***
Crocodylia - Alligatoridae
e Spectacled Caiman
d Brillenkaiman

1423 ***Caiman latirostris***
Crocodylia - Alligatoridae
e Broad-snouted Spectacled Caiman; Broad-nosed Caiman; Broad-mouthed Caiman; Broad-snouted

Caiman
d Breitschnauzenkaiman
f Caïman à museau large; Caïman à large museau

1424 ***Caiman yacari***
Crocodylia - Alligatoridae
e Paraguay Caiman; Yacare; Yacare Caiman
d Südlicher Krokodilkaiman; Krokodilkaiman
f Jacaré

1425 ***Caimanops***
Squamata - Agamidae
e Caiman Agamas; Five Keel Agamas

1426 ***Caimanops amphiboluroides***
Squamata - Agamidae
e Caiman Agama; Mulga Dragon (ANZ); Five Keel Agama

1427 ***Calabaria***
Squamata - Boidae
e African Burrowing Pythons; Small African Pythons; Burrowing Pythons; Calabar Ground Pythons; West African Pythons
d Erdpythone; Erdpythonschlangen
f Calabarias
i Calabarie

1428 ***Calabaria reinhardtii***
Squamata - Boidae
e African Burrowing Python; Small African Python; Calabar Ground Python; Burrowing Python; West African Python; Calabar Burrowing Boa
d Erdpython
f Calabaria; Calabaria de Reinhardt
i Calabaria di Reinhardt; Calabaria

1429 ***Calamaria***
Squamata - Colubridae
e Dwarf Snakes; Reed Snakes
d Zwergnattern; Zwergschlangen

1430 ***Calamaria bicolor***
Squamata - Colubridae
e Two-coloured Dwarf Snake

1431 ***Calamaria bitorques***
Squamata - Colubridae
e Luzon Dwarf Snake

1432 ***Calamaria buchi***
Squamata - Colubridae
e Dalat Dwarf Snake

1433 ***Calamaria gervaisi***
Squamata - Colubridae
e Philippine Dwarf Snake

1434 ***Calamaria leucogaster***
Squamata - Colubridae
e Ampat Lawang Dwarf Snake

1435 ***Calamaria linnaei***
Squamata - Colubridae
e Linné's Dwarf Snake; Linnaeus's Dwarf Snake
d Linnés Zwergschlange

1436 ***Calamaria lowi***
Squamata - Colubridae
e Lowe's Dwarf Snake

1437 ***Calamaria lumbricoidea***
Squamata - Colubridae
e Variable Dwarf Snake

1438 ***Calamaria margaritophora***
Squamata - Colubridae
e Bleeker's Dwarf Snake

1439 ***Calamaria pavimentata***
Squamata - Colubridae
e Collared Dwarf Snake; Collared Reed Snake

1440 ***Calamaria pavimentata miyarai***
Squamata - Colubridae
e Miyara's Collared Snake

1441 ***Calamaria schlegelii***
Squamata - Colubridae
e White-headed Dwarf Snake

1442 ***Calamaria septentrionalis***
Squamata - Colubridae
e Hong Kong Dwarf Snake

1443 ***Calamaria sumatrana***
Squamata - Colubridae
e Sumatra Dwarf Snake

1444 ***Calamaria virgulata***
Squamata - Colubridae
e Boie's Dwarf Snake

1445 ***Calambdontophis***
Squamata - Colubridae
e Tropical Forest Snakes

1446 ***Calambdontophis paucidens***
Squamata - Colubridae
e Tropical Forest Snake

1447 ***Calamorhabdium***
Squamata - Colubridae
e Iridescent Snakes

1448 ***Calamorhabdium acuticeps***
Squamata - Colubridae
e Sulawesi Iridescent Snake

1449 ***Calamorhabdium kukenthali***
Squamata - Colubridae
e Batjan Iridescent Snake

1450 ***Caledoniscus***
Squamata - Scincidae
e Caledonian Skinks

1451 ***Caledoniscus austrocaledonicus***
Squamata - Scincidae
e Caledonian Skink

1452 ***Caledoniscus orestes***
Squamata - Scincidae
e Sadler's Caledonian Skink

1453 ***Callagur***
Testudines - Emydidae
e Painted Terrapins; River Turtle; Callagurs
d Callagurschildkröten

1454 ***Callagur borneoensis***
Testudines - Emydidae
e Painted Terrapin; River Turtle; Biuku; Painted Batagur; Painted Terrapin
d Callagurschildkröte
f Émyde peinte de Bornéo

1455 ***Calliophis***
Squamata - Elapidae
e Oriental Coral Snakes; Coral Snakes
d Schmuckottern; Orientalische Korallenottern

1456 ***Calliophis bibroni***
Squamata - Elapidae
e Bibron's Coral Snake

1457 ***Calliophis gracilis***
Squamata - Elapidae
e Gray's Coral Snake

1458 ***Calliophis kelloggi***
Squamata - Elapidae
e Kellogg's Coral Snake

1459 ***Calliophis kelloggi iwasakii***
Squamata - Elapidae
e Iwasaki's Coral Snake

1460 ***Calliophis maculiceps***
Squamata - Elapidae
e Small-spotted Coral Snake
d Südostasiatische Korallenotter

1461 ***Calliophis melanurus***
Squamata - Elapidae
e Indian Coral Snake; Slender Coral Snake (ISC)

1462 ***Calliophis nigrescens***
Squamata - Elapidae
e Black Coral Snake

1463 ***Calliophis sauteri***
Squamata - Elapidae
e Oriental Coral Snake

1464 ***Callisaurus***
Squamata - Iguanaidae
e Zebratail Iguanas; Gridiron-tailed Lizards; Zebra-tailed Lizards
d Gitterschwanzleguane
f Lézards à queue zébrée; Callisaures

1465 ***Callisaurus draconoides***
Squamata - Iguanaidae
e Zebratail Lizard; Gridiron-tailed Lizard; Common Gridiron-tailed Lizard; Zebra-tailed Lizard
d Gitterschwanzleguan; Zebraschwanzleguan
f Lézard-dragon; Callisaure draconoïde

1466 ***Callisaurus draconoides crinitus***
Squamata - Iguanaidae
e Viscaiño Zebra-tailed Lizard

1467 ***Callisaurus draconoides draconoides***
Squamata - Iguanaidae
e Common Zebra-tailed Lizard

1468 ***Callisaurus draconoides myurus***
Squamata - Iguanaidae
e Nevada Zebratail Lizard; Nevada Zebra-tailed Lizard; Northern Zebra-tailed Lizard

1469 ***Callisaurus draconoides rhodostictus***
Squamata - Iguanaidae
e Mojave Zebratail Lizard; Western Zebra-tailed Lizard

1470 ***Callisaurus draconoides ventralis***
Squamata - Iguanaidae
e Arizona Zebratail Lizard; Eastern Zebra-tailed Lizard

1471 ***Callopistes***
Squamata - Teiidae
e False Monitors

1472 ***Callopistes flavipunctatus***
Squamata - Teiidae
e False Monitor
d Warran-Teju

1473 ***Callopistes maculatus***
Squamata - Teiidae
e Spotted False Monitor; Dwarf Tegu
d Chile-Teju

1474 ***Calloselasma***
Squamata - Viperidae
e Malayan Pit Vipers
d Malayische Mokassinottern; Malayen-Mokassinottern; Malayische Mokassinschlangen

1475 ***Calloselasma rhodostoma***
Squamata - Viperidae
e Malayan Pit Viper
d Malayische Mokassinotter; Malayen-Mokassinotter; Malayische Mokassinschlange; Malayische Mokassingrubenotter; Malayen-Mokassinschlange

1476 ***Calodactylodes***
Squamata - Gekkonidae
e Indian Geckos

1477 ***Calodactylodes aureus***
Squamata - Gekkonidae
e Beddome's Golden Gecko; Golden Gecko (ISC)

1478 ***Calodactylodes illingworthi***
Squamata - Gekkonidae
e Illingworth's Gecko; Sri Lankan Golden Gecko

1479 ***Calotes***
Squamata - Agamidae
e Varied Lizards; Variable Lizards; Caloteses (ISC); Beauty Lizards; Bloodsuckers
d Schönechsen
f Galéotes

1480 ***Calotes andamanensis***
Squamata - Agamidae
e Andaman Lizard
d Andaman-Schönechse
f Galéote d'Andaman

1481 ***Calotes calotes***
Squamata - Agamidae
e Common Tree Lizard; Sawback Agama; Sawback Agamid; Tree Lizard; Bloodsucker; Southern Green Calotes (ISC); Common Green Forest Lizard
d Schönechse; Sägerückenagame
f Galéote commun

1482 ***Calotes ceylonensis***
Squamata - Agamidae
e Ceylon Bloodsucker; Sri Lankan Forest Lizard
d Ceylon-Schönechse
f Galéote de Ceylon

1483 ***Calotes emma***
Squamata - Agamidae
e Thailand Bloodsucker
d Hinterindische Schönechse; Emmas Schönechse; Gehörnter Calotes

1484 ***Calotes grandisquamis***
Squamata - Agamidae
e Big-scaled Variable Lizard

1485 ***Calotes jerdoni***
Squamata - Agamidae
e Jerdon's Bloodsucker; Jerdon's Calotes (ISC); Eastern Green Calotes

1486 ***Calotes khakiensis***
Squamata - Agamidae
e Anderson's Bloodsucker

1487 ***Calotes kinabaluensis***
Squamata - Agamidae
e Sabah Bloodsucker

1488 ***Calotes kingdonwardi***
Squamata - Agamidae
e Kindon-Ward's Bloodsucker

1489 ***Calotes liocephalus***
Squamata - Agamidae
e Lionhead Agama

1490 ***Calotes liolepis***
Squamata - Agamidae
e Sri Lanka Agama

1491 ***Calotes maria***
Squamata - Agamidae
e Khasi Hills Agama

1492 ***Calotes medogenis***
Squamata - Agamidae
e Medog Bloodsucker

1493 ***Calotes mystaceus***
Squamata - Agamidae

e Indo-Chinese Agama; Indo-Chinese Tree Agama
d Blaukopf-Calotes

1494 ***Calotes nemoricola***
Squamata - Agamidae
e Niligiri Hills Agama

1495 ***Calotes nigrigularis***
Squamata - Agamidae
e Black-throated Bloodsucker
d Hochlandagame

1496 ***Calotes nigrilabis***
Squamata - Agamidae
e Black-lipped Agama; Black-lipped Forest Lizard

1497 ***Calotes rouxi***
Squamata - Agamidae
e Bombay Agama; Red-headed Lizard; Forest Calotes (ISC); Forest Bloodsucker (ISC)
d Rotkopfagame
f Galéote de Roux

1498 ***Calotes versicolor***
Squamata - Agamidae
e Variable Agama; Bloodsucker (ISC); Varied Lizard; Indian Bloodsucker; Indian Garden Lizard (ISC); Common Garden Lizard (ISC); Bloodsucker Indian Lizard
d Verschiedenfarbige Schönechse; Blutsaugeragame; Indischer Blutsauger; Blutsauger; Indische Schönechse; Indische Schönechse

1499 ***Calyptommatus***
Squamata - Gymnophthalmidae
e Microteiids

1500 ***Calyptommatus liolepis***
Squamata - Gymnophthalmidae
e Greater Microteiid

1501 ***Calyptommatus nicterus***
Squamata - Gymnophthalmidae
e Rodrigues's Microteiid

1502 ***Calyptommatus sinecrachiatus***
Squamata - Gymnophthalmidae
e Lesser Microteiid

1503 ***Calyptotis***
Squamata - Scincidae
e Fossorial Skinks

1504 ***Calyptotis lepidorostrum***
Squamata - Scincidae
e Australian Fossorial Skink

1505 ***Calyptotis ruficauda***
Squamata - Scincidae
e Redtail Fossorial Skink

1506 ***Calyptotis scutirostris***
Squamata - Scincidae
e Brown Fossorial Skink

1507 ***Calyptotis temporalis***
Squamata - Scincidae
e Flecked Fossorial Skink

1508 ***Calyptotis thorntonensis***
Squamata - Scincidae
e Thornton Peak Fossorial Skink

1509 ***Candoia***
Squamata - Boidae
e Ground Boas; Pacific Boas
d Pazifik-Boas; Südsee-Boas
i Candoia

1510 ***Candoia aspera***
Squamata - Boidae
e New Guinea Viper Boa; New Guinea Ground Boa; Dwarf Boa; New Guinea Boa
d Pazifik-Boa; Neuguinea Boa

1511 ***Candoia bibroni***
Squamata - Boidae
e Pacific Island Boa; Bibron's Boa; Pacific Tree Boa
d Bibrons Boa; Pazifik-Boa
f Boa de Bibron; Boa des Loyautés

1512 ***Candoia carinata***
Squamata - Boidae
e Solomon Islands Ground Boa; Pacific Boa; Keeled Tree Boa; New Guinea Boa; Papuan Tree Boa; Halmerheran Tree Boa; Pacific Ground Boa
d Neiguinea-Boa

1513 ***Candoia carinata carinata***
Squamata - Boidae
e Pacific Tree Boa
d Pazifik-Boa

1514 ***Candoia carinata paulsoni***
Squamata - Boidae
e Pacific Ground Boa
d Pazifik-Boa

1515 ***Cantoria***
Squamata - Colubridae
e Water Snakes; Girard's Water Snakes

1516 ***Cantoria annulata***
Squamata - Colubridae
e Annulated Water Snake

1517 ***Cantoria violacea***
Squamata - Colubridae
e Cantor's Water Snake; Girard's Water Snake

1518 ***Caretta***
Testudines - Cheloniidae
e Loggerhead Turtles; Loggerheads; Loggerhead Sea Turtles
d Unechte Karettschildkröten; Unechte Karettenschildkröten
f Tortues caouane

1519 ***Caretta caretta***
Testudines - Cheloniidae
e Loggerhead Turtle; Loggerhead; Loggerhead Musk Turtle; Loggerhead Sea Turtle; Lanternback Turtle; Red-brown Loggerhead
d Unechte Schildkröte; Unechte Karettschildkröte; Unechte Karettenschildkröte; Unechte Karette
f Cauane; Karet (Ants); Caret; Couanne; Tortue caouanne
i Tartaruga caretta; Tartaruga caretta comune; Caretta comune; Tartaruga marina comune; Caretta; Caretta comune

1520 ***Caretta caretta caretta***
Testudines - Cheloniidae
e Atlantic Loggerhead
i Tartarugone comune

1521 **Carettochelyidae**
Testudines
e Pignnose Turtles; Pitted-shelled Turtles (ANZ); Fly River Turtles
d Neuguinea-Weichschildkröten; Papua-Weichschildkröten; Papua-Schildkröten
f Carettochélyidés; Tortues à carapace molle

1522 ***Carettochelys***
Testudines - Carettochelyidae
e Pignose Turtles; Fly River Turtle; Pig-nosed Turtles; Pitted-shelled Turtles
d Neuguinea-Weichschildkröten; Papua-Schildkröten
f Tortues à carapace molle

1523 ***Carettochelys insculpta***
Testudines - Carettochelyidae
e Pignose Turtle; Pitted-shell Turtle (ANZ); New Guinea River Turtle; Pig-snouted River Turtle; New Guinean Plateless Turtle; River Turtle; New Guinean Pig-nosed Turtle; Fly River Turtle;

Pig-nosed Turtle
d Neuguinea-Weichschildkröte; Papua-Schildkröte
f Tortue à carapace molle de Nouvelle-Guinée; Tortue molle de Nouvelle-Guinée; Tortue sans écaille de Nouvelle-Guinée; Tortue à nez de cochon
i Tartaruga naso di porcello

1524 ***Carinatogecko***
Squamata - Gekkonidae
e Strand's Geckos

1525 ***Carinatogecko aspratilis***
Squamata - Gekkonidae
e Iranian Gecko

1526 ***Carinatogecko heteropholis***
Squamata - Gekkonidae
e Iraqui Gecko

1527 ***Carlia***
Squamata - Scincidae
e Carlias; Rainbow Skinks

1528 ***Carlia amax***
Squamata - Scincidae
e White-eared Carlia

1529 ***Carlia coensis***
Squamata - Scincidae
e Coen Carlia

1530 ***Carlia dogare***
Squamata - Scincidae
e Speckle-lipped Carlia

1531 ***Carlia fusca***
Squamata - Scincidae
e Indonesian Brown Carlia; Four-fingered Skink

1532 ***Carlia gracilis***
Squamata - Scincidae
e Copper Carlia; Slender Rainbow Skink

1533 ***Carlia jarnoldae***
Squamata - Scincidae
e Decorated Carlia

1534 ***Carlia johnstonei***
Squamata - Scincidae
e Woodland Carlia

1535 ***Carlia longipes***
Squamata - Scincidae
e Bronze Carlia

1536 ***Carlia munda***
Squamata - Scincidae
e Grey Carlia

1537 ***Carlia mundivensis***
Squamata - Scincidae
e Rock Carlia

1538 ***Carlia pectoralis***
Squamata - Scincidae
e Eastern Carlia; Open-litter Rainbow Skink

1539 ***Carlia rhomboidalis***
Squamata - Scincidae
e Speckled Carlia; Blue-throated Rainbow Skink
d Glanzschlankskink

1540 ***Carlia rimula***
Squamata - Scincidae
e Black Carlia

1541 ***Carlia rostralis***
Squamata - Scincidae
e Black-flecked Carlia; Black-throated Rainbow Skink

1542 ***Carlia rubrigularis***
Squamata - Scincidae

e Rainforest Carlia; Northern Red-throated Skink

1543 ***Carlia rufilatus***
Squamata - Scincidae
e Red-striped Carlia

1544 ***Carlia schmeltzii***
Squamata - Scincidae
e Olive Carlia; Robust Rainbow Skink

1545 ***Carlia scirtetis***
Squamata - Scincidae
e Cave Carlia

1546 ***Carlia storri***
Squamata - Scincidae
e Storr's Carlia; Brown Bicarinate Rainbow Skink

1547 ***Carlia tetradactyla***
Squamata - Scincidae
e Forest Carlia; Southern Rainbow Skink (ANZ)

1548 ***Carlia triacantha***
Squamata - Scincidae
e Central Carlia

1549 ***Carlia vivax***
Squamata - Scincidae
e Brown-flecked Carlia; Tussock Rainbow Skink

1550 ***Carphodactylus***
Squamata - Gekkonidae
e Chameleon Geckos

1551 ***Carphodactylus laevis***
Squamata - Gekkonidae
e Chameleon Gecko
f Gecko caméléon

1552 ***Carphophis***
Squamata - Colubridae
e Worm Snakes; Ground Snakes
d Wurmschlangen; Wurmnattern
f Couleuvres-ver

1553 ***Carphophis amoenus***
Squamata - Colubridae
e Worm Snake; Ground Snake; Eastern Worm Snake
d Wurmschlange; Wurmnatter
f Couleuvre-ver

1554 ***Carphophis amoenus amoenus***
Squamata - Colubridae
e Eastern Worm Snake

1555 ***Carphophis amoenus helenae***
Squamata - Colubridae
e Midwest Worm Snake; Midwestern Worm Snake

1556 ***Carphophis amoenus vermis***
Squamata - Colubridae
e Western Worm Snake

1557 ***Casarea***
Squamata - Bolyeriidae
e Round Island Boas; Mauritius Boa; Round Island Ground Boas
d Round-Island-Boas; Mauritius-Boas; Rundinsel-Boas
f Boas de Maurice

1558 ***Casarea dussumieri***
Squamata - Bolyeriidae
e Round Island Boa; Round Island Ground Boa; Round Island Keel-scaled Boa; Round Island Burrowing Boa
d Round-Island-Boa; Mauritius-Boa; Rundinsel-Boa
f Boa de Maurice; Round Island Boa
i Boa carenato dell'isola di Round

1559 ***Causus***
Squamata - Viperidae
e Night Adders

d Krötenvipern; Nachtottern; Krötenottern
f Vipères nocturnes

1560 ***Causus bilineatus***
Squamata - Viperidae
e Two-lined Night Adder

1561 ***Causus defilippii***
Squamata - Viperidae
e Snouted Night Adder
d Defilippis Nachtotter

1562 ***Causus lichtensteini***
Squamata - Viperidae
e Lichtenstein's Night Adder
d Lichtensteins Nachtotter

1563 ***Causus maculatus***
Squamata - Viperidae
e Spotted Night Adder; West African Night Adder

1564 ***Causus resimus***
Squamata - Viperidae
e Green Night Adder; Velvety Green Night Adder
d Grüne Nachtotter; Grüne Krötenotter

1565 ***Causus rhombeatus***
Squamata - Viperidae
e Common Night Adder (CSA); Rhombic Night Adder; Demon Adder
d Krötenotter; Nachtotter; Pfeilotter
f Vipère nocturne; Vipère démon

1566 ***Ceiolaemus***
Squamata - Iguanaidae
e Spotted Lizards

1567 ***Ceiolaemus marmorata***
Squamata - Iguanaidae
e Burmeister's Spotted Lizard

1568 ***Celestus***
Squamata - Anguidae
e Lesser Galliwasps; Galliwasps
d Doppelzungenschleichen

1569 ***Celestus barbouri***
Squamata - Anguidae
e Barbour's Galliwasp; Limestone Forest Galliwasp (WI)
d Barbours Doppelzungenschleiche

1570 ***Celestus bivittatus***
Squamata - Anguidae
e Two-striped Galliwasp

1571 ***Celestus crusculus***
Squamata - Anguidae
e Garman's Galliwasp; Jamaican Galliwasp (WI)

1572 ***Celestus curtissi***
Squamata - Anguidae
e Curtis's Galliwasp; Hispaniolan Khaki Galliwasp
d Curtiss-Doppelzungenschleiche

1573 ***Celestus darlingtoni***
Squamata - Anguidae
e Darlington's Galliwasp; Hispaniolan Striped Galliwasp
d Darlingtons Doppelzungenschleiche

1574 ***Celestus duquesneyi***
Squamata - Anguidae
e Duquesney's Galliwasp; Blue-tailed Galliwasp (WI)

1575 ***Celestus enneagrammus***
Squamata - Anguidae
e Huaxteca Lesser Galliwasp

1576 ***Celestus fowleri***
Squamata - Anguidae
e Fowler's Galliwasp; Bromeliad Galliwasp (WI)

d Fowlers Doppelzungenchleiche

1577 ***Celestus hewardi***
Squamata - Anguidae
e Heward's Galliwasp; Three-streaked Galliwasp; Red-spotted Galliwasp (WI)
d Hewards Doppelzungenchleiche

1578 ***Celestus macrotus***
Squamata - Anguidae
e Thomas's Galliwasp; La Selle Galliwasp

1579 ***Celestus marcanoi***
Squamata - Anguidae
e Marcano's Galliwasp; Pico Duarte Galliwasp

1580 ***Celestus microblepharis***
Squamata - Anguidae
e Tiny Eyelid Galliwasp; Small-eyed Galliwasp (WI)

1581 ***Celestus montanus***
Squamata - Anguidae
e Mountain Lesser Galliwasp

1582 ***Celestus occiduus***
Squamata - Anguidae
e Sinking Galliwasp; Jamaican Giant Galliwasp (WI)

1583 ***Celestus rozellae***
Squamata - Anguidae
e Rozella's Lesser Galliwasp; Rozella's Galliwasp

1584 ***Celestus stenurus***
Squamata - Anguidae
e Cope's Galliwasp; Hispaniolan Keeled Galliwasp

1585 ***Cemophora***
Squamata - Colubridae
e Scarlet Snakes
d Scharlachnattern
f Couleuvres écarlates

1586 ***Cemophora coccinea***
Squamata - Colubridae
e Scarlet Snake; Red Snake
d Scharlachnatter
f Couleuvre écarlate; Vipère écarlate

1587 ***Cemophora coccinea coccinea***
Squamata - Colubridae
e Florida Scarlet Snake

1588 ***Cemophora coccinea copei***
Squamata - Colubridae
e Northern Scarlet Snake

1589 ***Cemophora lineri***
Squamata - Colubridae
e Texas Scarlet Snake

1590 ***Cerastes***
Squamata - Viperidae
e Horned Vipers
d Afrikanische Hornvipern; Hornvipern
f Cérastes; Vipères cornues; Vipères à cornes

1591 ***Cerastes cerastes***
Squamata - Viperidae
e Horned Viper; Desert Horned Viper; Sahara Horned Viper; Carpet Viper; Sand Viper; Greater Cerastes Viper
d Hornviper
f Vipère cornue; Vipère à cornes ; Vipère à cornes du Sahara
i Ceraste cornuta; Diavolo del deserto

1592 ***Cerastes gasperettii***
Squamata - Viperidae
e Arabian Horned Viper

1593 ***Cerastes vipera***
Squamata - Viperidae

e Common Sand Viper; Sand Viper; Sahara Sand Viper; Lesser Cerastes Viper
d Avicenna-Viper; Avicennas Viper
f Céraste d'Avicenne; Vipère de l'Erg; Vipère d'Avicenne; Vipère des sables

1594 ***Ceratophora***
Squamata - Agamidae
e Horned Agamas; Horned Agamids; Horned Lizards
d Hornagamen; Nashornagamen
f Cératophores

1595 ***Ceratophora aspera***
Squamata - Agamidae
e Sri Lanka Horned Agama
d Hornagame; Spitznase

1596 ***Ceratophora stoddartii***
Squamata - Agamidae
e Mountain Horned Agama; Horned Agamid; Stoddard's Horned Lizard
d Stachelnase
f Cératophore de Stoddard

1597 ***Ceratophora tennentii***
Squamata - Agamidae
e Rhinoceros Agama; Tennent's Horned Lizard

1598 ***Cerberus***
Squamata - Colubridae
e Dog-faced Water Snakes
d Hundskopfwassertrugnattern

1599 ***Cerberus microlepis***
Squamata - Colubridae
e Dog-faced Water Snake

1600 ***Cerberus rhynchops***
Squamata - Colubridae
e Bokadam; Bockadam (ANZ); Dog-faced Water Snake (ISC); Estuarine Dog-faced Water Snake
d Hundskopfwassertrugnatter

1601 ***Cercaspis***
Squamata - Colubridae
e Wolf Snakes; Ceylon Wolf Snakes

1602 ***Cercaspis carinatus***
Squamata - Colubridae
e Ceylon Wolf Snake

1603 ***Cercophis***
Squamata - Colubridae
e Golden Snakes

1604 ***Cercophis auratus***
Squamata - Colubridae
e Schlegel's Golden Snake

1605 ***Cercosaura***
Squamata - Teiidae
e Ocellated Tegus; Shade Lizards

1606 ***Cercosaura ocellata***
Squamata - Teiidae
e Ocellated Tegu

1607 ***Cercosaura ocellata bassleri***
Squamata - Teiidae
e Black Striped Shade Lizard

1608 ***Cerrophidion***
Squamata - Viperidae
e Pit Vipers
d Berggrubenottern; Hochlandvipern

1609 ***Cerrophidion barbouri***
Squamata - Viperidae
e Barbour's Montane Pit Viper

1610 ***Cerrophidion godmani***
Squamata - Viperidae
e Godman's Montane Pit Viper; Godman's Mountain Pit Viper
d Godman-Bergrubenotter

1611 ***Cerrophidion tzotzilorum***
Squamata - Viperidae
e Tzotzil Montane Pit Viper; Tzotzil Mountain Pit Viper

1612 ***Chalarodon***
Squamata - Iguanaidae
e Madagascar Iguanas
d Madagaskar-Leguane

1613 ***Chalarodon madagascariensis***
Squamata - Iguanaidae
e Madagascar Iguana; Spotted Lizard
d Madagaskar-Leguan

1614 ***Chalcides***
Squamata - Scincidae
e Cylindrical Skinks
d Walzenskinke
f Chalcides

1615 ***Chalcides armitagei***
Squamata - Scincidae
e Armitage's Cylindrical Skink

1616 ***Chalcides atlantis***
Squamata - Scincidae
e Pasteur's Cylindrical Skink

1617 ***Chalcides bedriagai***
Squamata - Scincidae
e Spanish Cylindrical Skink; Round-bodied Skink; Bedriaga's Skink; Spanish Barrel Skink
d Iberischer Walzenskink; Spanischer Walzenskink
f Seps de Bedriaga
i Gongilo di Bedriaga

1618 ***Chalcides chalcides***
Squamata - Scincidae
e Algerian Cylindrical Skink; Three-toed Skink
d Erzschleiche
f Seps tridactyle; Seps strié
i Luscengola; Fienarola

1619 ***Chalcides colosii***
Squamata - Scincidae
e Colosi's Cylindrical Skink; Riffian Skink
f Seps riffain; Seps du Rif
i Gongilo ocellato

1620 ***Chalcides ebneri***
Squamata - Scincidae
e Ebner's Cylindrical Skink; Ebner's Skink
d Ebners Walzenskink
f Seps d'Ebner

1621 ***Chalcides guentheri***
Squamata - Scincidae
e Günther's Cylidrical Skink
d Günthers Walzenskink

1622 ***Chalcides levitoni***
Squamata - Scincidae
e Leviton's Cylindrical Skink
d Arabische Walzenechse

1623 ***Chalcides mauritanicus***
Squamata - Scincidae
e Cylindrical Skink; Two-fingered Skink
f Chalcides maurétanique; Seps de Maurétanie

1624 ***Chalcides mionecton***
Squamata - Scincidae
e Morocco Cylindrical Skink; Mionecton Skink
d Marokkanische Erzschleiche
f Seps mionecton

1625 ***Chalcides moseri***
Squamata - Scincidae
e Moser's Cylindrical Skink
d Santorin-Walzenskink

1626 ***Chalcides ocellatus***
Squamata - Scincidae
e Cylindrical Skink; Ocellated Skink; Eyed Skink
d Gefleckter Walzenskink; Walzenskink
f Seps ocellé; Chalcide ocellé; Scinque ocellé; Gongyle ocellé
i Gongilo; Gongilo ocellato

1627 ***Chalcides pistaciae***
Squamata - Scincidae
e Pyrenean Cylindrical Skink
f Seps de Valverde
i Gongilo ispano-lusitanico

1628 ***Chalcides polylepis***
Squamata - Scincidae
e Many-scaled Cylindrical Skink
d Marokkanischer Walzenskink
f Seps à écailles nombreuses; Scinque ocellé; Gongyle ocellé

1629 ***Chalcides pulchellus***
Squamata - Scincidae
e Mocquard's Cylindrical Skink

1630 ***Chalcides ragazzii***
Squamata - Scincidae
e Ragazzi's Cylindrical Skink

1631 ***Chalcides sexlinetus***
Squamata - Scincidae
e Six-lined Cylindrical Skink
d Gestreifter Kanarenskink

1632 ***Chalcides thierryi***
Squamata - Scincidae
e Thierry's Cylindrical Skink

1633 ***Chalcides viridianus***
Squamata - Scincidae
e Canaryan Cylindrical Skink; Canaries Cylindrical Skink
d Kanaren-Walzenskink; Kanaren-Skink

1634 ***Chalcidoseps***
Squamata - Scincidae
e Thwaite's Skinks

1635 ***Chalcidoseps thwaitesi***
Squamata - Scincidae
e Thwaites's Skink

1636 ***Chamaeleo***
Squamata - Chamaeleonidae
e Chameleons
d Chamäleone; Eigentliche Chamäleone
f Caméléons

1637 ***Chamaeleo affinis***
Squamata - Chamaeleonidae
e Rüppell's Desert Chameleon
d Äthiopisches Chamäleon

1638 ***Chamaeleo africanus***
Squamata - Chamaeleonidae
e African Chameleon
d Afrikanisches Chamäleon; Basiliskenchamäleon
f Caméléon africain

1639 ***Chamaeleo anchietae***
Squamata - Chamaeleonidae
e Double-scaled Chameleon; Angola Chameleon
d Angola-Chamäleon
f Caméléon d'Angola

1640 ***Chamaeleo angeli***
Squamata - Chamaeleonidae
e Angel's Chameleon

1641 ***Chamaeleo angusticoronatus***
Squamata - Chamaeleonidae
e Narrow-crowned Chameleon

1642 ***Chamaeleo antimena***
Squamata - Chamaeleonidae
e White-lined Chameleon

1643 ***Chamaeleo arabicus***
Squamata - Chamaeleonidae
e Arabian Chameleon

1644 ***Chamaeleo balteatus***
Squamata - Chamaeleonidae
e Rainforest Chameleon

1645 ***Chamaeleo barbouri***
Squamata - Chamaeleonidae
e Barbour's Chameleon

1646 ***Chamaeleo belandaensis***
Squamata - Chamaeleonidae
e Belalanda Chameleon

1647 ***Chamaeleo bifidus***
Squamata - Chamaeleonidae
e Two-horned Chameleon
d Zweihornchamäleon
f Caméléon à nez forchu

1648 ***Chamaeleo bitaeniatus***
Squamata - Chamaeleonidae
e Two-lined Chameleon; Double-banded Chameleon
d Zweistreifenchamäleon; Streifenchamäleon
f Caméléon à deux bandes

1649 ***Chamaeleo boettgeri***
Squamata - Chamaeleonidae
e Boettger's Chameleon

1650 ***Chamaeleo brevicornis***
Squamata - Chamaeleonidae
e Short-horned Chameleon
d Kurzhornchamäleon
f Caméléon à capuchon

1651 ***Chamaeleo calcarifer***
Squamata - Chamaeleonidae
e Hillenius's Chameleon

1652 ***Chamaeleo calyptratus***
Squamata - Chamaeleonidae
e Veiled Chameleon
d Jemen-Chamäleon
f Caméléon casqué
i Camaleonte velato

1653 ***Chamaeleo campani***
Squamata - Chamaeleonidae
e Madagascar Forest Chameleon

1654 ***Chamaeleo capuroni***
Squamata - Chamaeleonidae
e Madagascar Chameleon
d Kurzhornchamäleon

1655 ***Chamaeleo carpenteri***
Squamata - Chamaeleonidae
e Ruwenzori Mountains Chameleon

1656 ***Chamaeleo cephalolepis***
Squamata - Chamaeleonidae
e Comoro Island Chameleon

1657 ***Chamaeleo chameleon***
Squamata - Chamaeleonidae
e Common Chameleon; Chameleon; European Chameleon; Mediterranean Chameleon; Indian Chameleon
d Europäisches Chamäleon; Gemeines Chamäleon; Gewöhnliches Chamäleon; Chamäleon
f Caméléon vulgaire; Caméléon commun; Caméléon ordinaire; Caméléon européen
i Camaleonte; Cameleonte comune

1658 ***Chamaeleo chapini***
Squamata - Chamaeleonidae
e Grey Chameleon

1659 ***Chamaeleo cristatus***
Squamata - Chamaeleonidae
e Crested Chameleon
d Kammchamäleon

1660 ***Chamaeleo cucullatus***
Squamata - Chamaeleonidae
e Hooded Chameleon
f Caméléon à capuchon

1661 ***Chamaeleo deremensis***
Squamata - Chamaeleonidae
e Wavy Chameleon; Usambara Three-horned Chameleon
d Usambara-Dreihornchamäleon

1662 ***Chamaeleo dilepis***
Squamata - Chamaeleonidae
e Flapneck Chameleon; Flap-necked Chameleon; Common Chameleon
d Lappenchamäleon
f Caméléon bilobé
i Camaleonte orecchiuto africano

1663 ***Chamaeleo dilepis dilepis***
Squamata - Chamaeleonidae
e Common Flap-necked Chameleon (CSA)

1664 ***Chamaeleo fallax***
Squamata - Chamaeleonidae
e Deceptive Chameleon

1665 ***Chamaeleo fischeri***
Squamata - Chamaeleonidae
e Fischer's Chameleon; Usambara Two-horned Chameleon; Kilimanjaro Two-horned Chameleon
d Fischers Chamäleon
f Caméléon de Fischer

1666 ***Chamaeleo fuelleborni***
Squamata - Chamaeleonidae
e Ngosi Volcano Chameleon; Poroto Three-horned Chameleon
d Dreihornlappenchamäleon

1667 ***Chamaeleo furcifer***
Squamata - Chamaeleonidae
e Forknose Chameleon
d Gabelchamäleon

1668 ***Chamaeleo gallus***
Squamata - Chamaeleonidae
e Blade Chameleon

1669 ***Chamaeleo gastrotaenia***
Squamata - Chamaeleonidae
e Perinet Chameleon; Short-nosed Chameleon

1670 ***Chamaeleo globifer***
Squamata - Chamaeleonidae
e Flat-casqued Chameleon; Will's Chameleon

1671 ***Chamaeleo goetzi***
Squamata - Chamaeleonidae
e Ilota Chameleon; Goetze's Chameleon; Tanganyika Chameleon
d Goetzi-Chamäleon

1672 ***Chamaeleo gracilis***
Squamata - Chamaeleonidae
e Graceful Chameleon; Spur-heeled Chameleon; Gracile Chameleon
d Fersenspornchamäleon; Zierliches Chamäleon; Fleckenchmäleon

1673 ***Chamaeleo guibei***
Squamata - Chamaeleonidae
e Northeastern Chameleon; Guibe's Chameleon

1674 ***Chamaeleo hoehnelii***
Squamata - Chamaeleonidae
e High-casqued Chameleon
d Helmchamäleon; Hoehnels Chamäleon

1675 ***Chamaeleo incornutus***
Squamata - Chamaeleonidae
e Poroto Mountain Chameleon; Ukinga Hornless Chameleon

1676 ***Chamaeleo jacksoni***
Squamata - Chamaeleonidae
e Jackson's Chameleon; Jackson's Three-horned Chameleon; Three-horned Chameleon
d Jacksons Dreihornchamäleon; Jacksons Chamäleon; Dreihornchamäleon; Ostafrikanisches Dreihornchamleon; Jackson-Chamäleon
f Caméléon de Jackson

1677 ***Chamaeleo jacksoni xanthlophus***
Squamata - Chamaeleonidae
e Yellow-crested Jackson's Chamaeleon
d Dreihornchamäleon; Ostafrikanisches Dreihornchamleon

1678 ***Chamaeleo jacksoni merumontana***
Squamata - Chamaeleonidae
e Dwarf Jackson's Chameleon; Mount Meru Chameleon

1679 ***Chamaeleo johnstoni***
Squamata - Chamaeleonidae
e Johnston's Chameleon; Three-horned Chameleon; Johnston's Three-horned Chameleon; Ruwenzori Three-horned Chameleon
d Johnston-Chamäleon

1680 ***Chamaeleo lateralis***
Squamata - Chamaeleonidae
e Jewelled Chameleon; Carpet Chameleon
d Teppichchamäleon
f Caméléon à bandes laterales

1681 ***Chamaeleo lateralis major***
Squamata - Chamaeleonidae
e Giant Jewelled Chameleon

1682 ***Chamaeleo laterispinis***
Squamata - Chamaeleonidae
e Uzungwe Mountain Chameleon; Spiny-sided Chameleon
d Seitenstachelchamäleon

1683 ***Chamaeleo linotus***
Squamata - Chamaeleonidae
e Maroantsetra Chameleon

1684 ***Chamaeleo malthe***
Squamata - Chamaeleonidae
e Yellow-green Chameleon

1685 ***Chamaeleo marsabitensis***
Squamata - Chamaeleonidae
e Tilbury's Chameleon
d Marsabit-Chamäleon

1686 ***Chamaeleo melleri***
Squamata - Chamaeleonidae
e Meller's Chameleon; Giant One-horned Chameleon
d Elefantenchamäleon; Elefantenohrchamäleon; Mellers Chamäleon; Meller-Chamäleon

1687 ***Chamaeleo minor***
Squamata - Chamaeleonidae
e South-central Chameleon

1688 ***Chamaeleo mlanjensis***
Squamata - Chamaeleonidae
e Mlanje Mountain Chameleon

1689 ***Chamaeleo monachus***
Squamata - Chamaeleonidae
e Monarch Chameleon

1690 ***Chamaeleo monoceras***
Squamata - Chamaeleonidae
e One-horned Chameleon

1691 ***Chamaeleo montium***
Squamata - Chamaeleonidae

e Cameroon Sailfin Chameleon; Mountain Chameleon
d Bergchamäleon; Zweihornbergchmäleon
f Caméléon à deux cornes rostrales

1692 ***Chamaeleo namaquensis***
Squamata - Chamaeleonidae
e Namaqua Chameleon
d Wüstenchamäleon
f Caméléon du désert; Caméléon namaquois

1693 ***Chamaeleo nasutus***
Squamata - Chamaeleonidae
e Bignose Chameleon
f Caméléon nasu

1694 ***Chamaeleo oshaughnessyi***
Squamata - Chamaeleonidae
e O'Shaughnessy's Chameleon

1695 ***Chamaeleo oustaleti***
Squamata - Chamaeleonidae
e Oustalet's Chameleon; Malagasy Giant Chameleon
d Riesenchamäleon
f Caméléon d'Oustalet

1696 ***Chamaeleo oweni***
Squamata - Chamaeleonidae
e Owen's Chameleon
d Dreihornchamäleon; Owens Dreihornchamäleon
f Caméléon à trois cornes; Caméléon d'Owen

1697 ***Chamaeleo pardalis***
Squamata - Chamaeleonidae
e Panther Chameleon
d Pantherchamäleon
f Caméléon panthère

1698 ***Chamaeleo parsonii***
Squamata - Chamaeleonidae
e Parson's Chameleon
d Parsons Chamäleon
f Caméléon de Parsons
i Camaleonte di Parson

1699 ***Chamaeleo petteri***
Squamata - Chamaeleonidae
e Petter's Chameleon

1700 ***Chamaeleo peyrieresi***
Squamata - Chamaeleonidae
e Brygoo's Chameleon

1701 ***Chamaeleo pfefferi***
Squamata - Chamaeleonidae
e Pfeffer's Chameleon

1702 ***Chamaeleo polleni***
Squamata - Chamaeleonidae
e Mayotte Island Chameleon

1703 ***Chamaeleo quadricornis***
Squamata - Chamaeleonidae
e Four-horned Chameleon
d Vierhornchamäleon
f Caméléon à quatre cornes

1704 ***Chamaeleo quilensis***
Squamata - Chamaeleonidae
e Bocage's Chameleon
d Südafrikanisches Lappenchamäleon

1705 ***Chamaeleo rhinoceratus***
Squamata - Chamaeleonidae
e Rhinoceros Chameleon

1706 ***Chamaeleo rudis***
Squamata - Chamaeleonidae
e Coarse Chameleon; Ruwenzori Side-striped Chameleon
d Zweistreifenchmäleon

1707 ***Chamaeleo ruspoli***
Squamata - Chamaeleonidae

e Lanza's Chameleon

1708 ***Chamaeleo senegalensis***
Squamata - Chamaeleonidae
e Senegal Chameleon
d Senegal-Chamäleon
f Caméléon du Sénégal

1709 ***Chamaeleo spinosus***
Squamata - Chamaeleonidae
e Prickly Chameleon; Rosette-nosed Chameleon

1710 ***Chamaeleo tavetanus***
Squamata - Chamaeleonidae
e Mount Kilimanjaro Two-horned Chamaeleon

1711 ***Chamaeleo tempeli***
Squamata - Chamaeleonidae
e Tanzania Mountain Chameleon; Tubercle-nosed Chameleon

1712 ***Chamaeleo tigris***
Squamata - Chamaeleonidae
e Tiger Chameleon; Seychelles Chameleon
d Tigerchamäleon
f Caméléon tigré

1713 ***Chamaeleo tsaratananensis***
Squamata - Chamaeleonidae
e Tsaratanan Chameleon

1714 ***Chamaeleo tuzentae***
Squamata - Chamaeleonidae
e Ambiky Chameleon

1715 ***Chamaeleo verrucosus***
Squamata - Chamaeleonidae
e Warty Chameleon
d Warzenchamäleon
f Caméléon verruqueux

1716 ***Chamaeleo werneri***
Squamata - Chamaeleonidae
e Werner's Chameleon; Uzungwe Three-horned Chameleon
d Werners Dreihornchmäleon

1717 ***Chamaeleo wiedersheimi***
Squamata - Chamaeleonidae
e Mount Lefo Chameleon

1718 ***Chamaeleo willsii***
Squamata - Chamaeleonidae
e Canopy Chameleon

1719 ***Chamaeleo xenorhinus***
Squamata - Chamaeleonidae
e Strangenose Chameleon; Single Welded-horn Chameleon

1720 ***Chamaeleo zeylanicus***
Squamata - Chamaeleonidae
e Indian Chameleon

1721 ***Chamaeleolis***
Squamata - Iguanaidae
e Crested False Chameleons
d Falsche Chamäleone

1722 ***Chamaeleolis barbatus***
Squamata - Iguanaidae
e Garrido's False Chameleon
d Falsches Chamäleon

1723 ***Chamaeleolis chamaeleonides***
Squamata - Iguanaidae
e False Chameleon; Short-bearded Anole
d Kubanisches Falsches Chamäleon; Falsches Chamäleon
f Anolis caméléonide

1724 ***Chamaeleolis porcus***
Squamata - Iguanaidae
e Cope's False Chameleon
d Falsches Chamäleon

1725 Chamaeleonidae
Squamata
e Chameleons; True Chameleons; Old World Chameleons
d Chamäleone
f Caméléonidés; Caméléons

1726 *Chamaelinorops*
Squamata - Iguanaidae
e Mimic Chameleons

1727 *Chamaelinorops barbouri*
Squamata - Iguanaidae
e Mimic Chameleon

1728 *Chamaelycus*
Squamata - Colubridae
e Banded Snakes

1729 *Chamaelycus christyi*
Squamata - Colubridae
e Christy's Banded Snake

1730 *Chamaelycus fasciatus*
Squamata - Colubridae
e African Banded Snake

1731 *Chamaelycus parkeri*
Squamata - Colubridae
e Parker's Banded Snake

1732 *Chamaesaura*
Squamata - Cordylidae
e Snake-Lizards; Grass Lizards (CSA)
d Schlangengürtelechsen
f Chamésaures

1733 *Chamaesaura aenea*
Squamata - Cordylidae
e Transvaal Snake-Lizard; Transvaal Grass Lizard (CSA)
d Transvaal-Schlangenechse
f Chamésaure du Transvaal

1734 *Chamaesaura anguina*
Squamata - Cordylidae
e Cape Snake-Lizard; Cape Grass Lizard (CSA); Anguine Lizard
d Kap-Schlangengürtelechse
f Chamésaure du Cap

1735 *Chamaesaura macrolepis*
Squamata - Cordylidae
e Largescale Snake-Lizard; Largescale Grass Lizard (CSA); Large-scaled Snake-Lizard (CSA)
d Großschuppige Schlangengürtelechse
f Chamésaure macrolépide

1736 *Chapinophis*
Squamata - Colubridae
e Guatemalan Snakes

1737 *Chapinophis xanthocheilus*
Squamata - Colubridae
e Guatemalan Snake

1738 *Charina*
Squamata - Boidae
e Rubber Boas
d Gummiboas; Zwergboas
f Boas cautchouc

1739 *Charina bottae*
Squamata - Boidae
e Rubber Boa; Ball Snake; Rubber Snake (NA); Two-headed Snake (NA)
d Nordamerikanische Gummiboa; Nordamerikanische Sandboa; Gummiboa
f Boa caoutchouc
i Boa di gomma

1740 *Charina bottae bottae*
Squamata - Boidae
e Pacific Rubber Boa; Northern Rubber Boa

1741 ***Charina bottae umbratica***
Squamata - Boidae
e Southern Rubber Boa

1742 ***Charina bottae utahensis***
Squamata - Boidae
e Rocky Mountain Rubber Boa

1743 **Chelidae**
Testudines
e Side-necked Turtles; Austro-American Side-necked Turtles; Crooked Heads; Snake-necked Turtles
d Schlangenhalsschildkröten
f Chélyidés; Tortues à cou de serpent; Chélidés
i Tartarughe al collo di serpente

1744 ***Chelodina***
Testudines - Chelidae
e Snakeneck Turtles; Australian Snake-necked Turtles; Snake-necked Turtles
d Schlangenhalsschildkröten; Australische Schlangenhalsschildkröten
f Chélodines

1745 ***Chelodina expansa***
Testudines - Chelidae
e Giant Snakeneck Turtle; Broad-shelled River Turtle; Giant Snake-necked Turtle; Broad-shelled Tortoise
d Riesen-Schlangenhalsschildkröte
i Tartaruga gigante collo di serpente

1746 ***Chelodina longicollis***
Testudines - Chelidae
e Common Snakeneck Turtle; Eastern Snake-necked Turtle; Long-necked Tortoise (ANZ); Australian Snake-necked Turtle; Australian Eastern Snake-necked Turtle; Common Snakeneck
d Glattrücken-Schlangenhalsschildkröte; Australische Schlangenhalsschildkröte; Glattrückige Schlangenhalsschildkröte
f Tortue à cou de serpent; Chélodine à cou long
i Testuggine collo lungho d'Australia; Tartaruga lungocollo comune

1747 ***Chelodina mccordi***
Testudines - Chelidae
e McCord's Snakeneck Turtle
d McCords Schlangenhalsschildkröte
i Tartaruga collo di serpente di Roti

1748 ***Chelodina novaeguineae***
Testudines - Chelidae
e New Guinea Snakeneck Turtle; New Guinea Snake-necked Turtle; New Guinea Snake-necked Tortoise; New Guinea Snakeneck Tortoise
d Neuguinea-Schlangenhalsschildköte
f Chélodine de la Nouvelle-Guinée

1749 ***Chelodina oblonga***
Testudines - Chelidae
e Narrow-breasted Snakeneck Turtle; Oblong Turtle (ANZ); Narrow-breasted Snake-necked Turtle; Oblong Tortoise; Oblong Snake-necked Turtle
d Schmalbrust-Schlangenhalsschildkröte

1750 ***Chelodina parkeri***
Testudines - Chelidae
e Parker's Snakeneck Turtle; Parker's Snake-necked Turtle
d Parkers Schlangenhalsschildkröte

1751 ***Chelodina pritchardi***
Testudines - Chelidae
e Pritchard's Snakeneck Turtle

1752 ***Chelodina reimanni***
Testudines - Chelidae
e Reimann's Snakeneck Turtle

d Reimanns Schlangenhalsschildkröte

1753 ***Chelodina rugosa***
Testudines - Chelidae
e Northern Australian Snakeneck Turtle; Northern Snake-necked Turtle (ANZ); North Australian Snake-necked Turtle; Northern Long-necked Tortoise; Northern Side-necked Turtle
d Rauhe Schlangenhalsschildkröte

1754 ***Chelodina siebenrocki***
Testudines - Chelidae
e Siebenrock's Snakeneck Turtle; Siebenrock's Snake-necked Turtle
d Siebenrock-Schlangenhalsschildkröte; Siebenrocks Schlangenhalsschildkröte
i Tartaruga collo di serpente di Siebenrock

1755 ***Chelodina steindachneri***
Testudines - Chelidae
e Steindachner's Snakeneck Turtle; Flat-shelled Turtle (ANZ); Dinner-plate Turtle; Steindachner's Snake-necked Turtle; Steindachner's Tortoise; Dinner-plate Tortoise; Plated Snake-necked Turtle
d Teller-Schlangenhalsschildkröte
i Tartaruga di Steindachner

1756 ***Chelonia***
Testudines - Cheloniidae
e Green Turtles; Turtles; Green Sea Turtles
d Meeresschildkröten; Suppenschildkröten
f Chélonées; Chélones

1757 ***Chelonia agassizii***
Testudines - Cheloniidae
e Pacific Green Sea Turtle; Black Turtle; Pacific Green Turtle
d Schwarze Suppenschildkröte
f Chélonée franche du Pacifique; Chélone franche du Pacifique
i Tartaruga verde di Agassiz

1758 ***Chelonia mydas***
Testudines - Cheloniidae
e Green Turtle; Green Sea Turtle; Edible Turtle; Indian Ocean Green Turtle; Common Green Turtle; Honu; Black Sea Turtle
d Grüne Meeresschildkröte; Gewöhnliche Suppenschildkröte; Suppenschildkröte
f Tortue verte; Tortue franche; Tortue à soupe; Tortue comestible; Chélone franche; Chélonée franche; Tortue blanc (Ants)
i Tartaruga verde; Tartaruga franca; Tartaruga mida; Mida

1759 ***Chelonia mydas mydas***
Testudines - Cheloniidae
e Atlantic Green Turtle

1760 **Cheloniidae**
Testudines
e Marine Turtles; Sea Turtles
d Seeschildkröten; Meeresschildkröten; Eigentliche Meeresschildkröten
f Tortues marines; Tortues de mer; Chéloniidées
i Tartarughe marine

1761 ***Chelosania***
Squamata - Agamidae
e Chameleon Dragons; Chameleon Agamas

1762 ***Chelosania brunnea***
Squamata - Agamidae
e Chameleon Dragon; Chameleon Agama

1763 ***Chelus***
Testudines - Chelidae
e Matamata Turtles; Matamatas
d Fransenschildkröten; Matamatas

f Chélydes; Matamatas

1764 ***Chelus fimbriata***
Testudines - Chelidae
e Matamata Turtle; Matamata; Amazon Matamata
d Fransenschildkröte
f Matamata
i Matamata

1765 ***Chelydra***
Testudines - Chelydridae
e Common Snapping Turtles; Snapping Turtles; Loggerhead Snapping Turtles
d Schnappschildkröten
f Chélydres

1766 ***Chelydra serpentina***
Testudines - Chelydridae
e Snapping Turtle; Common Snapping Turtle; Alligator-tailed Turtle; Snapper (ZZ); Mud Turtle (NA)
d Schnappschildkröte
f Chélydre serpentine; Tortue hargneuse; Tortue happante; Tortue vorace d'eau douce; Tortue happeuse
i Testugine azzanatrice; Tartaruga azzannatrice

1767 ***Chelydra serpentina acutirostris***
Testudines - Chelydridae
e South American Snapping Turtle; Ecuadorian Snapping Turtle
d Südamerikanische Schnappschildkröte
f Tortue happeuse d'Amérique du Sud

1768 ***Chelydra serpentina osceola***
Testudines - Chelydridae
e Florida Snapping Turtle
d Florida-Schnappschildkröte
f Tortue happeuse de Floride

1769 ***Chelydra serpentina rossignoni***
Testudines - Chelydridae
e Central American Snapping Turtle
d Mittelamerikanische Schnappschildkröte
f Tortue happeuse d'Amérique Centrale

1770 ***Chelydra serpentina serpentina***
Testudines - Chelydridae
e Common Snapping Turtle (NA); Eastern Snapping Turtle
d Nordamerikanische Schnappschildkröte
f Tortue happeuse commune

1771 **Chelydridae**
Testudines
e Snapping Turtles; Loggerhead Snapping Turtles
d Alligatorschildkröten; Kaimanschildkröten; Schnappschildkröten
f Tortues harpantes; Tortues voraces
i Tartarughe azzannatrici

1772 ***Chersina***
Testudines - Testudinidae
e Bowsprit Tortoises
d Schnabelbrustschildkröten

1773 ***Chersina angulata***
Testudines - Testudinidae
e South African Bowsprit Tortoise; Angulate Tortoise (CSA); Bowsprit Tortoise
d Afrikanische Schnabelbrustschildkröte; Schnabelbrustschildkröte
f Tortue angulée
i Chersina; Testuggine del Sudafrica

1774 ***Chersodromus***
Squamata - Colubridae
e Earth Runners

1775 ***Chersodromus liebmanni***
Squamata - Colubridae
e Liebmann's Earth Runner

1776 ***Chersodromus rubriventris***
Squamata - Colubridae
e Redbelly Earth Runner

1777 ***Chilomeniscus***
Squamata - Colubridae
e Sand Snakes; Banded Burrowing Snakes
d Sandschlangen; Bändersandnattern
f Couleuvres arénicoles

1778 ***Chilomeniscus cinctus***
Squamata - Colubridae
e Banded Sand Snake; Burrowing Sand Snake
d Gebänderte Sandschlange; Bändersandnatter
f Couleuvre arénicole rayée

1779 ***Chilomeniscus punctatissimus***
Squamata - Colubridae
e Island Burrowing Sand Snake; Espiritu Santo Island Sand Snake

1780 ***Chilomeniscus savagei***
Squamata - Colubridae
e Savage's Sand Snake; Cerralvo Island Sand Snake

1781 ***Chilomeniscus stramineus***
Squamata - Colubridae
e Spotted Sand Snake; Bandless Sand Snake (NA); Variable Sand Snake

1782 ***Chilorhinophis***
Squamata - Colubridae
e Two-headed Snakes

1783 ***Chilorhinophis butleri***
Squamata - Colubridae
e Butler's Two-headed Snake

1784 ***Chilorhinophis carpenteri***
Squamata - Colubridae
e Liwale Two-headed Snake

1785 ***Chilorhinophis gerardi***
Squamata - Colubridae
e Gerard's Two-headed Snake; Gerrard's Black-and-yellow Burrowing Snake (CSA); Tanganyika Two-headed Snake; Gerard's Striped Burrowing Snake (CSA)

1786 ***Chinemys***
Testudines - Emydidae
e Chinese Pond Turtles; Redneck Pond Turtles
d China-Sumpfschildkröten
f Chinémydes

1787 ***Chinemys nigricans***
Testudines - Emydidae
e Chinese Pond Turtle; Asian Yellow Pond Turtle; Red-necked Pond Turtle; Chinese Red-necked Pond Turtle
d Chinesische Rothalsschildkröten
f Chinémyde noirâtre
i Tartaruga cinese palustra collo rosso

1788 ***Chinemys reevesii***
Testudines - Emydidae
e Chinese Pond Turtle; Reeves's Turtle; Chinese Three-keeled Pond Turtle; Reeves's Pond Turtle
d Chinesische Dreikielschildkröte
f Chinémyde de Reeves
i Tartaruga cinese palustra

1789 ***Chionactis***
Squamata - Colubridae
e Shovelnose Snakes; Shovel-nosed Snakes; Shovel-nosed Ground Snakes
d Schaufelnasenschlangen; Schaufelnasennattern;

Shaufelnasenwühlnattern
f Couleuvres à nez spatulé

1790 ***Chionactis occipitalis***
Squamata - Colubridae
e Western Shovelnose Snake; Western Shovel-nosed Snake; Desert Burrowing Snake
d Westliche Schaufelnasenschlange; Shaufelnasennatter; Westliche Shaufelnasenwühlnatter
f Couleuvre à nez spatulé de l'Ouest

1791 ***Chionactis occipitalis annulata***
Squamata - Colubridae
e Colorado Desert Shovelnose Snake; Colorado Desert Shovel-nosed Snake; Banded Burrowing Snake
d Westliche Schaufelnasenwühlnatter

1792 ***Chionactis occipitalis klauberi***
Squamata - Colubridae
e Tucson Shovelnose Snake; Tucson Shovel-nosed Snake

1793 ***Chionactis occipitalis occipitalis***
Squamata - Colubridae
e Mojave Shovelnose Snake; Mojave Shovel-nosed Snake; Mojave Ringed Snake; Western Shovel-nosed Snake

1794 ***Chionactis occipitalis talpina***
Squamata - Colubridae
e Nevada Shovelnose Snake; Nevada Shovel-nosed Snake

1795 ***Chionactis palarostris***
Squamata - Colubridae
e Sonoran Shovelnose Snake; Sonoran Shovel-nosed Snake; Sonora Shovel-nosed Snake
d Sonora-Schaufelnasenschlange
f Couleuvre à nez spatulé du Sonora

1796 ***Chionactis palarostris organica***
Squamata - Colubridae
e Sonoran Shovelnose Snake; Sonoran Shovel-nosed Snake; Organ-pipe Shovel-nosed Snake; Organ-pipe Shovelnose Snake; Sonora Shovel-nosed Snake

1797 ***Chionactis saxatilis***
Squamata - Colubridae
e Mountain Shovelnose Snake; Mountain Shovel-nosed Snake

1798 ***Chirindia***
Squamata - Amphisbaenidae
e South African Worm Lizards; Pink Round-headed Worm Lizards; Sharp-snouted Worm Lizards

1799 ***Chirindia bushbyi***
Squamata - Amphisbaenidae
e Mozambique Worm Lizard

1800 ***Chirindia everbecki***
Squamata - Amphisbaenidae
e Mbanja Worm Lizard

1801 ***Chirindia langi***
Squamata - Amphisbaenidae
e Lang's Worm Lizard; Msimbit Worm Lizard (CSA)

1802 ***Chirindia mpwapwaensis***
Squamata - Amphisbaenidae
e Mpwapwa Worm Lizard

1803 ***Chirindia orientalis***
Squamata - Amphisbaenidae
e Mikindani Worm Lizard

1804 ***Chirindia rondoensis***
Squamata - Amphisbaenidae
e Nichingidi Worm Lizard
d Tanzania-Doppelschleiche

1805 ***Chirindia swynnertoni***
Squamata - Amphisbaenidae
e Swynnerton's Worm Lizard; Swynnerton's Round-headed Worm Lizard; Swynnerton's Round-snouted Amphisbaenian (CSA)

1806 ***Chironius***
Squamata - Colubridae
e Sipos; Whip Snakes
d Sipos

1807 ***Chironius bicarinatus***
Squamata - Colubridae
e Two-headed Sipo

1808 ***Chironius carinatus***
Squamata - Colubridae
e Sipo; Brown Keeled Racer
d Sipo

1809 ***Chironius cinnamomeus***
Squamata - Colubridae
e Surinam Sipo

1810 ***Chironius cochranae***
Squamata - Colubridae
e Cochran's Sipo

1811 ***Chironius exoletus***
Squamata - Colubridae
e Linnaeus's Sipo; Green Keeled Racer; Common Whip Snake

1812 ***Chironius flavolineatus***
Squamata - Colubridae
e Boettger's Sipo

1813 ***Chironius flavopictus***
Squamata - Colubridae
e Werner's Sipo

1814 ***Chironius foveatus***
Squamata - Colubridae
e Coastal Sipo

1815 ***Chironius fuscus***
Squamata - Colubridae
e Brown Sipo; Olive Whip Snake

1816 ***Chironius grandisquamis***
Squamata - Colubridae
e Ecuador Sipo; Black Keeled Racer

1817 ***Chironius holochlorus***
Squamata - Colubridae
e Peru Sipo

1818 ***Chironius laevicollis***
Squamata - Colubridae
e Brazilian Sipo

1819 ***Chironius melas***
Squamata - Colubridae
e Black Sipo

1820 ***Chironius monticola***
Squamata - Colubridae
e Mountain Sipo

1821 ***Chironius multiventris***
Squamata - Colubridae
e South American Sipo

1822 ***Chironius multiventris multiventris***
Squamata - Colubridae
e Long-tailed Whip Snake

1823 ***Chironius pyrrhopogon***
Squamata - Colubridae
e Wied's Sipo

1824 ***Chironius quadricarinatus***
Squamata - Colubridae
e Central Sipo

1825 ***Chironius schlueteri***
Squamata - Colubridae
e Schlueter's Sipo

1826 ***Chironius scurrulus***
Squamata - Colubridae
e Wagler's Sipo; Rusty Whip Snake
f Serpent agouti (Ants)

1827 ***Chironius spixii***
Squamata - Colubridae
e Spix's Sipo

1828 ***Chironius vincenti***
Squamata - Colubridae
e Vincent's Sipo; St. Vincent Racer (WI)

1829 ***Chitra***
Testudines - Trionychidae
e Narrowhead Softshells; Striped Narrow-headed Soft-shelled Turtles; Southeast Asian Narrow-headed Softshell Turtle
d KurzkopfwWeichschildkröten; Chitraweichschildkröten

1830 ***Chitra chitra***
Testudines - Trionychidae
e Nutaphand's Narrowhead Softshell
d Kurzkopfweichschildkröte

1831 ***Chitra indica***
Testudines - Trionychidae
e Narrowhead Softshell; Narrow-headed Soft-shelled Turtle; Chitra Turtle (ISC); Narrow-headed Softshell Turtle
f Trionyx à rayures

1832 ***Chlamydosaurus***
Squamata - Agamidae
e Frilled Lizards; Frilled Dragons
d Kragenechsen; Kragenagamen
f Chlamydosaures; Lézards à collerete
i Clamidosauri

1833 ***Chlamydosaurus kingii***
Squamata - Agamidae
e Australian Frilled Lizard; Frilled Dragon; Frilled Lizard (ANZ);

Bicycle Lizard
d Kragenagame; Kragenechse
f Chlamydosaure de King; Lézard à collerette
i Clamidosauro; Clamidosauro ornato; Lucertola dal collare

1834 ***Chondrodactylus***
Squamata - Gekkonidae
e Whorled Sand Geckos; Giant Ground Geckos (CSA); Sand Geckos
d Sandgeckos
f Chondrodactyles

1835 ***Chondrodactylus anguilifer***
Squamata - Gekkonidae
e Whorled Sand Gecko; Giant Ground Gecko (CSA); Sand Gecko
d Sandgecko
f Gecko de Namibie

1836 ***Chondrodactylus anguilifer namibensis***
Squamata - Gekkonidae
d Kornfingergecko

1837 ***Christinus***
Squamata - Gekkonidae
e Australian Clawed Geckos

1838 ***Christinus guentheri***
Squamata - Gekkonidae
e Günther's Island Gecko
d Lord-Howe-Insel-Blattfingergecko

1839 ***Christinus marmoratus***
Squamata - Gekkonidae
e Marbled Gecko
d Südaustralischer Blattfingergecko
f Gecko marbré

1840 ***Chrysemys***
Testudines - Emydidae
e Painted Turtles
d Zierschildkröten
f Tortues friandes de soleil
i Tartarughe palustre

1841 ***Chrysemys picta***
Testudines - Emydidae
e Painted Turtle
d Zierschildkröte
f Tortue peinte; Chrysémyde peinte
i Tartaruga palustre dipinta

1842 ***Chrysemys picta belli***
Testudines - Emydidae
e Western Painted Turtle
f Tortue peinte de l'Ouest

1843 ***Chrysemys picta dorsalis***
Testudines - Emydidae
e Southern Painted Turtle
d Rückenstreifenzierschildkröte
f Chrysémyde du Sud

1844 ***Chrysemys picta marginata***
Testudines - Emydidae
e Midland Painted Turtle
d Gebänderte Zierschildkröte
f Chrysémyde du Nord

1845 ***Chrysemys picta picta***
Testudines - Emydidae
e Eastern Painted Turtle
f Chrysémyde peinte de l'Est

1846 ***Chrysopelea***
Squamata - Colubridae
e Flying Tree Snakes; Flying Snakes; Tree Snakes
d Schmuckbaumnattern; Schmuckbaumschlangen
f Couleuvres volantes

1847 ***Chrysopelea ornata***
Squamata - Colubridae
e Golden Tree Snake; Golden Flying Snake (ISC); Flying Snake; Black Tree Snake; Golden Gliding Snake; Ornate Flying Snake
d Gewöhnliche Schmuckbaumnatter; Goldschlange; Schmuckbaumschlange
f Couleuvre volante; Serpent volant

1848 ***Chrysopelea paradisi***
Squamata - Colubridae
e Paradise Tree Snake; Paradise Flying Snake; Flying Snake; Garden Flying Snake
d Paradis-Schmuckbaumnatter; Paradis-Schlange
f Serpent volant

1849 ***Chrysopelea pelias***
Squamata - Colubridae
e Barred Tree Snake; Banded Flying Snake; Twin-barred Tree Snake
d Gebänderte Schmuckbaumnatter

1850 ***Chrysopelea rhodopleurum***
Squamata - Colubridae
e Moluccan Flying Snake; Moluccas Flying Snake

1851 ***Chrysopelea taprobanica***
Squamata - Colubridae
e Indian Flying Snake

1852 ***Claudius***
Testudines - Kinosternidae
e Narrow-bridged Musk Turtles; Narrow-bridged Mud Turtles
d Schlammschildkröten; Großkopfschlammschildkröten

1853 ***Claudius angustatus***
Testudines - Kinosternidae
e Narrow-bridged Musk Turtle; Narrow-bridged Mud Turtle
d Großkopfschlammschildkröte
f Staurotype nain
i Tartaruga del muschio a piastrone piccolo

1854 ***Clelia***
Squamata - Colubridae
e Mussuranas; Clelias (WI)

d Mussuranas
f Mussuranas

1855 ***Clelia bicolor***
Squamata - Colubridae
e Two-coloured Mussurana

1856 ***Clelia clelia***
Squamata - Colubridae
e Mussurana; Black Mussurana; Common Mussurana; Windward Clelia (WI); Mussurana Snake
d Mussurana
f Mussurana

1857 ***Clelia equatoriana***
Squamata - Colubridae
e Equatorial Mussurana; Southern Mussurana

1858 ***Clelia errabunda***
Squamata - Colubridae
e Underwood's Mussurana; St. Lucia Clelia (WI)

1859 ***Clelia occipitolutea***
Squamata - Colubridae
e Brazilian Mussurana

1860 ***Clelia rustica***
Squamata - Colubridae
e Culebra (NA)

1861 ***Clelia scytalina***
Squamata - Colubridae
e Mexican Snake-eater; Mexican Mussurana; Brown Mussurama

1862 ***Clemmys***
Testudines - Emydidae
e Pond Turtles; American Freshwater Turtles; American Pond Turtles
d Amerikanische Wasserschildkröten; Wasserschildkröten; Neuweltliche Wasserschildkröten
f Émydes; Clemmydes

1863 ***Clemmys guttata***
Testudines - Emydidae
e Spotted Turtle; Spotted Pond Turtle; Spotted Black Pond Turtle
d Tropfenschildkröte
f Tortue ponctuée (Qué); Clemmyde à gouttelettes
i Testuggine palustre punteggiata; Tartaruga macchiata

1864 ***Clemmys insculpta***
Testudines - Emydidae
e Wood Turtle
d Waldbachschildkröte
f Clemmyde sculpté; Tortue des bois nord-américanine; Tortue des bois (Qué)
i Testuggine palustre scolpita; Tartaruga di bosco

1865 ***Clemmys marmorata***
Testudines - Emydidae
e Western Pond Turtle; Pacific Pond Turtle; California Pond Turtle; Pacific Turtle
d Pazifik-Wasserschildkröte
f Émyde de l'Ouest; Clemmyde marbrée
i Testuggine palustre marmorizata

1866 ***Clemmys marmorata marmorata***
Testudines - Emydidae
e Northwestern Pond Turtle; Northern Pacific Pond Turtle

1867 ***Clemmys marmorata pallida***
Testudines - Emydidae
e Southwestern Pond Turtle; Western Pond Turtle; Southern Pacific Pond Turtle

1868 ***Clemmys muhlenbergii***
Testudines - Emydidae
e Bog Turtle; Mühlenberg's Turtle
d Mühlenbergs Wasserschildkröte; Mühlenberg-Schildkröte
f Émyde de Mühlenberg; Clemmyde

de Mühlenberg; Tortue de Muhlenberg
i Testuggine dei pantani

1869 ***Clonophis***
Squamata - Colubridae
e Kirtland's Snakes; Kirtland's Water Snakes
d Lianennattern; Kirtlands Wassernattern
f Serpents de Kirtland; Couleuvres de Kirtland

1870 ***Clonophis kirtlandii***
Squamata - Colubridae
e Kirtland's Snake; Kirtland's Water Snake
d Lianennatter; Kirtlands Wassernatter
f Serpent de Kirtland

1871 ***Cnemaspis***
Squamata - Gekkonidae
e Small-toed Geckos; Dwarf Geckos (ISC)

1872 ***Cnemaspis affinis***
Squamata - Gekkonidae
e Stoliczka's Gecko

1873 ***Cnemaspis africana***
Squamata - Gekkonidae
e African Gecko; Usambara Forest Gecko

1874 ***Cnemaspis argus***
Squamata - Gekkonidae
e Dring's Gecko

1875 ***Cnemaspis barbouri***
Squamata - Gekkonidae
e Barbour's Gecko

1876 ***Cnemaspis beddomei***
Squamata - Gekkonidae
e Beddome's Gecko

1877 ***Cnemaspis boiei***
Squamata - Gekkonidae
e Boie's Gecko

1878 ***Cnemaspis boulengerii***
Squamata - Gekkonidae
e Boulenger's Gecko

1879 ***Cnemaspis dickersoni***
Squamata - Gekkonidae
e Dickerson's Gecko

1880 ***Cnemaspis dilepsis***
Squamata - Gekkonidae
e Two-scaled Gecko

1881 ***Cnemaspis elgonensis***
Squamata - Gekkonidae
e Loveridge's Gecko

1882 ***Cnemaspis gigas***
Squamata - Gekkonidae
e Perret's Gecko

1883 ***Cnemaspis goanensis***
Squamata - Gekkonidae
e Sharma's Gecko

1884 ***Cnemaspis indica***
Squamata - Gekkonidae
e Nilgiri Hills Gecko

1885 ***Cnemaspis jerdonii***
Squamata - Gekkonidae
e Jerdon's Indian Gecko; Punjab Snake-eyed Lacerta

1886 ***Cnemaspis kandiana***
Squamata - Gekkonidae
e Kelaart's Gecko

1887 ***Cnemaspis kendalli***
Squamata - Gekkonidae
e Kendall's Gecko; Kendall's Day

Gecko

1888 ***Cnemaspis koehleri***
Squamata - Gekkonidae
e Koehler's Gecko

1889 ***Cnemaspis kummnoli***
Squamata - Gekkonidae
e Trang Province Gecko

1890 ***Cnemaspis lavolineata***
Squamata - Gekkonidae
e Dew-lined Gecko

1891 ***Cnemaspis littoralis***
Squamata - Gekkonidae
e Indian Forest Gecko

1892 ***Cnemaspis mysoriensis***
Squamata - Gekkonidae
e Jerdon's Forest Gecko

1893 ***Cnemaspis nairi***
Squamata - Gekkonidae
e Kerala Gecko

1894 ***Cnemaspis nigridius***
Squamata - Gekkonidae
e Borneo Black Gecko

1895 ***Cnemaspis occidentalis***
Squamata - Gekkonidae
e Western Gecko

1896 ***Cnemaspis ornata***
Squamata - Gekkonidae
e Beddome's Ornate Gecko

1897 ***Cnemaspis petrodroma***
Squamata - Gekkonidae
e Nigeria Crag Gecko

1898 ***Cnemaspis podihuna***
Squamata - Gekkonidae
e Deraniyagala's Gecko

1899 ***Cnemaspis quattuorseriata***
Squamata - Gekkonidae
e Sternfeld's Gecko

1900 ***Cnemaspis siamensis***
Squamata - Gekkonidae
e Siamese Gecko; Green-eyed Gecko

1901 ***Cnemaspis sisparensis***
Squamata - Gekkonidae
e Theobald's Gecko

1902 ***Cnemaspis spinicollis***
Squamata - Gekkonidae
e Cameroon Collared Gecko

1903 ***Cnemaspis timoriensis***
Squamata - Gekkonidae
e Timor Gecko

1904 ***Cnemaspis tropidogaster***
Squamata - Gekkonidae
e Marked Belly Gecko

1905 ***Cnemaspis uzungwae***
Squamata - Gekkonidae
e Tanzania Gecko

1906 ***Cnemaspis wyandensis***
Squamata - Gekkonidae
e Wynaad Gecko

1907 ***Cnemidophorus***
Squamata - Teiidae
e Whiptails; Whiptail Lizards; Racerunners
d Renntejus; Rennechsen
f Lézards queue-en-fouet; Cnémidophores; Lézards agiles; Lézards courreurs; Lézards arc-en-ciel; Lézards queue-en-fouet et agiles

1908 ***Cnemidophorus alpinus***
Squamata - Teiidae
e Alpine Whiptail; Whiptail Lizard

1909 ***Cnemidophorus angusticeps***
Squamata - Teiidae
e Yucatan Whiptail

1910 ***Cnemidophorus arubensis***
Squamata - Teiidae
e Aruba Whiptail

1911 ***Cnemidophorus bacatus***
Squamata - Teiidae
e San Pedro Nolasco Whiptail; San Pedro Nolasco Island Whiptail

1912 ***Cnemidophorus burti***
Squamata - Teiidae
e Giant Spotted Whiptail; Canyon Spotted Whiptail (NA)
f Queue-en-fouet tacheté des canyons

1913 ***Cnemidophorus burti stictogrammus***
Squamata - Teiidae
e Giant Whiptail; Giant Spotted Whiptail (NA)

1914 ***Cnemidophorus burti xanthonotus***
Squamata - Teiidae
e Redback Whiptail; Red-backed Whiptail (NA)

1915 ***Cnemidophorus calidipes***
Squamata - Teiidae
e Rio Tepalcatepic Valley Whiptail

1916 ***Cnemidophorus canus***
Squamata - Teiidae
e Salsipuedes Island Whiptail

1917 ***Cnemidophorus carmenensis***
Squamata - Teiidae
e Carmen Island Blue-throated Whiptail

1918 ***Cnemidophorus catalinensis***
Squamata - Teiidae
e Santa Catalina Island Whiptail

1919 ***Cnemidophorus celeripes***
Squamata - Teiidae
e San Jose Island Whiptail

1920 ***Cnemidophorus ceralbensis***
Squamata - Teiidae
e Cerralvo Island Whiptail

1921 ***Cnemidophorus communis***
Squamata - Teiidae
e Colima Giant Whiptail; Giant Whiptail

1922 ***Cnemidophorus costatus***
Squamata - Teiidae
e Western Mexico Whiptail; Western Mexican Whiptail

1923 ***Cnemidophorus cozumelae***
Squamata - Teiidae
e Cozumel Racerunner; Cozumel Whiptail

1924 ***Cnemidophorus cryptus***
Squamata - Teiidae
e Cryptic Racerunner

1925 ***Cnemidophorus danheimae***
Squamata - Teiidae
e San Jose Island Blue-throated Whiptail

1926 ***Cnemidophorus deppei***
Squamata - Teiidae
e Blackbelly Racerunner; Black-bellied Racerunner; Seven-lined Racerunner
d Siebenstreifenrennechse
f Cnémidophore à sept raies

1927 ***Cnemidophorus dixoni***
Squamata - Teiidae
e Gray-checkered Whiptail (NA); Grey-chequered Whiptail
f Queue-en-fouet à carreaux gris

1928 ***Cnemidophorus espiritensis***
Squamata - Teiidae
e Espiritu Santo Island Orange-throated Whiptail

1929 ***Cnemidophorus estebannensis***
Squamata - Teiidae
e San Esteban Island Whiptail

1930 ***Cnemidophorus exsanguis***
Squamata - Teiidae
e Chihuahan Spotted Whiptail; Chihuahuan Whiptail; Chihuaha Whiptail; Chihuahua Spotted Whiptail
d Blasser Rennteju
f Queue-en-fouet tacheté de Chihuahua

1931 ***Cnemidophorus flagellicaudus***
Squamata - Teiidae
e Gila Spotted Whiptail
f Queue-en-fouet tacheté de Gila

1932 ***Cnemidophorus franciscensis***
Squamata - Teiidae
e San Francisco Island Blue-throated Whiptail

1933 ***Cnemidophorus gramivagrus***
Squamata - Teiidae
e McCrystal's Whiptail; Llanos Whiptail

1934 ***Cnemidophorus gularis***
Squamata - Teiidae
e Easstern Spotted Whiptail
d Rotkehlrennteju

1935 ***Cnemidophorus gularis gularis***
Squamata - Teiidae
e Texas Spotted Whiptail

1936 ***Cnemidophorus guttatus***
Squamata - Teiidae
e Mexican Racerunner
f Queue-en-fouet du Texas

1937 ***Cnemidophorus hyperythrus***
Squamata - Teiidae
e Orange Whiptail; Orange-throated Whiptail (NA); Orangethroat Whiptail
d Orangekehlige Rennechse
f Queue-en-fouet à gorge orange

1938 ***Cnemidophorus hyperythrus beldingi***
Squamata - Teiidae
e Belding's Orange Whiptail; Belding Orange-throated Whiptail (NA)

1939 ***Cnemidophorus hyperythrus hyperythrus***
Squamata - Teiidae
e Cape Orange-throated Whiptail
d Orangekehlige Rennechse
f Queue-en-fouet à gorge orange

1940 ***Cnemidophorus inornatus***
Squamata - Teiidae
e Little Striped Whiptail; Seven-striped Whiptail
d Schlichter Rennteju
f Queue-en-fouet rayé

1941 ***Cnemidophorus inornatus arizonae***
Squamata - Teiidae
e Arizona Striped Whiptail

1942 ***Cnemidophorus inornatus gypsi***
Squamata - Teiidae
e Little White Whiptail

1943 ***Cnemidophorus inornatus heptagrammus***
Squamata - Teiidae

e Trans-Pecos Striped Whiptail

1944 ***Cnemidophorus inornatus juniperus***
Squamata - Teiidae
e Woodland Striped Whiptail

1945 ***Cnemidophorus inornatus llanuras***
Squamata - Teiidae
e Plains Striped Whiptail

1946 ***Cnemidophorus labialis***
Squamata - Teiidae
e Baja California Whiptail

1947 ***Cnemidophorus lacertoides***
Squamata - Teiidae
e Bibron's Whiptail

1948 ***Cnemidophorus laredoensis***
Squamata - Teiidae
e Laredo Striped Whiptail
f Queue-en-fouet rayé de Laredo

1949 ***Cnemidophorus lemniscatus***
Squamata - Teiidae
e Rainbow Whiptail; Dotted Racerunner; Spotted Racerunner; Rainbow Whiptail
d Tüpfelrennechse
f Cnémidophore ponctué

1950 ***Cnemidophorus lineatissimus***
Squamata - Teiidae
e Many-striped Whiptail; Many-lined Whiptail

1951 ***Cnemidophorus longicaudus***
Squamata - Teiidae
e Longtail Whiptail; Lagartija Cola Roja (NA)

1952 ***Cnemidophorus marmoratus***
Squamata - Teiidae
e Marbled Whiptail

1953 ***Cnemidophorus marmoratus marmoratus***
Squamata - Teiidae
e Western Marbled Whiptail

1954 ***Cnemidophorus marmoratus reticuloriens***
Squamata - Teiidae
e Eastern Marbled Whiptail

1955 ***Cnemidophorus martyris***
Squamata - Teiidae
e San Pedro Martir Whiptail; San Pedro Martir Island Whiptail

1956 ***Cnemidophorus maximus***
Squamata - Teiidae
e Cape Region Whiptail; Giant Whiptail (NA)

1957 ***Cnemidophorus mexicanus***
Squamata - Teiidae
e Mexican Whiptail

1958 ***Cnemidophorus montaguae***
Squamata - Teiidae
e Giant Whiptail

1959 ***Cnemidophorus murinus***
Squamata - Teiidae
e Laurent's Whiptail; Blausanna Whiptail

1960 ***Cnemidophorus neomexicanus***
Squamata - Teiidae
e New Mexico Whiptail; New Mexican Whiptail (NA)
f Queue-en-fouet du Nouveau-Mexique

1961 ***Cnemidophorus ocellifer***
Squamata - Teiidae
e Spix's Whiptail; Calango (NA)

1962 ***Cemidophorus pai***
Squamata - Teiidae

e Pai Striped Whiptail

1963 ***Cnemidophorus opatae***
Squamata - Teiidae
e Opata Whiptail

1964 ***Cnemidophorus parvisocius***
Squamata - Teiidae
e Mexican Pygmy Whiptail

1965 ***Cnemidophorus pictus***
Squamata - Teiidae
e Monserrate Island Blue-throated Whiptail

1966 ***Cnemidophorus pseudolemniscatus***
Squamata - Teiidae
e Colee's Racerunner

1967 ***Cnemidophorus rodecki***
Squamata - Teiidae
e Rodeck's Whiptail

1968 ***Cnemidophorus sacki***
Squamata - Teiidae
e Sack's Spotted Whiptail; Blue-bellied Racerunner; Spotted Whiptail
d Fleckenrennechse
f Cnémidophore tacheté

1969 ***Cnemidophorus septemvittatus***
Squamata - Teiidae
e Plateau Spotted Whiptail; Mexican Plateau Spotted Whiptail

1970 ***Cnemidophorus septemvittatus septemvittatus***
Squamata - Teiidae
e Bif Bend Spotted Whiptail

1971 ***Cnemidophorus sexlineatus***
Squamata - Teiidae
e Six-lined Racerunner (NA); Six-lined Lizard; Prairie Lined Racerunner
d Sechsstreifenrennechse; Sechsstreifenrennteju
f Lézard agile à six lignes; Cnémidophore à six raies

1972 ***Cnemidophorus sexlineatus sexlineatus***
Squamata - Teiidae
e Six-lined Racerunner; Eastern Six-lined Racerunner

1973 ***Cnemidophorus sexlineatus stephensae***
Squamata - Teiidae
e Texas Yellowhead Racerunner; Yellow-headed Racerunner; Texas Yellow-headed Racerunner
d Westliche Rennechse; Getigerter Rennteju

1974 ***Cnemidophorus sexlineatus viridis***
Squamata - Teiidae
e Prairie Racerunner

1975 ***Cnemidophorus sonorae***
Squamata - Teiidae
e Sonoran Spotted Whiptail
f Queue-en-fouet tacheté du Sonora

1976 ***Cnemidophorus tessselatus***
Squamata - Teiidae
e Common Checkered Whiptail (NA); Colorado Chequered Whiptail; Checkered Whiptail (NA)
d Würfelrennechse
f Queue-en-fouet carrelé du Colorado

1977 ***Cnemidophorus tigris***
Squamata - Teiidae
e Western Whiptail; Marbled Whiptail; Tiger Whiptail
d Westliche Rennechse
f Queue-en-fouet tigré

1978 ***Cnemidophorus tigris gracilis***
Squamata - Teiidae
e Southern Whiptail; Arizona Desert Whiptail

1979 ***Cnemidophorus tigris multiscutatus***
Squamata - Teiidae
e Coastal Whiptail; Tiger Whiptail; Cedros Island Whiptail

1980 ***Cnemidophorus tigris mundus***
Squamata - Teiidae
e California Whiptail

1981 ***Cnemidophorus tigris punctilineatus***
Squamata - Teiidae
e Sonoran Tiger Whiptail

1982 ***Cnemidophorus tigris rubidus***
Squamata - Teiidae
e Reddish Whiptail

1983 ***Cnemidophorus tigris septentrionalis***
Squamata - Teiidae
e Northern Whiptail; Painted Desert Whiptail; Plateau Tiger Whiptail

1984 ***Cnemidophorus tigris stejnegeri***
Squamata - Teiidae
e Coastal Whiptail

1985 ***Cnemidophorus tigris tigris***
Squamata - Teiidae
e Great Basin Whiptail

1986 ***Cnemidophorus tigris vividus***
Squamata - Teiidae
e South Coronado Island Whiptail

1987 ***Cnemidophorus uniparens***
Squamata - Teiidae
e Desert Grassland Whiptail
f Queue-en-fouet du désert semi-aride

1988 ***Cnemidophorus vanzoi***
Squamata - Teiidae
e Vanzo's Whiptail; Maria Islands Whiptail (WI)

1989 ***Cnemidophorus velox***
Squamata - Teiidae
e Plateau Striped Whiptail; Plateau Whiptail (NA)
f Queue-en-fouet rayé des plateaux

1990 ***Cnemidophorus vittatus***
Squamata - Teiidae
e Slender Whiptail

1991 ***Cnemidophorus xanthonotus***
Squamata - Teiidae
e Redback Whiptail; Red-backed Whiptail (NA)

1992 ***Coeranoscinus***
Squamata - Scincidae
e Fawn Skinks

1993 ***Coeranoscinus frontalis***
Squamata - Scincidae
e Chocolate Burrowing Skink

1994 ***Coeranoscinus reticulatus***
Squamata - Scincidae
e Fawn Skink

1995 ***Coleodactylus***
Squamata - Gekkonidae
e Amazon Geckos

1996 ***Coleodactylus amazonicus***
Squamata - Gekkonidae
e Amazon Gecko

1997 ***Coleodactylus brachystoma***
Squamata - Gekkonidae
e Goias Gecko

1998 ***Coleodactylus guimaraesi***
Squamata - Gekkonidae
e Rondonian Gecko

1999 ***Coleodactylus meridionalis***
Squamata - Gekkonidae
e Meridian Gecko

2000 ***Coleodactylus septentrionalis***
Squamata - Gekkonidae
e Ilha Marca Gecko; Ilha di Maraca Leaf-glitter Gecko

2001 ***Coleonyx***
Squamata - Eublepharidae
e Banded Geckos
d Wüstengeckos; Krallengeckos
f Geckos-léopard; Coléonyx

2002 ***Coleonyx brevis***
Squamata - Eublepharidae
e Texas Banded Gecko
d Krallengecko; Texas-Krallengecko
f Gecko-léopard du Texas

2003 ***Coleonyx elegans***
Squamata - Eublepharidae
e Yucatan Banded Gecko
d Mexikanischer Krallengecko; Mittelamerikanischer Krallengecko

2004 ***Coleonyx fasciatus***
Squamata - Eublepharidae
e Black Banded Gecko

2005 ***Coleonyx gypsicolus***
Squamata - Eublepharidae
e San Marcos Island Barefoot Gecko

2006 ***Coleonyx mitratus***
Squamata - Eublepharidae
e Central American Banded Gecko
d Belize-Krallengecko; Krallengecko
f Gecko de Nicaragua

2007 ***Coleonyx reticulatus***
Squamata - Eublepharidae
e Reticulate Banded Gecko; Reticulated Gecko; Texas Banded Gecko
f Gecko-léopard réticulée; Coléonyx réticulé

2008 ***Coleonyx switaki***
Squamata - Eublepharidae
e Barefoot Banded Gecko; Barefoot Gecko; Switak's Banded Gecko
f Gecko-léopard de Baja

2009 ***Coleonyx switaki switaki***
Squamata - Eublepharidae
e Peninsular Banded Gecko; Peninsular Barefoot Gecko

2010 ***Coleonyx variegatus***
Squamata - Eublepharidae
e Western Banded Lizard; Western Banded Gecko; Banded Gecko (NA)
d Gebänderter Krallengecko; Gebänderter Wüstengecko
f Gecko-léopard de l'Ouest; Coléonyx varié

2011 ***Coleonyx variegatus abbotti***
Squamata - Eublepharidae
e San Diego Banded Gecko

2012 ***Coleonyx variegatus bogerti***
Squamata - Eublepharidae
e Tucson Banded Gecko

2013 ***Coleonyx variegatus utahensis***
Squamata - Eublepharidae
e Utah Banded Gecko

2014 ***Coleonyx variegatus variegatus***
Squamata - Eublepharidae
e Desert Banded Gecko; Tucson Banded Gecko

2015 ***Collorhabdium***
Squamata - Colubridae
e Reed Snakes; Williamson's Reed Snakes

2016 ***Collorhabdium williamsoni***
Squamata - Colubridae
e Williamson's Reed Snake

2017 ***Colobodactylus***
Squamata - Teiidae
e Amaral's Teiids

2018 ***Colobodactylus dalcyanus***
Squamata - Teiidae
e Vanzolini's Teiid

2019 ***Colobodactylus taunayi***
Squamata - Teiidae
e Taunay Teiid

2020 ***Colobosaura***
Squamata - Teiidae
e Colobosauras

2021 ***Colobosaura kraepelini***
Squamata - Teiidae
e Kraepelin's Colobosaura

2022 ***Colobosaura landii***
Squamata - Teiidae
e Colobosaura

2023 ***Colobosaura mentalis***
Squamata - Teiidae
e Amaral's Colobosaura

2024 ***Colobosaura modesta***
Squamata - Teiidae
e Bahia Colobosaura

2025 ***Colobosauroides***
Squamata - Teiidae
e Cuha's Teiid

2026 ***Colobosauroides caerensis***
Squamata - Teiidae
e Cuha's Teiid

2027 ***Coloptychon***
Squamata - Anguidae
e Lateral-fold Lizards; Alligator Lizards

2028 ***Coloptychon rhombifer***
Squamata - Anguidae
e Peters's Lateral-fold Lizard; Peters's Alligator Lizard

2029 ***Colopus***
Squamata - Gekkonidae
e Kalahari Geckos; Kalahari Ground Geckos (CSA)

2030 ***Colopus wahlbergi***
Squamata - Gekkonidae
e Wahlberg's Kalahari Gecko; Kalahari Ground Gecko (CSA)
f Gecko terrestre du Kalahari

2031 ***Coluber***
Squamata - Colubridae
e Common Racers (NA); Racers; Whip Snakes; North American Racers
d Zornnattern; Pfeilnattern; Bunte Zornnattern; Eigentliche Nattern
f Coleuvres agiles; Couleuvres

2032 ***Coluber bholanathi***
Squamata - Colubridae
e Sharma's Racer

2033 ***Coluber brevis***
Squamata - Colubridae
e Short Racer

2034 ***Coluber constrictor***
Squamata - Colubridae
e Eastern Racer; Racer (NA); Northern Racer; Black Snake; Black Racer; North American Racer; American Racer
d Schwarznatter
f Couleuvre agile noire; Couleuvre noire; Couleuvre agile

2035 ***Coluber constrictor anthicus***
Squamata - Colubridae
e Buttermilk Racer

2036 ***Coluber constrictor constrictor***
Squamata - Colubridae
e Northern Black Racer; Black Racer (NA)

2037 ***Coluber constrictor etheridgei***
Squamata - Colubridae
e Tan Racer

2038 ***Coluber constrictor flaviventris***
Squamata - Colubridae
e Yellowbelly Racer; Eastern Yellowbelly Racer

2039 ***Coluber constrictor foxi***
Squamata - Colubridae
e Blue Racer

2040 ***Coluber constrictor helvigularis***
Squamata - Colubridae
e Brownchin Racer; Brown-chinned Racer; Everglades Racer

2041 ***Coluber constrictor latrunculus***
Squamata - Colubridae
e Blackmask Racer; Black-masked Racer

2042 ***Coluber constrictor oaxaca***
Squamata - Colubridae
e Mexican Racer

2043 ***Coluber constrictor paludicola***
Squamata - Colubridae
e Everglades Racer

2044 ***Coluber constrictor priapus***
Squamata - Colubridae
e Southern Black Racer

2045 ***Coluber cypriensis***
Squamata - Colubridae
e Cyprus Racer
d Zypern-Natter; Zypern Schlanknatter
f Couleuvre de Chypre
i Colubro cipriota

2046 ***Coluber elegantissimus***
Squamata - Colubridae
e Elegant Racer
d Prachtzornnatter
f Couleuvre fouet élégante

2047 ***Coluber florulentus***
Squamata - Colubridae
e Flowered Racer
d Ägyptische Zornnatter
f Couleuvre algérienne; Couleuvre d'Égypte

2048 ***Coluber gracilis***
Squamata - Colubridae
e Graceful Racer

2049 ***Coluber hippocrepis***
Squamata - Colubridae
e Horseshoe Racer; Horseshoe Snake; Horseshoe Whip Snake
d Hufeisennatter
f Couleuvre fer-à-cheval
i Colubro ferro di cavallo; Colubro sardo

2050 ***Coluber insularis***
Squamata - Colubridae
e Saro Island Racer

2051 ***Coluber jugularis***
Squamata - Colubridae
e Fire Racer; Large Whip Snake; Caspian Whip Snake; Balkan Whip Snake; European Whip Snake; Persian Large Whip Snake
d Jochnatter; Pfeilnatter; Balkan-Natter; Kaspische Zornnatter; Kaspische Pfeilnatter; Persische Pfeilnatter

f Serpent-fouet; Couleuvre flèche; Couleuvre de la Caspienne; Couleuvre-fouet
i Colubro caspico; Colubro saltatore

2052 ***Coluber karelini***
Squamata - Colubridae
e Spotted Desert Racer

2053 ***Coluber laurenti***
Squamata - Colubridae
e Balkan Racer; Balkan Whip Snake
d Zornnatter; Balkan-Zornnatter; Pfeilnatter; Südenpfeilnatter
f Couleuvre des Balkans; Couleuvre de Laurenti
i Biacco minore; Biacco balcanico; Colubro dei Balcani

2054 ***Coluber manseri***
Squamata - Colubridae
e Manser's Black Racer

2055 ***Coluber messanai***
Squamata - Colubridae
e Schatti's Racer

2056 ***Coluber mormon***
Squamata - Colubridae
e Western Yellowbelly Racer; Western Yellow-bellied Racer; Western Racer

2057 ***Coluber najadum***
Squamata - Colubridae
e Ghamcheh Racer; Dahl's Whip Snake; Slender Racer
d Schlanknatter; Steignatter
f Serpent-fouet; Couleuvre fouet; Couleuvre de Dahl; Couleuvre à coup tacheté
i Colubro saetta

2058 ***Coluber nummifer***
Squamata - Colubridae
e Asian Racer; Reuss's Whip Snake
d Reuss-Zornnatter; Münzennatter
f Couleuvre à monnaie; Couleuvre à nummulites
i Colubro nummifero

2059 ***Coluber ravergieri***
Squamata - Colubridae
e Ravergier's Snake; Spotted Whip Snake; Mountain Racer
d Ravergieris Zornnatter; Bunte Zornnatter; Transkaspische Zornnatter
f Couleuvre fouet de Ravergier; Couleuvre de Ravergier; Couleuvre magnifique
i Colubro di Ravergier

2060 ***Coluber rhodorachis***
Squamata - Colubridae
e Braid Snake; Cliff Racer
d Jahns Pfeilnatter
f Couleuvre de Jan

2061 ***Coluber rogersi***
Squamata - Colubridae
e Rogers's Racer
d Anderson-Pfeilnatter; Andersons Pfeilnatter
f Couleuvre d'Anderson

2062 ***Coluber sinai***
Squamata - Colubridae
e Sinai Racer
d Sinai-Zornnatter
f Couleuvre du Sinaï

2063 ***Coluber smithi***
Squamata - Colubridae
e Smith's Racer; Southern Flowered Snake
d Smiths Natter; Smiths Pfeilnatter
f Couleuvre fouet de Smith; Couleuvre de Smith

2064 ***Coluber socotrae***
Squamata - Colubridae
e Socotran Racer

2065 ***Coluber spinalis***
Squamata - Colubridae
e Slender Racer

2066 ***Coluber thomasi***
Squamata - Colubridae
e Thomas's Semi-banded Racer

2067 ***Coluber variabilis***
Squamata - Colubridae
e Variable Snake

2068 ***Coluber ventromaculatus***
Squamata - Colubridae
e Hardwicke's Rat Snake; Glossy-bellied Racer; Gray's Rat Snake (ISC)
d Gefleckte Zornnatter
f Couleuvre tachetée
i Colubro di Gray

2069 ***Coluber viridiflavus***
Squamata - Colubridae
e Yellow-green Racer; Green Whip Snake; European Whip Snake; Dark Green Whip Snake; Dark Green Racer; Western Whip Snake; Dark Green Snake
d Gelbgrüne Zornnatter; Gelbgrüne Pfeilnatter; Zornnatter; Pfeilnatter
f Couleuvre verte-et-jaune; Zamenis
i Biacco; Colubro comune; Colubri verde-e-giallo; Biacco maggiore; Milordo; Frustone; Biàcc

2070 ***Coluber viridiflavus carbonarius***
Squamata - Colubridae
e Black Whip Snake
d Schwarze Pfeilnatter; Karbonarnatter; Karbonarschlange
i Colubro carbonaro

2071 **Colubridae**
Squamata
e Colubrid Snakes; Typical Snakes; Colubrids
d Nattern; Echte Nattern
f Colubridés; Couleuvres

2072 ***Compsophis***
Squamata - Colubridae
e Bighead Snakes

2073 ***Compsophis albiventris***
Squamata - Colubridae
e Madagascar Bighead Snake

2074 ***Coniophanes***
Squamata - Colubridae
e Black-striped Snakes; Black-banded Snakes; Debris Snakes
f Couleuvres à raies noires

2075 ***Coniophanes alvarezi***
Squamata - Colubridae
e Alvarez's Debris Snake

2076 ***Coniophanes andresensis***
Squamata - Colubridae
e Andresen Snake

2077 ***Coniophanes bipunctatus***
Squamata - Colubridae
e Two-spotted Snake; Two-spotted Debris Snake

2078 ***Coniophanes dromiciformis***
Squamata - Colubridae
e Peters's Running Snake

2079 ***Coniophanes fissidens***
Squamata - Colubridae
e Yellowbelly Snake; White-lipped Debris Snake

2080 ***Coniophanes flangivirgatus***
Squamata - Colubridae
e Peters's Snake

2081 ***Coniophanes imperialis***
Squamata - Colubridae

e Black-striped Snake; Black-banded Snake; Imperial Snake; Black-striped Debris Snake; Regal Black-striped Snake
f Couleuvre à raies noires

2082 ***Coniophanes imperialis imperialis***
Squamata - Colubridae
e Black-striped Snake; Tamaulpin Black Striped Snake

2083 ***Coniophanes joanae***
Squamata - Colubridae
e Joan's Snake; Joan's Debris Snake

2084 ***Coniophanes lateritius***
Squamata - Colubridae
e Stripeless Snake

2085 ***Coniophanes meridanus***
Squamata - Colubridae
e Peninsula Stripeless Snake; Yucatan Debris Snake

2086 ***Coniophanes piceivittis***
Squamata - Colubridae
e Cope's Black-striped Snake; Striped Debris Snake

2087 ***Coniophanes quinquevittatus***
Squamata - Colubridae
e File-striped Snake; Five-lined Debris Snake

2088 ***Coniophanes schmidti***
Squamata - Colubridae
e Faded Black-striped Snake; Schmidt's Debris Snake

2089 ***Conolophus***
Squamata - Iguanaidae
e Galapagos Land Iguanas; Land Iguanas
d Drusenköpfe
f Iguanes terrestres

2090 ***Conolophus pallidus***
Squamata - Iguanaidae
e Santa Fe Land Iguana; Barrington Land Iguana; Barrington Island Land Iguana; Barrington Island Iguana
d Santa-Fé-Drüsenkopf

2091 ***Conolophus subcristatus***
Squamata - Iguanaidae
e Galapagos Land Iguana; Land Iguana
d Galapagos-Drüsenkopf; Galapagos-Landleguan
f Iguane terrestre des Galapagos

2092 ***Conophis***
Squamata - Colubridae
e Road Guarders
d Trugnattern; Streifennattern

2093 ***Conophis biserialis***
Squamata - Colubridae
e Two-lined Mexican Road Guarder; Two-lined Mexican Earth Snake

2094 ***Conophis lineatus***
Squamata - Colubridae
e Road Guarder; Common Road Guarder
d Mexikanische Trugnatter

2095 ***Conophis nasus***
Squamata - Colubridae
e Largenose Road Guarder; Largenose Earth Snake

2096 ***Conophis pulcher***
Squamata - Colubridae
e Beautiful Road Guarder

2097 ***Conophis vittatus***
Squamata - Colubridae
e Striped Road Guarder; Mexican Road Guarder

2098 ***Contia***
Squamata - Colubridae
e Sharptail Snakes; Sharp-tailed Snakes
d Dornschwanzschlangen
f Couleuvres à queue pointue; Couleuvres à queue fine

2099 ***Contia tenuis***
Squamata - Colubridae
e Sharp-tailed Snake; Sharptail Snake; Purple-tailed Snake
d Dornschwanzschlange
f Couleuvre à queue pointue; Couleuvre à queue fine

2100 ***Cophosaurus***
Squamata - Iguanaidae
e Earless Lizards; Greater Earless Lizards
d Texanische Taubiguane; Taubiguane
f Cophosaures; Lézards gros sans oreilles; Iguanes sourds

2101 ***Cophosaurus texanus***
Squamata - Iguanaidae
e Greater Earless Lizard
d Texanischer Taubiguan; Taubiguan
f Cophosaure du Texas; Iguane sourd

2102 ***Cophosaurus texanus scitulus***
Squamata - Iguanaidae
e Southwestern Earless Lizard; Greater Earless Lizard; Chihuahuan Earless Lizard; Chihuahuan Greater Earless Lizard

2103 ***Cophosaurus texanus texanus***
Squamata - Iguanaidae
e Texas Earless Lizard

2104 ***Cophoscincopus***
Squamata - Scincidae
e Keeled Water Skinks
d Wasserskinke

2105 ***Cophoscincopus durus***
Squamata - Scincidae
e Keeled Water Skink
d Wasserskink

2106 ***Cophotis***
Squamata - Agamidae
e Earless Agamas; Deaf Agamids; Deaf Agamas
d Taubagamen; Baumagamen

2107 ***Cophotis ceylanica***
Squamata - Agamidae
e Ceylon Deaf Agama
d Ceylon-Taubagame; Baumagame; Taubagame

2108 ***Cophotis sumatrana***
Squamata - Agamidae
e Sumatra Deaf Agama
d Sumatra-Taubagame

2109 ***Corallus***
Squamata - Boidae
e Tree Boas
d Hundskopfboas; Windeschlangen; Hundskopfschlinger
f Serpents canines

2110 ***Corallus annulatus***
Squamata - Boidae
e Annulated Boa; Ringed Tree Boa; Costa Rica Boa; Annulated Tree Boa
d Geringelte Baumboa; Ringelboa

2111 ***Corallus caninus***
Squamata - Boidae
e Emerald Tree Boa; Emerald Boa; Dog-headed Tree Boa; Green Tree Boa
d Grüne Hundskopfboa; Echte Hundskopfboa; Grüner Hundskopfschlinger; Hundskopfschlinger; Surinam-Hundskopfschlinger; Hundskopfboa; Smaragdboa

f Boa émeraude; Boa canine; Serpent à perroquets; Boa canin (Ants); Faux jacquot (Ants)
i Boa smeraldino

2112 ***Corallus enhydris***
Squamata - Boidae
e Amazon Tree Boa; Garden Tree Boa; Yellow Tree Boa
d Gartenboa; Hundskopfboa; Cooksche Boa; Baumboa
f Boa arboricole (Ants); Boa de Cook
i Boa dei giardini

2113 ***Corallus enhydris cooki***
Squamata - Boidae
e Cook's Garden Tree Boa; Cook's Tree Boa; St. Vincent Boa (WI)
d Cooks Gartenboa

2114 ***Corallus ruschenbergii***
Squamata - Boidae
e Ruschenberg's Tree Boa

2115 **Cordylidae**
Squamata
e Girdle-tailed Lizards; Spinytail Lizards
d Gürtelechsen; Gürtelschwänze; Gürtelschweife
f Cordylidés

2116 ***Cordylosaurus***
Squamata - Cordylidae
e Blue-black Plated Lizards; Dwarf Plated Lizards (CSA)
d Blauschwarze Schildechsen
f Cordylosaures

2117 ***Cordylosaurus subtesselatus***
Squamata - Cordylidae
e Blue-black Plated Lizard; Dwarf Plated Lizard (CSA)
d Blauschwarze Schildechse

2118 ***Cordylus***
Squamata - Cordylidae
e Spinytail Lizards; Girdled Lizards (CSA); Spiny-tailed Lizards (CSA); Club-tailed Lizards
d Gürtelschweife; Echte Gürtelschweife
f Cordyles
i Cordili

2119 ***Cordylus caeroleopunctatus***
Squamata - Cordylidae
e Blue-spotted Spinytail Lizard; Blue-spotted Girdled Lizard (CSA)
d Blaufleckiger Gürtelschweif
f Cordyle ponctué de bleu

2120 ***Cordylus campbelli***
Squamata - Cordylidae
e Campbell's Spinytail Lizard; Campbell's Girdled Lizard (CSA)

2121 ***Cordylus cataphractus***
Squamata - Cordylidae
e Armadillo Spinytail Lizard; Armadillo Girdled Lizard (CSA); Armadillo Lizard; Armadillo Girdle-tailed Lizard
d Panzergürtelschweif
f Cordyle cataphracté; Zonure cataphracté; Armadillo

2122 ***Cordylus cordylus***
Squamata - Cordylidae
e Common Spinytail Lizard; Cape Girdled Lizard (CSA); Common Cape Girdled Lizard; Rough-sealed Girdled Lizard
d Zwerggürtelschweif; Gewöhnlicher Gürtelschweif
f Cordyle vulgaire

2123 ***Cordylus giganteus***
Squamata - Cordylidae
e Giant Spinytail Lizard; Giant Girdled Lizard (CSA); African Armadillo Lizard; Sungazer (CSA); Lord Derby's Girdled Lizard; Giant

Sungazer
d Riesengürtelschweif; Südafrikanischer Sonnengucker
f Cordyle géant; Zonure géante

2124 ***Cordylus lawrenci***
Squamata - Cordylidae
e Lawrence's Girdled Lizard

2125 ***Cordylus macropholis***
Squamata - Cordylidae
e Coastal Spinytail Lizard; Large-scaled Girdled Lizard (CSA)

2126 ***Cordylus mclachlani***
Squamata - Cordylidae
e McLachlan's Spinytail Lizard; McLachlan's Girdled Lizard (CSA)

2127 ***Cordylus namaquensis***
Squamata - Cordylidae
e Namaqua Spinytail Lizard; Namaqua Girdled Lizard (CSA)

2128 ***Cordylus peersi***
Squamata - Cordylidae
e Hewitt's Spinytail Lizard; Peers's Girdled Lizard (CSA)

2129 ***Cordylus polyzonus***
Squamata - Cordylidae
e African Spinytail Lizard; Karoo Girdled Lizard (CSA)
f Cordyle polyzoné; Zonure polyzoné

2130 ***Cordylus pustulatus***
Squamata - Cordylidae
e Herero Girdled Lizard

2131 ***Cordylus rhodesianus***
Squamata - Cordylidae
e Zimbabwe Girdled Lizard; Rhodesian Girdled Lizard (CSA); Manica Girdled Lizard (CSA); Montane Girdled Lizard; Montane Spiny-tailed Lizard (CSA)

2132 ***Cordylus tasmani***
Squamata - Cordylidae
e Tasman's Girdled Lizard
d Tanzania-Gürtelschweif

2133 ***Cordylus tropidosternum***
Squamata - Cordylidae
e Tropical Spinytail Lizard; Tropical Girdled Lizard (CSA); Tropical Spiny-tailed Lizard (CSA); Eastern Spiny-tailed Lizard; Savanna Spiny-tailed Lizard (CSA)
d Zwerggürtelschweif; Gürtelschweif
f Cordyle épineuse; Lézard Sungazer

2134 ***Cordylus tropidosternum jonesii***
Squamata - Cordylidae
e Lowveld Girdled Lizard; Tropical Spinytail Lizard
i Cordilo

2135 ***Cordylus tropidosternum tropidosternum***
Squamata - Cordylidae
e Tropical Girdled Lizard

2136 ***Cordylus ukingensis***
Squamata - Cordylidae
e Ukinga Spinytail Lizard; Ukinga Spiny-tailed Lizard

2137 ***Cordylus vittifer***
Squamata - Cordylidae
e Reichenow's Spinytail Lizard; Transvaal Girdled Lizard (CSA)
d Rostbrauner Gürtelschweif

2138 ***Cordylus warreni***
Squamata - Cordylidae
e Warren's Spinytail Lizard; Warren's Girdled Lizard; Warren's Spiny-tailed Lizard (CSA); Zonure
d Warrens Gürtelschweif
f Cordyle de Warren

2139 ***Cordylus warreni barbertonensis***
Squamata - Cordylidae
e Barberton's Dark-girdled Lizard

2140 ***Cordylus warreni breyeri***
Squamata - Cordylidae
e Breyer's Dark-girdled Lizard

2141 ***Cordylus warreni depressus***
Squamata - Cordylidae
e Flat Dark-girdled Lizard

2142 ***Cordylus warreni laevigatus***
Squamata - Cordylidae
e Zoutspansberg Dark-girdled Lizard

2143 ***Cordylus warreni perkoensis***
Squamata - Cordylidae
e Olifant's River Dark-girdled Lizard

2144 ***Cordylus warreni vandami***
Squamata - Cordylidae
e Van Dams Dark-girdled Lizard

2145 ***Coronella***
Squamata - Colubridae
e Smooth Snakes; European Smooth Snakes
d Glattnattern; Schlingnatttern
f Coronelles; Couleuvres-lézard

2146 ***Coronella austriaca***
Squamata - Colubridae
e Smooth Snake; European Smooth Snake
d Glattnatter; Schlingnatter; Haselnatter
f Couleuvre lisse; Coronelle lisse; Coronelle légère; Coronelle
i Colubro liscio; Colubro austriaco; Coronella austriaca

2147 ***Coronella girondica***
Squamata - Colubridae
e Southern Smooth Snake
d Gironde-Glattnatter; Gironde-Natter; Girondische Glattnatter; Girondische Schlingnatter
f Couleuvre bordelaise; Couleuvre girondine; Coronelle bordelaise; Couleuvre coronelle girondine; Coronelle girondine
i Colubro di Riccioli; Colubro del Riccioli; Colubro girondina; Colubro dai Riccioli

2148 ***Corucia***
Squamata - Scincidae
e Solomon Islands Skinks
d Winkelschwanzskinke
f Scinques géants

2149 ***Corucia zebrata***
Squamata - Scincidae
e Solomon Islands Skink; Monkey-tailed Skink ; Prehensile-tailed Giant Skink; Solomons Giant Skink
d Wickelschwanzskink; Salomonen-Winkelschwanzskink; Salomonen-Riesenskink
f Scinque géant; Scinque à queue préhensile des Îles Salomons

2150 ***Coryophophylax***
Squamata - Agamidae
e Subcristate Angleheads; Nicobar Forest Dragons
d Kammwinkelkopfagamen

2151 ***Coryophophylax subcristatus***
Squamata - Agamidae
e Subcristate Anglehead; Nicobar Forest Dragon
d Kammwinkelkopfagame

2152 ***Corytophanes***
Squamata - Iguanaidae
e Helmeted Iguanas; Helmeted Iguanids; Helmeted Lizards; Helmeted Basilisks
d Helmleguane
f Corytophanes

2153 ***Corytophanes cristatus***
Squamata - Iguanaidae
e Smooth Helmeted Iguana; Elegant Helmeted Lizard; Helmeted Iguana; Helmeted Lizard; Smooth Helmeted Basilisk; Helmeted Basilisk; Smooth-head Helmeted Basilisk
d Helmleguan
f Corytophane à crête; Caméléon des forêts; Iguane à casque

2154 ***Corytophanes hernandezii***
Squamata - Iguanaidae
e Hernandez's Helmeted Iguana; Hernandez's Helmeted Lizard; Hernandez's Helmeted Basisk
d Hernandez-Helmleguan
f Corytophane caméléopside

2155 ***Corytophanes percarinatus***
Squamata - Iguanaidae
e Keeled Helmeted Iguana; Keeled Helmeted Lizard; Keeled Helmeted Basilisk
d Guatemala-Helmleguan

2156 ***Cosymbotus***
Squamata - Gekkonidae
e Asian House Geckos
d Asiatische Hausgeckos

2157 ***Cosymbotus craspedotus***
Squamata - Gekkonidae
e Mocquard's House Gecko

2158 ***Cosymbotus platyurus***
Squamata - Gekkonidae
e Asian House Gecko; Frilled House Gecko (ISC); Flat-tailed Gecko; Flat-tailed House Gecko (NA); Asian Flat-tailed Gecko
d Thailändischer Hausgecko

2159 ***Crenadactylus***
Squamata - Gekkonidae
e Clawless Geckos

2160 ***Crenadactylus ocellatus***
Squamata - Gekkonidae
e Ocellated Clawless Gecko; Clawless Gecko (ANZ)

2161 ***Cricosaura***
Squamata - Xantusiidae
e Cuban Night Lizards
d Spitzschnauzenskinke; Kuba-Nachtechsen
f Cricosaures

2162 ***Cricosaura typica***
Squamata - Xantusiidae
e Cuban Night Lizard
d Eigentlicher Spitzschnauzenskink; Kuba-Nachtechse
f Cricosaure typique

2163 ***Crisantophis***
Squamata - Colubridae
e Dunn's Road Guarders

2164 ***Crisantophis nevermanni***
Squamata - Colubridae
e Dunn's Road Guarder; Southern Road Guarder

2165 ***Crocodilurus***
Squamata - Teiidae
e Crocodile Teiids; Crocodile Tegus
d Krokodilschwanzechsen
f Crocodilures

2166 ***Crocodilurus lacertinus***
Squamata - Teiidae
e Crocodile Teiid; Crocodile Tegu
d Krokodilschwanzechse
f Crocodilure lézardet

2167 **Crocodylia**
d Krokodilier
f Crocodiliens

2168 **Crocodylidae**
Squamata
e True Crocodiles; Crocodiles
d Krokodile; Echte Krokodile; Panzerechsen
f Crocodilidés
i Coccodrilli e alligatori

2169 ***Crocodylus***
Crocodylia - Crocodylidae
e Crocodiles
d Echte Krokodile; Panzerechsen
f Crocodiles
i Coccodrilli

2170 ***Crocodylus acutus***
Crocodylia - Crocodylidae
e American Crocodile; Common Crocodile (NA); Central American Alligator
d Spitzkrokodil
f Crocodile américain; Crocodile à museau pointu

2171 ***Crocodylus cataphractus***
Crocodylia - Crocodylidae
e African Slender-snouted Crocodile; Western African Crocodile; West African Long-snouted Crocodile; Long-snouted Crocodile; Slender-snouted Crocodile
d Panzerkrokodil
f Crocodile à museau allongé; Crocodile noir; Faux gavial africain; Crocodile à nuque cuirassée; Crocodile camus africain; Crocodile à museau de gavial

2172 ***Crocodylus intermedius***
Crocodylia - Crocodylidae
e Orinoco Crocodile
d Orinoko-Krokodil
f Crocodile de l'Orénoque; Crocodile de l'Amazon

2173 ***Crocodylus johnstoni***
Crocodylia - Crocodylidae
e Australian Freshwater Crocodile; Freshwater Crocodile (ANZ); Australian Crocodile; Johnston River Crocodile; Johnston's Crocodile
d Australien-Krokodil
f Crocodile d'Australie; Crocodile de Johnston

2174 ***Crocodylus moreletti***
Crocodylia - Crocodylidae
e Morelet's Crocodile
d Beulenkrokodil
f Crocodile de Morelet

2175 ***Crocodylus niloticus***
Crocodylia - Crocodylidae
e Nile Crocodile; Nilotic Crocodile; African Crocodile
d Nil-Krokodil
f Crocodile du Nil; Crocodile

2176 ***Crocodylus novaeguineae***
Crocodylia - Crocodylidae
e Freshwater Crocodile; New Guinea Crocodile; New Guinean Crocodile
d Neuguinea-Krokodil
f Crocodile de la Nouvelle Guinée

2177 ***Crocodylus novaeguineae mindorensis***
Crocodylia - Crocodylidae
e Philippine Crocodile; Mindoro Crocodile
d Mindoro-Krokodil
f Crocodile de Mindoro

2178 ***Crocodylus palustris***
Crocodylia - Crocodylidae
e Mugger Crocodile; Mugger; Snub-nosed Crocodile; Marsh Crocodile (ISC)
d Sumpfkrokodil; Stumpfkrokodil
f Crocodile des marais; Crocodile indien des marais

2179 ***Crocodylus porosus***
Crocodylia - Crocodylidae
e Saltwater Crocodile; Man-eating Saltwater Crocodile; Estuarine Crocodile; Indo-Pacific Crocodile
d Leistenkrokodil
f Crocodile marin; Crocodile des estuaires; Crocodile à double crète

2180 ***Crocodylus rhombifer***
Crocodylia - Crocodylidae
e Cuban Crocodile
d Rautenkrokodil; Kuba-Krokodil
f Crocodile rhombifère; Crocodile rhombifère du Cuba; Crocodile du Cuba

2181 ***Crocodylus siamensis***
Crocodylia - Crocodylidae
e Siamese Crocodile; Estuarine Crocodile; East Indian Saltwater Crocodile
d Siam-Krokodil
f Crocodile siamois; Crocodile de Siam

2182 ***Crossobamon***
Squamata - Gekkonidae
e Comb-toed Geckos
d Wüstengeckos; Kammzehengeckos

2183 ***Crossobamon eversmanni***
Squamata - Gekkonidae
e Comb-toed Gecko
d Kammzehengecko

2184 ***Crossobamon orientalis***
Squamata - Gekkonidae
e Sind Gecko

2185 ***Crotalus***
Squamata - Viperidae
e Rattlesnakes
d Klapperschlangen
f Crotales; Crotales maillés; Serpents à sonnettes
i Crotali; Vipere dalle fossette; Serpenti a sonagli; Mocassini

2186 ***Crotalus adamanteus***
Squamata - Viperidae
e Eastern Diamondback Rattlesnake; Eastern Diamondback (NA); Highland Ratlesnake (NA); Diamondback (NA); Eastern Diamond Rattlesnake; Eastern Diamon-backed Rattlesnake
d Diamantklapperschlange; Östliche Diamantklapperschlange
f Crotale diamantin de l'Est; Crotale diamant; Crotale diamantin
i Crotalo diamantino; Crotalo diamantino orientale

2187 ***Crotalus angelensis***
Squamata - Viperidae
e Angel Island Speckled Rattlesnake

2188 ***Crotalus aquilus***
Squamata - Viperidae
e Queretaran Dusky Rattlesnake

2189 ***Crotalus atrox***
Squamata - Viperidae
e Western Diamondback Rattlesnake; Western Diamondback (NA); Western Diamond Rattlesnake; Texas Rattlesnake; Western Diamond-backed Rattlesnake
d Texas-Klapperschlange; Westliche Diamantklapperschlange
f Crotale diamantin de l'Ouest; Crotale du Texas
i Crotalo diamantino occidentale

2190 ***Crotalus basiliscus***
Squamata - Viperidae
e Mexican West Coast Rattlesnake; Basilisc Rattlesnake; Oaxacan Rattlesnake
d Basiliskenklapperschlange; Mexikanische Westküsten-Klapperschlange

2191 ***Crotalus catalinensis***
Squamata - Viperidae
e Santa Catalina Island Rattlesnake; Santa Catalina Rattlesnake
d Santa-Catalina-Klapperschlange

2192 ***Crotalus cerastes***
Squamata - Viperidae
e Sidewinder Rattlesnake; Sidewinder; Horned Rattlesnake (NA); American Sidewinder; Mojave Desert Sidewinder
d Gehörnte Klapperschlange; Seitenwinder-Klappershlange; Seitenwinder
f Crotale cornu; Serpent à sonnettes cornu
i Crotalo ceraste

2193 ***Crotalus cerastes cerastes***
Squamata - Viperidae
e Mojave Desert Sidewinder
d Mojave-Seitenwinder-Klapperschlange

2194 ***Crotalus cerastes cercobombus***
Squamata - Viperidae
e Sonoran Sidewinder; Sonoran Desert Sidewinder
d Sonora-Seitenwinder-Klapperschlange

2195 ***Crotalus cerastes laterorepens***
Squamata - Viperidae
e Colorado Desert Sidewinder
d Colorado-Seitenwinder-Klapperschlange

2196 ***Crotalus durissus***
Squamata - Viperidae
e Neotropical Rattlesnake; Cascabel Rattlesnake; Cascabel; Tropical Rattlesnake; Cascaval; Tropical Rattlesnake; South American Rattlesnake
d Schauerklapperschlange; Tropische Klapperschlange; Cascabel-Klapperschlange; Schreckensklapperschlange; Schauerklapperschlange; Cascaval
f Crotale des tropiques; Cascabel; Crotale diamantin; Serpent à sonnette; Crotale; Crotale sud-américain
i Cascavel; Serpente a sonagli tropicale

2197 ***Crotalus durissus culminatus***
Squamata - Viperidae
d Nordwestliche Schauerklapperschlange

2198 ***Crotalus durissus cumanensis***
Squamata - Viperidae
e Columbian Rattlesnake
d Kolumbianische Klapperschlange

2199 ***Crotalus durissus durissus***
Squamata - Viperidae
d Schauerklapperschlange

2200 ***Crotalus durissus terrificus***
Squamata - Viperidae
e South American Rattlesnake
d Südamerikanische Klapperschlange; Eigentliche Cascaval

2201 ***Crotalus durissus totonacus***
Squamata - Viperidae
e Totonacan Rattlesnake
d Totonaca-Klapperschlange

2202 ***Crotalus durissus tzabcan***
Squamata - Viperidae
e Yucatan Rattlesnake
d Yucatan-Schauerklapperschlange

2203 ***Crotalus durissus unicolor***
Squamata - Viperidae
e Aruba Rattlesnake
d Aruba-Klapperchlange

2204 ***Crotalus enyo***
Squamata - Viperidae
e Baja California Rattlesnake (NA); Lower California Rattlesnake
d Nieder-Kalifornische Klapperschlange

2205 ***Crotalus estebanensis***
Squamata - Viperidae
e San Esteban Island Rattlesnake

2206 ***Crotalus exsul***
Squamata - Viperidae
e Cedros Island Diamond Rattlesnake; Red Diamond Rattlesnake; Red Diamond Rattler (NA); Cedros Island Rattler; Cedros Island Rattlesnake

2207 ***Crotalus exsul exsul***
Squamata - Viperidae
e Peninsular Red Diamond Rattlesnake

2208 ***Crotalus horridus***
Squamata - Viperidae
e Timber Rattlesnake; Canebrake Rattlesnake; Seminole Rattler (NA); Swamp Rattler (NA); Banded Rattlesnake
d Waldklapperschlange
f Crotale des bois; Crotale

2209 ***Crotalus horridus atricaudatus***
Squamata - Viperidae
e Canebrake Rattlesnake
d Waldklapperschlange
f Crotale des bois

2210 ***Crotalus intermedius***
Squamata - Viperidae
e Mexican Smallhead Rattlesnake; Tatalcan Small-headed Rattlesnake

2211 ***Crotalus lannomi***
Squamata - Viperidae
e Autlan Rattlesnake

2212 ***Crotalus lepidus***
Squamata - Viperidae
e Rock Rattlesnake; Mottled Rock Rattlesnake; Green Rattlesnake
d Felsenklapperschlange
f Crotale rupestre; Crotale des rochers

2213 ***Crotalus lepidus klauberi***
Squamata - Viperidae
e Banded Rock Rattlesnake; New Mexico Banded Rock Rattlesnake
d Gestreifte Felsenklapperschlange

2214 ***Crotalus lepidus lepidus***
Squamata - Viperidae
e Mottled Rock Rattlesnake
d Felsenklapperschlange

2215 ***Crotalus lepidus maculosus***
Squamata - Viperidae
e Durango Rock Rattlesnake
d Durango-Felsenklapperschlange

2216 ***Crotalus lepidus morulus***
Squamata - Viperidae
e Tamaulipas Rock Rattlesnake
d Tamaulipas-Felsenklapperschlange

2217 ***Crotalus lorenzoensis***
Squamata - Viperidae
e San Lorenzo Island Rattlesnake

2218 ***Crotalus mitchelli***
Squamata - Viperidae
e Speckled Rattlesnake; Mitchell's Rattlesnake; San Lucan Speckled Rattlesnake
d Gefleckte Klapperschlange
f Crotale de Mitchell

2219 ***Crotalus mitchelli mitchelli***
Squamata - Viperidae
e San Lucan Speckled Rattlesnake
d San-Lucas Gefleckte Klapperschlange

2220 ***Crotalus mitchelli pyrrhus***
Squamata - Viperidae
e Southwestern Speckled Rattlesnake
d Südwestliche Gefleckte Klapperschlange

2221 ***Crotalus mitchelli stephensi***
Squamata - Viperidae
e Panamint Rattlesnake

2222 ***Crotalus molossus***
Squamata - Viperidae
e Black-tailed Rattlesnake; Blacktail Rattlesnake
d Schwarzschwanzklapperschlange
f Crotale ä queue noire

2223 ***Crotalus molossus molossus***
Squamata - Viperidae
e Northern Black-tailed Rattlesnake (NA); Blacktail Rattlesnake
d Schwarzschwanzklapperschlange

2224 ***Crotalus molossus nigrescens***
Squamata - Viperidae
e Mexican Black-tailed Rattlesnake
d Mexikanische Schwarzschwanzklapperschlange

2225 ***Crotalus muertensis***
Squamata - Viperidae
e El Muerto Island Speckled Rattlesnake

2226 ***Crotalus polystictus***
Squamata - Viperidae
e Mexican Lancehead Rattlesnake; Mexican Lance-headed Rattlesnake
d Mexikanische Lanzenkopfklapperschlange

2227 ***Crotalus pricei***
Squamata - Viperidae
e Twin-spotted Rattlesnake; Twin-spot Rattlesnake; Arizona Twin-spotted Rattlesnake; Price's Rattlesnake
d Prices Klapperschlange
f Crotale de Price

2228 ***Crotalus pricei pricei***
Squamata - Viperidae
e Western Twin-spotted Rattlesnake
d Prices Klapperschlange

2229 ***Crotalus pusillus***
Squamata - Viperidae
e Tancitaran Dusky Rattlesnake
d Tancitaro-Klapperschlange

2230 ***Crotalus ruber***
Squamata - Viperidae
e Red Diamond Rattlesnake; Red Rattlesnake
d Rote Diamantklapperschlange
f Crotale diamantin rouge
i Diamante rosso

2231 ***Crotalus ruber lucanensis***
Squamata - Viperidae
e Cape Red Rattlesnake

2232 ***Crotalus ruber ruber***
Squamata - Viperidae
e Northern Red Rattlesnake
d Rote Diamantklapperschlange
f Crotale diamantin rouge

2233 ***Crotalus scutulatus***
Squamata - Viperidae
e Mojave Rattlesnake; Mojave Green Rattlesnake; Mojave Diamond Rattlesnake
d Mojave-Klapperschlange
f Crotale du Mojave
i Crotalo di Mojave

2234 ***Crotalus scutulatus scutulatus***
Squamata - Viperidae
e Mojave Green Rattlesnake; Mojave Rattlesnake
d Mojave-Klapperschlange

2235 ***Crotalus stejnegeri***
Squamata - Viperidae
e Longtail Rattlesnake; Long-tailed Rattlesnake

2236 ***Crotalus tigris***
Squamata - Viperidae
e Tiger Rattlesnake; Tiger Rattler
d Tigerklapperschlange
f Crotale tigré

2237 ***Crotalus tortugensis***
Squamata - Viperidae
e Tortuga Island Rattlesnake; Tortuga Island Diamondback Rattlenake; Tortuga Island Diamond Rattlesnake

2238 ***Crotalus transversus***
Squamata - Viperidae
e Cross-banded Mountain Rattlesnake

2239 ***Crotalus triseriatus***
Squamata - Viperidae
e Mexican Dusky Rattlesnake; Spotted Rattlesnake; Central Plateau Dusky Rattlesnake
d Dunkle Mexikanische Klapperschlange

2240 ***Crotalus triseriatus aquilus***
Squamata - Viperidae
e Queretaran Dusky Rattlesnake

2241 ***Crotalus triseriatus triseriatus***
Squamata - Viperidae
e Central Plateau Dusky Rattlesnake

2242 ***Crotalus unicolor***
Squamata - Viperidae
e Aruba Island Rattlesnake

2243 ***Crotalus vegrandis***
Squamata - Viperidae
e Uracoan Rattlesnake
d Uracoa-Klapperschlange; Venezolanische Klapperschlanage

2244 ***Crotalus viridis***
Squamata - Viperidae
e Western Rattlesnake; Prairie Rattlesnake
d Prärie-Klapperschlange; Prärien-Klapperschlange
f Crotale des prairies; Crotale de l'ouest
i Crotalo verde; Crotalo del Pacifico

2245 ***Crotalus viridis abyssus***
Squamata - Viperidae
e Grand Canyon Rattlesnake; Plains Rattlesnake

2246 ***Crotalus viridis caliginis***
Squamata - Viperidae
e Los Coronados Island Rattlesnake

2247 ***Crotalus viridis cerberus***
Squamata - Viperidae
e Arizona Black Rattlesnake
d Schwarze Arizona-Klapperschlange

2248 ***Crotalus viridis concolor***
Squamata - Viperidae
e Midget Faded Rattlesnake; New Mexico Midget Faded Rattlesnake

2249 ***Crotalus viridis helleri***
Squamata - Viperidae
e Southern Pacific Rattlesnake
d Südliche Prairie-Klapperschlange

2250 ***Crotalus viridis lutosus***
Squamata - Viperidae
e Great Basin Rattlesnake

2251 ***Crotalus viridis nuntius***
Squamata - Viperidae
e Hopi Rattlesnake
d Hopi-Prairie-Klapperschlanage

2252 ***Crotalus viridis oreganus***
Squamata - Viperidae

e Northern Pacific Rattlesnake; Pacific Rattlesnake; Blacktail Rattlesnake; Pacific Rattler (NA)
d Nordpazifische Klapperschlange; Nördliche Prairie-Klapperschlange

2253 ***Crotalus viridis viridis***
Squamata - Viperidae
e Prairie Rattlesnake
d Prairie-Klapperschlange
f Crotale des prairies

2254 ***Crotalus willardi***
Squamata - Viperidae
e Ridge-nosed Rattlesnake; Ridgenose Rattlesnake; Willard's Rattlesnake
d Willard-Klapperschlange; Willards Klapperschlange; Kantenkopfklapperschlange
f Crotale de Willard

2255 ***Crotalus willardi meridionalis***
Squamata - Viperidae
e Southern Ridge-nosed Rattlesnake
d Südliche Willard-Klapperschlange

2256 ***Crotalus willardi obscurus***
Squamata - Viperidae
e New Mexico Ridgenose Rattlesnake; Animas Ridge-nosed Rattlesnake; New Mexico Ridge-nosed Rattlesnake
d Neumexiko Willard-Klapperschlange

2257 ***Crotalus willardi silus***
Squamata - Viperidae
e Chihuahan Ridge-nosed Rattlesnake
d Chihuaha-Willard-Klapperschlange

2258 ***Crotalus willardi willardi***
Squamata - Viperidae
e Arizona Ridgenose Rattlesnake; Ridgenose Rattlesnake; Arizona Ridge-nosed Rattlesnake; Willards Rattlesnake
d Willard-Klapperschlange

2259 ***Crotaphopeltis***
Squamata - Colubridae
e Tropical Water Snakes; Herald Snakes; Red-lipped Snakes (CSA); Yellow-lipped Snakes; Herald's Snakes

2260 ***Crotaphopeltis acarina***
Squamata - Colubridae
e Roman's Tropical Snake

2261 ***Crotaphopeltis barotseensis***
Squamata - Colubridae
e Barotse Water Snake

2262 ***Crotaphopeltis hotamboeia***
Squamata - Colubridae
e Herald Snake (CSA); Red-lipped Snake (CSA); Black-templed Cat Snake (CSA); White-lipped Snake; Irrascible Snake
d Weißlippenschlange

2263 ***Crotaphopeltis tornieri***
Squamata - Colubridae
e Werner's Cat Snake

2264 ***Crotaphytus***
Squamata - Iguanaidae
e Collared Lizards
d Halsbandleguane
f Crotaphytinés; Lézards à colliers

2265 ***Crotaphytus bicinctores***
Squamata - Iguanaidae
e Mojave Black-collared Lizard; Great Basin Collared Lizard

2266 ***Crotaphytus collaris***
Squamata - Iguanaidae
e Collared Lizard; Common Collared Lizard; Eastern Collared Lizard
d Halsbandleguan; Kugelechse
f Lézard à colliers

2267 ***Crotaphytus collaris auriceps***
Squamata - Iguanaidae
e Yellowhead Collared Lizard; Yellow-headed Collared Lizard

2268 ***Crotaphytus collaris baileyi***
Squamata - Iguanaidae
e Western Collared Lizard

2269 ***Crotaphytus collaris collaris***
Squamata - Iguanaidae
e Eastern Collared Lizard
d Halsbandleguan; Kugelechse

2270 ***Crotaphytus collaris fuscus***
Squamata - Iguanaidae
e Chihuahuan Collared Lizard
d Brauner Halsbandleguan

2271 ***Crotaphytus dickersonae***
Squamata - Iguanaidae
e Dickerson's Collared Lizard

2272 ***Crotaphytus insularis***
Squamata - Iguanaidae
e Black-collared Lizard; Desert Collared Lizard (NA); Angel Island Collared Lizard
d Wüstenhalsbandleguan
f Lézard à colliers noirs

2273 ***Crotaphytus insularis bicinctores***
Squamata - Iguanaidae
e Great Basin Collared Lizard

2274 ***Crotaphytus insularis vestigium***
Squamata - Iguanaidae
e Baja California Collared Lizard

2275 ***Crotaphytus nebrius***
Squamata - Iguanaidae
e Sonoran Collared Lizard

2276 ***Crotaphytus reticulatus***
Squamata - Iguanaidae
e Reticulate Collared Lizard
f Lézard à collier réticulé

2277 ***Crotaphytus vestigium***
Squamata - Iguanaidae
e Baja Black-collared Lizard; Baja California Collared Lizard

2278 ***Cryophis***
Squamata - Colubridae
e Forest Snakes

2279 ***Cryophis hallbergi***
Squamata - Colubridae
e Hallberg's Cloud Forest Snake

2280 ***Cryptagama***
Squamata - Agamidae
e Hidden Agamas; Fringe-lip Agamas; Gravel Dragons (ANZ); Large-eared Forest Dragons

2281 ***Cryptagama aurita***
Squamata - Agamidae
e Hidden Agama; Gravel Dragon (ANZ); Large-eared Forest Dragon; Fringe-lip Agama

2282 ***Cryptoblepharus***
Squamata - Scincidae
e Snake-eyed Skinks; Coastal Skinks (CSA)s; Island-dwellers; Rock Dwellers

2283 ***Cryptoblepharus boutonii***
Squamata - Scincidae
e Snake-eyed Skink; Bouton's Skink (CSA); Bouton's Island-dweller
d Inselskink

2284 ***Cryptoblepharus carnabyi***
Squamata - Scincidae
e Grey Snake-eyed Skink; Carnaby's Wall Skink (ANZ)

2285 ***Cryptoblepharus eximius***
Squamata - Scincidae
e Pygmy Snake-eyed Skink

2286 ***Cryptoblepharus fuhni***
Squamata - Scincidae
e Fuhn's Snake-eyed Skink; Fuhn's Island-dweller

2287 ***Cryptoblepharus litoralis***
Squamata - Scincidae
e Green-black Snake-eyed Skink; Supralittoral Shining Snake-eyed Skink; Supralittoral Shining Skink

2288 ***Cryptoblepharus megastictus***
Squamata - Scincidae
e Metallic Snake-eyed Skink; Spotted Snake-eyed Skink

2289 ***Cryptoblepharus plagiocephalus***
Squamata - Scincidae
e Australian Snake-eyed Skink; Australian Island-dweller; Callose-palmed Shining Skink (NA); Callose-palmed Shining Snake-eyed Skink (NA)

2290 ***Cryptoblepharus poecilopleurus***
Squamata - Scincidae
e Snake-eyed Skink; Mottled Snake-eyed Skink

2291 ***Cryptoblepharus virgatus***
Squamata - Scincidae
e Silver Snake-eyed Skink; Wall Skink; Cream-striped Shining Skink

2292 ***Cryptolycus***
Squamata - Colubridae
e Dwarf Wolf Snakes; Cryptis Wolf Snakes

2293 ***Cryptolycus nanus***
Squamata - Colubridae
e Dwarf Wolf Snake; Cryptic Wolf Snake (CSA)

2294 ***Cryptophidion***
Squamata - Colubridae
e Secret Snakes

2295 ***Cryptophidion annamensis***
Squamata - Colubridae
e Vietnam Secret Snake

2296 ***Cryptoposcincus***
Squamata - Scincidae
e Secret Skinks

2297 ***Cryptoposcincus minimus***
Squamata - Scincidae
e Mocquard's Secret Skink

2298 ***Ctenoblepharis***
Squamata - Iguanaidae
e Desert Iguanas

2299 ***Ctenoblepharis adspersus***
Squamata - Iguanaidae
e Coastal Desert Iguana

2300 ***Ctenophorus***
Squamata - Agamidae
e Australian Dragons
d Bartagemen

2301 ***Ctenophorus caudicinctus***
Squamata - Agamidae
e Ringtail Dragon; Ring-tailed Dragon (ANZ); Ring-tailed Bicycle Dragon (ANZ)
d Ringenschwanzbartagame

2302 ***Ctenophorus clayi***
Squamata - Agamidae
e Clay's Dragon; Black-shouldered Ground Dragon (ANZ)

2303 ***Ctenophorus cristatus***
Squamata - Agamidae
e Crested Dragon; Crested Bicycle Dragon (ANZ)
d Kammbartagame

2304 ***Ctenophorus decresii***
Squamata - Agamidae
e Kangaroo Island Dragon; Tawny Dragon (ANZ); Tawny Crevice Dragon (ANZ)

2305 ***Ctenophorus femoralis***
Squamata - Agamidae
e Exmouth Gulf Dragon; Long-tailed Sand Dragon (ANZ)

2306 ***Ctenophorus fionni***
Squamata - Agamidae
e Eyre Peninsula Dragon; Peninsula Crevice Dragon (ANZ); Peninsula Dragon (ANZ)

2307 ***Ctenophorus fordi***
Squamata - Agamidae
e Mallee Dragon; Mallee Sand Dragon (ANZ)
d Fords Bartagame
f Amphibolure de Ford

2308 ***Ctenophorus gibba***
Squamata - Agamidae
e Gibba Plains Dragon; Bulldust Ground Dragon (ANZ)

2309 ***Ctenophorus inermis***
Squamata - Agamidae
e De Vis's Bloodsucker

2310 ***Ctenophorus isolepis***
Squamata - Agamidae
e Military Dragon; Military Sand Dragon (ANZ)
d Soldatenagame

2311 ***Ctenophorus maculatus***
Squamata - Agamidae
e Spotted Dragon; Spotted Agama; Spotted Sand Dragon (ANZ)
d Gefleckte Bodenagame
f Amphibolure tacheté

2312 ***Ctenophorus maculosus***
Squamata - Agamidae
e Lake Eyre Dragon (ANZ); Lake Eyre Agama; Small Spotted Dragon; Saltlake Ground Dragon (ANZ)
d Lake-Eyre-Agame
f Amphibolure deLac Eyrie

2313 ***Ctenophorus mckenziei***
Squamata - Agamidae
e McKenzie's Dragon; Dwarf Bicycle Dragon (ANZ)

2314 ***Ctenophorus nuchalis***
Squamata - Agamidae
e Central Netted Dragon; Central Netted Ground Dragon (ANZ)
d Netzagame

2315 ***Ctenophorus ornatus***
Squamata - Agamidae
e Ornate Dragon; Ornate Crevice Dragon (ANZ); Painted Dragon; Painted Ground Dragon (ANZ)
d Schmuckbartagame; Gemalte Bartagame; Australische Bodenagame
f Amphibolure orné; Amphibolure bigarré

2316 ***Ctenophorus reticulatus***
Squamata - Agamidae
e Western Netted Dragon; Western Netted Ground Dragon (ANZ)
d Netzagame
f Amphibolure réticulé

2317 ***Ctenophorus rubens***
Squamata - Agamidae
e Reddening Sand Dragon

2318 ***Ctenophorus rufescens***
Squamata - Agamidae
e Red Dragon; Bicycle Lizard; Rusty Crevice Dragon (ANZ)
f Amphibolure rouget

2319 ***Ctenophorus salinarum***
Squamata - Agamidae
e Saltbush Dragon; Saltpan Ground Dragon (ANZ)

2320 ***Ctenophorus scutulatus***
Squamata - Agamidae
e Lozenge-marked Dragon; Lozenge-marked Bicycle Dragon (ANZ)

2321 ***Ctenophorus tjantjalua***
Squamata - Agamidae
e Tjantjalu Dragon

2322 ***Ctenophorus vadnappa***
Squamata - Agamidae
e Houston's Dragon; Red-barred Crevice Dragon (ANZ)

2323 ***Ctenophorus yinnietharra***
Squamata - Agamidae
e Yinnietharra Dragon; Yinnietharra Crevice Dragon (ANZ)

2324 ***Ctenosaura***
Squamata - Iguanaidae
e Greater Spinytail Iguanas; Spiny-tailed Iguanas
d Schwarze Leguane; Schwarzleguane

2325 ***Ctenosaura acanthura***
Squamata - Iguanaidae
e Northeastern Spinytail Iguana; Eastern Spinytail Iguana
d Schwarzer Leguan
f Iguane noir du Mexique

2326 ***Ctenosaura alfredschmidti***
Squamata - Iguanaidae
e Campeche Spinytail Iguana

2327 ***Ctenosaura bakeri***
Squamata - Iguanaidae
e Baker's Spinytail Iguana; Utila Spinytail Iguana
d Utila-Schwarzleguan; Utila-Leguan

2328 ***Ctenosaura clarki***
Squamata - Iguanaidae
e Balsas Spinytail Iguana; Balsas Armed Lizard

2329 ***Ctenosaura defensor***
Squamata - Iguanaidae
e Yucatan Spinytail Iguana

2330 ***Ctenosaura flavidorsalis***
Squamata - Iguanaidae
e Yellow-backed Spinytail Iguana

2331 ***Ctenosaura hemilopha***
Squamata - Iguanaidae
e Cape Spinytail Iguana; Cape Spiny-tailed Iguana; Short-crested Spiny-tailed Iguana; Cape Iguana; Northern False Iguana; Spiny-tailed Iguana; Peninsular Spinytail Iguana (NA)
f Iguane à queue épineuse-et-crête courte; Iguane commun à queue épineuse

2332 ***Ctenosaura hemilopha hemilopha***
Squamata - Iguanaidae
e Peninsular Spiny-tailed Iguana; Iguana

2333 ***Ctenosaura hemilopha macrolopha***
Squamata - Iguanaidae
e Sonoran Spiny-tailed Iguana

2334 ***Ctenosaura oedirhina***
Squamata – Iguanaidae
e De Queiroz's Spinytail Iguana; Roatan Spinytail Iguana

2335 ***Ctenosaura pectinata***
Squamata - Iguanaidae

e Mexican Spinytail Iguana; Western Spiny-tailed Iguana; Spiny-tailed Iguana; Spinytail Iguana
d Schwarzleguan
f Iguane à queue épineuse-et-crête courte

2336 ***Ctenosaura quinquecarinata***
Squamata - Iguanaidae
e Oaxacan Spinytail Iguana; Central American Armed Lizard; Five-keeled Spinytail Iguana
d Kurzkopfkielschwanzleguan

2337 ***Ctenosaura similis***
Squamata - Iguanaidae
e Black Iguana; Black Spiny-tailed Iguana; Common Spinytail Iguana; Black Spinytail Iguana
d Schwarzer Leguan

2338 ***Ctenotus***
Squamata - Scincidae
e Ctenotuses
d Streifenskinke

2339 ***Ctenotus alacer***
Squamata - Scincidae
e Black Ctenotus

2340 ***Ctenotus alleni***
Squamata - Scincidae
e Allen's Ctenotus

2341 ***Ctenotus allotropis***
Squamata - Scincidae
e Mallee Ctenotus

2342 ***Ctenotus angusticeps***
Squamata - Scincidae
e Olive-grey Ctenotus

2343 ***Ctenotus arcanus***
Squamata - Scincidae
e Olive-brown Ctenotus

2344 ***Ctenotus ariadnae***
Squamata - Scincidae
e Dunes Ctenotus; Ariadna's Ctenotus

2345 ***Ctenotus arnhemensis***
Squamata - Scincidae
e Western Arnhem Ctenotus

2346 ***Ctenotus astarte***
Squamata - Scincidae
e Crevice Ctenotus

2347 ***Ctenotus atlas***
Squamata - Scincidae
e Porcupinegrass Ctenotus

2348 ***Ctenotus australis***
Squamata - Scincidae
e Australian Striped Ctenotus

2349 ***Ctenotus borealis***
Squamata - Scincidae
e Boreal Ctenotus

2350 ***Ctenotus brachyonyx***
Squamata - Scincidae
e Arid Ctenotus; Murray Striped Skink (ANZ)

2351 ***Ctenotus brooksi***
Squamata - Scincidae
e Brooks's Ctenotus; Brooks's Striped Skink (ANZ)

2352 ***Ctenotus calurus***
Squamata - Scincidae
e Eight-striped Ctenotus

2353 ***Ctenotus capricorni***
Squamata - Scincidae
e Capricorn Ctenotus

2354 ***Ctenotus catnifer***
Squamata - Scincidae

e Heath Ctenotus

2355 ***Ctenotus coggeri***
Squamata - Scincidae
e Cogger's Ctenotus

2356 ***Ctenotus colleti***
Squamata - Scincidae
e Collett's Ctenotus

2357 ***Ctenotus decaneurus***
Squamata - Scincidae
e Ten-striped Ctenotus

2358 ***Ctenotus delli***
Squamata - Scincidae
e White-dotted Ctenotus

2359 ***Ctenotus dux***
Squamata - Scincidae
e Desert Ctenotus; Chief Ctenotus

2360 ***Ctenotus ehmanni***
Squamata - Scincidae
e Ehmann's Ctenotus

2361 ***Ctenotus essingtonii***
Squamata - Scincidae
e Essington's Ctenotus

2362 ***Ctenotus eurydice***
Squamata - Scincidae
e Woodland Ctenotus

2363 ***Ctenotus eutaenius***
Squamata - Scincidae
e Townsville Ctenotus

2364 ***Ctenotus fallens***
Squamata - Scincidae
e Coastal Ctenotus

2365 ***Ctenotus gagudju***
Squamata - Scincidae
e Alligator River Ctenotus

2366 ***Ctenotus gemmula***
Squamata - Scincidae
e Southern Ctenotus

2367 ***Ctenotus grandis***
Squamata - Scincidae
e Grand Ctenotus

2368 ***Ctenotus greeri***
Squamata - Scincidae
e Greer's Ctenotus

2369 ***Ctenotus hanloni***
Squamata - Scincidae
e Hanlon's Ctenotus

2370 ***Ctenotus hebetior***
Squamata - Scincidae
e Grass Ctenotus

2371 ***Ctenotus helenae***
Squamata - Scincidae
e Helena's Ctenotus

2372 ***Ctenotus hilli***
Squamata - Scincidae
e Hill's Ctenotus

2373 ***Ctenotus impar***
Squamata - Scincidae
e Eleven-striped Ctenotus

2374 ***Ctenotus ingrami***
Squamata - Scincidae
e Ingram's Ctenotus

2375 ***Ctenotus inornatus***
Squamata - Scincidae
e Sandstone Ctenotus

2376 ***Ctenotus joanae***
Squamata - Scincidae

e Newcastle Waters Ctenotus

2377 ***Ctenotus kurnbudj***
Squamata - Scincidae
e Eucalypt Ctenotus

2378 ***Ctenotus labillardieri***
Squamata - Scincidae
e Labillardière's Ctenotus; Labilliardière's Skink
d Labilliardières Skink
f Scinque de Labilliardière

2379 ***Ctenotus lancelini***
Squamata - Scincidae
e Lancelin Ctenotus
d Lancelin-Streifenskink

2380 ***Ctenotus lateralis***
Squamata - Scincidae
e Interior Ctenotus

2381 ***Ctenotus leae***
Squamata - Scincidae
e Red Sand Ctenotus

2382 ***Ctenotus leonhardii***
Squamata - Scincidae
e Leonhard's Ctenotus

2383 ***Ctenotus mastigura***
Squamata - Scincidae
e Iridescent Ctenotus

2384 ***Ctenotus militaris***
Squamata - Scincidae
e Military Ctenotus

2385 ***Ctenotus mimetes***
Squamata - Scincidae
e Southwestern Ctenotus

2386 ***Ctenotus monticola***
Squamata - Scincidae
e Atherton Tableland Ctenotus

2387 ***Ctenotus pallescens***
Squamata - Scincidae
e Hummockgrass Ctenotus

2388 ***Ctenotus pantherinus***
Squamata - Scincidae
e Swift Ctenotus

2389 ***Ctenotus piankai***
Squamata - Scincidae
e White-bellied Ctenotus

2390 ***Ctenotus pulchellus***
Squamata - Scincidae
e Queensland Ctenotus

2391 ***Ctenotus quattuordecimlineatus***
Squamata - Scincidae
e Many-lined Ctenotus

2392 ***Ctenotus quinkan***
Squamata - Scincidae
e Cooktown-Laura Ctenotus

2393 ***Ctenotus rawlinsoni***
Squamata - Scincidae
e Rawlinson's Ctenotus

2394 ***Ctenotus regius***
Squamata - Scincidae
e Royal Ctenotus; Regal Striped Skink (ANZ)

2395 ***Ctenotus robustus***
Squamata - Scincidae
e Robust Ctenotus; Large Striped Skink (ANZ)

2396 ***Ctenotus rubicundus***
Squamata - Scincidae
e Pilbara Ctenotus

2397 ***Ctenotus rutilans***
Squamata - Scincidae
e Scrub Ctenotus

2398 ***Ctenotus saxatilis***
Squamata - Scincidae
e Rock Ctenotus

2399 ***Ctenotus schevilli***
Squamata - Scincidae
e Black Soil Ctenotus

2400 ***Ctenotus schomburgkii***
Squamata - Scincidae
e Variable Ctenotus

2401 ***Ctenotus septenarius***
Squamata - Scincidae
e Seven-striped Ctenotus

2402 ***Ctenotus serotinus***
Squamata - Scincidae
e Gravel Ctenotus

2403 ***Ctenotus serventyi***
Squamata - Scincidae
e Plains Ctenotus

2404 ***Ctenotus severus***
Squamata - Scincidae
e Mulga Ctenotus

2405 ***Ctenotus spaldingi***
Squamata - Scincidae
e Spalding's Ctenotus

2406 ***Ctenotus storri***
Squamata - Scincidae
e Storr's Ctenotus

2407 ***Ctenotus strauchii***
Squamata - Scincidae
e Strauch's Ctenotus

2408 ***Ctenotus striaticeps***
Squamata - Scincidae
e Striated Ctenotus

2409 ***Ctenotus taeniolatus***
Squamata - Scincidae
e Coppertail Ctenotus; Copper-tailed Skink (ANZ)

2410 ***Ctenotus tanamiensis***
Squamata - Scincidae
e Tanami Desert Ctenotus

2411 ***Ctenotus tantillus***
Squamata - Scincidae
e Northern Ctenotus

2412 ***Ctenotus uber***
Squamata - Scincidae
e Sandy Ctenotus

2413 ***Ctenotus uber orientalis***
Squamata - Scincidae
e Uber Striped Skink

2414 ***Ctenotus vertebralis***
Squamata - Scincidae
e Escarpment Ctenotus

2415 ***Ctenotus xenopleura***
Squamata - Scincidae
e West Australian Ctenotus

2416 ***Ctenotus youngsoni***
Squamata - Scincidae
e Youngson's Ctenotus

2417 ***Ctenotus zastictus***
Squamata - Scincidae
e Shark Bay Ctenotus

2418 ***Ctenotus zebrilla***
Squamata - Scincidae
e Highland Ctenotus

2419 ***Cuora***
Testudines - Emydidae
e Asian Box Turtles; Malayan Box Turtles; Box Turtles; Old World Box Turtles
d Scharnierschildkröten
f Tortues-boîte; Tortues tabatières; Tortues tabatières d'Asie

2420 ***Cuora amboinensis***
Testudines - Emydidae
e South East Asia Box Turtle; Amboina Box Turtle; Malayan Box Turtle (ISC); East Indian Box Turtle
d Amboina-Scharnierschildkröte
f Tortue-boîte de Malaisie; Tortue-boîte d'Asie orientale
i Tartaruga scatola malese

2421 ***Cuora amboinensis amboiensis***
Testudines - Emydidae
e Amboina Box Turtle
f Tortue-boîte d'Amboine

2422 ***Cuora amboinensis couro***
Testudines - Emydidae
f Tortue-boîte de la Sonde

2423 ***Cuora amboinensis kamaroma***
Testudines - Emydidae
e Thailand Box Turtle
f Tortue-boîte de Thaïlande

2424 ***Cuora aurocapitata***
Testudines - Emydidae
e Yellow-headed Box Turtle; Yellow-margined Box Turtle; Golden-headed Box Turtle
d Goldkopfscharnierschildkröte
f Tortue-boîte à tête jaune
i Tartaruga scatola dalla testa gialla

2425 ***Cuora flavomarginata***
Testudines - Emydidae
e Yellow-margined Box Turtle; Snake-eating Turtle; Common Yellow-margined Box Turtle; Chinese Box Turtle
d Gelbrandscharnierschildkröte
f Tortue-boîte à bords jaunes
i Tartaruga scatola cinese a bordi gialli

2426 ***Cuora flavomarginata evelynae***
Testudines - Emydidae
e Ryu-Kyu Yellow-margined Tortoise
f Tortue-boîte à bords jaunes des Ryu-Kyu

2427 ***Cuora flavomarginata flavomarginata***
Testudines - Emydidae
e Taiwan Yellow-margined Box Turtle
f Tortue-boîte à bords jaunes de Taïwan

2428 ***Cuora flavomarginata sinensis***
Testudines - Emydidae
e Chinese Yellow-margined Box Turtle
f Tortue-boîte à bords jaunes du Chine

2429 ***Cuora galbinifrons***
Testudines - Emydidae
e Indo-Chinese Box Turtle; Vietnam Box Turtle; Flowerback Box Turtle
d Südostasiaitische Scharnierschildkröte; Hinterindische Scharnierschildkröte
f Tortue-boîte à front jaune
i Tartaruga scatola indocinese

2430 ***Cuora mccordi***
Testudines - Emydidae
e McCord's Box Turtle
d McCords Scharnierschildkröte
f Tortue-boîte de McCord
i Tartaruga scatola di McCord

2431 ***Cuora pani***
Testudines - Emydidae
e Pan's Box Turtle
d Pans Scharnierschildkröte
f Tortue-boîte de Pan

i Tartaruga scatola cinese

2432 ***Cuora trifasciata***
Testudines - Emydidae
e Chinese Three-striped Box Turtle; Three-banded Box Turtle; Three-lined Box Turtle; Three-striped Box Turtle
d Dreistreifenscharnierschildkröte
f Tortue-boîte à trois bandes
i Tartaruga scatola dalle tre fasce

2433 ***Cuora yunnanensis***
Testudines - Emydidae
e Yunnan Box Turtle
d Yunnan-Scharnierschildkröte
f Tortue-boîte du Yunnan

2434 ***Cuora zhoui***
Testudines - Emydidae
e Zhou's Box Turtle; Pale-headed Box Turtle
d Zhous Scharnierschildkröte

2435 ***Cyclagras***
Squamata - Colubridae
e Water Cobras; Brazilian False Water Cobras
d Brasilianische Glanznattern

2436 ***Cyclagras gigas***
Squamata - Colubridae
e Brazilian Water Cobra; False Water Cobra; South American River Snake; Brazilian False Water Cobra
d Brasilianische Glanznatter; Falsche Wasserkobra; Brasilianische Glattnatter

2437 ***Cyclanorbis***
Testudines - Trionychidae
e Sub-Saharan Flapshell Turtles
d Mittelafrikanische Klappenweichschildkröten

2438 ***Cyclanorbis elegans***
Testudines - Trionychidae
e Nubian Flapshell Turtle; Nubian Soft-shelled Turtle
d Rückenfleckenweichschildkröte

2439 ***Cyclanorbis senegalensis***
Testudines - Trionychidae
e Senegal Flapshell Turtle; Senegal Soft-shelled Turtle
d Senegal-Weichschildkröte
f Trionyx à clapets du Sénégal

2440 ***Cyclemys***
Testudines - Emydidae
e Asian Leaf Turtles
d Dornschildkröten
f Cyclemmydes

2441 ***Cyclemys dentata***
Testudines - Emydidae
e Asian Leaf Turtle; Leaf Turtle; Garo and Khasi Hill Tortoise
d Malayische Dornschildkröte
f Cyclemmyde dentelée
i Tartaruga foglia asiatica

2442 ***Cyclemys tcheponensis***
Testudines - Emydidae
e Stripeneck Leaf Turtle; Stripe-necked Leaf Turtle
d Laos-Dornschildkröte; Indochinesische Dornschildkröte
i Tartaruga foglia dal collo striato

2443 ***Cyclocorus***
Squamata - Colubridae
e Lined Snakes

2444 ***Cyclocorus lineatus***
Squamata - Colubridae
e Reinhardt's Lined Snake

2445 ***Cycloderma***
Testudines - Trionychidae
e Central African Flapshell Turtles

d Westafrikanische Klappenweichschildkröten

2446 ***Cycloderma aubryi***
Testudines - Trionychidae
e Aubry's Flapshell Turtle; Aubry's Soft-shelled Turtle; West African Flapshell Turtle
d Rotrückige Klappenweichschildkröte
f Trionyx à clapets d'Aubry

2447 ***Cycloderma frenatum***
Testudines - Trionychidae
e Zambezi Flapshell Turtle; Zambezi Soft-shelled Turtle; Zambezi Soft-shelled Terrapin (CSA)
d Graurückige Klappenweichschildkröte
f Trionyx à clapets du Zambézie

2448 ***Cyclodina***
Squamata - Scincidae
e New Zealand Skinks

2449 ***Cyclodina aenea***
Squamata - Scincidae
e Girard's Skink

2450 ***Cyclodina alani***
Squamata - Scincidae
e Alan's Skink

2451 ***Cyclodina macgregori***
Squamata - Scincidae
e MacGregor's New Zealand Skink

2452 ***Cyclodina oliveri***
Squamata - Scincidae
e Oliver's New Zealand Skink

2453 ***Cyclodina ornata***
Squamata - Scincidae
e Gray's Ornate Skink

2454 ***Cyclodina whitakeri***
Squamata - Scincidae
e Whitaker's New Zealand Skink

2455 ***Cyclodomorphus***
Squamata - Scincidae
e Pale-throated Skinks

2456 ***Cyclodomorphus branchialis***
Squamata - Scincidae
e Lesser Pale-throated Skink

2457 ***Cyclodomorphus casuarinae***
Squamata - Scincidae
e Oak Skink
d Kasuarinen-Blauzungenskink

2458 ***Cyclodomorphus maxima***
Squamata - Scincidae
e Greater Pale-throated Skink

2459 ***Cyclophiops***
Squamata - Colubridae
e Green Snakes; Asian Green Snakes
d Asiatische Grünschlangen

2460 ***Cyclophiops doriae***
Squamata - Colubridae
e Doria's Green Snake

2461 ***Cyclophiops major***
Squamata - Colubridae
e South China Green Snake; Chinese Green Snake
d Chinesische Grünnatter

2462 ***Cyclophiops multicinctus***
Squamata - Colubridae
e North China Green Snake

2463 ***Cyclura***
Squamata - Iguanaidae
e Rhinoceros Iguanas
d Wirtelschwanzleguane; Nashornleguane

f Cyclures; Iguanes à cornes

2464 ***Cyclura carinata***
Squamata - Iguanaidae
e Turks Island Iguana; Mayaguana; Rock Iguana; Southern Bahamas Rock Iguana
d Turks-Caicos-Wirtelschwanzleguan; Wirtelschwanz; Türks-und-Caicos-Leguan

2465 ***Cyclura collei***
Squamata - Iguanaidae
e Jamaica Iguana; Jamaican Rock Iguana (WI)

2466 ***Cyclura cornuta***
Squamata - Iguanaidae
e Rhinoceros Iguana; Corned Iguana; Hispaniolan Rock Iguana ; Rhinoceros Rock Iguana
d Nashornleguan; Mona-Nashornleguan
f Iguane cornu; Iguane rhinocéros; Iguane à cornes

2467 ***Cyclura cyclura***
Squamata - Iguanaidae
e Exuma Island Ground Iguana; Allen Cays Rock Iguana; Northern Bahamas Rock Iguana

2468 ***Cyclura nubila***
Squamata - Iguanaidae
e Cayman Islands Ground Iguana; Cuban Ground Iguana ; Rock Iguana; Cuban Iguana; Cuban Rock Iguana (WI)
d Kuba-Wirtelschwanzleguan

2469 ***Cyclura nubila lewisi***
Squamata - Iguanaidae
e Grand Cayman Ground Iguana; Grand Cayman Iguana; Blue Rock Iguana (WI)

2470 ***Cyclura pinguis***
Squamata - Iguanaidae
e Anegada Ground Iguana; Anegada Rock Iguana
d Anegada-Leguan; Anegada-Wirtelschwanzleguan

2471 ***Cyclura ricordi***
Squamata - Iguanaidae
e Ricord's Ground Iguana; Banded Rock Agama
d Hispaniola-Leguan

2472 ***Cyclura rileyi***
Squamata - Iguanaidae
e White Cay Ground Iguana; Watling Island Ground Iguana; Central Bahamas Rock Iguana (WI)

2473 ***Cyclura stejnegeri***
Squamata - Iguanaidae
e Mona Rock Iguana

2474 ***Cylindrophis***
Squamata - Uropeltidae
e Island Pipe Snakes; Cylinder Snakes; Pipe Snakes; Asian Pipe Snakes; Oriental Pipe Snakes; Earth Snakes
d Walzenschlangen; Rollschlangen
f Serpents-tube tachetés
i Serpenti cilindrici

2475 ***Cylindrophis aruensis***
Squamata - Uropeltidae
e Aru Island Pipe Snake; Aru Cylinder Snake
d Aru-Walzenschlange

2476 ***Cylindrophis boulengeri***
Squamata - Uropeltidae
e Boulenger's Pipe Snake

2477 ***Cylindrophis isolepis***
Squamata - Uropeltidae
e Jampea Island Pipe Snake

2478 ***Cylindrophis lineatus***
Squamata - Uropeltidae
e Blanford's Pipe Snake; Striped Assam Pipe Snake; Blanford's Earth Snake
d Blanfords Rollschlange

2479 ***Cylindrophis maculatus***
Squamata - Uropeltidae
e Spotted Pipe Snake; Ceylonese Cylinder Snake; Sri Lankan Pipe Snake
d Ceylon-Walzenschlange; Gefleckte Warzenschlange
f Serpent-tube tacheté

2480 ***Cylindrophis melanotus***
Squamata - Uropeltidae
e Black Pipe Snake

2481 ***Cylindrophis opisthorhodus***
Squamata - Uropeltidae
e Island Pipe Snake

2482 ***Cylindrophis rufus***
Squamata - Uropeltidae
e Red Pipe Snake; Red Cylinder Snake; Red-tailed Pipe Snake; Malayan Pipe Snake
d Rote Walzenschlange; Rotschwanzrollschlange; Korallenrollschlange
f Serpent à deux têtes; Tête-double; Queue-double; Serpent double-tête
i Cilindrofide rosso

2483 ***Cynisca***
Squamata - Amphisbaenidae
e West African Worm Lizards

2484 ***Cynisca bifrontalis***
Squamata - Amphisbaenidae
e French Congo Worm Lizard

2485 ***Cynisca degrysi***
Squamata - Amphisbaenidae
e Sierra Leone Worm Lizard

2486 ***Cynisca feae***
Squamata - Amphisbaenidae
e Ugly Worm Lizard

2487 ***Cynisca haughi***
Squamata - Amphisbaenidae
e Haugh's Worm Lizard

2488 ***Cynisca kraussi***
Squamata - Amphisbaenidae
e Ghana Worm Lizard

2489 ***Cynisca lamottei***
Squamata - Amphisbaenidae
e Guinea Worm Lizard

2490 ***Cynisca leonina***
Squamata - Amphisbaenidae
e Los Archipelago Worm Lizard

2491 ***Cynisca leucura***
Squamata - Amphisbaenidae
e Coast Worm Lizard

2492 ***Cynisca liberiensis***
Squamata - Amphisbaenidae
e Liberia Worm Lizard

2493 ***Cynisca mulleri***
Squamata - Amphisbaenidae
e Muller's Worm Lizard

2494 ***Cynisca oligopholis***
Squamata - Amphisbaenidae
e Cassine River Worm Lizard

2495 ***Cynisca schaeferi***
Squamata - Amphisbaenidae
e Cameroon Worm Lizard

2496 ***Cyrtodactylus***
Squamata - Gekkonidae

e Bow-fingered Geckos; Naked-fingered Geckos
d Bogenfinger; Nacktfingergeckos; Bodenfingergeckos
f Cyrtodactyles

2497 ***Cyrtodactylus aguasanensis***
Squamata - Gekkonidae
e Mindanao Bow-fingered Gecko

2498 ***Cyrtodactylus angularis***
Squamata - Gekkonidae
e Angulated Bow-fingered Gecko

2499 ***Cyrtodactylus annulatus***
Squamata - Gekkonidae
e Annulated Bow-fingered Gecko

2500 ***Cyrtodactylus baluensis***
Squamata - Gekkonidae
e Balu Bow-fingered Gecko

2501 ***Cyrtodactylus biordinis***
Squamata - Gekkonidae
e Guadalcanal Bow-fingered Gecko

2502 ***Cyrtodactylus brevipalmatus***
Squamata - Gekkonidae
e Kampuchea Bow-fingered Gecko

2503 ***Cyrtodactylus cavernicolus***
Squamata - Gekkonidae
e Sarawak Bow-fingered Gecko

2504 ***Cyrtodactylus chitralensis***
Squamata - Gekkonidae
e Pakistan Bow-fingered Gecko

2505 ***Cyrtodactylus condorensis***
Squamata - Gekkonidae
e Pulo Condore Bow-fingered Gecko

2506 ***Cyrtodactylus consobrinoide***
Squamata - Gekkonidae
e Cousin Bow-fingered Gecko

2507 ***Cyrtodactylus consobrinus***
Squamata - Gekkonidae
e Peters's Bow-fingered Gecko
d Gebänderter Bogenfinger

2508 ***Cyrtodactylus darmandvillei***
Squamata - Gekkonidae
e D'Armandville Bow-fingered Gecko

2509 ***Cyrtodactylus dattanensis***
Squamata - Gekkonidae
e Khan's Bow-fingered Gecko

2510 ***Cyrtodactylus derongo***
Squamata - Gekkonidae
e Derongo Bow-fingered Gecko

2511 ***Cyrtodactylus deveti***
Squamata - Gekkonidae
e Moluccan Bow-fingered Gecko

2512 ***Cyrtodactylus elok***
Squamata - Gekkonidae
e Malaysia Bow-fingered Gecko

2513 ***Cyrtodactylus fasciolatus***
Squamata - Gekkonidae
e Striped Bow-fingered Gecko

2514 ***Cyrtodactylus feae***
Squamata - Gekkonidae
e Ugly Bow-fingered Gecko

2515 ***Cyrtodactylus fraenatus***
Squamata - Gekkonidae
e Sri Lanka Bow-fingered Gecko

2516 ***Cyrtodactylus fumosus***
Squamata - Gekkonidae
e Sulawesi Bow-fingered Gecko

2517 ***Cyrtodactylus gubernatoris***
Squamata - Gekkonidae
e Navigator Bow-fingered Gecko

2518 ***Cyrtodactylus himalayensis***
Squamata - Gekkonidae
e Himalaya Bow-fingered Gecko

2519 ***Cyrtodactylus indusoani***
Squamata - Gekkonidae
e Punjab Bow-fingered Gecko

2520 ***Cyrtodactylus ingeri***
Squamata - Gekkonidae
e Sabah Bow-fingered Gecko

2521 ***Cyrtodactylus intermedius***
Squamata - Gekkonidae
e Intermediate Bow-fingered Gecko
d Zimtfarbener Bogenfinger; Kleiner Bogenfinger

2522 ***Cyrtodactylus irregularis***
Squamata - Gekkonidae
e Irregular Bow-fingered Gecko

2523 ***Cyrtodactylus jellesmae***
Squamata - Gekkonidae
e Kabena Bow-fingered Gecko

2524 ***Cyrtodactylus khasiensis***
Squamata - Gekkonidae
e Assam Bow-fingered Gecko

2525 ***Cyrtodactylus khirmanensis***
Squamata - Gekkonidae
e Nikolsky's Bow-fingered Gecko

2526 ***Cyrtodactylus laevigatus***
Squamata - Gekkonidae
e Komodo Bow-fingered Gecko

2527 ***Cyrtodactylus lateralis***
Squamata - Gekkonidae
e Sumatra Bow-fingered Gecko

2528 ***Cyrtodactylus lawderanus***
Squamata - Gekkonidae
e Stoliczka's Bow-fingered Gecko; Lawder's Ground Gecko

2529 ***Cyrtodactylus loriae***
Squamata - Gekkonidae
e Boulenger's Bow-fingered Gecko

2530 ***Cyrtodactylus louisiadensis***
Squamata - Gekkonidae
e Ringtail Bow-fingered Gecko; Ring-tailed Gecko (ANZ); Louisiades Bow-fingered Gecko
d Bänderschwanz-Nacktfinger

2531 ***Cyrtodactylus malayanus***
Squamata - Gekkonidae
e Borneo Bow-fingered Gecko

2532 ***Cyrtodactylus malcolmsmithi***
Squamata - Gekkonidae
e Malcolm's Bow-fingered Gecko

2533 ***Cyrtodactylus mansarulus***
Squamata - Gekkonidae
e Jammu Bow-fingered Gecko

2534 ***Cyrtodactylus marmoratus***
Squamata - Gekkonidae
e Marbled Bow-fingered Gecko
f Gymnodactyle marbré

2535 ***Cyrtodactylus matsuii***
Squamata - Gekkonidae
e Hikida Bow-fingered Gecko

2536 ***Cyrtodactylus medongensis***
Squamata - Gekkonidae
e Medog Bow-fingered Gecko

2537 ***Cyrtodactylus minikanus***
Squamata - Gekkonidae

e False Bow-fingered Gecko

2538 ***Cyrtodactylus mintoni***
Squamata - Gekkonidae
e Minton's Bow-fingered Gecko

2539 ***Cyrtodactylus novaeguinea***
Squamata - Gekkonidae
e New Guinea Bow-fingered Gecko

2540 ***Cyrtodactylus oldhami***
Squamata - Gekkonidae
e Odham's Bow-fingered Gecko

2541 ***Cyrtodactylus papuensis***
Squamata - Gekkonidae
e Papua Bow-fingered Gecko

2542 ***Cyrtodactylus peguensis***
Squamata - Gekkonidae
e Thai Bow-fingered Gecko
d Augenfleck-Bogenfinger; Gefleckter Krallengecko

2543 ***Cyrtodactylus philippinicus***
Squamata - Gekkonidae
e Philippine Bow-fingered Gecko

2544 ***Cyrtodactylus pubisculus***
Squamata - Gekkonidae
e Inger's Bow-fingered Gecko

2545 ***Cyrtodactylus pulchellus***
Squamata - Gekkonidae
e Beautiful Bow-fingered Gecko; Beautiful Bent-toed Gecko
d Malayischer Bogenfinger; Bändergecko
f Gymnodtyle gentil

2546 ***Cyrtodactylus quadrivirgatus***
Squamata - Gekkonidae
e Taylor's Bow-fingered Gecko

2547 ***Cyrtodactylus redimiculus***
Squamata - Gekkonidae
e Palawan Bow-fingered Gecko

2548 ***Cyrtodactylus rohtasfortai***
Squamata - Gekkonidae
e Azad Bow-fingered Gecko

2549 ***Cyrtodactylus rubidus***
Squamata - Gekkonidae
e Red Bow-fingered Gecko; Bent-toed Gecko (ISC)

2550 ***Cyrtodactylus sadleiri***
Squamata - Gekkonidae
e Sadleir's Bow-fingered Gecko

2551 ***Cyrtodactylus sermowaiensis***
Squamata - Gekkonidae
e De Rooij's Bow-fingered Gecko

2552 ***Cyrtodactylus stoliczkai***
Squamata - Gekkonidae
e Frontier Bow-fingered Gecko; Karakorum Rock Gecko

2553 ***Cyrtodactylus sworderi***
Squamata - Gekkonidae
e Johore Bow-fingered Gecko

2554 ***Cyrtodactylus tibetinus***
Squamata - Gekkonidae
e Tibetan Bow-fingered Gecko

2555 ***Cyrtodactylus variegatus***
Squamata - Gekkonidae
e Variegated Bow-fingered Gecko

2556 ***Cyrtodactylus wetariensis***
Squamata - Gekkonidae
e Wetar Bow-fingered Gecko

2557 ***Cyrtodactylus yoshii***
Squamata - Gekkonidae

e Yoshi's Bow-fingered Gecko

2558 ***Cyrtopodion***
Squamata - Gekkonidae
e Bowfoot Geckos

2559 ***Cyrtopodion agamuroides***
Squamata - Gekkonidae
e Nikolsky's Gecko

2560 ***Cyrtopodion basoglui***
Squamata - Gekkonidae
f Cyrtopodion de Basoglu
i Geco di Basoglu

2561 ***Cyrtopodion elongatus***
Squamata - Gekkonidae
e Yangihissar Gecko

2562 ***Cyrtopodion kachhensis***
Squamata - Gekkonidae
e Kachh Gecko; Kutch Gecko

2563 ***Cyrtopodion kachhensis kachhensis***
Squamata - Gekkonidae
e Warty Rock Gecko (ISC)

2564 ***Cyrtopodion montiumsalsorum***
Squamata - Gekkonidae
e Salt Range Gecko

2565 ***Cyrtopodion scabrum***
Squamata - Gekkonidae
e Roughtail Gecko; Keeled Rock Gecko; Rough-tailed Gecko (NA)
d Krokodilgecko
f Cyrtopodion rugueux
i Geco scabro

2566 ***Cyrtopodion watsoni***
Squamata - Gekkonidae
e Watson's Gecko

D

2567 ***Daboia***
Squamata - Viperidae
e Mid-East Vipers
d Mittelasiatische Ottern

2568 ***Daboia mauritanica***
Squamata - Viperidae
e Morocco Viper; Moorish Viper
d Atlas-Otter
f Vipère de l'Atlas; Vipère de Maurétanie

2569 ***Daboia palaestinae***
Squamata - Viperidae
e Palestine Viper
d Palästina-Viper
f Vipère de Palestine

2570 ***Daboia raddei***
Squamata - Viperidae
e Caucasus Viper; Russian Viper; Radde's Viper; Armenian Viper; Radde's Rock Viper
d Armenische Bergotter; Raddes Otter; Radde-Viper
f Vipère de Radde; Vipère d'Arménie
i Vipera di Radde

2571 ***Daboia xanthina***
Squamata - Viperidae
e Ottoman Viper; Coastal Viper; Rock Viper; Mountain Viper
d Bergotter; Kleinasiatische Bergotter
f Vipère d'Asie mineure; Vipère ottomane; Vipère montagnarde
i Vipera ottomana; Vipera di Xanto

2572 ***Darlingtonia***
Squamata - Colubridae
e Ground Snakes

2573 ***Darlingtonia haetiana***
Squamata - Colubridae
e Haitian Ground Snake; Montane Ground Snake

2574 ***Dasia***
Squamata - Scincidae
e Dasias; Tree Skinks
d Baumskinke

2575 ***Dasia griffini***
Squamata - Scincidae
e Griffin's Dasia

2576 ***Dasia grisea***
Squamata - Scincidae
e Grey Dasia
d Brauner Baumskink

2577 ***Dasia haliana***
Squamata - Scincidae
e Ceylonese Dasia
d Ceylon-Baumskink

2578 ***Dasia olivacea***
Squamata - Scincidae
e Olive Dasia; Emerald Skink
d Olivfarbener Baumskink; Getüpfelter Smaragdskink

2579 ***Dasia semicincta***
Squamata - Scincidae
e Peters's Dasia

2580 ***Dasia subcaerulea***
Squamata - Scincidae
e Boulenger's Dasia

2581 ***Dasypeltis***
Squamata - Colubridae

e Egg-eating Snakes; African Egg-eating Snakes; Egg-eaters (CSA)
d Afrikanische Eierschlangen
f Serpents mangeurs d'oeufs; Mangeurs d'oeufs

2582 ***Dasypeltis atra***
Squamata - Colubridae
e African Egg-eating Snake

2583 ***Dasypeltis fasciata***
Squamata - Colubridae
e Western Forest Egg-eating Snake; Central African Egg-eating Snake; Western Forest Egg-eater
d Gestreifte Eierschlange

2584 ***Dasypeltis inornata***
Squamata - Colubridae
e Brown Egg-eating Snake (CSA); Southern Brown Egg-eating Snake; Southern Brown Egg-eater (CSA); Brown Egg-eater
d Braune Eierschlange

2585 ***Dasypeltis medici***
Squamata - Colubridae
e East African Egg-eating Snake; East African Egg-eater (CSA); Bianconi's Egg-eater; Eastern Egg-eater (CSA); Rufous Egg-eater (CSA)

2586 ***Dasypeltis scabra***
Squamata - Colubridae
e Common Egg-eating Snake (CSA); Common Rhombic Egg-eater; Rhombic Egg-eater (CSA); Common Egg-eater (CSA); African Egg-eating Snake; Egg-eating Snake
d Gewöhnliche Eierschlange; Eierschlange; Afrikanische Eierschlange
f Serpent mangeur d'oeufs; Couleuvre mangeuse d'oeufs
i Mangiatore d'uova

2587 ***Davewakeum***
Squamata - Scincidae
e Miriam's Skinks

2588 ***Davewakeum miriamae***
Squamata - Scincidae
e Miriam's Skink

2589 ***Deinagkistrodon***
Squamata - Viperidae
e Hundred-pace Vipers
d Chinesische Nasenottern

2590 ***Deinagkistrodon acutus***
Squamata - Viperidae
e Hundred-pace Viper; Chinese Moccasin; Chinese Copperhead; Sharp-nosed Pit Viper; Hundred-pacer; Chinese Sharp-nosed Viper
d Chinesische Nasenotter

2591 ***Deirochelys***
Testudines - Emydidae
e Chicken Turtles
d Langhalsschmuckschildkröten
f Tortues reticulées; Déirochélydes

2592 ***Deirochelys reticularia***
Testudines - Emydidae
e Chicken Turtle
d Langhalsschmuckschildkröte
f Tortue reticulée; Déirochélyde réticulaire; Émyde réticulée
i Testuggine reticolata

2593 ***Deirochelys reticularia chrysea***
Testudines - Emydidae
e Florida Chicken Turtle
f Émyde réticulée de Floride

2594 ***Deirochelys reticularia miaria***
Testudines - Emydidae
e Western Chicken Turtle
f Émyde réticulée de l'Ouest

2595 ***Deirochelys reticularia reticularia***
Testudines - Emydidae
e Eastern Chicken Turtle
f Émyde réticulée de l'Est

2596 ***Delma***
Squamata - Gekkonidae
e Flapped Scalyfoots; Flapped Geckos
d Glattschuppige Flossenfüße

2597 ***Delma australis***
Squamata - Gekkonidae
e Australian Scalyfoot; Australian Gecko; Southern Legless Lizard (ANZ)

2598 ***Delma borea***
Squamata - Gekkonidae
e Black-banded Scalyfoot; Black-banded Gecko

2599 ***Delma butleri***
Squamata - Gekkonidae
e Butler's Scalyfoot; Butler's Gecko; Butler's Legless Lizard (ANZ)

2600 ***Delma elegans***
Squamata - Gekkonidae
e Elegant Scalyfoot; Elegant Gecko

2601 ***Delma fraseri***
Squamata - Gekkonidae
e Fraser's Scalyfoot; Fraser's Gecko
d Frasers Flossenfuß

2602 ***Delma grayii***
Squamata - Gekkonidae
e Gray's Scalyfoot; Gray's Gecko
d Grays Flossenfuß

2603 ***Delma haroldi***
Squamata - Gekkonidae
e Harold's Scalyfoot; Harold's Gecko

2604 ***Delma impar***
Squamata - Gekkonidae
e Victoria Scalyfoot; Victoria Gecko; Striped Legless Skink (ANZ)

2605 ***Delma inornata***
Squamata - Gekkonidae
e Unadorned Scalyfoot; Unadorned Gecko; Olive Legless Lizard (ANZ)
d Schmuckloser Flossenfuß

2606 ***Delma labialis***
Squamata - Gekkonidae
e Yellow-lipped Scalyfoot; Yellow-lipped Gecko

2607 ***Delma molleri***
Squamata - Gekkonidae
e Southern Scalyfoot; Southern Gecko

2608 ***Delma nasuta***
Squamata - Gekkonidae
e Northern Scalyfoot; Northern Gecko

2609 ***Delma pax***
Squamata - Gekkonidae
e Hamersley Scalyfoot; Hamersley Gecko

2610 ***Delma plebia***
Squamata - Gekkonidae
e Grey Scalyfoot; Grey Gecko

2611 ***Delma tincta***
Squamata - Gekkonidae
e Dyed Scalyfoot; Dyed Gecko
d Gemalter Flossenfuß

2612 ***Delma torquata***
Squamata - Gekkonidae
e Collared Scalyfoot; Collared Gecko

2613 ***Demansia***
Squamata - Elapidae
e Venomous Whip Snakes; Whip Snakes; Brown Snakes; Australian

Brown Snakes
d Peitschenschlangen; Braunschlangen
f Serpents jaunes

2614 ***Demansia atra***
Squamata - Elapidae
e Black Whip Snake
d Schwarze Peitschenotter

2615 ***Demansia calodera***
Squamata - Elapidae
e Calodera Whip Snake

2616 ***Demansia olivacea***
Squamata - Elapidae
e Olive Whip Snake
d Olive Peitschenotter

2617 ***Demansia papuensis***
Squamata - Elapidae
e Papua Whip Snake

2618 ***Demansia psammophis***
Squamata - Elapidae
e Yellow-faced Whip Snake
d Gelbkopfpeitschenschlange; Gelbkopfbraunschlange

2619 ***Demansia rufescens***
Squamata - Elapidae
e Storr's Whip Snake

2620 ***Demansia simplex***
Squamata - Elapidae
e Woodland Whip Snake

2621 ***Demansia torquata***
Squamata - Elapidae
e Collared Whip Snake
d Halsbandpeitschenotter

2622 ***Demansia vestigiata***
Squamata - Elapidae
e De Vis's Whip Snake

2623 ***Dendragama***
Squamata - Agamidae
e Tree Agamas; Sumatran Tree Dragons
d Baumagamen
f Dendragames

2624 ***Dendragama boulengeri***
Squamata - Agamidae
e Boulenger's Tree Agama; Sumatran Tree Dragon
f Dendragame de Boulenger

2625 ***Dendrelaphis***
Squamata - Colubridae
e Bronzebacks; Bronzeback Snakes; Bronze-backed Snakes
d Bronzenattern

2626 ***Dendrelaphis bifrenalis***
Squamata - Colubridae
e Boulenger's Bronzeback; Boulenger's Bronzeback Snake

2627 ***Dendrelaphis calligaster***
Squamata - Colubridae
e Northern Bronzeback; Northern Bronzeback Snake; Northern Tree Snake (ANZ)
d Nördliche Bronzenatter

2628 ***Dendrelaphis caudolineatus***
Squamata - Colubridae
e Gray's Bronzeback; Gray's Bronzeback Snake

2629 ***Dendrelaphis caudolineolatus***
Squamata - Colubridae
e Striped Bronzeback; Striped Bronzeback Snake

2630 ***Dendrelaphis cyanochloris***
Squamata - Colubridae
e Wall's Bronzeback; Wall's Bronzeback Snake; Blue-green Bronzeback Snake

2631 ***Dendrelaphis formosus***
Squamata - Colubridae
e Elegant Bronzeback; Elegant Bronzeback Snake

2632 ***Dendrelaphis gastrostictus***
Squamata - Colubridae
e New Guinea Bronzeback; New Guinea Bronzeback Snake

2633 ***Dendrelaphis gorei***
Squamata - Colubridae
e Gore's Bronzeback; Gore's Bronzeback Snake

2634 ***Dendrelaphis grandoculis***
Squamata - Colubridae
e Southern Bronzeback; Southern Bronzeback Snake

2635 ***Dendrelaphis humayuni***
Squamata - Colubridae
e Tiwari's Bronzeback; Tiwari's Bronzeback Snake

2636 ***Dendrelaphis lorentzi***
Squamata - Colubridae
e Lorentz's Bronzeback; Lorentz's Bronzeback Snake

2637 ***Dendrelaphis oliveri***
Squamata - Colubridae
e Oliver's Bronzeback; Oliver's Bronzeback Snake

2638 ***Dendrelaphis papuensis***
Squamata - Colubridae
e Papua Bronzeback; Papua Bronzeback Snake

2639 ***Dendrelaphis pictus***
Squamata - Colubridae
e Gmelin's Bronzeback; Gmelin's Bronzeback Snake; Indonesian Bronzeback; Eastern Bronzeback; Indo-Malayan Bronzeback; Painted Tree Snake; Common Bronzeback; Painted Bronzeback
d Gewöhnliche Bronzenatter; Gemeine Bronzenatter

2640 ***Dendrelaphis punctulatus***
Squamata - Colubridae
e Common Bronzeback; Common Bronzeback Snake; Common Tree Snake (ANZ); Green Tree Snake

2641 ***Dendrelaphis salmonis***
Squamata - Colubridae
e Günther's Bronzeback; Günther's Bronzeback Snake

2642 ***Dendrelaphis subocularis***
Squamata - Colubridae
e Mountain Bronzeback; Mountain Bronzeback Snake

2643 ***Dendrelaphis tristis***
Squamata - Colubridae
e Daudin's Bronzeback; Daudin's Bronzeback Snake; Seba's Bronzeback; Indian Bronzeback; Common Indian Bronzeback (ISC); Common Indian Tree Snake (ISC)

2644 ***Dendroaspis***
Squamata - Elapidae
e Mambas
d Mambas
f Mambas; Mambas rouge-et-verts
i Mambe

2645 ***Dendroaspis angusticeps***
Squamata - Elapidae
e Mamba; Common Mamba; Eastern Green Mamba (CSA); East African Green Mamba; Green Mamba (CSA); South African Mamba
d Mamba; Blattgrüne Mamba; Gewöhnliche Mamba; Ostafrikanische Grüne Mamba
f Mamba noir-et-vert
i Mamba verde

2646 ***Dendroaspis jamesoni***
Squamata - Elapidae
e Jameson's Mamba
d Jameson-Mamba; Jamesons Mamba
f Mamba de Jameson

2647 ***Dendroaspis polylepis***
Squamata - Elapidae
e Black Mamba; Brown Mamba (CSA); Black-mouthed Mamba (CSA)
d Schwarze Mamba; Schwarzmamba
f Mamba noir
i Mamba nero

2648 ***Dendroaspis viridis***
Squamata - Elapidae
e Green Mamba; Western Green Mamba; West African Green Mamba
d Grüne Mamba; Westafrikanische Grüne Mamba
f Mamba vert; Serpent de bananier
i Mamba verde

2649 ***Dendrolycus***
Squamata - Colubridae
e Rainforest Snakes

2650 ***Dendrolycus elapoides***
Squamata - Colubridae
e Cameroon Rainforest Snake

2651 ***Dendrophidion***
Squamata - Colubridae
e Forest Racers

2652 ***Dendrophidion bivittatus***
Squamata - Colubridae
e Forest Racer

2653 ***Dendrophidion boshelli***
Squamata - Colubridae
e Boshell's Forest Racer

2654 ***Dendrophidion brunneus***
Squamata - Colubridae
e Günther's Forest Racer

2655 ***Dendrophidion clarkii***
Squamata - Colubridae
e Clark's Forest Racer

2656 ***Dendrophidion dendrophis***
Squamata - Colubridae
e Olive Forest Racer; Common Forest Racer

2657 ***Dendrophidion nuchalis***
Squamata - Colubridae
e Peters's Forest Racer; Three-coloured Forest Racer

2658 ***Dendrophidion paucicarinatum***
Squamata - Colubridae
e Cope's Forest Racer; Cloud Forest Racer

2659 ***Dendrophidion percarinatum***
Squamata - Colubridae
e South American Forest Racer; Brown Forest Racer

2660 ***Dendrophidion vinitor***
Squamata - Colubridae
e Barred Forest Racer

2661 ***Denisonia***
Squamata - Elapidae
e Ornamental Snakes; Banded Snakes; Australian Copperheads
d Bänderottern; Ornamentschlangen

2662 ***Denisonia devisi***
Squamata - Elapidae
e De Vis's Banded Snake
d De Vis-Bänderotter

2663 ***Denisonia maculata***
Squamata - Elapidae
e Ornamental Snake

d Ornament-Bänderotter

2664 Dermatemydidae
Testudines
e Central American River Turtles; River Turtles; Tabasco Turtles
d Tabasco-Schildkröten
f Dermatémydidés; Tortues de Tabasco

2665 *Dermatemys*
Testudines - Dermatemydidae
e Central American River Turtles
d Tabasco-Schildkröten
f Dermatémydes

2666 *Dermatemys mawii*
Testudines - Dermatemydidae
e Central American River Turtle; Tabasco Turtle
d Tabasco-Schildkröte
f Dermatémyde de Mawe; Tortue de Tababasco
i Dermatemide di fiume

2667 Dermochelyidae
Testudines
e Leatherback Turtles; Leathery Turtles (ANZ); Leatherbacks
d Lederschildkröten; Meeresschildkröten
f Tortues marines; Dermochélyidés; Tortues-Luths; Tortues cuir
i Sfargide; Tartarughe liuto

2668 *Dermochelys*
Testudines - Dermochelyidae
e Leatherback Turtles; Leatherback Sea Turtles; Leatherbacks
d Lederschildkröten
f Tortues-luth; Tortues-cuir

2669 *Dermochelys coriacea*
Testudines - Dermochelyidae
e Leatherback Turtle; Leatherback Sea Turtle; Leathery Turtle (ANZ); Luth Turtle (ANZ); Luth; Trunk Turtle; Leatherback
d Lederschildkröte; Luth
f Bataklin (Ants); Tortue-luth; Tortue cuir; Luth
i Tartaruga liuto; Sfargide; Sfargide coriacea; Dermochelide coriacea; Grande sfargide

2670 *Dermochelys coriacea coriacea*
Testudines - Dermochelyidae
e Atlantic Leatherback; Leatherback; Marine Leatherback; Leathery Turtle

2671 *Dermochelys coriacea schlegeli*
Testudines - Dermochelyidae
e Pacific Leatherback Sea Turtle

2672 *Diadophis*
Squamata - Colubridae
e Ringneck Snakes; Ring-necked Snakes
d Ringhalsnattern; Halsbandnattern
f Couleuvres à collier

2673 *Diadophis punctatus*
Squamata - Colubridae
e Ringneck Snake; Ring-necked Snake; Southern Ring-necked Snake; Eastern Ringneck Snake
d Ringhalsnatter; Halsbandnatter; Halsringnatter
f Couleuvre à collier

2674 *Diadophis punctatus acricus*
Squamata - Colubridae
e Key Ringneck Snake; Key Ring-necked Snake

2675 *Diadophis punctatus amabilis*
Squamata - Colubridae
e Pacific Ringneck Snake; Pacific Ring-necked Snake; California Ring-necked Snake (NA); Western Ring-necked Snake (NA); Red-bellied Snake (NA); Western Ringneck
d Halsbandnatter

2676 ***Diadophis punctatus anthonyi***
Squamata - Colubridae
e Todos Santos Island Ringneck Snake; Todos Santos Island Ring-necked Snake

2677 ***Diadophis punctatus arnyi***
Squamata - Colubridae
e Prairie Ringneck Snake; Prairie Ring-necked Snake

2678 ***Diadophis punctatus edwardsii***
Squamata - Colubridae
e Northern Ringneck Snake; Northern Ring-necked Snake

2679 ***Diadophis punctatus modestus***
Squamata - Colubridae
e San Bernardino Ringneck Snake; San Bernardino Ring-necked Snake

2680 ***Diadophis punctatus occidentalis***
Squamata - Colubridae
e Northwestern Ringneck Snake; Northwestern Ring-necked Snake; Ring-necked Snake

2681 ***Diadophis punctatus pulchellus***
Squamata - Colubridae
e Coralbelly Ringneck Snake; Coralbelly Ring-necked Snake; Coral-bellied Ringneck Snake; Coral-bellied ring-necked Sn ake

2682 ***Diadophis punctatus punctatus***
Squamata - Colubridae
e Southern Ringneck Snake; Southern Ring-necked Snake
f Couleuvre à collier

2683 ***Diadophis punctatus regalis***
Squamata - Colubridae
e Regal Ringneck Snake; Regal Ring-necked Snake; Sonoran Ring-necked Snake

2684 ***Diadophis punctatus similis***
Squamata - Colubridae
e San Diego Ringneck Snake; San Diego Ring-necked Snake

2685 ***Diadophis punctatus stictogenys***
Squamata - Colubridae
e Mississippi Ringneck Snake; Mississippi Ring-necked Snake

2686 ***Diadophis punctatus vandenburghi***
Squamata - Colubridae
e Monterey Ringneck Snake; Monterey Ring-necked Snake

2687 ***Diaphorolepis***
Squamata - Colubridae
e Frog-eating Snakes

2688 ***Diaphorolepis laevis***
Squamata - Colubridae
e Colombian Frog-eating Snake

2689 ***Diaphorolepis wagneri***
Squamata - Colubridae
e Ecuador Frog-eating Snake; Wagner's Frog-eating Snake

2690 **Dibamidae**
Squamata
e Blind Lizards; Wormlike Lizards; Blind Skinks; Dibamids
d Schlangenschleichen
f Dibamidés

2691 ***Dibamus***
Squamata - Dibamidae
e Blind Skinks; Dibamid Lizards
d Schlangenschleichen
f Dibames

2692 ***Dibamus alfredi***
Squamata - Dibamidae
e Alfred's Blind Skink; Alfred's Dibame
d Alfreds Schlangenschleiche

2693 ***Dibamus bogadeki***
Squamata - Dibamidae
e Bogadek's Blind Skink
f Dibame

2694 ***Dibamus bourreti***
Squamata - Dibamidae
e Bourret's Blind Skink; Bourret's Dibame
d Bourrets Schlangenschleiche
f Dibame de Bourret

2695 ***Dibamus celebensis***
Squamata - Dibamidae
e Sulawesi Blind Skink

2696 ***Dibamus greeri***
Squamata - Dibamidae
e Greer's Blind Skink

2697 ***Dibamus leucurus***
Squamata - Dibamidae
e White Blind Skink

2698 ***Dibamus montanus***
Squamata - Dibamidae
e Mountain Blind Skink

2699 ***Dibamus novaeguinea***
Squamata - Dibamidae
e New Guinea Blind Skink; New Guinea Dibame
d Neuguinea-Schlangenschleiche

2700 ***Dibamus seramensis***
Squamata - Dibamidae
e Seram Blind Skink

2701 ***Dibamus smithi***
Squamata - Dibamidae
e Smith's Blind Skink

2702 ***Dibamus taylori***
Squamata - Dibamidae
e Taylor's Blind Skink

2703 ***Dicrodon***
Squamata - Teiidae
e Desert Tegus
d Wüstentejus

2704 ***Dicrodon guttulatum***
Squamata - Teiidae
e Peru Desert Tegu

2705 ***Dicrodon heterolepis***
Squamata - Teiidae
e Ecuador Desert Tegu

2706 ***Dicrodon holmbergi***
Squamata - Teiidae
e Holmberg's Desert Tegu

2707 ***Dinodon***
Squamata - Colubridae
e Big-tooth Snakes
d Großnattern; Großzahnnattern

2708 ***Dinodon flavozonatus***
Squamata - Colubridae
e Big-tooth Snake

2709 ***Dinodon gammiei***
Squamata - Colubridae
e Gammie's Wolf Snake

2710 ***Dinodon orientalis***
Squamata - Colubridae
e Oriental Big-tooth Snake; Oriental Odd-tooth Snake

2711 ***Dinodon rosozonatum***
Squamata - Colubridae
e Rose Big-tooth Snake
d Großzahnnatter

2712 ***Dinodon rufozonatum***
Squamata - Colubridae

e Red-banded Snake; Red-banded Wolf Snake

2713 ***Dinodon rufozonatum rufozonatum***
Squamata - Colubridae
e Red-banded Odd-tooth Snake

2714 ***Dinodon rufozonatum walli***
Squamata - Colubridae
e Sakishima Odd-tooth Snake

2715 ***Dinodon semicarinatum***
Squamata - Colubridae
e Loo-choo Big-tooth Snake; Ryuku Odd-tooth Snake

2716 ***Dinodon septentrionalis***
Squamata - Colubridae
e White-banded Big-tooth Snake; White-banded Wolf Snake

2717 ***Diplodactylus***
Squamata - Gekkonidae
e Spinytail Geckos; Dyed Geckos
d Stachelschwanzgeckos; Australische Geckos

2718 ***Diplodactylus alboguttatus***
Squamata - Gekkonidae
e White Spotted; Collared Gecko

2719 ***Diplodactylus assimilis***
Squamata - Gekkonidae
e Grey Spinytail Gecko

2720 ***Diplodactylus byrnei***
Squamata - Gekkonidae
e Byrne's Gecko

2721 ***Diplodactylus ciliaris***
Squamata - Gekkonidae
e Spinytail Gecko; Spiny-tailed Gecko (ANZ); Northern Spiny-tailed Gecko
d Stachelschwanzgecko

2722 ***Diplodactylus conspicillatus***
Squamata - Gekkonidae
e Remarkable Gecko; Fat-tailed Diplodactylus (ANZ)

2723 ***Diplodactylus damaeus***
Squamata - Gekkonidae
e Deer Gecko; Beaded Gecko (ANZ)

2724 ***Diplodactylus elderi***
Squamata - Gekkonidae
e Jewelled Gecko
d Juwelengecko

2725 ***Diplodactylus fulleri***
Squamata - Gekkonidae
e Lake Disappointment Gecko

2726 ***Diplodactylus furcosus***
Squamata - Gekkonidae
e Forked Gecko

2727 ***Diplodactylus galeatus***
Squamata - Gekkonidae
e Kluge's Helmeted Gecko

2728 ***Diplodactylus granariensis***
Squamata - Gekkonidae
e Pilbara Gecko

2729 ***Diplodactylus immaculatus***
Squamata - Gekkonidae
e Immaculate Gecko

2730 ***Diplodactylus intermedius***
Squamata - Gekkonidae
e Eastern Spinytail Gecko; Eastern Spiny-tailed Gecko (ANZ)

2731 ***Diplodactylus jeanae***
Squamata - Gekkonidae
e Jean's Spinytail Gecko

2732 ***Diplodactylus kenneallyi***
Squamata - Gekkonidae
e Kenneally's Gecko

2733 ***Diplodactylus maini***
Squamata - Gekkonidae
e Australian Woodland Gecko

2734 ***Diplodactylus mcmillani***
Squamata - Gekkonidae
e McMillan's Spinytail Gecko

2735 ***Diplodactylus michaelseni***
Squamata - Gekkonidae
e Michaelsen's Spinytail Gecko

2736 ***Diplodactylus mitchelli***
Squamata - Gekkonidae
e Mitchell's Gecko

2737 ***Diplodactylus occultus***
Squamata - Gekkonidae
e Alligator River Gecko

2738 ***Diplodactylus ornatus***
Squamata - Gekkonidae
e Gray's Ornate Gecko

2739 ***Diplodactylus polyophthalmus***
Squamata - Gekkonidae
e Many-eyed Gecko

2740 ***Diplodactylus pulcher***
Squamata - Gekkonidae
e Beautiful Gecko

2741 ***Diplodactylus rankini***
Squamata - Gekkonidae
e Rankin's Spinytail Gecko

2742 ***Diplodactylus savagei***
Squamata - Gekkonidae
e Spinifex Gecko

2743 ***Diplodactylus spinigerus***
Squamata - Gekkonidae
e Western Spinytail Gecko; Western Spiny-tailed Gecko (ANZ)
d Westlicher Stachelschwanzgecko; Australischer Dornschwanzgecko

2744 ***Diplodactylus squarrosus***
Squamata - Gekkonidae
e Heath Gecko

2745 ***Diplodactylus steindachneri***
Squamata - Gekkonidae
e Steindachner's Gecko

2746 ***Diplodactylus stenodactylus***
Squamata - Gekkonidae
e Long-fingered Gecko

2747 ***Diplodactylus strophurus***
Squamata - Gekkonidae
e Silver Spinytail Gecko

2748 ***Diplodactylus taeniatus***
Squamata - Gekkonidae
e White Spinytail Gecko

2749 ***Diplodactylus taenicauda***
Squamata - Gekkonidae
e Goldentail Gecko; Golden-tailed Gecko (ANZ)

2750 ***Diplodactylus tessellatus***
Squamata - Gekkonidae
e Tessellated Gecko
d Geziegelter Stachelschwanzgecko; Mosaikdoppelfingergecko

2751 ***Diplodactylus vittatus***
Squamata - Gekkonidae
e Wood Gecko
d Steingecko

2752 ***Diplodactylus wellingtonae***
Squamata - Gekkonidae

e Wellington's Spinytail Gecko

2753 ***Diplodactylus williamsi***
Squamata - Gekkonidae
e Williams's Spinytail Gecko
d Williams-Stachelgecko; Williams Stachelschwanzgecko

2754 ***Diplodactylus wilsoni***
Squamata - Gekkonidae
e Wilson's Gecko

2755 ***Diplodactylus wombeyi***
Squamata - Gekkonidae
e Wombey's Gecko

2756 ***Diploglossus***
Squamata - Anguidae
e Greater Galliwasps; Galliwasp
d Gallwespenschleichen
f Diploglosses

2757 ***Diploglossus anelpistus***
Squamata - Anguidae
e San Cristobal Galliwasp; Alta Gracia Giant Galliwasp

2758 ***Diploglossus bilobatus***
Squamata - Anguidae
e O'Shaughnessy's Galliwasp

2759 ***Diploglossus carraui***
Squamata - Anguidae
e Comedero Galliwasp

2760 ***Diploglossus costatus***
Squamata - Anguidae
e Common Galliwasp; Hispaniolan Smooth Galliwasp
d Hispaniola-Schleiche

2761 ***Diploglossus cyanochloris***
Squamata - Anguidae
e Blue-green Galliwasp

2762 ***Diploglossus delasagra***
Squamata - Anguidae
e Cuban Galliwasp; Cuban Pale-necked Galliwasp

2763 ***Diploglossus fasciatus***
Squamata - Anguidae
e Banded Galliwasp

2764 ***Diploglossus garridoi***
Squamata - Anguidae
e Cuban Small-eared Galliwasp

2765 ***Diploglossus legnotus***
Squamata - Anguidae
e Campbell's Galliwasp

2766 ***Diploglossus lessonae***
Squamata - Anguidae
e Brazilian Galliwasp

2767 ***Diploglossus maculata***
Squamata - Anguidae
e Spotted Galliwasp

2768 ***Diploglossus microcephalus***
Squamata - Anguidae
e Smallhead Galliwasp

2769 ***Diploglossus microlepis***
Squamata - Anguidae
e Small-lipped Galliwasp

2770 ***Diploglossus millepunctatus***
Squamata - Anguidae
e Dotted Galliwasp
d Weißpunkt-Gallwespenschleiche; Malpelo-Gallwespenschleiche

2771 ***Diploglossus monotropis***
Squamata - Anguidae
e Kuhl's Galliwasp

2772 ***Diploglossus montiserrati***
Squamata - Anguidae
e Montserrat Galliwasp

2773 ***Diploglossus montisilvestris***
Squamata - Anguidae
e Panama Galliwasp

2774 ***Diploglossus owenii***
Squamata - Anguidae
e Owen's Galliwasp
f Diploglosse d'Owen

2775 ***Diploglossus pleei***
Squamata - Anguidae
e Puerto Rico Galliwasp; Plee's Diploglossus; Puerto Rican Galliwasp
f Diploglosse de Plee

2776 ***Diploglossus warreni***
Squamata - Anguidae
e Warren's Galliwasp; Hispaniolan Giant Galliwasp
d Warrens Doppelzungenschleiche

2777 ***Diplolaemus***
Squamata - Iguanaidae
e Patagonia Iguanas

2778 ***Diplolaemus bibronii***
Squamata - Iguanaidae
e Bibron's Iguana

2779 ***Diplolaemus darwinii***
Squamata - Iguanaidae
e Darwin's Iguana; Patagonian Lizard
d Patagonischer Leguan; Darwins Eidechse; Patagonischer Erdleguan

2780 ***Diplolaemus leopardinus***
Squamata - Iguanaidae
e Leopard Iguana

2781 ***Diporiphora***
Squamata - Agamidae
e Two-pored Agamas; Two-pored Dragons; Two-pored Lizards; Two-lined Dragons (ANZ)
d Australische Bodenagamen

2782 ***Diporiphora albilabris***
Squamata - Agamidae
e Kimberley Plateau Two-pored Dragon; White-lipped Two-line Dragon (ANZ)

2783 ***Diporiphora australis***
Squamata - Agamidae
e Australian Two-pored Dragon; Eastern Two-line Dragon (ANZ); Two-lined Dragon

2784 ***Diporiphora bennettii***
Squamata - Agamidae
e Bennett's Two-pored Dragon; Robust Two-line Dragon (ANZ)

2785 ***Diporiphora bennettii arnhemica***
Squamata - Agamidae
e Arnhem Land Two-line Dragon

2786 ***Diporiphora bilineata***
Squamata - Agamidae
e Two-lined Dragon; Two-pored Lizard; Northern Two-line Dragon (ANZ)

2787 ***Diporiphora convergens***
Squamata - Agamidae
e Crystal Creek Dragon; Crystal Creek Two-line Dragon (ANZ)

2788 ***Diporiphora lalliae***
Squamata - Agamidae
e Storr's Two-pored Dragon; Lally's Two-line Dragon (ANZ)

2789 ***Diporiphora linga***
Squamata - Agamidae
e Coastal Two-pored Dragon;

Pink Two-line Dragon

2790 ***Diporiphora magna***
Squamata - Agamidae
e Northern Two-pored Dragon; Yellow-sided Two-line Dragon

2791 ***Diporiphora pindan***
Squamata - Agamidae
e Pindan Two-pored Dragon; Pindan Two-line Dragon

2792 ***Diporiphora reginae***
Squamata - Agamidae
e Queen Two-pored Dragon; Plain-backed Two-line Dragon (ANZ)

2793 ***Diporiphora superba***
Squamata - Agamidae
e Exalted Two-pored Dragon; Superb Two-line Dragon (ANZ)
d Australische Baumagame

2794 ***Diporiphora valens***
Squamata - Agamidae
e Pilbara Two-pored Dragon; Pilbara Two-line Dragon (ANZ)

2795 ***Diporiphora winneckei***
Squamata - Agamidae
e Winnecke's Two-pored Dragon; Cane Grass Dragon Lizard; Cane Grass Two-line Dragon

2796 ***Dipsadoboa***
Squamata - Colubridae
e Tree Snakes; Cat-eyed Tree Snakes (CSA)

2797 ***Dipsadoboa aulica***
Squamata - Colubridae
e Court Tree Snake; Cross-barred Tree Snake (CSA); Marbled Tree Snake (CSA); Cross-barred Snake (CSA)

2798 ***Dipsadoboa aulicus***
Squamata - Colubridae
e Cross-barred Snake; Günther's Cross-barred Snake

2799 ***Dipsadoboa duchesnii***
Squamata - Colubridae
e Boulenger's Brown Tree Snake

2800 ***Dipsadoboa shrevei***
Squamata - Colubridae
e Shreve's Tree Snake

2801 ***Dipsadoboa unicolor***
Squamata - Colubridae
e Günther's Green Tree Snake

2802 ***Dipsadoboa werneri***
Squamata - Colubridae
e Werner's Green Tree Snake

2803 ***Dipsas***
Squamata - Colubridae
e Snail-eaters; Thirst Snakes
d Amerikanische Schneckennattern; Dickkopfnattern; Schneckennattern

2804 ***Dipsas albifrons***
Squamata - Colubridae
e Sauvage's Snail-eater

2805 ***Dipsas articulata***
Squamata - Colubridae
e American Snail-eater; Caribbean Snail-eater

2806 ***Dipsas bicolor***
Squamata - Colubridae
e Two-coloured Snail-eater

2807 ***Dipsas boettgeri***
Squamata - Colubridae
e Boettgers Snail-eater

2808 ***Dipsas brevifacies***
Squamata - Colubridae

e Snail-eating Thirst Snake; Short-facied Snail-eater

2809 ***Dipsas catesbyi***
Squamata - Colubridae
e Catesby's Snail-eater; Ornate Thirst Snake

2810 ***Dipsas copei***
Squamata - Colubridae
e Cope's Snail-eater

2811 ***Dipsas elegans***
Squamata - Colubridae
e Elegant Snail-eater

2812 ***Dipsas ellipsifera***
Squamata - Colubridae
e Boulenger's Snail-eater

2813 ***Dipsas gaigeae***
Squamata - Colubridae
e Gaige's Thirst Snake

2814 ***Dipsas gracilis***
Squamata - Colubridae
e Graceful Snail-eater

2815 ***Dipsas incerta***
Squamata - Colubridae
e Jan's Snail-eater

2816 ***Dipsas indica***
Squamata - Colubridae
e Neotropical Snail-eater; Common Slug-eating Snake

2817 ***Dipsas indica indica***
Squamata - Colubridae
e Big-headed Thirst Snake

2818 ***Dipsas infrenalis***
Squamata - Colubridae
e Rosen's Snail-eater

2819 ***Dipsas latifasciata***
Squamata - Colubridae
e Striped Snail-eater

2820 ***Dipsas latifrontalis***
Squamata - Colubridae
e Venezuela Snail-eater

2821 ***Dipsas longicaudata***
Squamata - Colubridae
e Longhead Snail-eater

2822 ***Dipsas maxillaris***
Squamata - Colubridae
e Werner's Snail-eater; Tabasco Snail-eater

2823 ***Dipsas neivai***
Squamata - Colubridae
e Bahia Snail-eater

2824 ***Dipsas oreas***
Squamata - Colubridae
e Ecuador Snail-eater

2825 ***Dipsas pavonina***
Squamata - Colubridae
e Northern Snail-eater; Ringed Thirst Snake

2826 ***Dipsas perijaensis***
Squamata - Colubridae
e Aleman's Snail-eater

2827 ***Dipsas peruana***
Squamata - Colubridae
e Peru Snail-eater

2828 ***Dipsas poecilolepis***
Squamata - Colubridae
e Sao Paolo Snail-eater

2829 ***Dipsas polylepis***
Squamata - Colubridae

e Many-scaled Snail-eater

2830 ***Dipsas pratti***
Squamata - Colubridae
e Pratt's Snail-eater

2831 ***Dipsas sanctijoannis***
Squamata - Colubridae
e Tropical Snail-eater

2832 ***Dipsas schunkii***
Squamata - Colubridae
e Schunk's Snail-eater

2833 ***Dipsas temporalis***
Squamata - Colubridae
e Temporal Snail-eater; Snail-eater

2834 ***Dipsas tenuissima***
Squamata - Colubridae
e Tropical Snail-eater; Pacific Snail-eater

2835 ***Dipsas variegata***
Squamata - Colubridae
e Variegated Snail-eater

2836 ***Dipsas vermiculata***
Squamata - Colubridae
e Vermiculate Snail-eater

2837 ***Dipsas viguieri***
Squamata - Colubridae
e Bocourt's Snail-eater

2838 ***Dipsina***
Squamata - Colubridae
e Dwarf Beaked Snakes

2839 ***Dipsina multimculata***
Squamata - Colubridae
e Dwarf Beaked Snake

2840 ***Dipsosaurus***
Squamata - Iguanaidae
e Desert Iguanas; Crested Lizards
d Wüstenleguane
f Dipsosaures

2841 ***Dipsosaurus dorsalis***
Squamata - Iguanaidae
e Desert Iguana (NA); Keeled-back Iguana; Pygmy Iguana; Northern Crested Lizard
d Wüstenleguan
f Iguane du désert; Dipsosaure du désert

2842 ***Dipsosaurus dorsalis dorsalis***
Squamata - Iguanaidae
e Desert Iguana; Northern Desert Iguana (NA)

2843 ***Dispholidus***
Squamata - Colubridae
e Boomslangs
d Boomslangen; Trugnattern
f Boomslangs

2844 ***Dispholidus typus***
Squamata - Colubridae
e Boomslang (CSA); Back-fanged Tree Snake; Tree Snake (CSA)
d Grüne Boomslang; Trugnatter; Boomslang
f Serpent d'arbre du Cap; Boomslang
i Boomslang

2845 ***Dispholidus typus kivuensis***
Squamata - Colubridae
e Boomslang

2846 ***Disteira***
Squamata - Elapidae
e Common Sea Snake

2847 ***Disteira kingii***
Squamata - Elapidae
e King Sea Snake

2848 ***Disteira major***
Squamata - Elapidae
e Common Sea Snake

2849 ***Disteira nigrocinctus***
Squamata - Elapidae
e Black-banded Sea Snake

2850 ***Disteira wallii***
Squamata - Elapidae
e Wall's Sea Snake

2851 ***Ditaxodon***
Squamata - Colubridae
e Hensel's Snakes

2852 ***Ditaxodon taeniatus***
Squamata - Colubridae
e Hensel's Snake

2853 ***Dityophis***
Squamata - Colubridae
e Günther's Racers

2854 ***Dityophis vivax***
Squamata - Colubridae
e Günther's Racer

2855 ***Dogania***
Testudines - Trionychidae
e Malayan Softshells
d Malayen-Weichschildkröten

2856 ***Dogania subplana***
Testudines - Trionychidae
e Malayan Softshell; Malayan Soft-shelled Turtle; Malayan Soft-shell Turtle; Redcheek Softshell-turtle
d Malayen-Weichschildkröte
f Trionyx de Malaisie

2857 ***Dracaena***
Squamata - Teiidae
e Caiman Lizards
d Krokodiltejus

2858 ***Dracaena guianensis***
Squamata - Teiidae
e Guyana Caiman Lizard; Guianan Caiman Lizard; Norther Caiman Lizard
d Guayana-Krokodilteju; Krokodilteju
f Téju caiman

2859 ***Dracaena paraguayensis***
Squamata - Teiidae
e Paraguay Caiman Lizard; Paraguyan Caiman Lizard
d Paraguay-Krokodilteju

2860 ***Draco***
Squamata - Agamidae
e Flying Dragons; Asiatic Flying Lizards; Dracos
d Flugdrachen; Flattereidechsen
f Dragons volants

2861 ***Draco affinis***
Squamata - Agamidae
e Bartlett's Flying Dragon

2862 ***Draco biaro***
Squamata - Agamidae
e Lazell's Flying Dragon

2863 ***Draco blanfordii***
Squamata - Agamidae
e Blanford's Flying Dragon; Blanford's Dragon
d Blanfords Flugdrache
f Dragon de Blanford

2864 ***Draco dussumieri***
Squamata - Agamidae
e Indian Flying Dragon; Indian Dragon; Flying Lizard; Draco (ISC); Gliding Lizard (ISC); South Indian Flying Lizard (ISC)
d Indischer Flugdrache

f Dragon de Dussumier; Dragon d'Inde

2865 ***Draco fimbriatus***
Squamata - Agamidae
e Fringed Flying Dragon
f Dragon frangé

2866 ***Draco formosus***
Squamata - Agamidae
e Formosa Flying Dragon

2867 ***Draco haematopogon***
Squamata - Agamidae
e Red-barbed Flying Dragon; Red-barbed Dragon
d Rotbartflugdrache
f Dragon barbe-rouge

2868 ***Draco jarechii***
Squamata - Agamidae
e Jarechi's Flying Dragon

2869 ***Draco lineatus***
Squamata - Agamidae
e Lined Flying Dragon
d Linealflugdrache
f Dragon rayé; Dragon de Daudin

2870 ***Draco maculatus***
Squamata - Agamidae
e Spotted Flying Dragon; Spotty Dragon
d FleckfFlugdrache
f Dragon tacheté

2871 ***Draco maximus***
Squamata - Agamidae
e Great Flying Dragon

2872 ***Draco melanopogon***
Squamata - Agamidae
e Black-barbed Flying Dragon; Black-barbed Dragon
d Schwarzbartflugdrache
f Dragon barbe-noire

2873 ***Draco norvilli***
Squamata - Agamidae
e Alcock's Flying Dragon

2874 ***Draco obscurus***
Squamata - Agamidae
e Veiled Flying Dragon

2875 ***Draco punctatus***
Squamata - Agamidae
e Dotted Flying Dragon

2876 ***Draco quinquefasciatus***
Squamata - Agamidae
e Five-lined Flying Dragon; Five-lined Dragon
d Fünfstreifenflugdrache
f Dragon à cinq bandes

2877 ***Draco spilopterus***
Squamata - Agamidae
e Philippine Flying Dragon

2878 ***Draco taeniopterus***
Squamata - Agamidae
e Thai Flying Dragon; Flying Lizard; Flying Dragon
f Dragon taenioptère

2879 ***Draco volans***
Squamata - Agamidae
e Common Flying Dragon
d Gewöhnlicher Flugdrache; Gemeiner Flugdrache
f Dragon volant; Dragon de Timor

2880 ***Drepanoides***
Squamata - Colubridae
e Black-collared Snakes; Egg-eaters

2881 ***Drepanoides anomalus***
Squamata - Colubridae
e Black-collared Snake; Amazon Egg-eater

2882 ***Dromicodryas***
Squamata - Colubridae
e Striped Snakes

2883 ***Dromicodryas bernieri***
Squamata - Colubridae
e Bernier's Striped Snake

2884 ***Dromicodryas quadrilineatus***
Squamata - Colubridae
e Four-striped Snake

2885 ***Dromicus***
Squamata - Colubridae
e Island Racers; Ground Snakes

2886 ***Dromicus calliaemus***
Squamata - Colubridae
e Jamaica Island Racer; Jamaican Red Ground Snake (WI)

2887 ***Dromicus exiguum***
Squamata - Colubridae
e Virgin Island Racer; Puerto Rican Ground Snake

2888 ***Dromicus funereum***
Squamata - Colubridae
e Cope's Island Racer; Jamaican Black Ground Snake (WI)

2889 ***Dromicus polylepis***
Squamata - Colubridae
e Many-scaled Island Racer; Jamaican Long-tailed Ground Snake (WI)

2890 ***Dromophis***
Squamata - Colubridae
e Olympic Snakes
d Sumpfschlange

2891 ***Dromophis lineatus***
Squamata - Colubridae
e Lined Olympic Snake; Striped Swamp Snake (CSA); Buff-striped Grass Snake (CSA); Lined Swamp Snake (CSA)
d Gestreifte Sumpfschlange

2892 ***Dromophis praeornatus***
Squamata - Colubridae
e African Swamp Snake

2893 ***Dryadophis***
Squamata - Colubridae
e Forest Racers

2894 ***Dryadophis dorsalis***
Squamata - Colubridae
e Bocourt's Tropical Racer; Brown Highland Racer

2895 ***Dryadophis melanolomus***
Squamata - Colubridae
e Dark Tropical Racer; Olive Racer; Lizard-eater

2896 ***Dryadophis pleei***
Squamata - Colubridae
e Plee's Forest Racer; Plee's Racer

2897 ***Drymarchon***
Squamata - Colubridae
e Indigo Snakes; Speckled Snakes; Cribos
d Indigoschlangen; Indigonattern
f Serpents indigos

2898 ***Drymarchon corais***
Squamata - Colubridae
e Indigo Snake; Gopher Snake; Cribo; Florida Indigo Snake; Western Indigo Snake
d Indigoschlange; Indigonatter; Blaue Bullennatter
f Serpent indigo
i Serpente indaco

2899 ***Drymarchon corais corais***
Squamata - Colubridae

e Yellowtail Cribo; Common Cribo
d Indigonatter

2900 ***Drymarchon corais couperi***
Squamata - Colubridae
e Eastern Indigo Snake; Indigo Snake (NA)
d Östliche Indigonatter

2901 ***Drymarchon corais erebennus***
Squamata - Colubridae
e Texax Indigo Snake
d Texas-Indigonatter

2902 ***Drymarchon corais melanurus***
Squamata - Colubridae
e Blacktail Cribo
d Indigonatter

2903 ***Drymarchon corais rubidus***
Squamata - Colubridae
e Mexico Redtail Cribo
d Mexikanische Indigonatter

2904 ***Drymarchon corais unicolor***
Squamata - Colubridae
e Central American Cribo
d Mittelamerikanischer Cribo

2905 ***Drymobius***
Squamata - Colubridae
e Speckled Racers; Slender Ground Snakes; Tropical Racers; Central American Racers; Racers; Neotropiccal Racers
d Rennn-Nattern
f Couleuvres coureuses

2906 ***Drymobius chloroticus***
Squamata - Colubridae
e Green Highland Racer

2907 ***Drymobius margaritiferus***
Squamata - Colubridae
e Speckled Racer; Green-spotted Snake; Mexican Speckled Snake; Central American Speckled Racer
d Perlnatter; Gesprenklte Bodenschlange; Juwelennatter
f Couleuvre coureuse mouchetée; Couleuvre tachetée

2908 ***Drymobius margaritiferus margaritiferus***
Squamata - Colubridae
e Northern Speckled Racer
d Perlnatter

2909 ***Drymobius melanotropis***
Squamata - Colubridae
e Black Forest Racer; Green Racer

2910 ***Drymobius rhombifer***
Squamata - Colubridae
e Esmeralda Racer; Diamondback Racer; Blotched Racer

2911 ***Drymoluber***
Squamata - Colubridae
e Woodland Racers; Glossy Racers

2912 ***Drymoluber brazili***
Squamata - Colubridae
e Brazilian Woodland Racer; Common Glossy Racer

2913 ***Drymoluber dichrous***
Squamata - Colubridae
e Northern Woodland Racer

2914 ***Dryocalamus***
Squamata - Colubridae
e Bridal Snakes; Wood-Reed Snakes
d Waldrohrschlangen

2915 ***Dryocalamus davisoni***
Squamata - Colubridae
e Blanford's Bridal Snake

2916 ***Dryocalamus gracilis***
Squamata - Colubridae

e Scarce Bridal Snake

2917 ***Dryocalamus nympha***
Squamata - Colubridae
e Vellore Bridal Snake

2918 ***Dryocalamus subannulatus***
Squamata - Colubridae
e Malayan Bridal Snake

2919 ***Dryocalamus tristrigatus***
Squamata - Colubridae
e Striped Bridal Snake

2920 ***Dryophiops***
Squamata - Colubridae
e Whip Snakes

2921 ***Dryophiops philippina***
Squamata - Colubridae
e Philippine Whip Snake

2922 ***Dryophiops rubescens***
Squamata - Colubridae
e Red Whip Snake; Keel-bellied Whip Snake; Speckle-headed Tree Snake

2923 ***Drysdalia***
Squamata - Elapidae
e Australian Crowned Snakes; Crowned Snakes (ANZ)

2924 ***Drysdalia coronata***
Squamata - Elapidae
e Crowned Snake

2925 ***Drysdalia coronoides***
Squamata - Elapidae
e White-lipped Snake
d Weißlippenotter

2926 ***Drysdalia mastersi***
Squamata - Elapidae
e Masters's Snake
d Masters-Schlange; Masters-Otter

2927 ***Drysdalia rhodogaster***
Squamata - Elapidae
e Australian Rosebelly Snake

2928 ***Duberria***
Squamata - Colubridae
e Slug-eaters; African Snail-eating Snakes; Homalosoma Snakes; Slug-eater Snakes
d Afrikanische Schneckenfresser

2929 ***Duberria lutrix***
Squamata - Colubridae
e Slug-eater (CSA); Common Slug-eater (CSA); Abyssinia Slug-eater; Homalosoma Snake; African Slug-eater

2930 ***Duberria variegata***
Squamata - Colubridae
e Variegated Slug-eater; Spotted Slug-eater (CSA); Peters's Slug-eater; Variegated Snake

E

2931 ***Ebenavia***
Squamata - Gekkonidae
e Madagascar Clawless Geckos

2932 ***Ebenavia inunguis***
Squamata - Gekkonidae
e Madagascar Clawless Gecko

2933 ***Echananthera***
Squamata - Colubridae
e Brazilian Tropical Snakes

2934 ***Echananthera amoenus***
Squamata - Colubridae
e Brazilian Tropical Snake

2935 ***Echinosaura***
Squamata - Teiidae
e Rough Teiids
d Stacheltejus

2936 ***Echinosaura horrida***
Squamata - Teiidae
e Rough Teiid
d Stachelteju

2937 ***Echiopsis***
Squamata - Elapidae
e Bardick's Snakes
d Bardick-Schlangen

2938 ***Echiopsis curta***
Squamata - Elapidae
e Bardick's Snake; Bardick (ANZ)
d Bardick; Bardick-Schlange

2939 ***Echis***
Squamata - Viperidae
e Saw-scaled Vipers; Carpet Vipers
d Sandrasselottern
f Échides; Serpents de Cléopatre
i Echidi

2940 ***Echis carinatus***
Squamata - Viperidae
e Saw-scaled Viper; Phoorsa
d Sandrasselotter; Efa-Schlange
f Échide carénée; Vipère des pyramides; Vipère d'Égypte; Vipère à écailles carénées
i Echide carinato; Serpente squame a sega

2941 ***Echis carinatus leakeyi***
Squamata - Viperidae
e Eastern Saw-scaled Viper
d Östliche Sandrasselotter

2942 ***Echis carinatus ocellatus***
Squamata - Viperidae
e Western Saw-scaled Viper
d Westliche Sandrasselotter

2943 ***Echis coloratus***
Squamata - Viperidae
e Palestine Saw-scaled Viper; Burton's Painted Carpet Viper; Arabian Carpet Viper; Burton's Carpet Viper; Arabian Saw-scaled Viper
d Arabische Sandrasselotter
f Echide d'Arabie

2944 ***Echis hughesi***
Squamata - Viperidae
e Hughes's Saw-scaled Viper

2945 ***Echis jogeri***
Squamata - Viperidae
e Joger's Saw-scaled Viper

2946 ***Echis leucogaster***
Squamata - Viperidae
e Roman's Saw-scaled Viper; White-bellied Carpet Viper
f Echide à ventre blanc

2947 ***Echis megalocephalos***
Squamata - Viperidae
e Cherlin's Saw-scaled Viper

2948 ***Echis multisquamatus***
Squamata - Viperidae
e Iranian Saw-scaled Viper

2949 ***Echis ocellatus***
Squamata - Viperidae
e African Saw-scaled Viper

2950 ***Echis pyramidum***
Squamata - Viperidae
e Egyptian Saw-scaled Viper; East African Carpet Viver; Saw-scaled Viper; Carpet Viper
d Nordostafrikanische Sandrasselotter

2951 ***Echis sochurchi***
Squamata - Viperidae
e Stemmler's Saw-scaled Viper

2952 ***Egernia***
Squamata - Scincidae
e Spiny-tailed Skinks; Spinytail Skinks; Spiny Skinks; Egernias
d Stachelskinke; Dornschwanzskinke; Stachelschwanzskinke; Stachelechsen

2953 ***Egernia carinata***
Squamata - Scincidae
e Dark Spinytail Skink

2954 ***Egernia coventryi***
Squamata - Scincidae
e Coventry's Spintytail Skink; Coventry's Skink (ANZ); Swamp Skink (ANZ)

2955 ***Egernia cunninghami***
Squamata - Scincidae
e Cunningham's Skink
d Cunninghams Dornschwanzskink; Felsenskink; Plattschwänzchen; Cunninghams Skink
f Scinque à queue épineuse

2956 ***Egernia depressa***
Squamata - Scincidae
e Pygmy Spinytail Skink; Pygmy Spiny-tailed Skink (ANZ)
d Plattschwanzstachelskink

2957 ***Egernia douglasi***
Squamata - Scincidae
e Douglas's Skink
d Baumstachelskink

2958 ***Egernia formosa***
Squamata - Scincidae
e Formosa Skink

2959 ***Egernia frerei***
Squamata - Scincidae
e Major Skink

2960 ***Egernia hosmeri***
Squamata - Scincidae
e Hosmer's Skink
d Hosmer-Stachelskink; Hosmers Skink

2961 ***Egernia inornata***
Squamata - Scincidae
e Desert Skink
d Wüstenstachelskink

2962 ***Egernia kingii***
Squamata - Scincidae
e King's Skink

2963 ***Egernia kintorei***
Squamata - Scincidae
e Great Desert Skink

2964 ***Egernia luctuosa***
Squamata - Scincidae
e Peters's Swamp Skink
d Trauerskink

2965 ***Egernia major***
Squamata - Scincidae
e Land Mullet
d Großer Stachelskink; Stachelskink

2966 ***Egernia margaretae***
Squamata - Scincidae
e Black-spotted Skink

2967 ***Egernia mcpheei***
Squamata - Scincidae
e McPhee's Skink

2968 ***Egernia modesta***
Squamata - Scincidae
e Pale-spotted Skink

2969 ***Egernia multiscutata***
Squamata - Scincidae
e Coastal Dune Skink; Heath Skink (ANZ)

2970 ***Egernia napoleonis***
Squamata - Scincidae
e Napoleon Skink

2971 ***Egernia pilbarensis***
Squamata - Scincidae
e Pilbara Skink

2972 ***Egernia pulchra***
Squamata - Scincidae
e Keel-scaled Skink

2973 ***Egernia richardi***
Squamata - Scincidae
e Richard's Alligator River Skink

2974 ***Egernia rugosa***
Squamata - Scincidae
e Yakka Skink

2975 ***Egernia saxatilis***
Squamata - Scincidae
e Black Rock Skink
d Felsenstachelskink; Brauner Felsenskink; Schwarzer Felsenskink

2976 ***Egernia slateri***
Squamata - Scincidae
e Slater's Dull Skink

2977 ***Egernia stokesii***
Squamata - Scincidae
e Gidgee Skink
d Stokes-Stachelskink; Dornschwanzskink; Stachelskink

2978 ***Egernia striata***
Squamata - Scincidae
e Elliptical-eyed Skink
d Gestreifte Stachelechse; Nachtskink

2979 ***Egernia striolata***
Squamata - Scincidae
e Tree Skink
d Baumskink

2980 ***Egernia whitii***
Squamata - Scincidae
e White's Skink
d White-Stachelskink

2981 ***Eirenis***
Squamata - Colubridae
e Dwarf Racers
d Zwergnattern
f Couleuvres nains
i Irene

2982 ***Eirenis arabica***
Squamata - Colubridae
e Arabian Dwarf Racer

2983 ***Eirenis barani***
Squamata - Colubridae
e Baran Dwarf Racer
d Baran-Zwergnatter; Barans Zwergnatter
f Couleuvre naine de Baran
i Irene di Baran

2984 ***Eirenis collaris***
Squamata - Colubridae
e Collared Dwarf Racer; Collared Dwarf Snake
d Halsbandzwergnatter
f Couleuvre naine à collier
i Irene dal collare

2985 ***Eirenis coronella***
Squamata - Colubridae
e Crowned Dwarf Racer; Crowned Peace Snake
d Krönchenzwergnatter
f Couleuvre naine couronnée
i Irene coronella

2986 ***Eirenis coronelloides***
Squamata - Colubridae
e Sinai Dwarf Racer

2987 ***Eirenis eiselti***
Squamata - Colubridae
e Eiselt's Dwarf Racer
d Eiselts Zwergnatter
f Couleuvre naine d'Eiselt
i Irene di Eiselt

2988 ***Eirenis modestus***
Squamata - Colubridae
e Asia Minor Dwarf Racer; Ring-headed Dwarf Snake
d KopfbindenZwergnatter
f Couleuvre naine à tête rayée; Couleuvre naine d'Asie mineure
i Irene modesta

2989 ***Eirenis punctatolineatus***
Squamata - Colubridae
e Boettger's Dwarf Racer; Dotted Dwarf Racer; Dotted Dwarf Snake
d Armenische Zwergnatter; Punktenzwergnatter
f Couleuvre naine à points alignés; Couleuvre naine d'Arménie; Couleuvre naine ponctuée
i Irene lineopuctata

2990 ***Eirenis rothii***
Squamata - Colubridae
e Roth's Dwarf Racer; Roth's Snake
d Roths Zwergnatter
f Couleuvre naine de Roth
i Irene di Roth

2991 ***Elachistodon***
Squamata - Colubridae
e Indian Egg-eating Snakes
d Indische Eierschlangen
f Couleuvres de Westermann

2992 ***Elachistodon westermanni***
Squamata - Colubridae
e Indian Egg-eating Snake; Westermann's Snake
d Indische Eierschlange
f Couleuvre de Westermanm

2993 ***Elaphe***
Squamata - Colubridae
e Rat Snakes
d Kletternattern; Rattenschlangen; Hühnerschlangen
f Couleuvres ratiers; Élaphes
i Elafi

2994 ***Elaphe bairdi***
Squamata - Colubridae
e Baird's Rat Snake; Baird's Pilot

Snake
d Bairds Kletternatter
i Elafe di Baird

2995 ***Elaphe bimaculata***
Squamata - Colubridae
e Twin-spotted Rat Snake; Chinese Leopard Snake; Chinese Twin-spotted Rat Snake
d Chinesische Zweifleckkatter; Zweifleckkatter; Zweifleckkletternatter; Zweistreifennatter; Chinesische Leopardnatter

2996 ***Elaphe cantoris***
Squamata - Colubridae
e Eastern Trinket Snake; Cantor's Rat Snake; Ring-tailed Dhaman

2997 ***Elaphe carinata***
Squamata - Colubridae
e Taiwan Stink Snake; Stink Snake; Stinking Goddess; China King Rat Snake; King Rat Snake; Keeled Rat Snake
d Stinknatter; Gekielte Kletternatter
i Serpente dei ratti reale cinese

2998 ***Elaphe carinata carinata***
Squamata - Colubridae
e Chinese Keeled Rat Snake

2999 ***Elaphe carinata yonaguniensis***
Squamata - Colubridae
e Yonaguni Keeled Rat Snake

3000 ***Elaphe climacophora***
Squamata - Colubridae
e Japanese Rat Snake; Albino Rat Snake
d Inselnatter; Japanische Inselkletternatter; Inselkletternatter; Japan-Kletternatter

3001 ***Elaphe conspicillata***
Squamata - Colubridae
e Red Japanese Rat Snake; Japanese Forest Rat Snake; Burrowing Rat Snake; Japanese Wood Snake
d Japanische Walnatter

3002 ***Elaphe davidi***
Squamata - Colubridae
e Père David's Rat Snake
d Davids Kletternatter

3003 ***Elaphe dione***
Squamata - Colubridae
e Steppe Rat Snake; Dione Snake; Pallas's Coluber; Steppes Rat Snake; Dione's Rat Snake
d Dione-Natter; Steppennatter; Kletternatter
f Couleuvre Dione; Couleuvre de Dioné; Elaphe de Dioné
i Colubro Dione

3004 ***Elaphe emoryi***
Squamata - Colubridae
e Great Plains Rat Snake; Emory's Rat Snake; Spotted Chicken Snake (NA); Emory's Pilot Snake
d Prairie-Kornnatter; Graue Kornnatter

3005 ***Elaphe emoryi emoryi***
Squamata - Colubridae
e Northern Plains Rat Snake

3006 ***Elaphe erythrura***
Squamata - Colubridae
e Reddish Rat Snake; Philippine Rat Snake
d Philippinen Kletternatter

3007 ***Elaphe flavirufa***
Squamata - Colubridae
e Yellow-red Rat Snake; Red-yellow Rat Snake
d Mexikanische Nachtnatter

3008 ***Elaphe flavirufa pardalina***
Squamata - Colubridae

d Mexikanische Nachtnatter

3009 ***Elaphe flavolineata***
Squamata - Colubridae
e Black Copper Rat Snake; Common Malayan Rat Snake; Common Malayan Racer; Yellow-striped Snake
d Gelbstreifennatter

3010 ***Elaphe frenata***
Squamata - Colubridae
e Assam Green Trinket Snake
d Zügelnatter

3011 ***Elaphe gloydi***
Squamata - Colubridae
e Eastern Fox Snake
d Östliche Fuchsnatter

3012 ***Elaphe guttata***
Squamata - Colubridae
e Corn Snake; Red Rat Snake; Red Chicken Snake (NA); Red House Snake; Spotted Racer
d Kornnatter
f Élaphe des blés; Couleuvre à gouttelettes; Serpent des blés
i Serpente del grano; Elafe scarlatta

3013 ***Elaphe guttata guttata***
Squamata - Colubridae
e Corn Snake
d Kornnatter
f Serpent des blés; Couleuvre à gouttelettes

3014 ***Elaphe guttata meahllmorum***
Squamata - Colubridae
e Southwestern Rat Snake; South Plains Rat Snake

3015 ***Elaphe guttata rosacea***
Squamata - Colubridae
d Rosa Kornnatter

3016 ***Elaphe helenae***
Squamata - Colubridae
e Common Trinket Snake; Trinket Snake (ISC); Helena's Rat Snake; Daudin's Rat Snake; Trinket Rat Smake
d Indische Schmucknatter; Indische Schmuckbaumnatter; Indische Schönnatter

3017 ***Elaphe hodgsonii***
Squamata - Colubridae
e Himalaya Trinket Snake; Hodgson's Rat Snake
d Hodgsons Kletternatter

3018 ***Elaphe hohenackeri***
Squamata - Colubridae
e Trans-Caucasian Rat Snake; Caucasian Snake
d Transkaukasische Kletternatter
f Couleuvre de Hohenacker; Élaphe de Hohenacker
i Colubro di Hohenacker

3019 ***Elaphe janseni***
Squamata - Colubridae
e Jansen's Rat Snake
d Celebes-Schwarzschwanznatter

3020 ***Elaphe japonica***
Squamata - Colubridae
e Northern Japanese Rat Snake

3021 ***Elaphe leonardi***
Squamata - Colubridae
e Burmese Rat Snake
d Leonards Kletternatter

3022 ***Elaphe longissima***
Squamata - Colubridae
e Aesculapian Rat Snake; Aesculapian Snake
d Äskulapnatter
f Couleuvre d'Esculape; Elaphe d'Esculape

i Saettone; Colubro di Eusculapio; Serpente di Esculapio; Saettone di Esculapio; Frustone

3023 ***Elaphe longissima longissima***
Squamata - Colubridae
e Aesculapian Snake
d Äskulapnatter

3024 ***Elaphe mandarina***
Squamata - Colubridae
e Mandarin Rat Snake
d Mandarinnatter; Mandarinschlange
i Serpente dei ratti mandarino; Serpente mandarino

3025 ***Elaphe moellendorfi***
Squamata - Colubridae
e Red-headed Rat Snake; Flower Snake
d Blumennatter; Moellendorfs Kletternatter; Moellendorf's Blumennatter; Chinesische Kletternatter

3026 ***Elaphe obsoleta***
Squamata - Colubridae
e Eastern Rat Snake; Red Rat Snake; Common Rat Snake; Pilot Black Snake (NA); Mountain Black Snake (NA); Black Chicken Snake (NA); Rat Snake; American Rat Snake
d Grüne Erdnatter; Pilotnatter; Kükennatter; Erdnatter; Schwarze Rattenschlange
f Élaphe ratier; Couleuvre obscure; Serpent ratier
i Elafe americano; Serpente pilota nero

3027 ***Elaphe obsoleta lindheimerii***
Squamata - Colubridae
e Texas Rat Snake
d Texas-Kükennatter; Erdnatter; Lindheimers Erdnatter; Pilotnatter; Schwarze Rattennatter

3028 ***Elaphe obsoleta obsoleta***
Squamata - Colubridae
e Black Rat Snake; American Rat Snake
d Pilotnatter; Schwarze Rattennatter; Schwarze Erdnatter; Schwarze Pilotnatter
f Couleuvre obscure; Serpen ratier noire

3029 ***Elaphe obsoleta quadrivittata***
Squamata - Colubridae
e Yellow Rat Snake; Four-lined Chicken Snake; Common Chicken Snake (NA); Chicken Snake (NA); Four-lined Rat Snake
d Kükennatter; Gelbe Erdnatter; Gelbe Erdnatter
f Serpent-ratier jaune

3030 ***Elaphe obsoleta rossalleni***
Squamata - Colubridae
e Everglades Rat Snake
d Everglades-Kükenatter

3031 ***Elaphe obsoleta spiloides***
Squamata - Colubridae
e Gray Rat Snake (NA); Spotted Chicken Snake
d Graue Erdnatter; Graue Pilotnatter
f Serpent-ratier gris

3032 ***Elaphe perlacea***
Squamata - Colubridae
e Szechwan Rat Snake
d Perlennatter

3033 ***Elaphe persica***
Squamata - Colubridae
e Persia Rat Snake
d Persische Kletternatter

3034 ***Elaphe phaescens***
Squamata - Colubridae
e Yucatan Rat Snake

3035 ***Elaphe porphyracea***
Squamata - Colubridae
e Black-banded Trinket Snake; Broad-barred Rat Snake; Purple-spotted Rat Snake; Red Mountain Rat Snake
i Elafe rosso montano

3036 ***Elaphe porphyracea porphyracea***
Squamata - Colubridae
d Rote Bambusnatter

3037 ***Elaphe prasina***
Squamata - Colubridae
e Green Trinket Snake; Green Tree Racer; Green Tree Rat Snake; Small Green Rat Snake; Green Rat Snake
d Grüne Strauchnatter; Grüne Baumnatter

3038 ***Elaphe quadrivirgata***
Squamata - Colubridae
e Japanese Four-lined Rat Snake; Japanese Four-lined Snake
d Japanische Vierstreifennatter; Rotaugennatter

3039 ***Elaphe quatuorlineata***
Squamata - Colubridae
e Four-lined Rat Snake; Four-rayed Snake; Four-lined Snake
d Vierstreifennatter
f Élaphe à quatre raies; Couleuvre à quatre raies; Couleuvre à quatre bandes; Couleuvre à quatre lignes
i Cervone

3040 ***Elaphe quatuorlineata sauromates***
Squamata - Colubridae
e Eastern Four-lined Rat Snake
d Östliche Vierstreifennatter

3041 ***Elaphe radiata***
Squamata - Colubridae
e Radiated Rat Snake; Copper-headed Racer; Copper-headed Rat Snake; Stripe-necked Rat Snake; Asian Copperhed; Copperhead (ISC); Copperhead Rat Snake
d Strahlennatter
f Serpent-ratier à tête cuivrée
i Serpente dei ratti raggiato

3042 ***Elaphe rufodorsata***
Squamata - Colubridae
e Frog-eating Rat Snake; Rayed Coluber
d Rlotrückennatter

3043 ***Elaphe scalaris***
Squamata - Colubridae
e Ladder Snake; Ladder Rat Snake
d Treppennatter
f Couleuvre à échelons; Bisse
i Colubro scalare; Colubro bilineato

3044 ***Elaphe schrenckii***
Squamata - Colubridae
e Russian Rat Snake; Amur Rat Snake; Siberian Rat Snake; Great Black Coluber; Manchurian Black Water Snake
d Amur-Natter; Liebesnatter; Russische Rattenschlange; Sibirische Rattenschlange

3045 ***Elaphe situla***
Squamata - Colubridae
e Leopard Rat Snake ; Leopard Snake; European Leopard Snake; Asia-minor Chicken Snake
d Leopardnatter
f Couleuvre léopard; Couleuvre léopardine
i Colubro leopardino

3046 ***Elaphe subradiata***
Squamata - Colubridae
e Sunda Rat Snake; Indonesian Rat Snake
d Indonesische Kletternatter

3047 ***Elaphe taeniura***
Squamata - Colubridae
e Taiwan Beauty Snake; Beauty Snake; Beauty Rat Snake; Banded Coluber; Stripe-tailed Rat Snake; Striped Racer; Cave Racer
d Schönnatter; Streifenkletternatter; Streifennatter; Schönkletternatter; Streifenschwanznatter
i Serpente dalla coda a strisce

3048 ***Elaphe taeniura friesi***
Squamata - Colubridae
e Taiwan Beauty Snake
d Taiwan-Schönnatter
f Serpent-ratier de Chine

3049 ***Elaphe taeniura grabowskyi***
Squamata - Colubridae
e Borneo Blue Beauty

3050 ***Elaphe taeniura gridleyi***
Squamata - Colubridae
e Cave-dwelling Rat Snake

3051 ***Elaphe taeniura schmackeri***
Squamata - Colubridae
e Sakishima Beauty Snake

3052 ***Elaphe triaspis***
Squamata - Colubridae
e Cope's Rat Snake; Green Rat Snake (NA); Mexican Green Snake
d Grüne Rattennatter; Rattennatter
f Élaphe verte

3053 ***Elaphe triaspis intermedia***
Squamata - Colubridae
e Western Green Rat Snake

3054 ***Elaphe vulpina***
Squamata - Colubridae
e Fox Snake; Western Fox Snake
d Fuchsnatter
f Élaphe vulpine; Couleuvre renardine; Couleuvre jaune

3055 ***Elaphe vulpina vulpina***
Squamata - Colubridae
e Western Fox Snake
d Fuchsnatter; Westliche Fuchsnatter

3056 **Elapidae**
Squamata
e Cobras, Kraits and Sea Snakes; Cobras; Coral Snakes (NA); Front-fanged Snakes
d Giftnattern; Nattern
f Élapidés

3057 ***Elapognathus***
Squamata - Elapidae
e Little Brown Snakes
d Kleine Braunottern

3058 ***Elapognathus minor***
Squamata - Elapidae
e Little Brown Snake; Little Desert Snake
d Kleine Braunotter

3059 ***Elapoides***
Squamata - Colubridae
e Brown Mountain Racers

3060 ***Elapoides fuscus***
Squamata - Colubridae
e Brown Mountain Racer

3061 ***Elapomorphus***
Squamata - Colubridae
e Diadem Snakes

3062 ***Elapomorphus bilineatus***
Squamata - Colubridae
e Two-lined Lizard-eating Snake

3063 ***Elapomorphus dimidiatus***
Squamata - Colubridae
e Jan's Diadem Snake

3064 ***Elapomorphus lemniscatus***
Squamata - Colubridae
e Duméril's Diadem Snake

3065 ***Elapomorphus lepidus***
Squamata - Colubridae
e Minas Gerais Snake

3066 ***Elapomorphus mertensi***
Squamata - Colubridae
e Mertens's Lizard-eating Snake

3067 ***Elapomorphus nasutus***
Squamata - Colubridae
e Gomes's Lizard-eating Snake

3068 ***Elapomorphus punctatus***
Squamata - Colubridae
e Dotted Diadem Snake

3069 ***Elapomorphus quinquelineatus***
Squamata - Colubridae
e Radde's Lizard-eating Snake Snake

3070 ***Elapomorphus spegazzinii***
Squamata - Colubridae
e Spegazzini's Diadem Snake

3071 ***Elapomorphus tricolor***
Squamata - Colubridae
e Tricolor Lizard-eating Snake

3072 ***Elapomorphus wuchereri***
Squamata - Colubridae
e Wucherer's Lizard-eating Snake

3073 ***Elapotinus***
Squamata - Colubridae
e Jan's Snakes

3074 ***Elapotinus picteti***
Squamata - Colubridae
e Jan's Snake

3075 ***Elapsoidea***
Squamata - Elapidae
e Venomous Garter Snakes; Garter Snakes
d Afrikanische Korallenschlangen; Afrikanische Bänderkorallenschlangen

3076 ***Elapsoidea chelazzii***
Squamata - Elapidae
e Somali Garter Snake

3077 ***Elapsoidea guentheri***
Squamata - Elapidae
e Günther's Garter Snake; Western Garter Snake (CSA)

3078 ***Elapsoidea laticincta***
Squamata - Elapidae
e Werner's Garter Snake

3079 ***Elapsoidea loveridgei***
Squamata - Elapidae
e Loveridge's Garter Snake

3080 ***Elapsoidea nigra***
Squamata - Elapidae
e Black Garter Snake

3081 ***Elapsoidea semiannulata***
Squamata - Elapidae
e Half-banded Garter Snake (CSA); Southern Garter Snake

3082 ***Elapsoidea semiannulata boulengeri***
Squamata - Elapidae
e Boulenger's Garter Snake; Half-banded Garter Snake (CSA)

3083 ***Elapsoidea semiannulata semiannulata***
Squamata - Elapidae
e Angola Garter Snake

3084 ***Elapsoidea sundevallii***
Squamata - Elapidae
e African Garter Snake
d Südafrikanische Strumpfbandnatter; Afrikanische Termitenotter

3085 ***Elapsoidea sundevallii longicauda***
Squamata - Elapidae
e Sundevall's Garter Snake (CSA)

3086 ***Elapsoidea sundevallii media***
Squamata - Elapidae
e Sundevall's Garter Snake (CSA)

3087 ***Elapsoidea sundevallii sundevallii***
Squamata - Elapidae
e Sundevall's Garter Snake (CSA)

3088 ***Elgaria***
Squamata - Anguidae
e Western Alligator Lizards; Alligator Lizards
d Alligatorschleichen; Krokodilschleichen
f Lézards-alligator du Pacifique

3089 ***Elgaria cedronensis***
Squamata - Anguidae
e Cedros Alligator Lizard; Cedros Island Alligator Lizard

3090 ***Elgaria coerulea***
Squamata - Anguidae
e Northern Alligator Lizard; Long-tailed Lizard; California Plated Lizard
d Nördliche Krokodilschleiche
f Lézard-alligator du Nord

3091 ***Elgaria coerulea coerulea***
Squamata - Anguidae
e San Francisco Alligator Lizard

3092 ***Elgaria coerulea palmeri***
Squamata - Anguidae
e Sierra Alligator Lizard

3093 ***Elgaria coerulea principes***
Squamata - Anguidae
e Northern Alligator Lizard; Northwestern Alligator Lizard (NA)

3094 ***Elgaria coerulea shastensis***
Squamata - Anguidae
e Shasta Alligator Lizard

3095 ***Elgaria kingii***
Squamata - Anguidae
e Madrean Alligator Lizard (NA); Sonoran Alligator Lizard; Arizona Alligator Lizard (NA)
d Mexikanische Alligatorschleiche; Arizona-Alligatorschleiche
f Lézard-alligator d'Arizona

3096 ***Elgaria kingii nobilis***
Squamata - Anguidae
e Madrean Alligator Lizard; Arizona Alligator Lizard
f Lézard-alligator d'Arizona

3097 ***Elgaria multicarinata***
Squamata - Anguidae
e California Alligator Lizard; Southern Alligator Lizard (NA); Santa Cruz Island Alligator Lizard
d Südliche Alligatorchleiche; Südliche Krokodilschleiche
f Lézard-alligator du Sud; Gerrhonote multicaréné; Lézard alligator; Lézard crocodile

3098 ***Elgaria multicarinata multicarinata***
Squamata - Anguidae
e Southern Alligator Lizard (NA); California Alligator Lizard; California Southern Alligator Lizard

3099 ***Elgaria multicarinata scincicauda***
Squamata - Anguidae
e Oregon Alligator Lizard; Skink-tailed Plated Lizard

3100 ***Elgaria multicarinata webbii***
Squamata - Anguidae
e San Diego Alligator Lizard

3101 ***Elgaria panamintina***
Squamata - Anguidae
e Panamint Alligator Lizard
f Lézard-alligator panamint

3102 ***Elgaria parva***
Squamata - Anguidae
e Pygmy Alligator Lizard

3103 ***Elgaria paucicarinata***
Squamata - Anguidae
e San Lucan Alligator Lizard

3104 ***Elseya***
Testudines - Chelidae
e Australia Snapping Turtles; Australian Snapping Turtles
d Elseya-Schildkröten

3105 ***Elseya dentata***
Testudines - Chelidae
e Northern Australia Snapping Turtle; Northern Snapping Turtle (ANZ); Elseya; Snapping Tortoise; White-headed Turtle
d Elseya-Schildkröte
i Tartaruga azzannatrice del Australia settentrionale

3106 ***Elseya latisternum***
Testudines - Chelidae
e Serrated Snapping Turtle; Saw-shelled Turtle (ANZ); Sawshell Tortoise
d Breitbrust-Elseya-Schildkröte; Breitbrustspitzkopfschildkröte
i Tartaruga azzannatrice a carapace dentellato

3107 ***Elseya macrurus***
Testudines - Chelidae
e Cann's Snapping Turtle

3108 ***Elseya novaeguineae***
Testudines - Chelidae
e New Guinea Snapping Turtle; New Guinea Sideneck Turtle; New Guinea Snapper
d Rückenflecken-Elseya-Schildkröte; Rückenfleckenspitzkopfschildkröte
i Tartaruga azzannatrice della Nova Guinea

3109 ***Emmochliophis***
Squamata - Colubridae
e Pichinda Snakes

3110 ***Emmochliophis fugleri***
Squamata - Colubridae
e Pichinda Snake

3111 ***Emoia***
Squamata - Scincidae
e Emo Skinks; Emoias
d Küstenschlankskinke

3112 ***Emoia acrocarinata***
Squamata - Scincidae
e Pointed Emo Skink

3113 ***Emoia adspersa***
Squamata - Scincidae
e Steindachner's Emo Skink

3114 ***Emoia aenea***
Squamata - Scincidae
e Bronze Emo Skink

3115 ***Emoia ahli***
Squamata - Scincidae
e Ahl's Emo Skink

3116 ***Emoia aneityumensis***
Squamata - Scincidae
e Medway's Emo Skink

3117 ***Emoia arnoensis***
Squamata - Scincidae

e Emo Skink

3118 ***Emoia atrocostata***
Squamata - Scincidae
e Mangrove Skink; Mangrove Emo Skink; Green Tree Skink; Black-sided Emo Skink
d Mangrovenskink

3119 ***Emoia aurulenta***
Squamata - Scincidae
e Brown's Emo Skink

3120 ***Emoia battersbyi***
Squamata - Scincidae
e Battersby's Emo Skink

3121 ***Emoia baudini***
Squamata - Scincidae
e Baudin's Emo Skink

3122 ***Emoia bismarkensis***
Squamata - Scincidae
e Bismark Emo Skink

3123 ***Emoia boettgeri***
Squamata - Scincidae
e Boettger's Emo Skink

3124 ***Emoia bogerti***
Squamata - Scincidae
e Bogert's Emo Skink

3125 ***Emoia brongersmai***
Squamata - Scincidae
e Brongersma's Emo Skink

3126 ***Emoia caeruleocauda***
Squamata - Scincidae
e Pacific Bluetail Skink; Pacific Blue-tailed Skink

3127 ***Emoia callistica***
Squamata - Scincidae
e New Guinea Emo Skink

3128 ***Emoia campbelli***
Squamata - Scincidae
e Montane Emo Skink

3129 ***Emoia coggeri***
Squamata - Scincidae
e Cogger's Emo Skink

3130 ***Emoia concolor***
Squamata - Scincidae
e Fiji Green Emo Skink

3131 ***Emoia cyanogaster***
Squamata - Scincidae
e Teal Emo Skink

3132 ***Emoia cyanura***
Squamata - Scincidae
e Bluetail Emo Skink; Copper-tailed Skink; Blue-tailed Emo Skink
d Blauschwanzskink; Blauschwanzschlankskink

3133 ***Emoia cyclops***
Squamata - Scincidae
e Cyclops Emo Skink

3134 ***Emoia digul***
Squamata - Scincidae
e Digul Emo Skink

3135 ***Emoia erronan***
Squamata - Scincidae
e Common Emo Skink

3136 ***Emoia flavigularis***
Squamata - Scincidae
e Yellow-throated Emo Skink

3137 ***Emoia guttata***
Squamata - Scincidae
e Allison's Emo Skink

3138 ***Emoia impar***
Squamata - Scincidae
e Azure-tailed Skink

3139 ***Emoia iriannensis***
Squamata - Scincidae
e Irian Emo Skink

3140 ***Emoia isolata***
Squamata - Scincidae
e Isolated Emo Skink

3141 ***Emoia jakati***
Squamata - Scincidae
e Kopstein's Emo Skink

3142 ***Emoia jamur***
Squamata - Scincidae
e Jamur Emo Skink

3143 ***Emoia klossi***
Squamata - Scincidae
e Kloss's Emo Skink

3144 ***Emoia kordoana***
Squamata - Scincidae
e Meyer's Emo Skink

3145 ***Emoia kuekenthali***
Squamata - Scincidae
e Kueckenthal's Emo Skink

3146 ***Emoia laobaoensis***
Squamata - Scincidae
e Bourret's Emo Skink

3147 ***Emoia lawesi***
Squamata - Scincidae
e Günther's Emo Skink

3148 ***Emoia layaltiensis***
Squamata - Scincidae
e Roux's Emo Skink

3149 ***Emoia longicauda***
Squamata - Scincidae
e Longtail Emo Skink; Long-tailed Emo Skink

3150 ***Emoia loveridgei***
Squamata - Scincidae
e Loveridge's Emo Skink

3151 ***Emoia maculata***
Squamata - Scincidae
e Spotted Emo Skink

3152 ***Emoia maxima***
Squamata - Scincidae
e Great Emo Skink

3153 ***Emoia mivarti***
Squamata - Scincidae
e Boulenger's Emo Skink

3154 ***Emoia montana***
Squamata - Scincidae
e Mountain Emo Skink

3155 ***Emoia murphyi***
Squamata - Scincidae
e Murphy's Emo Skink

3156 ***Emoia nativittatis***
Squamata - Scincidae
e Island Emo Skink

3157 ***Emoia nigra***
Squamata - Scincidae
e Black Emo Skink; Black Skink
d Lackschwarzer Südsee-Skink

3158 ***Emoia nigromarginata***
Squamata - Scincidae
e Black-bordered Emo Skink

3159 ***Emoia obscura***
Squamata - Scincidae

e Obscure Emo Skink

3160 ***Emoia orbita***
Squamata - Scincidae
e Eyelid Emo Skink

3161 ***Emoia pallidiceps***
Squamata - Scincidae
e De Vis's Emo Skink

3162 ***Emoia paniai***
Squamata - Scincidae
e Coastal Emo Skink

3163 ***Emoia parkeri***
Squamata - Scincidae
e Parker's Emo Skink

3164 ***Emoia physicae***
Squamata - Scincidae
e Slender Emo Skink

3165 ***Emoia physicina***
Squamata - Scincidae
e Five-toed Emo Skink

3166 ***Emoia ponapea***
Squamata - Scincidae
e Kiester's Emo Skink

3167 ***Emoia popei***
Squamata - Scincidae
e Pope's Emo Skink

3168 ***Emoia pseudocyanura***
Squamata - Scincidae
e False Bluetail Emo Skink

3169 ***Emoia pseudopallidiceps***
Squamata - Scincidae
e Arboreal Emo Skink

3170 ***Emoia reimschisseli***
Squamata - Scincidae
e Reimschissel's Emo Skink

3171 ***Emoia rennelensis***
Squamata - Scincidae
e Bright Emo Skink

3172 ***Emoia ruficauda***
Squamata - Scincidae
e Redtail Emo Skink

3173 ***Emoia rufilabialis***
Squamata - Scincidae
e Red-lipped Emo Skink

3174 ***Emoia samoensis***
Squamata - Scincidae
e Samoa Emo Skink

3175 ***Emoia sanfordi***
Squamata - Scincidae
e Sanford's Emo Skink

3176 ***Emoia schmidti***
Squamata - Scincidae
e Schmidt's Emo Skink

3177 ***Emoia similis***
Squamata - Scincidae
e Dunn's Emo Skink; Dunn's Emoia

3178 ***Emoia slevini***
Squamata - Scincidae
e Slevin's Emo Skink

3179 ***Emoia sorex***
Squamata - Scincidae
e Sorex Emo Skink

3180 ***Emoia submetallica***
Squamata - Scincidae
e Macleay's Emo Skink

3181 ***Emoia taumakoensis***
Squamata - Scincidae

e Taumako Emo Skink

3182 ***Emoia tetrataenia***
Squamata - Scincidae
e Four-striped Emo Skink

3183 ***Emoia tropidolepis***
Squamata - Scincidae
e Forest Emo Skink

3184 ***Emoia trossula***
Squamata - Scincidae
e Gibbons's Emo Skink

3185 ***Emoia veracunda***
Squamata - Scincidae
e Tropical Emo Skink

3186 **Emydidae**
Testudines
e Emydid Turtles; Pond Turtles; Marsh Turtles; Freshwater Turtles; Pond and Box Turtles
d Sumpfschildkröten
f Tortues subaquatiques; Tortues bourbeuses; Tortues d'eau douce; Tortues fluviatiles; Tortues des marais; Émidés
i Emididi

3187 ***Emydocephalus***
Squamata - Elapidae
e Turtlehead Sea Snakes
d Schildkrötenköpfige Seeschlangen

3188 ***Emydocephalus annulatus***
Squamata - Elapidae
e Ringed Turtlehead Sea Snake; Egg-eating Sea Snake
d Eierfressende Seeschlange; Schildkrötenköpfige Seeschlange

3189 ***Emydocephalus ijimae***
Squamata - Elapidae
e Turtlehead Sea Snake; Ijima's Turtle-headed Sea Snake

3190 ***Emydoidea***
Testudines - Emydidae
e Blanding's Turtles; American Pond Turtles
d Amerikanische Sumpfschildkröten
f Tortues mouchetées; Tortues de Blanding

3191 ***Emydoidea blandingii***
Testudines - Emydidae
e Blanding's Turtle; American Pond Turtle
d Amerikanische Sumpfschildkröte
f Tortue mouchetée (Qué); Tortue de Blanding; Emyde de Blanding
i Tartaruga di Blanding

3192 ***Emydura***
Testudines - Chelidae
e Australian River Turtles; Sharp-snouted Turtles; Australian Mollusc-eating Turtles; Australian Short-necked Turtles
d Spitzkopfschildkröten; Spitzkopfschlangenhalsschildkröten
f Émydures

3193 ***Emydura australis***
Testudines - Chelidae
e Australian Bighead Turtle; Australian Big-headed Turtle; Northwestern Red-faced Tortoise; Australian Sharp-snouted Turtle
d Australische Spitzkopfschildkröte

3194 ***Emydura krefftii***
Testudines - Chelidae
e Krefft's River Turtle; Krefft's Turtle; Krefft's Sharp-snouted Turtle; Krefft's Tortoise
d Krefft-Spitzkopfschildkröte; Kreffts Spitzkopfschildkröte
f Émydure de Krefft

3195 ***Emydura macquarrii***
Testudines - Chelidae
e Murray River Turtle; Murray Turtle; Macquarr's Murray Turtle; Macquarr's Murray River Turtle; Macquarr's Tortoise; Murray River Short-necked Turtle
d Murray-River-Spitzkopfschildkröte; Breitrandspitzkopfschildkröte
f Émydure de Macquarr
i Tartaruga capogrosso

3196 ***Emydura signata***
Testudines - Chelidae
e Brisbane Shortneck Turtle; Brisbane Short-necked Turtle; Brisbane Shortneck Tortoise

3197 ***Emydura subglobosa***
Testudines - Chelidae
e Redbelly Shortneck Turtle; Painted Shortneck Turtle; Red-bellied Short-necked Turtle
d Rotbauchspitzkopfschildkröte; Rundliche Spitzkopfschildkröte
i Tartaruga orecchie rosse al collo corto

3198 ***Emydura victoriae***
Testudines - Chelidae
e Victoria Shortneck Turtle; Victoria Short-necked Turtle

3199 ***Emys***
Testudines - Emydidae
e European Pond Turtles; Freshwater Turtles
d Eigentliche Sumpfschildkröten
f Tortues des marais; Tortues bourbeuses; Tortues boueuses; Tortues jaunes; Cistudes

3200 ***Emys marmorata***
Testudines - Emydidae
e Western Pond Turtle
d Europäische Sumpfschildkröten; Eigentliche Sumpfschildkröten

3201 ***Emys orbicularis***
Testudines - Emydidae
e European Pond Turtle; Swamp Turtle; European Swamp Terrapin; European Swamp Tortoise; European Pond Terrapin; Common European Pond Turtle
d Europäische Sumpfschildkröte; Teichschildkröte
f Tortue bourbeuse; Cistude d'Europe; Cistude des marais; Tortue boueuse; Tortue jaune; Tortue des marais
i Testuggine palustre; Testuggine d'acqua; Tartaruga palustre europea; Tartaruga palustre; Galana d'acqua; Tartuca stizziata; Tartaruga d'acqua comune; Emide europea; Testuggine d'acqua comune; Tartaruga palustre comune; Testuggine palustre europea; Testuggine acquatica; Tartuca do Sciumi

3202 ***Emys orbicularis capolongoi***
Testudines - Emydidae
e Corsican Pond Turtle; Korsische Sumpfschildkröte
d Korsische Sumpfschildkröte

3203 ***Emys orbicularis colchia***
Testudines - Emydidae
e Colchic Pond Turtle; Kolchische Sumpfschildkröte
d Kolchische Sumpfschildkröte

3204 ***Emys orbicularis eiselti***
Testudines - Emydidae
e Eiselt's; Eiselts Sumpfschildkröte
d Eiselts Sumpfschildkröte

3205 ***Emys orbicularis fritsjuergenobsti***
Testudines - Emydidae
e Eastern Spanish Pond Turtle; Ostspanische Sumpfschildkröte
d Ostspanische Sumpfschildkröte

3206 ***Emys orbicularis galloitalica***
Testudines - Emydidae

e Mediterranean Pond Turtle; Mittelländische Sumpfschildkröte
d Mittelländische Sumpfschildkröte

3207 ***Emys orbicularis hellenica***
Testudines - Emydidae
e East Mediterranenan Pond Turtle; Ostmediterrane Sumpfschildkröte
d Ostmediterrane Sumpfschildkröte

3208 ***Emys orbicularis hispanica***
Testudines - Emydidae
e Iberian Pond Turtle; Iberische Sumpfschildkröte
d Iberische Sumpfschildkröte

3209 ***Emys orbicularis lanzai***
Testudines - Emydidae
e Corsican Pond Turtle; Korsische Sumpfschildkröte
d Korsische Sumpfschildkröte

3210 ***Emys orbicularis luteofusca***
Testudines - Emydidae
e Central Anatolian Pond Turtle; Zentralanatolische Sumpfschildkröte
d Zentralanatolische Sumpfschildkröte

3211 ***Emys orbicularis occidentalis***
Testudines - Emydidae
e North African Pond Turtle; Nordafrikanische Sumpfschildkröte
d Nordafrikanische Sumpfschildkröte
f Cistude d'Europe

3212 ***Emys orbicularis orbicularis***
Testudines - Emydidae
e Pontic Pond Turtle; Pontische Sumpfschildkröte
d Pontische Sumpfschildkröte

3213 ***Emys orbicularis persica***
Testudines - Emydidae
e Iranian Pond Turtle; Persische Sumpfschildkröte
d Persische Sumpfschildkröte

3214 ***Enhydrina***
Squamata - Elapidae
e Beaked Sea Snakes

3215 ***Enhydrina schistosa***
Squamata - Elapidae
e Beaked Sea Snake; Common Beaked Sea Snake; Common Sea Snake; Hook-nosed Sea Snake (ISC); Valakadiyan (ISC)
d Schnabelseeschlange

3216 ***Enhydrina zweifeli***
Squamata - Elapidae
e Zweifel's Sea Snake; Zweifel's Beaked Sea Snake

3217 ***Enhydris***
Squamata - Colubridae
e Rainbow Water Snakes; Water Snakes (CSA)
d Orientalische Wassertrugnattern; Choury-Schlangen
f Couleuvres aquatiques

3218 ***Enhydris albomaculatua***
Squamata - Colubridae
e White-spotted Water Snake

3219 ***Enhydris alternans***
Squamata - Colubridae
e Reuss's Water Snake

3220 ***Enhydris bennetti***
Squamata - Colubridae
e Bennett's Water Snake

3221 ***Enhydris bocourti***
Squamata - Colubridae
e Bocourts Water Snake
f Couleuvre aquatique

3222 ***Enhydris chinensis***
Squamata - Colubridae
e Chinese Water Snake

d Chinesiche Wassertrugnatter

3223 ***Enhydris doriae***
Squamata - Colubridae
e Sarawak Water Snake

3224 ***Enhydris dussumieri***
Squamata - Colubridae
e Dussumier's Water Snake

3225 ***Enhydris enhydris***
Squamata - Colubridae
e Rainbow Water Snake; Schneider's Water Snake; Schneider's Smooth Water Snake (ISC)

3226 ***Enhydris indica***
Squamata - Colubridae
e West Malaysia Water Snake

3227 ***Enhydris innominata***
Squamata - Colubridae
e Tay Minh Water Snake

3228 ***Enhydris jagori***
Squamata - Colubridae
e Striped Water Snake; Jagor's Water Snake

3229 ***Enhydris longicauda***
Squamata - Colubridae
e Longhead Water Snake

3230 ***Enhydris maculosa***
Squamata - Colubridae
e Blanford's Spotted Water Snake

3231 ***Enhydris matanensis***
Squamata - Colubridae
e Boulenger's Water Snake

3232 ***Enhydris pahangensis***
Squamata - Colubridae
e Pahang Water Snake

3233 ***Enhydris pakistanica***
Squamata - Colubridae
e Sind River Snake

3234 ***Enhydris plumbea***
Squamata - Colubridae
e Yellowbelly Water Snake; Rice-paddy Snake; Yellow-bellied Water Snake; Plumbeous Water Snake
d Reisfeld-Wassssertrugnatter

3235 ***Enhydris polylepis***
Squamata - Colubridae
e Macleay's Water Snake
d Maclays Wassertrugnatter

3236 ***Enhydris punctata***
Squamata - Colubridae
e Spotted Water Snake

3237 ***Enhydris sieboldi***
Squamata - Colubridae
e Siebold's Water Snake

3238 ***Enhydris smithi***
Squamata - Colubridae
e Smith's Water Snake; Smith's African Water Snake
d Afrikanische Wassernatter

3239 ***Enuliophis***
Squamata - Colubridae
e Colombian Longtail Snakes

3240 ***Enuliophis sclateri***
Squamata - Colubridae
e Colombian Longtail Snake; White-headed Snake

3241 ***Enulius***
Squamata - Colubridae
e Mexican Longtail Snakes

3242 ***Enulius bifoveatus***
Squamata - Colubridae

e Guanaja Longtail Snake

3243 ***Enulius flavitorques***
Squamata - Colubridae
e Pacific Longtail Snake; Common Longtail Snake

3244 ***Enulius oligostichus***
Squamata - Colubridae
e Mexican Longtail Snake

3245 ***Enulius roatanensis***
Squamata - Colubridae
e Roatan Longtail Snake

3246 ***Enyalioides***
Squamata - Iguanaidae
e Dwarf Iguanas; Wood Lizards

3247 ***Enyalioides cofanorum***
Squamata - Iguanaidae
e Duellman's Dwarf Iguana; Cofan Wood Lizard

3248 ***Enyalioides heterolepis***
Squamata - Iguanaidae
e Bocourt's Dwarf Iguana

3249 ***Enyalioides laticeps***
Squamata - Iguanaidae
e Guichenot's Dwarf Iguana; Amazon Wood Lizard

3250 ***Enyalioides microlepis***
Squamata - Iguanaidae
e Tiny-scale Dwarf Iguana; Brown Wood Lizard

3251 ***Enyalioides oshaughnessyi***
Squamata - Iguanaidae
e O'Shaughnessy's Dwarf Iguana

3252 ***Enyalioides palpebralis***
Squamata - Iguanaidae
e Boulenger's Dwarf Iguana; Horned Wood Lizard

3253 ***Enyalioides praestabilis***
Squamata - Iguanaidae
e Dwarf Iguana; Blue-spotted Wood Lizard

3254 ***Enyalius***
Squamata - Iguanaidae
e Fathead Anoles; Brazilian Lizards
d Brasilianische Leguane
f Enyales

3255 ***Enyalius bileneatus***
Squamata - Iguanaidae
e Two-lined Fathead Anole; Two-lined Brazilian Lizards
f Enyale à deux raies

3256 ***Enyalius brasiliensis***
Squamata - Iguanaidae
e Brazilian Fathead Anole; Brazilian Chameleon; Brazilian Tree Lizard

3257 ***Enyalius catenatus***
Squamata - Iguanaidae
e Wied's Fathead Anole
d Brasilianisches Chamäleon; Brasiliansicher Leguan
f Enyale rhombifère

3258 ***Enyalius iheringii***
Squamata - Iguanaidae
e Ihering's Fathead Anole

3259 ***Enyalius leechii***
Squamata - Iguanaidae
e Leech's Fathead Anole

3260 ***Enyalius perdites***
Squamata - Iguanaidae
e Jackson's Fathead Anole

3261 ***Eopleopus***
Squamata - Teiidae

e Gaudichaud's Eopleopuses

3262 ***Eopleopus guadichaudii***
Squamata - Teiidae
e Gaudichaud's Eopleopus

3263 ***Ephalophis***
Squamata - Elapidae
e Grey's Sea Snakes

3264 ***Ephalophis greyi***
Squamata - Elapidae
e Grey's Sea Snake

3265 ***Epicrates***
Squamata - Boidae
e Rainbow Boas; Slender Boas; Tree Boas
d Schlankboas; Glattlippenboas
f Boas; Boas arc-en-ciel; Serpents foulard (Ants)

3266 ***Epicrates angulifer***
Squamata - Boidae
e Cuban Boa; Pale-headed Tree Boa
d Kuba-Schlankboa; Kuba-Boa; Kubanische Schlankboa
f Boa de Cuba

3267 ***Epicrates cenchria***
Squamata - Boidae
e Rainbow Boa; Thick-necked Tree Boa; Common Rainbow Boa
d Bunte Regenbogenboa; Regenbogenboa
f Aboma; Arc-en-ciel; Boa arc-en-ciel; Serpent foulard (Ants)

3268 ***Epicrates cenchria alvarezie***
Squamata - Boidae
e Argentine Rainbow Boa
d Argentinische Regenbogenboa

3269 ***Epicrates cenchria cenchria***
Squamata - Boidae
e Red Rainbow Boa; Brazilian Rainbow Boa
d Rote Regenbogenboa; Brasilianische Regenbogenboa; Bunte Regenbogenboa

3270 ***Epicrates cenchria crassus***
Squamata - Boidae
e Rainbow Boa
d Regenbogenboa

3271 ***Epicrates cenchria maurus***
Squamata - Boidae
e Brown Rainbow Boa; Columbian Rainbow Boa
d Braune Regenbogenboa; Kolumbianische Regenbogenboa
f Boa arc-en-ciel de Colombie; Serpent foulard (Ants)

3272 ***Epicrates chrysogaster***
Squamata - Boidae
e Turks Island Boa; Southern Bahamas Boa (WI)
d Bahama-Schlankboa

3273 ***Epicrates exsul***
Squamata - Boidae
e Abaco Island Boa; Exsul Boa; Northern Bahamas Boa (WI)
d Abaco-Insel-Schlankboa; Great-Abaco-Schlankboa

3274 ***Epicrates fordi***
Squamata - Boidae
e Ford's Boa; Dominican Boa; Hispaniolan Desert Boa
d Fords Boa; Ford-Boa

3275 ***Epicrates gracilis***
Squamata - Boidae
e Hispaniola Boa; Hispaniolan Gracile Boa

3276 ***Epicrates inornatus***
Squamata - Boidae
e Puerto Rica Boa; Puerto Rican Boa; Yellow-headed Boa; Puerto Rican

Tree Boa ; Puerto Rican Giant Boa
d Puerto-Rico-Schlankboa; Puerto-Rico-Boa
f Boa de Porto Rico

3277 ***Epicrates monensis***
Squamata - Boidae
e Mona Island Boa; Virgin Islands Boa; Mona Boa; Tree Boa ; Puerto Rican Dwarf Boa
d Mona-Schlankboa
f Boa des Îles Vierges; Boa de l'île Mona

3278 ***Epicrates striatus***
Squamata - Boidae
e Haitian Boa; Bahama Boa (NA); Bimini Streaked Tree Boa; Hispaniolan Giant Boa (WI)
d Haiti-Boa; Silbergraue Schlankboa; Schlankboa

3279 ***Epicrates striatus ailurus***
Squamata - Boidae
e Cat Island Boa

3280 ***Epicrates striatus exagistus***
Squamata - Boidae
e Tiburon Peninsula Boa

3281 ***Epicrates striatus fosteri***
Squamata - Boidae
e Bimini Boa
d Bimini-Boa; Bimini-Schlankboa

3282 ***Epicrates striatus fowleri***
Squamata - Boidae
e Berry Island Boa

3283 ***Epicrates striatus mccraniei***
Squamata - Boidae
e Ragged Islands Boa

3284 ***Epicrates striatus striatus***
Squamata - Boidae
e Fischer's Boa; Haitian Boa
d Silbergraue Schlankboa

3285 ***Epicrates striatus strigulatus***
Squamata - Boidae
e New Providence Island Boa; Eleuthera Island Boa; Nassau Boa

3286 ***Epicrates striatus warreni***
Squamata - Boidae
e Ile de la Tortue Boa

3287 ***Epicrates subflavus***
Squamata - Boidae
e Jamaica Boa; Jamaican Boa (WI)
d Jamaika-Boa; Jamaika Schlankboa
f Boa de la Jamaïque

3288 ***Eremias***
Squamata - Lacertidae
e Racerunners; Desert Lacertas
d Wüstenrenner; Wüstenrennereidechsen; Buckelnasen
f Érémias

3289 ***Eremias acutirostris***
Squamata - Lacertidae
e Point-snouted Racerunner; Reticulate Desert Lacerta

3290 ***Eremias argus***
Squamata - Lacertidae
e Mongolia Racerunner; Mongolian Racerunner
d Mongolischer Wüstenrenner; Steppenrenner

3291 ***Eremias brenchleyi***
Squamata - Lacertidae
e Ordos Racerunner

3292 ***Eremias buechneri***
Squamata - Lacertidae
e Kaschar Racerunner

d Kaschgar-Wüstenrenner

3293 ***Eremias ercolinii***
Squamata - Lacertidae
e Lanza's Racerunner

3294 ***Eremias grammica***
Squamata - Lacertidae
e Reticulate Racerunner

3295 ***Eremias lineoocellata***
Squamata - Lacertidae
e Ocellated Racerunner; Ocellated Sand Lizard (CSA)
d Fleckenstreifenwüstenrenner

3296 ***Eremias lugubris***
Squamata - Lacertidae
e Mourning Racerunner; Dull Lizard; Black-and-yellow Sand Lizard (CSA)
f Lézard lugubre

3297 ***Eremias multiocellata***
Squamata - Lacertidae
e Multi-ocellated Racerunner
d Kleinfleckiger Wüstenrenner; Vierfleckiger Wüstenrenner

3298 ***Eremias przewalskii***
Squamata - Lacertidae
e Gobi Racerunner
d Przewalskis Wüstenrenner; Schachbrettwüstenrenner

3299 ***Eremias quadrifrons***
Squamata - Lacertidae
e Alachan Racerunner
d Alaschan-Wüstenrenner

3300 ***Eremias suphani***
Squamata - Lacertidae
e Basoglu Racerunner
d Türkischer Wüstenrenner
f Eremias de Süphan
i Eremia di Suphan

3301 ***Eremias velox***
Squamata - Lacertidae
e Rapid Racerunner; Rapid Fringe-toed Lizard; Racerunner
d Kaukasischer Wüstenrenner; Schneller Wüstenrenner
f Érémias véloce
i Eremia veloce

3302 ***Eremias vermiculata***
Squamata - Lacertidae
e Variegated Racerunner
d Zentralasien-Wüstenrenner

3303 ***Eremiascincus***
Squamata - Scincidae
e Sand Swimmers; Sand-swimming Skinks
d Sandschwimmer

3304 ***Eremiascincus fasciolatus***
Squamata - Scincidae
e Narrow-banded Sand Swimmer; Striped Sand-swimming Skink
d Schmalbändiger Sandschwimmer

3305 ***Eremiascincus richardsonii***
Squamata - Scincidae
e Broad-banded Sand Swimmer; Richardson's Sand-swimming Skink
d Breitbändiger Sandschwimmer

3306 ***Eretmochelys***
Testudines - Cheloniidae
e Hawksbill Turtles; Hawksbills; Hawksbill Sea Turtles
d Echte Karettschildkröten; Echte Karettenschildkröten
f Carets

3307 ***Eretmochelys imbricata***
Testudines - Cheloniidae
e Hawksbill Turtle; Hawksbill; Hawkbill; Tortoiseshell Turtle; Hawksbill Sea Turtle; Carey
d Echte Karettschildkröte; Echte Karettenschildkröte; Echte Karette

f Tortue marine à écailles; Caret; Tortue à écailles; Tortue tuilée; Tortue imbriquée; Tortue caret
i Tartaruga embricata

3308 ***Eretmochelys imbricata bissa***
Testudines - Cheloniidae
e Pacific Hawksbill; Pacific Hawksbill Sea Turtle

3309 ***Eretmochelys imbricata imbricata***
Testudines - Cheloniidae
e Atlantic Hawksbill; Hawksbill; Hawkbill; Tortoiseshell Turtle; Atlantic Hawksbill Sea Turtle

3310 ***Eridiphas***
Squamata - Colubridae
e Baja California Night Snakes

3311 ***Eridiphas marcosensis***
Squamata - Colubridae
e San Marcos Island Night Snake

3312 ***Eridiphas sleveni***
Squamata - Colubridae
e Baja California Night Snake

3313 ***Eridiphas sleveni sleveni***
Squamata - Colubridae
e Baja California Night Snake

3314 ***Eristicophis***
Squamata - Viperidae
e MacMahon's Vipers; MacMahon's Desert Vipers; Leaf-nosed Vipers
d MacMahon-Vipern
f Vipères de MacMahon

3315 ***Eristicophis macmahonii***
Squamata - Viperidae
e MacMahon's Viper; MacMahon's Desert Viper; Leaf-nosed Viper
d MacMahon-Viper
f Vipère de Macmahon; Vipère des sables

3316 ***Erpeton***
Squamata - Colubridae
e Tentacled Snakes; Tentacle Snakes
d Fühlerschlangen; Tentakelschlangen

3317 ***Erpeton tentaculatum***
Squamata - Colubridae
e Tentacled Snake; Tentacle Snake; Fishing Snake
d Fühlerschlange; Tentakelschlange
i Ngu-kradan; Serpente tentacolato

3318 ***Erymnochelys***
Testudines - Pelomedusidae
e Madagascar Bighead Turtles
d Madagaskar-Schienenschildkröten
f Podocnémides

3319 ***Erymnochelys madagascariensis***
Testudines - Pelomedusidae
e Madagascar Bighead Turtle; Madagascan Big-headed Side-necked Turtle
d Madagaskar-Schienenschildkröte
f Podocnémide de Madagascar; Podocnémis de Madagascar; Réré
i Podocnemide del Madagascar

3320 ***Erythrolamprus***
Squamata - Colubridae
e False Coral Snakes
d Falsche Korallennattern

3321 ***Erythrolamprus aesculapii***
Squamata - Colubridae
e Aesculapian False Coral Snake; Tropical American Coral Mimic (NA); Aesculapian Coral Snake; False Coral Snake
d Falsche Äsculap-Korallennatter
f Faux corail

3322 ***Erythrolamprus aesculapii aesculapii***
Squamata - Colubridae

e Common False Coral Snake

3323 ***Erythrolamprus bizona***
Squamata - Colubridae
e Jan's False Coral Snake; Double-banded False Coral Snake

3324 ***Erythrolamprus guentheri***
Squamata - Colubridae
e Günther's False Coral Snake; Pink-naped False Coral Snake

3325 ***Erythrolamprus mimus***
Squamata - Colubridae
e Mimic False Coral Snake; Single-banded False Coral Snake

3326 ***Erythrolamprus pseudocorallus***
Squamata - Colubridae
e False Coral Snake

3327 ***Eryx***
Squamata - Boidae
e Sand Boas; Earth Boas (ISC)
d Sandboas
f Éryx; Javelots; Boas des sables

3328 ***Eryx colubrinus***
Squamata - Boidae
e Egyptian Sand Boa
d Ägyptische Sandboa; Ägyptische Schlankboa
f Boa du sable; Boa des sables d'Égypte

3329 ***Eryx colubrinus loveridgi***
Squamata - Boidae
e Kenya Sand Boa; East African Sand Boa
d Kenya-Sandboa; Sandboa

3330 ***Eryx conicus***
Squamata - Boidae
e Rough-scaled Sand Boa; Rough-tailed Sand Boa; Russell's Sand Boa; Indian Sand Boa; Russell's Earth Boa (ISC); Common Sand Boa
d Gefleckte Sandboa; Bunte Sandboa
f Boa du sable; Boa aux écailles rugueuses; Boa des sables

3331 ***Eryx elegans***
Squamata - Boidae
e Central Asia Sand Boa

3332 ***Eryx jaculus***
Squamata - Boidae
e Caucasian Sand Boa; Javelin Sand Boa; Egyptian Sand Boa; European Sand Boa; Spotted Sand Boa; Turkish Sand Boa; Sand Boa
d Europäische Sandboa; Sandboa; Westliche Sandboa; Türkische Sandboa; Sandschlange
f Javelot; Éryx-javelot; Boa des sables occidental; Boa des sables européen; Boa-javelot
i Erice iaculo; Boa delle sabbie occidentale

3333 ***Eryx jayakari***
Squamata - Boidae
e Arabian Sand Boa; Jayakar's Sand Boa

3334 ***Eryx johnii***
Squamata - Boidae
e Brown Sand Boa; Indian Sand Boa; Indian Boa; Two-headed Snake; John's Earth Boa (ISC); John's Sand Boa (ISC)
d Indische Sandboa
f Éryx de John; Boa des sables; Boa de l'Inde; Boa des sables de l'Inde

3335 ***Eryx miliaris***
Squamata - Boidae
e Dwarf Sand Boa; Desert Sand Boa
d Wüstensandboa; Zwergsandboa; Östliche Sandboa
f Boa des sables oriental; Eryx miliaire
i Boa delle sabbie orientale

3336 ***Eryx muelleri***
Squamata - Boidae
e Müller's Sand Boa; Bosenborgo Sand Python
f Boa du sable

3337 ***Eryx somalicus***
Squamata - Boidae
e Somali Sand Boa; Somalian Sand Boa

3338 ***Eryx tataricus***
Squamata - Boidae
e Tartar Sand Boa; Tartary Sand Boa
d Tatarische Sandboa; Große Sandboa; Sandboa

3339 ***Etheridgeum***
Squamata - Colubridae
e Sumatra Etheridge Snakes

3340 ***Etheridgeum pulchrum***
Squamata - Colubridae
e Sumatra Etheridge Snake

3341 **Eublepharidae**
Squamata
e Eyelid Geckos
d Geckos; Lidgeckos
f Eublepharidés

3342 ***Eublepharis***
Squamata - Eublepharidae
e Eyelid Geckos; Fat-tailed Geckos; Leopard Geckos (ISC)
f Geckos-panthère

3343 ***Eublepharis angramainyu***
Squamata - Eublepharidae
e Iraqui Eyelid Gecko

3344 ***Eublepharis hardwickii***
Squamata - Eublepharidae
e Hardwicke's Eyelid Gecko

3345 ***Eublepharis macularius***
Squamata - Eublepharidae
e Leopard Gecko (ISC); Fat-tailed Gecko; Spotted Cat-eyed Gecko (ISC)
d Leopardgecko; Panthergecko
f Gecko-panthère à paupières mobiles; Gecko-léopard
i Geco leopardo

3346 ***Eublepharis turcmenicus***
Squamata - Eublepharidae
e Turkmenistan Eyelid Gecko; Turkmenistan Fat-tailed Gecko

3347 ***Eugongylus***
Squamata - Scincidae
e New Caledonia Skinks; Indo-Pacific Skinks

3348 ***Eugongylus albofasciolatus***
Squamata - Scincidae
e White-striped Cape Skink; White-striped Sheen Skink

3349 ***Eugongylus haraldmeieri***
Squamata - Scincidae
e New Caledonia Skink

3350 ***Eugongylus microlepis***
Squamata - Scincidae
e Tonga Skink

3351 ***Eugongylus rufescens***
Squamata - Scincidae
e Purple Cape Skink; Bar-lipped Sheen Skink

3352 ***Eugongylus sulaensis***
Squamata - Scincidae
e Sula Skink

3353 ***Eulamprus***
Squamata - Scincidae
e Five-fingered Skinks

3354 ***Eulamprus amplus***
Squamata - Scincidae
e Olive Five-fingered Skink

3355 ***Eulamprus heatwolei***
Squamata - Scincidae
e Heatwole's Five-fingered Skink; Yellow-bellied Water Skink (ANZ)

3356 ***Eulamprus kosciuskoi***
Squamata - Scincidae
e Alpine Water Skink

3357 ***Eulamprus leuraensis***
Squamata - Scincidae
e Blue Mountains Tree Skink

3358 ***Eulamprus luteilateralis***
Squamata - Scincidae
e Clarke Range Tree Skink

3359 ***Eulamprus quoyii***
Squamata - Scincidae
e Eastern Water Skink

3360 ***Eulamprus tympanum marniae***
Squamata - Scincidae
e Corangamite Water Skink

3361 ***Eumeces***
Squamata - Scincidae
e Eyelid Skinks; Long-legged Skinks; Long Skinks; Opaque-lidded Skinks; Skinks; Great Skinks (NA)
d Langbeinige Skinke
f Scinques lignées; Eumécès

3362 ***Eumeces afer***
Squamata - Scincidae
e Peters's Eyelid Skink

3363 ***Eumeces algerensis***
Squamata - Scincidae
e Algerian Skink; Northern Skink; Berber Skink; Algerian Orange-tailed Skink
d Berberskink
f Eumécès d'Algérie; Scinque d'Algérie; Scinque berber

3364 ***Eumeces altamirani***
Squamata - Scincidae
e Tepalcatepec Skink

3365 ***Eumeces anthracinus***
Squamata - Scincidae
e Coal Skink
f Scinque anthracite

3366 ***Eumeces anthracinus anthracinus***
Squamata - Scincidae
e Northern Coal Skink

3367 ***Eumeces anthracinus pluvialis***
Squamata - Scincidae
e Southern Coal Skink

3368 ***Eumeces barbouri***
Squamata - Scincidae
e Barbour's Eyelid Skink

3369 ***Eumeces bocourti***
Squamata - Scincidae
e Bocourt's Eyelid Skink
f Scinque de Bocourt

3370 ***Eumeces brevirostris***
Squamata - Scincidae
e Shortnose Skink

3371 ***Eumeces callicephalus***
Squamata - Scincidae
e Fair-headed Skink; Mountain Skink (NA)
f Scinque des montagnes

3372 ***Eumeces capito***
Squamata - Scincidae
e Gail's Skink

3373 ***Eumeces chinensis***
Squamata - Scincidae
e Chinese Skink
f Eumécès de Chine

3374 ***Eumeces colimensis***
Squamata - Scincidae
e Colima Skink

3375 ***Eumeces copei***
Squamata - Scincidae
e Cope's Skink

3376 ***Eumeces coreensis***
Squamata - Scincidae
e Smith's Skink

3377 ***Eumeces dugesii***
Squamata - Scincidae
e Duges's Skink

3378 ***Eumeces egregius***
Squamata - Scincidae
e Blue-tailed Mole Skink; Florida Skink; Mole Skink; Striped Red-tailed Skink

3379 ***Eumeces egregius egregius***
Squamata - Scincidae
e Florida Keys Mole Skink

3380 ***Eumeces egregius insularis***
Squamata - Scincidae
e Cedar Key Mole Skink

3381 ***Eumeces egregius lividus***
Squamata - Scincidae
e Bluetail Mole Skink; Blue-tailed Mole Skink

3382 ***Eumeces egregius onocrepis***
Squamata - Scincidae
e Pensinsula Mole Skink

3383 ***Eumeces egregius similis***
Squamata - Scincidae
e Northern Mole Skink

3384 ***Eumeces elegans***
Squamata - Scincidae
e Shanghai Elegant Skink

3385 ***Eumeces fasciatus***
Squamata - Scincidae
e Five-lined Skink; Common Five-lined Skink (NA)
d Fünfstreifenskink; Streifenskink
f Scinque pentaligne; Eumécès à bandes; Scinque à cinque lignes; Scinque à bandes

3386 ***Eumeces gilberti***
Squamata - Scincidae
e Gilbert's Skink; Greater Western Skink; Gilbert Skink
d Gilberts Skink
f Scinque de Gilbert

3387 ***Eumeces gilberti arizonensis***
Squamata - Scincidae
e Arizona Skink

3388 ***Eumeces gilberti cancellosus***
Squamata - Scincidae
e Variegated Skink

3389 ***Eumeces gilberti gilberti***
Squamata - Scincidae
e Greater Brown Skink

3390 ***Eumeces gilberti placerencis***
Squamata - Scincidae
e Northern Brown Skink

3391 ***Eumeces gilberti rubricaudatus***
Squamata - Scincidae
e Western Red-tailed Skink

3392 ***Eumeces inexpectatus***
Squamata - Scincidae
e Southeastern Five-lined Skink; Florida Five-lined Skink; Southern Five-lined Skink
d Fünfbindenskink
f Scinque pentaligne du Sud-ouest

3393 ***Eumeces kishinouyei***
Squamata - Scincidae
e Japanese Skink

3394 ***Eumeces lagunensis***
Squamata - Scincidae
e San Lucan Skink

3395 ***Eumeces lateralis***
Squamata - Scincidae
e Ground Skink

3396 ***Eumeces laticeps***
Squamata - Scincidae
e Broad-headed Skink; Broadhead Skink; Greater Five-lined Skink
d Breitkopfskink; Rotkopfskink
f Scinque à tête large; Eumécès à tête large; Scinque à queue bleue

3397 ***Eumeces latiscutatus***
Squamata - Scincidae
e Far Eastern Skink

3398 ***Eumeces livi***
Squamata - Scincidae
e Hikida's Skink

3399 ***Eumeces longirostris***
Squamata - Scincidae
e Longnose Skink

3400 ***Eumeces lynxe***
Squamata - Scincidae
e Oak Forest Skink

3401 ***Eumeces managuae***
Squamata - Scincidae
e Managua Skink; Pacific Giant Skink

3402 ***Eumeces marginatus***
Squamata - Scincidae
e Ousima Skink

3403 ***Eumeces multilineatus***
Squamata - Scincidae
e Chihuahuan Skink

3404 ***Eumeces multivirgatus***
Squamata - Scincidae
e Many-lined Skink
f Scinque multiligne

3405 ***Eumeces multivirgatus epipleurotes***
Squamata - Scincidae
e Many-lined Skink; Variable Skink

3406 ***Eumeces multivirgatus gaigeae***
Squamata - Scincidae
e Variable Skink

3407 ***Eumeces multivirgatus multivirgatus***
Squamata - Scincidae
e Northern Many-lined Skink

3408 ***Eumeces obsoletus***
Squamata - Scincidae
e Great Plains Skink; Brown Skink (NA); Sonoran Skink
d Rotkopfskink; Spitzkopfskink
f Scinque des plaines

3409 ***Eumeces ochoterenae***
Squamata - Scincidae
e Guerreran Skink

3410 ***Eumeces okadae***
Squamata - Scincidae
e Okada's Skink

3411 ***Eumeces parviauriculatus***
Squamata - Scincidae
e Northern Pygmy Skink

3412 ***Eumeces parvulus***
Squamata - Scincidae
e Southern Pygmy Skink

3413 ***Eumeces poonaensis***
Squamata - Scincidae
e Poona Skink

3414 ***Eumeces popei***
Squamata - Scincidae
e Pope's Skink

3415 ***Eumeces quadrilineatus***
Squamata - Scincidae
e Hong Kong Skink

3416 ***Eumeces schneideri***
Squamata - Scincidae
e Schneider's Skink; Schneider's Gold Skink; Orange-tailed Skink; Gold Skink
d Tüpfelskink; Fleckenskink
f Euméćès de Schneider; Scinque de Schneider
i Eumece di Schneider

3417 ***Eumeces schwartzei***
Squamata - Scincidae
e Schwartze's Skink; Yucatan Giant Skink

3418 ***Eumeces septentrionalis***
Squamata - Scincidae
e Prairie Skink; Northern Prairie Skink
f Scinque des prairies

3419 ***Eumeces septentrionalis obtusirostris***
Squamata - Scincidae
e Southern Prairie Skink

3420 ***Eumeces septentrionalis pallidus***
Squamata - Scincidae
e Pallid Skink

3421 ***Eumeces septentrionalis septentrionalis***
Squamata - Scincidae
e Northern Prairie Skink; Black-banded Skink
f Scinque des Prairies

3422 ***Eumeces skiltonianus***
Squamata - Scincidae
e Western Skink; Skilton's Skink; Common Western Skink
d Skilton Skink
f Scinque de l'Ouest

3423 ***Eumeces skiltonianus interparietalis***
Squamata - Scincidae
e Coronado Island Skink; Coronado Skink (NA)

3424 ***Eumeces skiltonianus skiltonianus***
Squamata - Scincidae
e Skilton Skink; Skilton's Skink

3425 ***Eumeces skiltonianus utahensis***
Squamata - Scincidae
e Great Basin Skink

3426 ***Eumeces stimsonii***
Squamata - Scincidae
e Stimson's Skink

3427 ***Eumeces sumichrasti***
Squamata - Scincidae
e Sumichrasts's Skink

3428 ***Eumeces taeniolatus***
Squamata - Scincidae
e Alpine Punjab Skink

3429 ***Eumeces tamdaoensis***
Squamata - Scincidae

e Vietnam Skink; Vietnamese Skink

3430 ***Eumeces tetragrammus***
Squamata - Scincidae
e Four-lined Skink; Four-striped Skink
f Scinque tétraligne

3431 ***Eumeces tetragrammus brevilineatus***
Squamata - Scincidae
e Short-lined Skink

3432 ***Eumeces tetragrammus tetragrammus***
Squamata - Scincidae
e Four-lined Skink; Long-lined Skink

3433 ***Eumecia***
Squamata - Scincidae
e Western Serpentiform Skinks; Serpentiform Skinks

3434 ***Eumecia anchietae***
Squamata - Scincidae
e Western Serpentiform Skink

3435 ***Eunectes***
Squamata - Boidae
e Anacondas; Water Boas
d Anakondas; Wasserboas
f Eunectes; Anacondas

3436 ***Eunectes barbouri***
Squamata - Boidae
e Barbour's Anaconda

3437 ***Eunectes deschauenseei***
Squamata - Boidae
e Dark-spotted Anaconda; Deschauensee's Anaconda
f Anaconda; Couleuvre d'eau (Ants)

3438 ***Eunectes murinus***
Squamata - Boidae
e Green Anaconda; Anaconda; Water Boa
d Große Anakonda; Anakonda; Grüne Anakonda
f Anaconda; Anaconda vert; Couleuvre d'eau (Ants)
i Anaconda; Anaconda verde

3439 ***Eunectes murinus murinus***
Squamata - Boidae
e Common Anaconda

3440 ***Eunectes notaeus***
Squamata - Boidae
e Yellow Anaconda; Southern Anaconda; Paraguay Anaconda
d Südanakonda; Gelbe Anakonda; Paraguay-Anakonda; Südliche Anakonda
f Anaconda jaune; Anaconda de Paraguay
i Anaconda giallo; Anaconda

3441 ***Euprepes***
Squamata - Scincidae
e Serpentiform Skinks

3442 ***Euprepes acutilabris***
Squamata - Scincidae
e Sharp-lipped Skink

3443 ***Euprepes aureopunctatus***
Squamata - Scincidae
e Gold-dotted Skink

3444 ***Euprepes bilineatus***
Squamata - Scincidae
e Grandidier's Two-lined Skink

3445 ***Euprepes binotatus***
Squamata - Scincidae
e Bocage's Serpentiform Skink

3446 ***Euprepes chaperi***
Squamata - Scincidae
e Serpentiform Skink

3447 ***Euprepes vittatus***
Squamata - Scincidae
e Banded Skink

3448 ***Eurydactylodes***
Squamata - Gekkonidae
e Prehensile-tail Geckos

3449 ***Eurydactylodes symmetricus***
Squamata - Gekkonidae
e Symmetrical Gecko

3450 ***Eurydactylodes vieillardi***
Squamata - Gekkonidae
e Bavay's Gecko

3451 ***Euspondylus***
Squamata - Teiidae
e Sun Tegus

3452 ***Euspondylus acutirostris***
Squamata - Teiidae
e Sharp-snouted Sun Tegu

3453 ***Euspondylus ampuedae***
Squamata - Teiidae
e Lancini's Sun Tegu

3454 ***Euspondylus guentheri***
Squamata - Teiidae
e Günther's Sun Tegu

3455 ***Euspondylus maculatus***
Squamata - Teiidae
e Spotted Sun Tegu

3456 ***Euspondylus phelpsi***
Squamata - Teiidae
e Phelps's Sun Tegu

3457 ***Euspondylus rahmi***
Squamata - Teiidae
e Rahm's Sun Tegu

3458 ***Euspondylus simonsii***
Squamata - Teiidae
e Simons's Sun Tegu

3459 ***Euspondylus spinalis***
Squamata - Teiidae
e Boulenger's Sun Tegu

3460 ***Evesia***
Squamata - Scincidae
e Bell's Skinks

3461 ***Evesia bellii***
Squamata - Scincidae
e Bell's Skink

3462 ***Exelencophis***
Squamata - Colubridae
e Maria Madre Snakes

3463 ***Exelencophis neslsoni***
Squamata - Colubridae
e Maria Madre Snake

3464 ***Exiliboa***
Squamata - Boidae
e Oaxacan Dwarf Boas; Mexican Dwarf Boas; Bogert's Boas
d Mexikanische Zwergboas

3465 ***Exiliboa placata***
Squamata - Boidae
e Oaxacan Dwarf Boa; Bogert's Boa
d Oaxaca-Zwergboa; Mexikanische Zwergboa

F

3466 ***Farancia***
Squamata - Colubridae
e Mud Snakes; Rainbow Mud Snakes; Horn Snakes; Rainbow Snakes
d Schlammnattern; Halsbandnattern; Regenbogennattern
f Couleuvres arc-en-ciel

3467 ***Farancia abacura***
Squamata - Colubridae
e Mud Snake; Horn Snake; Red-bellied Mud Snake
d Schlammnatter
f Couleuvre mosaïque; Couleuvre cornue

3468 ***Farancia abacura abacura***
Squamata - Colubridae
e Eastern Mud Snake; Mud Snake; Stinging Snake (NA); Horn Snake (NA)
d Östliche Schlammnatter

3469 ***Farancia abacura reinwardtii***
Squamata - Colubridae
e Western Mud Snake
d Westliche Schlammnatter

3470 ***Farancia erythrogramma***
Squamata - Colubridae
e Rainbow Snake; Rainbow Mud Snake
d Regenbogennatter
f Couleuvre arc-en-ciel

3471 ***Farancia erythrogramma erythrogramma***
Squamata - Colubridae
e Rainbow Snake; Common Rainbow Snake
d Regenbogennatter
f Couleuvre arc-en-ciel

3472 ***Farancia erythrogramma seminola***
Squamata - Colubridae
e South Florida Rainbow Snake; Southern Florida Rainbow Snake

3473 ***Feylinia***
Squamata - Scincidae
e Feylinias
d Afrikanische Schlangenechsen

3474 ***Feylinia boulengeri***
Squamata - Scincidae
e Boulenger's Feylinia

3475 ***Feylinia currori***
Squamata - Scincidae
e Western Forest Feylinia; Curror's Feylinia

3476 ***Feylinia elegans***
Squamata - Scincidae
e Elegant Feylinia

3477 ***Feylinia polylepis***
Squamata - Scincidae
e Many-scaled Feylinia

3478 ***Ficimia***
Squamata - Colubridae
e Hooknose Snakes; Mexican Hooknose Snakes; Hook-nosed Snakes; Mexican Hook-nosed Snakes; Eastsern Hook-nosed Snakes
d Hakennasennattern; Mexikanischen Hakennasennattern; Sichelnattern
f Couleuvres à nez crochu

3479 ***Ficimia olivacea***
Squamata - Colubridae
e Brown Hooknose Snake; Mexican Hooknose Snake; Mexican Hook-

nosed Snake
d Mexikanische Hakennasennatter

3480 ***Ficimia publia***
Squamata - Colubridae
e Blotched Hooknose Snake; Cope's Hook-nosed Snake; Blotched Hook-nosed Snake

3481 ***Ficimia ramirezi***
Squamata - Colubridae
e Ramirez's Hooknose Snake; Ramirez's Hook-nosed Snake

3482 ***Ficimia ruspator***
Squamata - Colubridae
e Guerreran Hooknose Snake

3483 ***Ficimia streckeri***
Squamata - Colubridae
e Mexican Hooknose Snake; Mexican Hook-nosed Snake; Strecker's Hook-nosed Snake; Tamaulipan Hooknose Snake; Tamaulipan Hook-nosed Snake
f Couleuvre à nez crochu du Mexique

3484 ***Ficimia variegata***
Squamata - Colubridae
e Tehuantepec Hooknose Snake

3485 ***Fimbrios***
Squamata - Colubridae
e Bearded Snakes
d Lippennattern

3486 ***Fimbrios klossi***
Squamata - Colubridae
e Bearded Snake; Kloss's Snake
d Lippennatter

3487 ***Fojia***
Squamata - Scincidae
e Fojia Skinks

3488 ***Fojia bumui***
Squamata - Scincidae
e Fojia Skink

3489 ***Fordonia***
Squamata - Colubridae
e Whitebelly Mangrove Snakes; Crab-eting Water Snakes
d Krebstrugnattern
f Serpents d'eau à ventre blanc

3490 ***Fordonia leucobalia***
Squamata - Colubridae
e Whitebelly Mangrove Snake; Crab-eating Water Snake; White-bellied Mangrove Snake (ANZ)
d Krebstrugnatter
f Serpent d'eau à ventre blanc

3491 ***Furina***
Squamata - Elapidae
e Naped Snakes; Red-naped Snakes

3492 ***Furina barnardi***
Squamata - Elapidae
e Yellow-naped Snake
d Gelbkragenotter

3493 ***Furina diadema***
Squamata - Elapidae
e Red-naped Snake; Diadem Snake
d Rothalsotter

3494 ***Furina dunmalli***
Squamata - Elapidae
e Dunmall's Snake
d Dunmalls Schlange

3495 ***Furina ornata***
Squamata - Elapidae
e Orange-naped Snake

3496 ***Furina tristis***
Squamata - Elapidae

e Brown-headed Snake
d Australische Braunkopfnatter

G

3497 ***Gallotia***
Squamata - Lacertidae
e Gallot's Lizards; Gallotias; Canarian Lizards; Canary Island Lizard
d Kanaren-Eidechsen

3498 ***Gallotia atlantica***
Squamata - Lacertidae
e Atlantic Lizard
d Atlantische Eidechse; Atlantische Kanareneidechse; Purpurarieneidechse

3499 ***Gallotia galloti***
Squamata - Lacertidae
e Gallot's Lizard; West Canaries Gallotia
d Kanaren-Eidechse
f Lézard de Gallot

3500 ***Gallotia simonyi***
Squamata - Lacertidae
e Simony's Lizard; Hierro Giant Lizard
d Hierro-Rieseneidechse
f Lézard géant de Hierro

3501 ***Gambelia***
Squamata - Iguanaidae
e Leopard Lizards
d Leopardleguane
f Lézards-léopard

3502 ***Gambelia copeii***
Squamata - Iguanaidae
e Cope's Leopard Lizard

3503 ***Gambelia sila***
Squamata - Iguanaidae
e Bluntnose Leopard Lizard; Blunt-nosed Leopard Lizard; San Joaquin Leopard Lizard
d San-Joaquin-Leopardleguan
f Lézard-léopard à nez court

3504 ***Gambelia wislizenii***
Squamata - Iguanaidae
e Longnose Leopard Lizard; Leopard Lizard; Common Leopard Lizard; Long-nosed Leopard Lizard (NA)
d Leopardleguan
f Lézard-léopard à nez long; Lézard de Wislizen

3505 ***Gambelia wislizenii maculosa***
Squamata - Iguanaidae
e Lahontan Basin Leopard Lizard

3506 ***Gambelia wislizenii punctata***
Squamata - Iguanaidae
e Pale Leopard Lizard; Small-spotted Leopard Lizard (NA)

3507 ***Gambelia wislizenii wislizeni***
Squamata - Iguanaidae
e Longnose Leopard Lizard; Large-spotted Leopard Lizard (NA)

3508 ***Gastropholis***
Squamata - Lacertidae
e Keelbelly Ground Lizards

3509 ***Gastropholis vittata***
Squamata - Lacertidae
e Keelbelly Ground Lizard; Keel-bellied Ground Lizard

3510 ***Gastropyxis***
Squamata - Colubridae
e Emerald Snakes; Keel-scaled Green Snakes
d Grünnattern

3511 ***Gastropyxis smaragdina***
Squamata - Colubridae
e Emerald Snake; Keel-scaled Green Snake
d Grünnattern

3512 **Gavialidae**
Squamata
e Gavials
d Gaviale
f Gavialidés

3513 ***Gavialis***
Squamata - Gavialidae
e Gharials; Gavials; Indian Gavials
d Ganges-Gaviale
f Gaviales du Gange

3514 ***Gavialis gangeticus***
Squamata - Gavialidae
e Gavial; Indian Gavial ; Gharial (ISC); Ganges Gharial; Long-snouted Crocodile (ISC); True Gharial; Indian Gharial
d Ganges-Gavial; Schnabelkrokodil; Mudela
f Gaviale du Gange; Gharial

3515 ***Geargas***
Squamata - Colubridae
e Tehuantepec Striped Snakes

3516 ***Geargas redimitus***
Squamata - Colubridae
e Tehuantepec Striped Snake

3517 ***Geckoella***
Squamata - Gekkonidae
e Indian Geckos; Ground Geckos (ISC)

3518 ***Geckoella collegalensis***
Squamata - Gekkonidae
e Beddome's Indian Gecko

3519 ***Geckoella deccanensis***
Squamata - Gekkonidae
e Gunther's Indian Gecko

3520 ***Geckoella jeyporensis***
Squamata - Gekkonidae
e Patinghe Indian Gecko

3521 ***Geckoella nebulosus***
Squamata - Gekkonidae
e Gunther'sMadras Gecko

3522 ***Geckoella triedrus***
Squamata - Gekkonidae
e Sri Lanka Gecko; Dark Bent-toed Gecko

3523 ***Geckoella yakhuna***
Squamata - Gekkonidae
e Northern Sri Lanka Gecko

3524 ***Geckolepis***
Squamata - Gekkonidae
e Large-scaled Geckos

3525 ***Geckolepis anomala***
Squamata - Gekkonidae
e Anomolous Gecko

3526 ***Geckolepis maculata***
Squamata - Gekkonidae
e Peters's Spotted Gecko

3527 ***Geckolepis modesta***
Squamata - Gekkonidae
e Modest Gecko

3528 ***Geckolepis petiti***
Squamata - Gekkonidae
e Angel's Gecko

3529 ***Geckolepis polylepis***
Squamata - Gekkonidae
e Many-scaled Gecko

3530 ***Geckolepis typica***
Squamata - Gekkonidae
e Grandidier's Gecko
d Schuppengecko

3531 ***Geckonia***
Squamata - Gekkonidae
e Helmethead Geckos; Helmeted Geckos
d Helmkopfgeckos; Helmgeckos
f Geckos casqués

3532 ***Geckonia chazaliae***
Squamata - Gekkonidae
e Helmethead Gecko; Helmeted Gecko
d Helmkopfgecko; Helmgecko
f Gecko casqué

3533 ***Gehyra***
Squamata - Gekkonidae
e Dtellas
d Pazifik-Geckos

3534 ***Gehyra angusticaudata***
Squamata - Gekkonidae
e Narrowhead Dtella

3535 ***Gehyra australis***
Squamata - Gekkonidae
e Northern Dtella
d Australischer Hausgecko

3536 ***Gehyra baliola***
Squamata - Gekkonidae
e Duméril's Dtella

3537 ***Gehyra barea***
Squamata - Gekkonidae
e Banda Island Dtella

3538 ***Gehyra borroloola***
Squamata - Gekkonidae
e Carpentaria Dtella

3539 ***Gehyra brevipalmata***
Squamata - Gekkonidae
e Ceres

3540 ***Gehyra butleri***
Squamata - Gekkonidae
e Butler's Dtella

3541 ***Gehyra catenata***
Squamata - Gekkonidae
e Chain Dtella

3542 ***Gehyra dubia***
Squamata - Gekkonidae
e Macleay's Dtella; Dubious Dtella

3543 ***Gehyra fehlmanni***
Squamata - Gekkonidae
e Fehlmann's Dtella

3544 ***Gehyra fenestra***
Squamata - Gekkonidae
e Mitchell's Dtella

3545 ***Gehyra intermedia***
Squamata - Gekkonidae
e Ryuku Dtella

3546 ***Gehyra interstitialis***
Squamata - Gekkonidae
e Oudeman's Dtella

3547 ***Gehyra kimberleyi***
Squamata - Gekkonidae
e Kimberley Dtella

3548 ***Gehyra lacerata***
Squamata - Gekkonidae
e Lacerated Dtella

3549 ***Gehyra lampei***
Squamata - Gekkonidae
e Irian Jaya Dtella

3550 ***Gehyra leopoldi***
Squamata - Gekkonidae
e Leopold Dtella

3551 ***Gehyra marginata***
Squamata - Gekkonidae
e Ternate Dtella

3552 ***Gehyra membranacruralis***
Squamata - Gekkonidae
e Port Moresby Dtella

3553 ***Gehyra minuta***
Squamata - Gekkonidae
e Barkly Tableland Dtella

3554 ***Gehyra montium***
Squamata - Gekkonidae
e Mountain Dtella

3555 ***Gehyra mutilata***
Squamata - Gekkonidae
e Stump-toed Dtella; Stump-toed Gecko; Sugar Lizard; Pacific Lizard; Eastern House Gecko (ISC); Four-clawed Gecko
d Pazifik-Gecko
f Gecko à doigts mutilés

3556 ***Gehyra nana***
Squamata - Gekkonidae
e Dwarf Dtella

3557 ***Gehyra occidentalis***
Squamata - Gekkonidae
e Western Dtella

3558 ***Gehyra oceanica***
Squamata - Gekkonidae
e Oceanic Dtella; Oceanic Gecko
d Südsee-Gecko

3559 ***Gehyra pamela***
Squamata - Gekkonidae
e Sandstone Dtella

3560 ***Gehyra papuana***
Squamata - Gekkonidae
e Papua Dtella

3561 ***Gehyra pilbara***
Squamata - Gekkonidae
e Pilbara Dtella

3562 ***Gehyra punctata***
Squamata - Gekkonidae
e Spotted Dtella

3563 ***Gehyra purpurascens***
Squamata - Gekkonidae
e Purple Dtella

3564 ***Gehyra robusta***
Squamata - Gekkonidae
e Robust Dtella

3565 ***Gehyra variegata***
Squamata - Gekkonidae
e Tree Dtella (ANZ); Variegated Dtella
f Gecko arboricole

3566 ***Gehyra vorax***
Squamata - Gekkonidae
e Voracious Dtella; Fiji Gecko
d Halmahera-Riesengecko

3567 ***Gehyra xenopus***
Squamata - Gekkonidae
e Foreign Dtella

3568 ***Gekko***
Squamata - Gekkonidae
e Pacific Geckos; Geckos; Tropical Asian Geckos (NA)
d Geckos

3569 ***Gekko athymus***
Squamata - Gekkonidae

e Brown's Gecko

3570 ***Gekko auriverrucosus***
Squamata - Gekkonidae
e Shanxi Gecko

3571 ***Gekko chinensis***
Squamata - Gekkonidae
e Gray's Chinese Gecko

3572 ***Gekko gecko***
Squamata - Gekkonidae
e Tokay Gecko; Asian Tokay; Tokay (ISC); Tuctoo (ISC)
d Tokeh; Tokee
f Tokay; Gecko tokay
i Geko tokai

3573 ***Gekko gecko gecko***
Squamata - Gekkonidae
e Common Tokay Gecko

3574 ***Gekko gigante***
Squamata - Gekkonidae
e Giant Gecko

3575 ***Gekko grossmanni***
Squamata - Gekkonidae
e Marbled Gecko
d Gepunkteter Gecko; Vvietnam Gepunkteter Gecko
f Gecko marbré

3576 ***Gekko hokouensis***
Squamata - Gekkonidae
e Kwangsi Gecko

3577 ***Gekko japonicus***
Squamata - Gekkonidae
e Schlegel's Japanese Gecko; Japanese Gecko
d Japan-Gecko
f Gecko japonais

3578 ***Gekko kikuchii***
Squamata - Gekkonidae
e Botel Gecko

3579 ***Gekko liboensis***
Squamata - Gekkonidae
e Guizhou's Gecko

3580 ***Gekko palmatus***
Squamata - Gekkonidae
e Palm Gecko

3581 ***Gekko palwanensis***
Squamata - Gekkonidae
e Palawan Gecko

3582 ***Gekko petricolus***
Squamata - Gekkonidae
e ThaikoThai Gecko
d Laos-Gecko; Gelbrückengecko

3583 ***Gekko porosus***
Squamata - Gekkonidae
e Taylor's Gecko

3584 ***Gekko romblon***
Squamata - Gekkonidae
e Philippine Gecko

3585 ***Gekko scabridus***
Squamata - Gekkonidae
e Yunnan Gecko

3586 ***Gekko smithii***
Squamata - Gekkonidae
e Smith's Green-eyed Gecko; Forest Gecko
d Smiths Grünaugengecko; Smiths Grünaugenriesengecko

3587 ***Gekko subpalmatus***
Squamata - Gekkonidae
e Southern Palm Gecko

3588 ***Gekko swinhonis***
Squamata - Gekkonidae
e Peking Gecko

3589 ***Gekko taibaiensis***
Squamata - Gekkonidae
e Mingtao's Gecko

3590 ***Gekko tawaensis***
Squamata - Gekkonidae
e Tawa Gecko

3591 ***Gekko tuberculatus***
Squamata - Gekkonidae
e Bumpy Gecko

3592 ***Gekko verreuxi***
Squamata - Gekkonidae
e Andaman Gecko

3593 ***Gekko vittatus***
Squamata - Gekkonidae
e Lined Gecko; White-lined Gecko
d Streifengecko

3594 ***Gekko yakuensis***
Squamata - Gekkonidae
e Yakushima Gecko

3595 **Gekkonidae**
Squamata
e Geckos; Geckos and Legless Lizards; Gekkonid Lizards; Typical Geckos (CSA)
d Geckos; Haftzeher; Geckonen
f Geckos; Geckotiens; Geckonidés

3596 ***Geocalamus***
Squamata - Amphisbaenidae
e Kenya Worm Lizards

3597 ***Geocalamus acutus***
Squamata - Amphisbaenidae
e Wedge-snouted Worm Lizard; Voi Wedge-snouted Worm Lizard

3598 ***Geocalamus modestus***
Squamata - Amphisbaenidae
e Mpwapwa Wedge-snouted Worm Lizard

3599 ***Geocalamus noltei***
Squamata - Amphisbaenidae
e Tanganyika Wedge-snouted Worm Lizard

3600 ***Geochelone***
Testudines - Testudinidae
e Typical Tortoises
d Schildkröten; Landschildkröten
f Tortues

3601 ***Geochelone carbonaria***
Testudines - Testudinidae
e Redfoot Tortoise; Red-footed Tortoise
d Köhlerschildkröte
f Tortue charbonnière; Tortue-terre (Ants)
i Testuggine carbonaria; Testuggine terrestre a zampe rosse; Testuggine a zampe rosse

3602 ***Geochelone chilensis***
Testudines - Testudinidae
e Chaco Tortoise; Chilean Tortoise; Argentine Land Tortoise
d Argentinische Landschildkröte
f Tortue d'Argentine
i Testuggine d'Argentine; Testuggine argentina

3603 ***Geochelone denticulata***
Testudines - Testudinidae
e South American Yellowfoot Tortoise; Yellow-footed Tortoise; Yellowfoot Tortoise; South American Yellow-footed Tortoise
d Waldschildkröte
f Tortue dentelée; Tortue terrestre denticulée; Tortue-terre (Ants)
i Tartaruga denticolata; Testuggine terrestre a zampe gialle

3604 ***Geochelone elegans***
Testudines - Testudinidae
e Indian Star Tortoise; Star Tortoise; Indian Starred Tortoise (ISC)
d Indische Sternschildkröte; Indische Strahlenschildkröte; Sternschildkröte
f Tortue élégante
i Tartaruga elegante; Tartaruga indiana stellata

3605 ***Geochelone gigantea***
Testudines - Testudinidae
e Aldabra Tortoise; Aldabran Giant Tortoise
d Seychellen-Riesenschildkröte; Elefantenschildkröte
f Tortue géante d'Aldabra; Tortue géante des Seychelles
i Testuggine gigante delle Seicelle; Testuggine gigante di Aldabra

3606 ***Geochelone nigra***
Testudines - Testudinidae
e Galapagos Tortoise; Galapagos Giant Tortoise
d Galapagos-Riesenschildkröte; Elefantenschildkröte
f Tortue géante des Galapagos
i Testuggine delle Galapagos; Testuggine gigante delle Galapagos

3607 ***Geochelone pardalis***
Testudines - Testudinidae
e Leopard Tortoise
d Pantherschildkröte
f Tortue léopard
i Testuggine leopardo

3608 ***Geochelone pardalis babcocki***
Testudines - Testudinidae
e Tropical Leopard Tortoise

3609 ***Geochelone platynota***
Testudines - Testudinidae
e Burmese Star Tortoise; Burmese Starred Tortoise
d Burma-Landschildkröte
f Tortue à dos plat

3610 ***Geochelone radiata***
Testudines - Testudinidae
e Radiated Tortoise; Madagascar Radiated Tortoise
d Madagassische Strahlenschildkröte; Strahlenschildkröte
f Tortue étoilée; Tortue rayonnée; Tortue étoilée de Madagascar
i Testugggine raggiata

3611 ***Geochelone sulcata***
Testudines - Testudinidae
e African Spurred Tortoise
d Spornschildkröte; Sporenschildkröte
f Tortue sillonée
i Tartaruga solcata; Testuggine africana

3612 ***Geochelone yniphora***
Testudines - Testudinidae
e Angonoka; Madagascan Tortoise; Ploughshare Tortoise; Angonoka Radiated Tortoise; Angulated Tortoise; Angulated Madagascar Tortoise
d Madagassische Schnabelbrustschildkröte
f Tortue à soc; Tortue de Madagascar
i Testuggine Angonoka

3613 ***Geoclemys***
Testudines - Emydidae
e Spotted Pond Turtles; Black Pond Turtles
d Strahlendreikielschildkröten
f Géoclémydes

3614 ***Geoclemys hamiltonii***
Testudines - Emydidae
e Spotted Pond Turtle; Black-breasted Pond Turtle; Black Pond Turtle; Spotted Black Terrapin (ISC); North Indian Freshwater Tortoise; Black Spotted Pond Turtle
d Indische Strahlendreikielschildkröte;

Strahlendreikielschildkröte
f Géoclémyde d'Hamilton; Tortue d'Hamilton
i Tartaruga palustre di Hamilton

3615 ***Geodipsas***
Squamata - Colubridae
e African Forest Snakes; Forest Snakes

3616 ***Geodipsas boulengeri***
Squamata - Colubridae
e Boulenger's Forest Snake

3617 ***Geodipsas depressiceps***
Squamata - Colubridae
e Dark Forest Snake

3618 ***Geodipsas heimi***
Squamata - Colubridae
e Heim's Forest Snake

3619 ***Geodipsas infralineata***
Squamata - Colubridae
e Lined Forest Snake

3620 ***Geodipsas procterae***
Squamata - Colubridae
e Loveridge's Forest Snake; Uluguru Forest Snake

3621 ***Geodipsas vauerocegae***
Squamata - Colubridae
e Usambara Forest Snake

3622 ***Geoemyda***
Testudines - Emydidae
e Leaf Turtles
d Erdschildkröten; Zackenerdschildkröten
f Géoémydes

3623 ***Geoemyda japonica***
Testudines - Emydidae
e Okinawa Black-breasted Leaf Turtle
f Géoémyde des Ryu-Kyu

3624 ***Geoemyda spengleri***
Testudines - Emydidae
e Chinese Black-breasted Leaf Turtle; Black-breasted Leaf Turtle; Chinese Leaf Turtle
d Zackenerdschildkröte; Spenglers Erdschildkröte; Chinesische Zackenerdschildkröte
f Géoémyde de Spengler
i Tartaruga foglia a petto nero

3625 ***Geomyersia***
Squamata - Scincidae
e Australian Island Skinks

3626 ***Geomyersia coggeri***
Squamata - Scincidae
e Cogger's Island Skink

3627 ***Geomyersia glabra***
Squamata - Scincidae
e Greer's Island Skink

3628 ***Geophis***
Squamata - Colubridae
e Latin American Earth Snakes

3629 ***Geophis anocularis***
Squamata - Colubridae
e Sierra Mije Earth Snake

3630 ***Geophis betaniensis***
Squamata - Colubridae
e Betanien Earth Snake

3631 ***Geophis bicolor***
Squamata - Colubridae
e Mexican Plateau Earth Snake

3632 ***Geophis blanchardi***
Squamata - Colubridae
e Blanchard's Earth Snake

3633 ***Geophis brachycephalus***
Squamata - Colubridae

e Colombian Earth Snake; Grey Earth Snake

3634 ***Geophis cancellatus***
Squamata - Colubridae
e Chiapas Earth Snake

3635 ***Geophis carinosus***
Squamata - Colubridae
e Keeled Earth Snake

3636 ***Geophis chalybeus***
Squamata - Colubridae
e Veracruz Earth Snake

3637 ***Geophis championi***
Squamata - Colubridae
e Panamanian Earth Snake

3638 ***Geophis damiani***
Squamata - Colubridae
e Damian's Earth Snake

3639 ***Geophis downsi***
Squamata - Colubridae
e Savage's Earth Snake; Downs's Earth Snake

3640 ***Geophis dubius***
Squamata - Colubridae
e Mesa del Sur Earth Snake

3641 ***Geophis duellmani***
Squamata - Colubridae
e Sierra Juarez Earth Snake

3642 ***Geophis dugesii***
Squamata - Colubridae
e Duges's Earth Snake

3643 ***Geophis dunni***
Squamata - Colubridae
e Dunn's Earth Snake; Banded Earth Snake

3644 ***Geophis fulvoguttatus***
Squamata - Colubridae
e Mertens's Earth Snake; Yellow-blotched Earth Snake

3645 ***Geophis godmani***
Squamata - Colubridae
e Godman's Earth Snake; Yellow-bellied Earth Snake
d Godman-Berggrubenotter

3646 ***Geophis hoffmanni***
Squamata - Colubridae
e Hoffmann's Earth Snake; Common Earth Snake

3647 ***Geophis immaculatus***
Squamata - Colubridae
e Downs's Earth Snake; Immaculate Earth Snake

3648 ***Geophis incomptus***
Squamata - Colubridae
e Sierra Coalcoman

3649 ***Geophis isthmicus***
Squamata - Colubridae
e Isthmian Earth Snake

3650 ***Geophis laticinctus***
Squamata - Colubridae
e Mesa Central Earth Snake

3651 ***Geophis laticollaris***
Squamata - Colubridae
e Wide-collar Earth Snake

3652 ***Geophis latifrontalis***
Squamata - Colubridae
e Potosi Earth Snake

3653 ***Geophis maculiferus***
Squamata - Colubridae
e Michoacan Earth Snake

3654 ***Geophis multitorques***
Squamata - Colubridae
e Highland Earth Snake

3655 ***Geophis nasalis***
Squamata - Colubridae
e Coffee Earth Snake

3656 ***Geophis nigrocinctus***
Squamata - Colubridae
e Black-banded Earth Snake

3657 ***Geophis omiltemanus***
Squamata - Colubridae
e Guerreran Earth Snake

3658 ***Geophis petersii***
Squamata - Colubridae
e Peters's Earth Snake

3659 ***Geophis pyburni***
Squamata - Colubridae
e Pyburn's Earth Snake

3660 ***Geophis rhodogaster***
Squamata - Colubridae
e Rosebelly Earth Snake

3661 ***Geophis rostralis***
Squamata - Colubridae
e Sierra Madre Earth Snake

3662 ***Geophis russatus***
Squamata - Colubridae
e Red Earth Snake

3663 ***Geophis ruthveni***
Squamata - Colubridae
e Ruthven's Earth Snake

3664 ***Geophis sallaei***
Squamata - Colubridae
e Sallae's Earth Snake

3665 ***Geophis semiannulatus***
Squamata - Colubridae
e Broken-ringed Earth Snake

3666 ***Geophis semidoliatus***
Squamata - Colubridae
e Coral Earth Snake

3667 ***Geophis sieboldi***
Squamata - Colubridae
e Siebold's Earth Snake

3668 ***Geophis talamancae***
Squamata - Colubridae
e Talamanca Earth Snake

3669 ***Geophis tarascae***
Squamata - Colubridae
e Tarascan Earth Snake

3670 ***Geophis zeledoni***
Squamata - Colubridae
e Zeledon's Earth Snake

3671 ***Gerarda***
Squamata - Colubridae
e Gerard's Water Snakes

3672 ***Gerarda prevostiana***
Squamata - Colubridae
e Gerard's Water Snake

3673 ***Gerrhonotus***
Squamata - Anguidae
e Eastern Alligator Lizards
d Alligatorschleichen; Krokodilschleichen
f Lézards-alligator du Texas; Lézards-caïman

3674 ***Gerrhonotus imbricatus***
Squamata - Anguidae
e Eastern Alligator Lizard
d Kiel-Krokodilschleiche

3675 ***Gerrhonotus infernalis***
Squamata - Anguidae
e Texas Alligator Lizard

3676 ***Gerrhonotus kufoi***
Squamata - Anguidae
e Lugo's Alligator Lizard

3677 ***Gerrhonotus liocephalus***
Squamata - Anguidae
e Texas Alligator Lizard; Mexican Plated Lizard
d Texas-Alligatorschleiche; Texas-Krokodilschleiche
f Lézard-alligator du Texas

3678 ***Gerrhosaurus***
Squamata - Cordylidae
e Plated Lizards; Rock Lizards
d Schildechsen; Eigentliche Schildechsen
f Gerrhosaures
i Gerrosauri

3679 ***Gerrhosaurus auritus***
Squamata - Cordylidae
e Kalahari Plated Lizard
d Kalahari-Schildechse

3680 ***Gerrhosaurus bulsi***
Squamata - Cordylidae
e Laurent's Plated Lizard

3681 ***Gerrhosaurus flavigularis***
Squamata - Cordylidae
e Yellow-throated Plated Lizard
d Gelbkehlige Schildechse; Rotflankenschildechse
f Gerrhosaure à gorge jaune

3682 ***Gerrhosaurus major***
Squamata - Cordylidae
e Rough-scaled Plated Lizard; Tawny African Great Lizard; Tawny Plated Lizard (CSA); Sudan Plated Lizard
d Sudan-Schildechse; Braune Schildechse; Rote Sudan-Schildechse
f Gerrhosaure africain
i Gerrosauro maggiore

3683 ***Gerrhosaurus multilineatus***
Squamata - Cordylidae
e Keeled Plated Lizard; Kalahari Plated Lizard (CSA); Many-lined Plated Lizard (CSA); Angolan Plated Lizard (CSA)

3684 ***Gerrhosaurus nigrolineatus***
Squamata - Cordylidae
e Black-lined Plated Lizard
d Gestreifte Schildechse

3685 ***Gerrhosaurus typicus***
Squamata - Cordylidae
e Karoo Plated Lizard; Namaqua Plated Lizard (CSA)
f Gerrhosaure typique

3686 ***Gerrhosaurus validus***
Squamata - Cordylidae
e Giant Plated Lizard; Rock Plated Lizard (CSA); Smith's Rock Plated Lizard
d Felsenschildechse
f Gerrhosaure de Smith; Gerrhosaure

3687 ***Glaphyromorphus***
Squamata - Scincidae
e Pygmy Tree Skinks

3688 ***Glaphyromorphus brongersmai***
Squamata - Scincidae
e Brongersma's Tree Skink

3689 ***Glaphyromorphus cracens***
Squamata - Scincidae
e Vine Skink

3690 ***Glaphyromorphus crassicaudus***
Squamata - Scincidae

e Spotted-flank Ground Skink

3691 ***Glaphyromorphus darwiniensis***
Squamata - Scincidae
e Darwin's Ground Skink

3692 ***Glaphyromorphus douglasi***
Squamata - Scincidae
e Douglas's Tree Skink

3693 ***Glaphyromorphus fuscicaudis***
Squamata - Scincidae
e Brown Tree Skink; Grey-tailed Skink

3694 ***Glaphyromorphus gracilipes***
Squamata - Scincidae
e Southwestern Tree Skink

3695 ***Glaphyromorphus isolepis***
Squamata - Scincidae
e Night Skink; Smooth-scaledSkink

3696 ***Glaphyromorphus mjobergi***
Squamata - Scincidae
e Atherton Tableland Skink

3697 ***Glaphyromorphus nigricaudis***
Squamata - Scincidae
e Blacktail Skink

3698 ***Glaphyromorphus pardalis***
Squamata - Scincidae
e Panther Burrowing Skink

3699 ***Glaphyromorphus pumilus***
Squamata - Scincidae
e Pygmy Tree Skink

3700 ***Glaphyromorphus punctatus***
Squamata - Scincidae
e Brown-spotted Skink

3701 ***Gnypetoscincus***
Squamata - Scincidae
e Keeled Queensland Skinks; Prickly Forest Skinks

3702 ***Gnypetoscincus queenslandiae***
Squamata - Scincidae
e Keeled Queensland Skink; Prickly Forest Skink

3703 ***Gomesophis***
Squamata - Colubridae
e Brazilian Burrowing Snakes

3704 ***Gomesophis brasiliensis***
Squamata - Colubridae
e Brazilian Burrowing Snake

3705 ***Gonatodes***
Squamata - Gekkonidae
e American Geckos; Streak Lizards; Bent-toed Geckos (NA)
d Mittelamerikanische Taggeckos
f Gonatodes

3706 ***Gonatodes albogularis***
Squamata - Gekkonidae
e Yellowhead Gecko; Yellow-headed Gecko; Padless Gecko; Neotropical Clawed Gecko (WI)
d Weißkehlgecko; Gelbkopfgecko
f Gecko à tête jaune; Gonatode à gorge blanche

3707 ***Gonatodes albogularis fuscus***
Squamata - Gekkonidae
e Dusky Yellow-headed Gecko
d Gelbkopfgecko

3708 ***Gonatodes annularis***
Squamata - Gekkonidae
e Annulated Gecko

3709 ***Gonatodes atillensis***
Squamata - Gekkonidae
e Antilles Gecko

3710 ***Gonatodes atricullaris***
Squamata - Gekkonidae
e Cajamarca Gecko

3711 ***Gonatodes caudiscutatus***
Squamata - Gekkonidae
e Shieldhead Gecko

3712 ***Gonatodes ceciliae***
Squamata - Gekkonidae
e Brilliant South American Gecko
d Buntgecko

3713 ***Gonatodes concinnatus***
Squamata - Gekkonidae
e O'Shaughnessy's Gecko; Collared Streak Lizard

3714 ***Gonatodes eladioi***
Squamata - Gekkonidae
e South American Gecko

3715 ***Gonatodes falconensis***
Squamata - Gekkonidae
e Estado Falcon Gecko

3716 ***Gonatodes hasemani***
Squamata - Gekkonidae
e Haseman's Gecko; Brown-headed Streak Lizard

3717 ***Gonatodes humeralis***
Squamata - Gekkonidae
e Trinidad Gecko; Amazon Streak Lizard

3718 ***Gonatodes ocellatus***
Squamata - Gekkonidae
e Eyespot Gecko

3719 ***Gonatodes petersi***
Squamata - Gekkonidae
e Peters's Gecko

3720 ***Gonatodes seigleri***
Squamata - Gekkonidae
e Estados Sucre Gecko

3721 ***Gonatodes taniae***
Squamata - Gekkonidae
e Estado Aragua Gecko

3722 ***Gonatodes tapajonicus***
Squamata - Gekkonidae
e Para Gecko

3723 ***Gonatodes vittatus***
Squamata - Gekkonidae
e Wiegmann's Striped Gecko; Streak Lizard
d Streifengecko

3724 ***Gongilomorphus***
Squamata - Scincidae
e Bojer's Skinks

3725 ***Gongilomorphus bojeri***
Squamata - Scincidae
e Bojer's Skink

3726 ***Gongylosoma***
Squamata - Colubridae
e Smooth Snakes

3727 ***Gongylosoma baliodeirus***
Squamata - Colubridae
e Boie's Smooth Snake

3728 ***Gongylosoma longicauda***
Squamata - Colubridae
e Peters's Smooth Snake

3729 ***Gongylosoma scripta***
Squamata - Colubridae
e Common Grey Ringneck

3730 ***Gongylus***
Squamata - Scincidae

e Thicktail Skinks

3731 ***Gongylus androvandii***
Squamata - Scincidae
e Androvandi's Skink

3732 ***Gongylus arborum***
Squamata - Scincidae
e Bavay's Skink

3733 ***Gongylus austrocaledonica***
Squamata - Scincidae
e Austro-Caledonian Skink

3734 ***Gongylus baudini***
Squamata - Scincidae
e Baudin's Skink

3735 ***Gongylus bougainvillii***
Squamata - Scincidae
e Bougainville's Skink

3736 ***Gongylus brachypoda***
Squamata - Scincidae
e Duméril's Skink

3737 ***Gongylus cartereti***
Squamata - Scincidae
e Carteret's Skink

3738 ***Gongylus coctei***
Squamata - Scincidae
e Bibron's Skink

3739 ***Gongylus concolor***
Squamata - Scincidae
e Brilliant Skink

3740 ***Gongylus crassicaudum***
Squamata - Scincidae
e Thicktail Skink

3741 ***Gonionotophis***
Squamata - Colubridae
e African Ground Snakes

3742 ***Gonionotophis brussauxi***
Squamata - Colubridae
e Mocquard's African Ground Snake

3743 ***Gonionotophis granti***
Squamata - Colubridae
e Grant's African Ground Snake

3744 ***Gonionotophis klingi***
Squamata - Colubridae
e Matschie's African Ground Snake

3745 ***Goniurosaurus***
Squamata - Eublepharidae
e Chinese Eyelid Geckos

3746 ***Goniurosaurus araneus***
Squamata - Eublepharidae
e Chinese Tiger Gecko

3747 ***Goniurosaurus kuroiwae***
Squamata - Eublepharidae
e Tokashiki Gecko

3748 ***Goniurosaurus lichtenfelderi***
Squamata - Eublepharidae
e Lichtenfelder's Gecko

3749 ***Goniurosaurus luii***
Squamata - Eublepharidae
e Chinese Cave Gecko; Chinese Leopard Gecko
f Gecko-léopard chinois; Gecko des caves chinois; Gecko des grottes chinois

3750 ***Gonocephalus***
Squamata - Agamidae
e Humphead Forest Dragons; Forest Dragons; Angleheads; Angle-headed Lizards
d Winkelkopfagamen
f Gonocéphales

3751 ***Gonocephalus abbotti***
Squamata - Agamidae
e Cochran's Forest Dragon

3752 ***Gonocephalus bellii***
Squamata - Agamidae
e Bell's Forest Dragon
d Bells Winkelkopfagame

3753 ***Gonocephalus beyschlagi***
Squamata - Agamidae
e Sumatra Forest Dragon

3754 ***Gonocephalus borneensis***
Squamata - Agamidae
e Borneo Forest Dragon

3755 ***Gonocephalus chamaeleontinus***
Squamata - Agamidae
e Chameleon Forest Dragon

3756 ***Gonocephalus denzeni***
Squamata - Agamidae
e Denzen's Forest Dragon

3757 ***Gonocephalus doriae***
Squamata - Agamidae
e Peters's Forest Dragon

3758 ***Gonocephalus grandis***
Squamata - Agamidae
e Giant Forest Dragon
d Große Winkelkopfagame

3759 ***Gonocephalus herveyi***
Squamata - Agamidae
e Hervey's Forest Dragon

3760 ***Gonocephalus interruptus***
Squamata - Agamidae
e Boulenger's Forest Dragon

3761 ***Gonocephalus klossi***
Squamata - Agamidae
e Kloss's Forest Dragon

3762 ***Gonocephalus lacunosus***
Squamata - Agamidae
e Manthey's Forest Dragon

3763 ***Gonocephalus liogaster***
Squamata - Agamidae
e Tropical Forest Dragon; Bornean Angle-headed Lizard
d Borneo-Winkelkopfagame
f Gonocéphale de Bornéo

3764 ***Gonocephalus megalepsis***
Squamata - Agamidae
e Bleeker's Forest Dragon

3765 ***Gonocephalus mjoeberbi***
Squamata - Agamidae
e Mjoeberb Forest Dragon

3766 ***Gonocephalus robinsonii***
Squamata - Agamidae
e Robinson's Forest Dragon

3767 ***Gonocephalus semperi***
Squamata - Agamidae
e Mindoro Forest Dragon

3768 ***Gonocephalus sophiae***
Squamata - Agamidae
e Negros Forest Dragon

3769 ***Gonyophis***
Squamata - Colubridae
e Rainbow Tree Snakes

3770 ***Gonyophis margaritatus***
Squamata - Colubridae
e Rainbow Tree Snake

3771 ***Gonyosoma***
Squamata - Colubridae
e Redtail Rat Snakes; Tree Racers;

Green Rat Snakes; Rat Snakes
d Spitzkopfnattern; Rotschwanznattern
i Colubri codarossa

3772 ***Gonyosoma oxycephalum***
Squamata - Colubridae
e Redtail Rat Snake; Tree Racer; Red-tailed Green Rat Snake; Red-tailed Rat Snake; Red-tailed Racer; Boie's Rat Snake; Rat-tailed Racer
d Spitzkopfnatter; Rotschwanznatter; Mangrovenrattenschlange
i Colubro codarossa

3773 ***Gopherus***
Testudines - Testudinidae
e Gopher Tortoises
d Gopherschildkröten
f Tortues fouisseuses
i Tartarughe del deserto

3774 ***Gopherus agassizii***
Testudines - Testudinidae
e Mojave Desert Tortoise (Mojave pop.); Sonoran Desert Tortoise (Sonora pop.); Desert Gopher (NA); Mojave Desert Tortoise; Californian Desert Tortoise; Desert Tortoise (NA); Western Desert Tortoise
d Kalifornische Gopherschildkröte; Agassiz-Gopherschildkröte; Wüstenschildkröte
f Tortue du désert
i Tartaruga del deserto; Testuggine del deserto

3775 ***Gopherus berlandieri***
Testudines - Testudinidae
e Berlandier's Tortoise; Texas Gopher Tortoise; Texas Tortoise
d Texas-Gopherschildkröte
f Tortue du Texas
i Tartaruga del Texas; Testuggine del deserto del Texas

3776 ***Gopherus flavomarginatus***
Testudines - Testudinidae
e Bolson Tortoise; Mexican Gopher Tortoise; Mexican Giant Gopher; Bolson Gopher Tortoise
d Bolson-Schildkröte; Mexikanische Gopherschildkröte; Gelbrandgopherschildkröte
f Tortue à bords jaunes
i Testuggine del deserto a margini gialli

3777 ***Gopherus polyphemus***
Testudines - Testudinidae
e Gopher Tortoise; Florida Gopher Tortoise
d Georgia-Gopherschildkröte; Gopherschildkröte
f Tortue gaufrée; Gophère polyphème
i Tartaruga del deserto; Tartaruga del deserto comune

3778 ***Graciliscincus***
Squamata - Scincidae
e Sadlier's Skinks

3779 ***Graciliscincus shonae***
Squamata - Scincidae
e Sadlier's Skink

3780 ***Graptemys***
Testudines - Emydidae
e Map Turtles; Map Turtles and Sawbacks
d Höckerschildkröten; Höckerschmuckschildkröten
f Tortues géographiques; Graptémydes

3781 ***Graptemys barbouri***
Testudines - Emydidae
e Barbour's Map Turtle; Barbour's Sawback Turtle; Barbour's Sawback
d Barbours Höckerschildkröte
f Tortue géographique de Barbour; Graptémyde de Barbour
i Tartaruga carta geografica di Barbour

3782 ***Graptemys caglei***
Testudines - Emydidae

e Cagle's Map Turtle
d Cagles Höckerschildkröte; Guadelupe-River-Höckerschildkröte
f Tortue géographique de Cagle; Graptémyde de Cagle

3783 ***Graptemys ernsti***
Testudines - Emydidae
e Ernst's Map Turtle; Escambia Map Turtle

3784 ***Graptemys flavimaculata***
Testudines - Emydidae
e Yellow-blotched Map Turtle; Yellow-blotched Sawback (NA); Yellow-blotched Sawback Map Turtle
d Gefleckte Höckerschildkröte
f Tortue géographique taches jaunes; Graptémyde à taches jaunes

3785 ***Graptemys geographica***
Testudines - Emydidae
e Common Map Turtle; Geographic Terrapin; Map Turtle (NA); Slider Turtle (NA); Slider (NA); Northern Map Turtle
d Landkartenhöckerschildkröte; Echte Landkartenhöckerschildkröte; Landkartenschildkröte
f Tortue géographique; Tortue à dessins géopgraphiques; Graptémyde géographique
i Tartaraguga carta geografica

3786 ***Graptemys gibbonsi***
Testudines - Emydidae
e Gibbons's Map Turtle; Pascagoula Map Turtle
d Pascagoula-Höckerschildkröte

3787 ***Graptemys nigrinoda***
Testudines - Emydidae
e Black-knobbed Map Turtle; Black-knobbed Sawback; Black-knobbed Sawback Map Turtle
d Schwarzkopfhöckerschildkröte
f Tortue géographique à nodules noirs; Graptémyde à points noirs
i Tartaruga dalla gobba nera

3788 ***Graptemys nigrinoda delticola***
Testudines - Emydidae
e Delta Map Turtle
f Graptémyde à points noirs du Sud

3789 ***Graptemys nigrinoda nigrinoda***
Testudines - Emydidae
e Black-knobbed Map Turtle
f Graptémyde à points noirs du Nord

3790 ***Graptemys oculifera***
Testudines - Emydidae
e Ringed Map Turtle; Ringed Sawback Turtle; Ringed Sawback
d Pfauenaugenhöckerschildkröte; Prachthöckerschildkröte; Pfauenaugehöckerschildkröte
f Tortue géographique ocellée; Graptémyde à ocelles
i Tartaruga carta geografica di Pearl

3791 ***Graptemys ouachitensis***
Testudines - Emydidae
e Ouachita Map Turtle
d Ouachita-Höckerschildkröte
f Tortue géographique Ouachita; Graptémyde de l'Ouatchita

3792 ***Graptemys ouatchitensis ouatchitensis***
Testudines - Emydidae
e Ouachita Map Turtle
f Tortue géographique Ouachita

3793 ***Graptemys ouatchitensis sabinensis***
Testudines - Emydidae
e Sabine Map Turtle
f Graptémyde de la Sabine

3794 ***Graptemys pseudogeographica***
Testudines - Emydidae

e False Map Turtle; Mississippi Sawback; Mississippi Sawback Turtle
d Falsche Landkartenschildkröte; Falsche Landkartenhöckerschildkröte
f Tortue pseudogéographique; Graptémyde pseudogéographique
i Tartaruga falsa carta geografica

3795 ***Graptemys pseudogeographica kohnii***
Testudines - Emydidae
e Mississippi Map Turtle; Mississippi Sawback; Mississippi Sawback Turtle
d Mississippi-Höckerschildkröte
f Tortue du Mississippi; Tortue à crête dorsale du Mississippi; Graptémyde du Sud
i Tartaraguga carta geografica

3796 ***Graptemys pseudogeographica pseudogeographica***
Testudines - Emydidae
e False Map Turtle
f Grapémyde du Nord

3797 ***Graptemys pulchra***
Testudines - Emydidae
e Alabama Map Turtle; Alabama Sawback; Alabama Sawback Turtle
d Schmuckhöckerschildkröte
f Tortue géographique d'Alabama; Graptémyde de l'Alabama

3798 ***Graptemys versa***
Testudines - Emydidae
e Texas Map Turtle; Texas Sawback; Texas Sawback Turtle
d Texas-Höckerschildkröte
f Tortue géographique du Texas; Graptémyde de Texas
i Tartaruga carta geografica del Texas

3799 ***Grayia***
Squamata - Colubridae
e African Water Snakes
d Afrikanische Wassernattern

3800 ***Grayia caesar***
Squamata - Colubridae
e Caesar's African Water Snake

3801 ***Grayia ornata***
Squamata - Colubridae
e Ornate African Water Snake

3802 ***Grayia tholloni***
Squamata - Colubridae
e Tholloni's African Water Snake; Thollonis Water Snake

3803 ***Gyalopion***
Squamata - Colubridae
e Desert Hooknose Snakes; Plateau Hooknose Snakes; Plateau Hook-nosed Snakes; Hook-nosed Snakes; Western Hook-nosed Snakes
d Mexikanische Hakennasennattern
f Couleuvres à nez crochu

3804 ***Gyalopion canum***
Squamata - Colubridae
e Western Hooknose Snake; Western Hook-nosed Snake; Pug-nosed Snake; Chihuahan Hook-nosed Snake; Chihuahan Hooknose Snake
d Westliche Hakennasennatter; Hakenschnauzennatter
f Couleuvre à nez crochu de l'Ouest

3805 ***Gyalopion quadrangularis***
Squamata - Colubridae
e Desert Hooknose Snake; Desert Hook-nosed Snake (NA); Thornscrub Hook-Nosed Snake; Thornscrub Hooknose Snake
f Couleuvre à nez crochu du désert

3806 ***Gyalopion quadrangularis desertorum***
Squamata - Colubridae
e Sonoran Hook-nosed Snake

3807 ***Gymnodactylus***
Squamata - Gekkonidae
e Naked-toed Geckos
d Nacktfingergeckos; Nacktfinger
f Gymnodactyles; Geckos à doigts nus

3808 ***Gymnodactylus geckoides***
Squamata - Gekkonidae
e Naked-toed Gecko
d Brasilianischer Nacktfingergecko

3809 ***Gymnodactylus guttulatus***
Squamata - Gekkonidae
e Brazilian Naked-toed Gecko

3810 **Gymnophthalmidae**
Squamata
e Spectacled Tegus
d Brillentegus; Microteiide Echsen
f Gymnophthalmidés

3811 ***Gymnophthalmus***
Squamata - Gymnophthalmidae
e Spectacled Tegus; Tegu Lizards
d Brillentejus

3812 ***Gymnophthalmus cryptus***
Squamata - Gymnophthalmidae
e Cryptic Spectacled Tegu

3813 ***Gymnophthalmus leucomystax***
Squamata - Gymnophthalmidae
e White Spectacled Tegu

3814 ***Gymnophthalmus lineatus***
Squamata - Gymnophthalmidae
e Striped Spectacled Tegu

3815 ***Gymnophthalmus pleei***
Squamata - Gymnophthalmidae
e Martinique Spectacled Tegu; Rough-scaled Worm Lizard (WI)

3816 ***Gymnophthalmus speciosus***
Squamata - Gymnophthalmidae
e Golden Spectacled Tegu; Central American Spectacled Microteid; Red-tailed Teiid
d Nördlicher Brillenteju

3817 ***Gymnophthalmus underwoodi***
Squamata - Gymnophthalmidae
e Underwood's Spectacled Tegu; Underwood's Spectacled Microteiid; Smooth-scaled Worm Lizard (WI)

H

3818 ***Haackgreerius***
Squamata - Scincidae
e Haacke-Greer's Skinks

3819 ***Haackgreerius miopus***
Squamata - Scincidae
e Haacke-Greer's Skink

3820 ***Haplocercus***
Squamata - Colubridae
e Ceylon Keelbacks
d Ceylon-Nattern

3821 ***Haplocercus ceylonensis***
Squamata - Colubridae
e Ceylon Keelback
d Ceylon-Nattern

3822 ***Hapsidophrys***
Squamata - Colubridae
e Black-lined Green Snakes

3823 ***Hapsidophrys lineatum***
Squamata - Colubridae
e Black-lined Green Snake

3824 ***Hardella***
Testudines - Emydidae
e Crowned River Turtles
d Diademschildkröten

3825 ***Hardella thurjii***
Testudines - Emydidae
e Crowned River Turtle; Brahminy River Turtle; Brahminy Terrapin (ISC); Kali Kauntha (ISC); Ganges Crowned River Turtle
d Diademschildkröte
f Émyde indienne à diadème du Gange
i Tartaruga di fiume incoronata

3826 ***Harpesaurus***
Squamata - Agamidae
e Nose-horned Lizards; Hook-nosed Lizards; Snout Agamas

3827 ***Harpesaurus beccarii***
Squamata - Agamidae
e Sumatra Nose-horned Lizard

3828 ***Harpesaurus ensicauda***
Squamata - Agamidae
e Nias Nose-horned Lizard

3829 ***Harpesaurus modigliani***
Squamata - Agamidae
e Modigliani's Nose-horned Lizard

3830 ***Harpesaurus thescelorhinos***
Squamata - Agamidae
e King's Nose-horned Lizard

3831 ***Harpesaurus tricinctus***
Squamata - Agamidae
e Java Nose-horned Lizard

3832 ***Helicops***
Squamata - Colubridae
e Mountain Keelbacks; Water Snakes
d Schielaugennattern; Scheelaugennattern

3833 ***Helicops angulatus***
Squamata - Colubridae
e Mountain Keelback; South American Water Snake

3834 ***Helicops carinicaudus***
Squamata - Colubridae
e Wied's Keelback
d Kielschwanznatter

3835 ***Helicops danieli***
Squamata - Colubridae
e Daniel's Keelback

3836 ***Helicops gomesi***
Squamata - Colubridae
e Sao Paolo Keelback

3837 ***Helicops hagmanni***
Squamata - Colubridae
e Hagmann's Keelback; Painted Watersnake

3838 ***Helicops hogei***
Squamata - Colubridae
e Hoge's Keelback

3839 ***Helicops leopardinus***
Squamata - Colubridae
e Leopard Keelback; Spotted Water Snake

3840 ***Helicops modestus***
Squamata - Colubridae
e Olive Keelback; Olivaceous Keelback

3841 ***Helicops pastazae***
Squamata - Colubridae
e Shreve's Keelback; Olive Water Snake

3842 ***Helicops petersi***
Squamata - Colubridae
e Spiral Keelback

3843 ***Helicops pictiventris***
Squamata - Colubridae
e Werner's Keelback

3844 ***Helicops polylepis***
Squamata - Colubridae
e Norman's Keelback; Günther's Keelback; Black-chinned Water Snake

3845 ***Helicops scalaris***
Squamata - Colubridae
e Ladder Keelback

3846 ***Helicops trivittatus***
Squamata - Colubridae
e Equatorial Keelback

3847 ***Helicops yacu***
Squamata - Colubridae
e Peru Keelback; Forest Water Snake

3848 ***Helminthophis***
Squamata - Anomalepididae
e Greater Blind Snakes; Greater Worm Snakes

3849 ***Helminthophis flavoterminatus***
Squamata - Anomalepididae
e Yellowtail Blind Snake; Yellowtail Worm Snake

3850 ***Helminthophis frontalis***
Squamata - Anomalepididae
e Costa Rica Blind Snake; Costa Rica Worm Snake; Pink-headed Blind Snake

3851 ***Helminthophis praeocularis***
Squamata - Anomalepididae
e Preocular Blind Snake; Preocular Worm Snake

3852 ***Heloderma***
Squamata - Helodermatidae
e Gila Monsters; Venomous Lizards; Beaded Lizards
d Gila-Krustenechsen; Krustenechsen
f Monstres de Gila; Hélodermes; Lézards vénéneux
i Elodermi; Gili

3853 ***Heloderma horridum***
Squamata - Helodermatidae
e Beaded Lizard; Escorpion (NA); Mexican Beaded Lizard (NA)

d Krustenechse; Escorpion; Südliche Krustenechse; Skorpionkrustenechse
f Héloderme granuleux; Héloderme mexicain; Héloderme hérissé
i Lucertola imperlata

3854 ***Heloderma suspectum***
Squamata - Helodermatidae
e Gila Monster
d Gila-Krustenechse; Nördliche Krusteneche; Gilatier
f Monstre de Gila
i Mostro di Gila; Gila

3855 ***Heloderma suspectum cinctum***
Squamata - Helodermatidae
e Banded Gila Monster
i Gila

3856 ***Heloderma suspectum suspectum***
Squamata - Helodermatidae
e Reticulated Gila Monster; Reticulate Gila Monster
i Gila

3857 **Helodermatidae**
Squamata
e Gila Monsters; Venomous Lizards (NA); Beaded Lizards
d Krustenechsen
f Monstres de Gila; Lézards venimeux; Hélodermatidés

3858 ***Helophis***
Squamata - Colubridae
e Sun Snakes

3859 ***Helophis schoutedeni***
Squamata - Colubridae
e Schouteden's Sun Snake

3860 ***Hemachatus***
Squamata - Elapidae
e Spitting Cobras; Ring-necked Cobras; Ringhals Snakes; Rinkhals Snakes (CSA); African Spitting Cobras
d Ringhalskobras; Ringhalsottern; Ringkobras; Südafrikanische Speikobras
i Cobra sputatori

3861 ***Hemachatus haemachatus***
Squamata - Elapidae
e Ringhals; Ringneck Spitting Cobra; Rinkhals (CSA); Ringhals Snake; Ring-necked Spitting Cobra
d Ringhalskobra; Ringhals
i Cobra sputatore

3862 ***Hemiaspis***
Squamata - Elapidae
e Grey Swamp Snakes; Swamp Snakes

3863 ***Hemiaspis dameli***
Squamata - Elapidae
e Grey Swamp Snake; Grey Snake (ANZ)

3864 ***Hemiaspis signata***
Squamata - Elapidae
e Blackbelly Swamp Snake; Black-bellied Swamp Snake (ANZ); Swamp Snake; Marsh Snake ; Black-bellied Marsh Snake
d Schwarzbäuchige Sumpfotter

3865 ***Hemibungarus***
Squamata - Elapidae
e Asian Coral Snakes

3866 ***Hemibungarus calligaster***
Squamata - Elapidae
e Philippine Coral Snake

3867 ***Hemibungarus japonicus***
Squamata - Elapidae
e Japanese Coral Snake

3868 ***Hemibungarus japonicus boettgeri***
Squamata - Elapidae
e Hai Coral Snake; Okinawa Coral

Snake

3869 ***Hemibungarus japonicus japonicus***
Squamata - Elapidae
e Hyan Coral Snake; Amami Coral Snake

3870 ***Hemibungarus macclellandi***
Squamata - Elapidae
e Asian Coral Snake; MacClelland's Coral Snake (ISC); Banded Coral Snake (ISC)

3871 ***Hemibungarus macclellandi iwasakii***
Squamata - Elapidae
e Iwasaki's Coral Snake

3872 ***Hemidactylus***
Squamata - Gekkonidae
e Leaf-toed Geckos; Tropical House Geckos (CSA); House Geckos (NA)
d Hausgeckos; Halbfinger; Halbfingergeckos; Halbzehergeckos; Halbzeher
f Hémidactyles
i Emidattili

3873 ***Hemidactylus agrius***
Squamata - Gekkonidae
e Country Leaf-toed Gecko

3874 ***Hemidactylus albopunctatus***
Squamata - Gekkonidae
e White Leaf-toed Gecko

3875 ***Hemidactylus anamallensis***
Squamata - Gekkonidae
e Anamally Hills Gecko

3876 ***Hemidactylus aporus***
Squamata - Gekkonidae
e Annobon Leaf-toed Gecko; Southern Forest Gecko (ISC)

3877 ***Hemidactylus arnoldi***
Squamata - Gekkonidae
e Arnold's Leaf-toed Gecko

3878 ***Hemidactylus barodanos***
Squamata - Gekkonidae
e Ethiopia Leaf-toed Gecko

3879 ***Hemidactylus bavazzanoi***
Squamata - Gekkonidae
e Somali Leaf-toed Gecko
f Gecko à bande de Somalie

3880 ***Hemidactylus bayonii***
Squamata - Gekkonidae
e Barboza's Leaf-toed Gecko

3881 ***Hemidactylus bouvieri***
Squamata - Gekkonidae
e Cape Verde Leaf-toed Gecko
d Bouvier-Halbzehergecko

3882 ***Hemidactylus bowringii***
Squamata - Gekkonidae
e Oriental Leaf-toed Gecko

3883 ***Hemidactylus brookii***
Squamata - Gekkonidae
e Brook's Leaf-toed Gecko; Brook's Gecko (ISC); Spotted Indian Gecko; House Lizard; Spotted Indian House Gecko (ISC); Northern Woodslave
d Afrikanischer Hausgecko
f Gecko des maisons asiatique

3884 ***Hemidactylus citernii***
Squamata - Gekkonidae
e Speedy Leaf-toed Gecko

3885 ***Hemidactylus curlei***
Squamata - Gekkonidae
e Northern Leaf-toed Gecko

3886 ***Hemidactylus depressus***
Squamata - Gekkonidae

e Sri Lanka Leaf-toed Gecko; Flattened Gecko

3887 ***Hemidactylus echinus***
Squamata - Gekkonidae
e Hedgehog Leaf-toed Gecko

3888 ***Hemidactylus fasciatus***
Squamata - Gekkonidae
e Banded Leaf-toed Gecko; Banded House Gecko; Common Pike-lined Skink
d Streifengecko
f Hémidactyle à bandes; Gecko à bandes; Gecko aux doigts en feuile

3889 ***Hemidactylus flaviviridis***
Squamata - Gekkonidae
e Indian Leaf-toed Gecko; Yellow-bellied House Gecko; Northern House Gecko (ISC); Yellow-spotted Gecko
d Indischer Hausgecko; Gelbgrüner Halbzehengecko

3890 ***Hemidactylus forbesii***
Squamata - Gekkonidae
e Socotra Leaf-toed Gecko

3891 ***Hemidactylus frenatus***
Squamata - Gekkonidae
e Common House Gecko; House Gecko; South Asian Waif Gecko; Cheechak; Bridled House Gecko; Southern House Gecko (ISC); Chit-Chat; Asian House Gecko
d Gewöhnlicher Halbzehengecko; Tschitschak; Asiatischer Hausgecko; Hausgecko
f Hémidactyle bridé; Gecko du Pacifique; Gecko des maisons; Margouillat domestique

3892 ***Hemidactylus funaiolii***
Squamata - Gekkonidae
e Kenya Leaf-toed Gecko; Archer's Post Gecko

3893 ***Hemidactylus garnotii***
Squamata - Gekkonidae
e Indo-Pacific Gecko; Indo-Pacific House Gecko; Fox Gecko; Garnot's House Gecko
d Indopazifik-Gecko; Indopazifischer Halbzehergecko
f Gecko renard; Hémidactyle de Garnot; Gecko de l'Indo-pacifique; Margouillat d'extérieur

3894 ***Hemidactylus giganteus***
Squamata - Gekkonidae
e Giant Leaf-toed Gecko

3895 ***Hemidactylus gracilis***
Squamata - Gekkonidae
e Graceful Leaf-toed Gecko

3896 ***Hemidactylus granchii***
Squamata - Gekkonidae
e Granchi's Leaf-toed Gecko

3897 ***Hemidactylus granti***
Squamata - Gekkonidae
e Grant's Leaf-toed Gecko

3898 ***Hemidactylus greefi***
Squamata - Gekkonidae
e Sao Tome Leaf-toed Gecko

3899 ***Hemidactylus homoeolepis***
Squamata - Gekkonidae
e Arabian Leaf-toed Gecko

3900 ***Hemidactylus intestinalis***
Squamata - Gekkonidae
e Nigeria Leaf-toed Gecko

3901 ***Hemidactylus isolepis***
Squamata - Gekkonidae
e Scaly Leaf-toed Gecko; Uniform-scaled Gecko

3902 ***Hemidactylus jubensis***
Squamata - Gekkonidae
e African Leaf-toed Gecko

3903 ***Hemidactylus karenorum***
Squamata - Gekkonidae
e Burmese Leaf-toed Gecko

3904 ***Hemidactylus klauberi***
Squamata - Gekkonidae
e Klauber's Leaf-toed Gecko

3905 ***Hemidactylus laevis***
Squamata - Gekkonidae
e Common Leaf-toed Gecko

3906 ***Hemidactylus laticaudatus***
Squamata - Gekkonidae
e Andersson's Leaf-toed Gecko

3907 ***Hemidactylus lemurinus***
Squamata - Gekkonidae
e Dhofar Leaf-toed Gecko

3908 ***Hemidactylus leschenaultii***
Squamata - Gekkonidae
e Leschenault's Leaf-toed Gecko; Bark Gecko (ISC)
f Hémidactyle de Leschenault; Gecko de Leschenault

3909 ***Hemidactylus longicephalus***
Squamata - Gekkonidae
e Longhead Leaf-toed Gecko; Long-headed Tropical House Gecko

3910 ***Hemidactylus mabouia***
Squamata - Gekkonidae
e Cosmopolitan Leaf-toed Gecko; Moreau's House Gecko; Moreau's Tropical House Gecko (CSA); Tasman's Tropical House Gecko; Moreau's Gecko; Tropical House Gecko (CSA); Southern Woodslave ; Amerafrican House Gecko; Common House Gecko (WI); Amerafrican House Gecko (NA)
d Brauner Halbzehengecko
f Hémidactyle mabouia; Gecko des maisons; Gecko tropical des maisons; Gecko tropical de Moreau; Mabouia de Jonnes

3911 ***Hemidactylus macropholis***
Squamata - Gekkonidae
e Longhead Leaf-toed Gecko
f Gecko de Boulanger

3912 ***Hemidactylus maculatus***
Squamata - Gekkonidae
e Spotted Leaf-toed Gecko; Rock Gecko (ISC)
d Gefleckter Halbfingergecko
f Hémidactyle tacheté

3913 ***Hemidactylus marmoratus***
Squamata - Gekkonidae
e Ryukyu Leaf-toed Gecko
d Marmorgecko

3914 ***Hemidactylus matschiei***
Squamata - Gekkonidae
e Togo Leaf-toed Gecko

3915 ***Hemidactylus megalops***
Squamata - Gekkonidae
e Parker's Leaf-toed Gecko

3916 ***Hemidactylus mercatorius***
Squamata - Gekkonidae
e Gray's Leaf-toed Gecko

3917 ***Hemidactylus modestus***
Squamata - Gekkonidae
e Moderate Leaf-toed Gecko
f Gecko de Tana

3918 ***Hemidactylus muriceus***
Squamata - Gekkonidae
e Guinea Leaf-toed Gecko

3919 ***Hemidactylus newtoni***
Squamata - Gekkonidae
e Newton's Leaf-toed Gecko

3920 ***Hemidactylus nigriventis***
Squamata - Gekkonidae
e Blackbelly Leaf-toed Gecko

3921 ***Hemidactylus ophiolepis***
Squamata - Gekkonidae
e Snakescale Leaf-toed Gecko

3922 ***Hemidactylus ophiolepoides***
Squamata - Gekkonidae
e Lanza's Leaf-toed Gecko

3923 ***Hemidactylus oxyrhinus***
Squamata - Gekkonidae
e Sharpnose Leaf-toed Gecko

3924 ***Hemidactylus palaichthus***
Squamata - Gekkonidae
e Antilles Leaf-toed Gecko; Maria Islands Leaf-toed Gecko (WI)

3925 ***Hemidactylus persicus***
Squamata - Gekkonidae
e Persia Leaf-toed Gecko; Persian Gecko

3926 ***Hemidactylus platycephalus***
Squamata - Gekkonidae
e Flathead Leaf-toed Gecko; Flat-headed Tropical House Gecko (CSA); Baobab Gecko (CSA); Tree Gecko
f Gecko des arbres

3927 ***Hemidactylus prashadi***
Squamata - Gekkonidae
e Bombay Leaf-toed Gecko

3928 ***Hemidactylus puccionii***
Squamata - Gekkonidae
e Zanzibar Leaf-toed Gecko

3929 ***Hemidactylus pumilio***
Squamata - Gekkonidae
e Pygmy Leaf-toed Gecko

3930 ***Hemidactylus reticulatus***
Squamata - Gekkonidae
e Reticulate Leaf-toed Gecko

3931 ***Hemidactylus richardsonii***
Squamata - Gekkonidae
e Richardson's Leaf-toed Gecko
f Gecko des forêts de Richardson

3932 ***Hemidactylus ruspolii***
Squamata - Gekkonidae
e Farm Leaf-toed Gecko
f Gecko du prince Ruspoli

3933 ***Hemidactylus scabriceps***
Squamata - Gekkonidae
e Annandale's Leaf-toed Gecko

3934 ***Hemidactylus sinaitus***
Squamata - Gekkonidae
e Sinai Leaf-toed Gecko

3935 ***Hemidactylus smithi***
Squamata - Gekkonidae
e Smith's Leaf-toed Gecko

3936 ***Hemidactylus somalicus***
Squamata - Gekkonidae
e Northern Somali Leaf-toed Gecko

3937 ***Hemidactylus squamulatus***
Squamata - Gekkonidae
e Tornier's Leaf-toed Gecko
f Gecko Nyika

3938 ***Hemidactylus stejnegeri***
Squamata - Gekkonidae
e Stejneger's Leaf-toed Gecko

3939 ***Hemidactylus subtriedrus***
Squamata - Gekkonidae
e Madras Leaf-toed Gecko

3940 ***Hemidactylus tanganicus***
Squamata - Gekkonidae
e Tanszania Leaf-toed Gecko

3941 ***Hemidactylus tasmani***
Squamata - Gekkonidae
e Tasmanian Leaf-toed Gecko
f Gecko tropical des maisons de Tasman

3942 ***Hemidactylus taylori***
Squamata - Gekkonidae
e Taylor's Leaf-toed Gecko

3943 ***Hemidactylus triedrus***
Squamata - Gekkonidae
e Dakota Leaf-toed Gecko; Termite Hill Gecko (ISC)
f Hémidactyle à tubercles triédres

3944 ***Hemidactylus tropidolepis***
Squamata - Gekkonidae
e Mocquards Leaf-toed Gecko
f Hecko de Ogden

3945 ***Hemidactylus turcicus***
Squamata - Gekkonidae
e Mediterranean Gecko; Turkish Gecko; Turkish Warty Gecko; Warty Gecko; Mediterranean House Gecko (NA)
d Europäischer Halbzehengecko; Europäischer Halbzeher; Scheibenfinger; Europäischer Halbfingergecko; Halbzeher; Europäischer Halbfinger; Europäischer Halbzehergecko
f Gecko verruqueux; Hémidactyle verruqueux; Gecko turc; Gecko méditerranéen
i Emidattilo turco; Geco verrucoso; Emidattilo verrucoso

3946 ***Hemidactylus turcicus turcicus***
Squamata - Gekkonidae
e Mediterranean Gecko; Turkish House Gecko
d Europäischer Halbzehengecko; Halbzeher; Europäischer Halbfinger
i Emidattilo turco

3947 ***Hemidactylus vietnamensis***
Squamata - Gekkonidae
e Vietnam Leaf-toed Gecko

3948 ***Hemidactylus yerburii***
Squamata - Gekkonidae
e Southern Leaf-toed Gecko

3949 ***Hemiergis***
Squamata - Scincidae
e Lined Skinks; Hemiergises

3950 ***Hemiergis decresiensis***
Squamata - Scincidae
e Cuvier's Three-toed Skink; Three-toed Skink (ANZ); Tridactyle Skink
f Hemiergis de Decres

3951 ***Hemiergis initialis***
Squamata - Scincidae
e Five-lined Skink

3952 ***Hemiergis millewae***
Squamata - Scincidae
e Orange-striped Skink; Millewa Skink (ANZ)

3953 ***Hemiergis peronii***
Squamata - Scincidae
e Yellowbelly Skink; Four-toed Skink (ANZ)
f Hemiergis de Peron

3954 ***Hemiergis quadrilinetum***
Squamata - Scincidae
e Golden Four-lined Skink

3955 ***Hemiphyllodactylus***
Squamata - Gekkonidae
e Gypsy Geckos
d Zigeunergeckos

3956 ***Hemiphyllodactylus hartertri***
Squamata - Gekkonidae
e Ewerner's Gypsy Gecko

3957 ***Hemiphyllodactylus insularis***
Squamata - Gekkonidae
e Insular Gypsy Gecko

3958 ***Hemiphyllodactylus typus***
Squamata - Gekkonidae
e Indo-Pacific Gypsy Gecko; Tree Gecko; Indo-Pacific Tree Gecko
d Zigeunergecko

3959 ***Hemiphyllodactylus typus aurantiacus***
Squamata - Gekkonidae
e Western Ghats Worm Gecko

3960 ***Hemiphyllodactylus typus typus***
Squamata - Gekkonidae
e Common Indo-Pacific Tree Gecko

3961 ***Hemiphyllodactylus yunnanensis***
Squamata - Gekkonidae
e Yunnan Gypsy Gecko

3962 ***Hemirhagerrhis***
Squamata - Colubridae
e Bark Snakes; Many-spotted Snakes

3963 ***Hemirhagerrhis kelleri***
Squamata - Colubridae
e Keller's Bark Snake; Stripe-bellied Snake

3964 ***Hemirhagerrhis nototaenia***
Squamata - Colubridae
e Bark Snake; Mopane Snake (CSA); Mopani Snake; Cape Many-spotted Snake

3965 ***Hemitheconyx***
Squamata - Eublepharidae
e Fat-tail Geckos; Fat-tailed Geckos
d Afrikanische Krallengeckos

3966 ***Hemitheconyx caudicinctus***
Squamata - Eublepharidae
e Fat-tail Gecko; West African Fat-tailed Gecko; Fat-tailed Gecko
d Afrikanischer Krallengecko
f Gecko à grosse queue; Gecko à queue grasse; Gecko à queue adipeuse
i Geko coda clava

3967 ***Hemitheconyx taylori***
Squamata - Eublepharidae
e Taylor's Fat-tail Gecko; Somali Fat-tailed Gecko

3968 ***Heosemys***
Testudines - Emydidae
e Forest Turtles
d Erdschildkröten; Asiatische Erdschildkröten
f Héosémydes

3969 ***Heosemys depressa***
Testudines - Emydidae
e Arakan Forest Turtle
d Flache Erdschildkröte
f Héosémyde de l'Arakan

3970 ***Heosemys grandis***
Testudines - Emydidae
e Giant Asian Pond Turtle; Orange-headed Temple Terrapin
d Riesenerdschildkröte
i Tartaruga palustre asiatica gigante

3971 ***Heosemys leytensis***
Testudines - Emydidae
e Philippine Pond Turtle; Leyte Pond Turtle

d Philippinen-Erdschildkröte
f Héosémyde de Leyte

3972 ***Heosemys silvatica***
Testudines - Emydidae
e Cochin Forest Leaf Turtle; Cochin Forest Cane Turtle; Kavalai Forest Turtle (ISC); Kerala Forest Terrapin (ISC)
d Gelbkopferdschildkröte
f Géoémyde de Cochin; Héosémyde de Cochin

3973 ***Heosemys spinosa***
Testudines - Emydidae
e Spiny Turtle; Ground-dwelling Turtle
d Stachelerdschildkröte
f Tortue épineuse; Héosémyde épineuse
i Tartaruga spinosa

3974 ***Heterodactylus***
Squamata - Teiidae
e Ground Teiids

3975 ***Heterodactylus imbricatus***
Squamata - Teiidae
e Rio de Janeiro Teiid

3976 ***Heterodactylus lundii***
Squamata - Teiidae
e Lund's Teiid

3977 ***Heterodon***
Squamata - Colubridae
e Hognose Snakes; Hog-nosed Snakes; North american Hog-nosed Snakes
d Hakennasennattern; Hakennattern; Schweinnasenschlangen
f Hétérodons
i Muso di porcello

3978 ***Heterodon nasicus***
Squamata - Colubridae
e Western Hognose Snake; Western Hog-nosed Snake; Texas Hog-nosed Snake (NA)
d Westliche Hakennatter
f Hétérodon de l'Ouest; Couleuvre à nez retroussé
i Muso di porcello occidentale; Serpente muso di porcello; Serpente naso di porco

3979 ***Heterodon nasicus gloydi***
Squamata - Colubridae
e Dusty Hognose Snake; Dusty Hog-nosed Snake
d Hakennatter

3980 ***Heterodon nasicus kennerlyi***
Squamata - Colubridae
e Mexican Hognose Snake; Mexican Hog-nosed Snake
d Mexikanische Hakennatter

3981 ***Heterodon nasicus nasicus***
Squamata - Colubridae
e Plains Hognose Snake; Plains Hog-nosed Snake; Western Hognose
d Westliche Hakennasennatter

3982 ***Heterodon platyrhinos***
Squamata - Colubridae
e Eastern Hognose Snake; Eastern Hog-nosed Snake; Common Hog-nosed Snake (NA); Puff Adder (NA); Spreading Adder (NA); Hissing Adder (NA); Blow Snake (NA); Blowing Viper (NA); Melanistic Eastern Hognose Snake; Eastern Hognose
d Östliche Hakennatter; Gewöhnliche Hakennatter; Südöstliche Hakennatter
f Hétérodon de l'Est; Serpent à groin; Couleuvre à nez plat
i Muso di porcello orientale

3983 ***Heterodon simus***
Squamata - Colubridae
e Southern Hognose Snake; Southern Hog-nosed Snake; Spreadhead (NA);

Blow Snake (NA); Puff Adder (NA); Hissing Adder (NA)
d Südliche Hakennatter
f Hétérodon du Sud

3984 ***Heteroliodon***
Squamata - Colubridae
e Madagascar Snakes

3985 ***Heteroliodon torquatus***
Squamata - Colubridae
e Boettger's Madagascar Snake

3986 ***Heteronotia***
Squamata - Gekkonidae
e Australian Forest Geckos

3987 ***Heteronotia binoei***
Squamata - Gekkonidae
e Bynoe's Gecko; Prickly Gecko
f Gecko de Bynoe

3988 ***Heteronotia planiceps***
Squamata - Gekkonidae
e Storr's Gecko

3989 ***Heteronotia spelea***
Squamata - Gekkonidae
e Desert Cave Gecko

3990 ***Heurnia***
Squamata - Colubridae
e Mamberamo River Snakes

3991 ***Heurnia ventrimaculata***
Squamata - Colubridae
e Mamberamo River Snake

3992 ***Hieremys***
Testudines - Emydidae
e Yellow-headed Temple Turtles; Annadale's Turtles
d Tempelschildkröten
f Hiérémydes

3993 ***Hieremys annadalei***
Testudines - Emydidae
e Yellow-headed Temple Turtle; Annandale's Turtle; Temple Turtle
d Tempelschildkröte
f Hiérémyde d'Annandale; Tortue des temples à tête jaune

3994 ***Holaspis***
Squamata - Lacertidae
e Sawtail Lizards; Blue-tailed Lizard (CSA)
d Sägeschwanzeidechsen

3995 ***Holaspis guentheri***
Squamata - Lacertidae
e Sawtail Lizard
d Sägeschwanzeidechse; Günthers Stacheleidechse

3996 ***Holbrookia***
Squamata - Iguanaidae
e Earless Lizards; Lesser Earless Lizards
d Taubleguane
f Lézards sans oreilles; Iguanes sourds

3997 ***Holbrookia lacerata***
Squamata - Iguanaidae
e Spot-tailed Earless Lizard; Blunt-tailed Earless Lizard
f Lézard sans oreilles lacéré

3998 ***Holbrookia lacerata lacerata***
Squamata - Iguanaidae
e Plateau Earless Lizard; Northern Spot-tailed Earless Lizards

3999 ***Holbrookia maculata***
Squamata - Iguanaidae
e Lesser Earless Lizard; Common esser Earless Lizard
d Gefleckter Taubleguan
f Lézard sans oreilles tacheté

4000 ***Holbrookia maculata approximans***
Squamata - Iguanaidae
e Speckled Earless Lizard

4001 ***Holbrookia maculata bunkeri***
Squamata - Iguanaidae
e Bunker's Earless Lizard

4002 ***Holbrookia maculata maculata***
Squamata - Iguanaidae
e Speckled Earless Lizard; Spotted Lizard (NA); Northern Earless Lizard; Great Plains Earless Lizard

4003 ***Holbrookia maculata perspicua***
Squamata - Iguanaidae
e Eastern Earless Lizard; Eastern Lesser Earless Lizard; Prairie Earless Lizard

4004 ***Holbrookia maculata pulchra***
Squamata - Iguanaidae
e Huachuca Earless Lizard; Northern Earless Lizard

4005 ***Holbrookia maculata ruthveni***
Squamata - Iguanaidae
e Bleached Earless Lizard

4006 ***Holbrookia maculata thermophila***
Squamata - Iguanaidae
e Western Earless Lizard

4007 ***Holbrookia propinqua***
Squamata - Iguanaidae
e Keeled Earless Lizard
f Lézard sans oreilles caréné

4008 ***Holbrookia propinqua propinqua***
Squamata - Iguanaidae
e Keeled Earless Lizard; Northern Keeled Earless Lizard

4009 ***Holbrookia subcaudalis***
Squamata - Iguanaidae
e Southern Earless Lizard; Southern Spot-tailed Earless Lizard

4010 ***Holbrookia texana***
Squamata - Iguanaidae
e Greater Earless Lizard
d Taubleguane

4011 ***Holodactylus***
Squamata - Eublepharidae
e Whole-toed Geckos

4012 ***Holodactylus africanus***
Squamata - Eublepharidae
e African Whole-toed Gecko
f Gecko nain à queue grasse; Gecko griffu africain

4013 ***Holodactylus cornii***
Squamata - Eublepharidae
e Scortecci's Whole-toed Gecko

4014 ***Hologerrhum***
Squamata - Colubridae
e Philippine Stripe-lipped Snakes

4015 ***Hologerrhum philippinum***
Squamata - Colubridae
e Philippine Stripe-lipped Snake

4016 ***Homalopsis***
Squamata - Colubridae
e Masked Water Snakes; Water Snakes; Dog-faced Water Snakes; Puff-faced Water Snakes
d Wassertrugnattern; Boa-Wassertrugnattern; Boa-Trugnattern
f Homalopsides

4017 ***Homalopsis buccata***
Squamata - Colubridae
e Masked Water Snake; Puff-faced Water Snake; Dog-faced Water Snake
d Boa-Wassertrugnatter; Boa-Trugnatter; Wassertrugnatter

f Homalopside joufflu

4018 ***Homonota***
Squamata - Gekkonidae
e Marked Geckos

4019 ***Homonota andicola***
Squamata - Gekkonidae
e Cei's Marked Gecko

4020 ***Homonota borellii***
Squamata - Gekkonidae
e Borelli's Marked Gecko

4021 ***Homonota darwinii***
Squamata - Gekkonidae
e Darwin's Marked Gecko; Prickly Gecko

4022 ***Homonota fasciatus***
Squamata - Gekkonidae
e Banded Marked Gecko

4023 ***Homonota gaudichaudii***
Squamata - Gekkonidae
e Chilean Marked Gecko

4024 ***Homonota horrida***
Squamata - Gekkonidae
e South American Marked Gecko

4025 ***Homonota penai***
Squamata - Gekkonidae
e Coquimbo Marked Gecko

4026 ***Homonota underwoodi***
Squamata - Gekkonidae
e Underwood's Marked Gecko

4027 ***Homonota uruguayensis***
Squamata - Gekkonidae
e Uruguay Marked Gecko

4028 ***Homonota whitii***
Squamata - Gekkonidae
e Argentine Marked Gecko

4029 ***Homopholis***
Squamata - Gekkonidae
e Madagascar Velvet Geckos; Velvet Geckos (CSA)

4030 ***Homopholis antongilensis***
Squamata - Gekkonidae
e Antongil Velvet Gecko

4031 ***Homopholis boivini***
Squamata - Gekkonidae
e Madagascar Velvet Gecko

4032 ***Homopholis fasciata***
Squamata - Gekkonidae
e Striped Velvet Gecko

4033 ***Homopholis mulleri***
Squamata - Gekkonidae
e Muller's Velvet Gecko

4034 ***Homopholis sakalava***
Squamata - Gekkonidae
e Grandidier's Velvet Gecko

4035 ***Homopholis wahlbergii***
Squamata - Gekkonidae
e Wahlberg's Velvet Gecko; Wahlberg's Velverty Gecko (CSA); Velvety Gecko (CSA); Wahlberg's Gecko (CSA)

4036 ***Homopus***
Testudines - Testudinidae
e Cape Tortoises; Padlopers (CSA)
d Flachschildkröten
f Homopodes
i Omopi

4037 ***Homopus areolatus***
Testudines - Testudinidae
e Beaked Cape Tortoise; Parrot-beaked

Tortoise; Parrot-beaked Cape Tortoise
d Areolen-Flachschildkröte
f Homopode aréolé
i Omopo arreolata; Testuggine del Capo areolato

4038 ***Homopus bergeri***
Testudines - Testudinidae
e Berger's Cape Tortoise; Nama Padloper (CSA); Berger's Padloper
d Bergers Flachschildkröte
f Homopode de Namibie
i Omopo di Berger

4039 ***Homopus boulengeri***
Testudines - Testudinidae
e Boulenger's Cape Tortoise; Donderweer Tortoise; Donderweer Skilpad (CSA); Klipskilpad (CSA); Rooisklipajie (CSA); Boulenger's Padloper (CSA); Karoo Padloper (CSA); Donderweer Tortoise (CSA)
d Boulenger-Flachschildkröte
f Homopode de Boulenger
i Omopo di Boulenger

4040 ***Homopus femorlis***
Testudines - Testudinidae
e Karoo Cape Tortoise; Karoo Tortoise; Great Padloper (CSA); Greater Padloper (CSA)
d Spornflachschildkröte
f Homopode à éperon

4041 ***Homopus signatus***
Testudines - Testudinidae
e Speckled Cape Tortoise; Speckled Padloper (CSA); Speckled Tortoise (CSA)
d Gesägte Flachschildkröte
f Homopode marqué
i Testuggine del Capo marcata

4042 ***Homopus signatus cafer***
Testudines - Testudinidae
e Southern Speckled Padloper

4043 ***Homopus signatus signatus***
Testudines - Testudinidae
e Speckled Padloper (CSA)

4044 ***Homoroselaps***
Squamata - Colubridae
e Harlequin Snakes; Dwarf Garter Snakes
d Afrikanische Korallenschlangen

4045 ***Homoroselaps dorsalis***
Squamata - Colubridae
e Striped Harlequin Snake; Striped Dwarf Garter Snake (CSA); Striped Garter Snake

4046 ***Homoroselaps lacteus***
Squamata - Colubridae
e Spotted Harlequin Snake; Southern Dwarf Garter Snake; Southern Garter Snake; Spotted Dwarf Garter Snake (CSA); Dwarf Garter Snake
d Afrikanische Korallenschlange

4047 ***Hoplocephalus***
Squamata - Elapidae
e Pale-headed Snakes; Broad-headed Snakes

4048 ***Hoplocephalus bitorquatus***
Squamata - Elapidae
e Pale-headed Snake
d Blaßkopfotter

4049 ***Hoplocephalus bungaroides***
Squamata - Elapidae
e Broadhead Snake; Broad-headed Snake; Yellow-spotted Snake
d Breitkopfotter; Gelbfleckenschlange; Breitkopfschlange

4050 ***Hoplocephalus stephensi***
Squamata - Elapidae
e Stephens's Banded Snake; Stephens's

Snake; Yellow-banded Snake; Banded Broad-headed Snake; Stephens Snake

4051 ***Hoplocercus***
Squamata - Iguanaidae
e Weapontails
d Stachelschwanzleguane

4052 ***Hoplocercus spinosus***
Squamata - Iguanaidae
e Weapontail; Prickletail Lizard
d Stachelschwanzleguan

4053 ***Hoplodactylus***
Squamata - Gekkonidae
e Sticky-toed Geckos

4054 ***Hoplodactylus chrysosireticus***
Squamata - Gekkonidae
e Golden Sticky-toed Gecko

4055 ***Hoplodactylus delcourti***
Squamata - Gekkonidae
e Delcourt's Sticky-toed Gecko
d Riesengecko

4056 ***Hoplodactylus duvauceli***
Squamata - Gekkonidae
e Northern Sticky-toed Gecko
d Duvaucels Gecko

4057 ***Hoplodactylus granulatus***
Squamata - Gekkonidae
e Gray's Sticky-toed Gecko
d Neuseeländischer Waldgecko

4058 ***Hoplodactylus kahutarae***
Squamata - Gekkonidae
e Whitaker's Sticky-toed Gecko

4059 ***Hoplodactylus maculatus***
Squamata - Gekkonidae
e Spotted Sticky-toed Gecko

4060 ***Hoplodactylus pacificus***
Squamata - Gekkonidae
e Pacific Sticky-toed Gecko
d Gewöhnlicher Baumgecko

4061 ***Hoplodactylus raikurae***
Squamata - Gekkonidae
e Stewart's Sticky-toed Gecko

4062 ***Hoplodactylus stephensi***
Squamata - Gekkonidae
e Stephens's Sticky-toed Gecko

4063 ***Hormonotus***
Squamata - Colubridae
e Uganda House Snakes

4064 ***Hormonotus modestus***
Squamata - Colubridae
e Uganda House Snake; Hallowell's House Snake

4065 ***Hydrablabes***
Squamata - Colubridae
e Small-eyed Snakes

4066 ***Hydrablabes periops***
Squamata - Colubridae
e Olive Small-eyed Snake

4067 ***Hydrablabes praefrontalis***
Squamata - Colubridae
e Mocquard's Small-eyed Snake

4068 ***Hydraethiops***
Squamata - Colubridae
e Blackbelly Snakes
d Schwarzbauchwassernattern

4069 ***Hydraethiops melanogaster***
Squamata - Colubridae
e Blackbelly Snake
d Schwarzbauchwassernatter

4070 ***Hydrelaps***
Squamata - Elapidae
e Darwin's Sea Snakes

4071 ***Hydrelaps darwiniensis***
Squamata - Elapidae
e Darwin's Sea Snake

4072 ***Hydrodynastes***
Squamata - Colubridae
e False Water Snakes
d Falsche Wasserkobras

4073 ***Hydrodynastes bicintus***
Squamata - Colubridae
e Hermann's Water Snake

4074 ***Hydrodynastes gigas***
Squamata - Colubridae
e Brazilian False Water Snake; False Water Cobra
d Brasilianische Glattnatter; Brasilianische Falsche Wasserkobra; Falsche Wasserkobra

4075 ***Hydromedusa***
Testudines - Chelidae
e South American Snakeneck Turtles; South American Side-necked Turtles
d Schlangenhalsschildkröten; Südamerikanische Schlangenhalsschildkröten
f Hydroméduses

4076 ***Hydromedusa maximiliani***
Testudines - Chelidae
e Brazilian Snakeneck Turtle; Maximilian's Snake-headed Turtle; Prince Maximilian's Turtle; Maximilian's Snake-necked Turtle; Brazil Snake-headed Turtle; Brazilian Snake-necked Turtle
d Brasilianische Schlangenhalsschildkröte
f Hydroméduse de Maximilian; Tortue à cou serpent de Brésil

4077 ***Hydromedusa tectifera***
Testudines - Chelidae
e South American Snakeneck Turtle; South American Side-necked Turtle; South American Snake-headed Turtle; Argentine Snake-headed Turtle; Cope's Turtle; South American Snake-necked Turtle; Uruguay Snake-necked Turtle
d Südamerikanische Schlangenhalsschildkröte; Argentinische Schlangenhalsschildkröte
f Hydroméduse à dos rugueux
i Tartaruga collo di serpente

4078 ***Hydromorphus***
Squamata - Colubridae
e Tropical Water Snakes

4079 ***Hydromorphus clarki***
Squamata - Colubridae
e Clark's Water Snake

4080 ***Hydromorphus concolor***
Squamata - Colubridae
e Costa Rica Water Snake; Common Water Snake

4081 ***Hydromorphus dunni***
Squamata - Colubridae
e Dunn's Water Snake; Panamanian Water Snake

4082 ***Hydrophis***
Squamata - Elapidae
e Asian Sea Snakes; Sea Snakes; Blue-banded Sea Snakes
d Ruderschlangen
f Hydrophides

4083 ***Hydrophis atriceps***
Squamata - Elapidae
e Southeast Asia Sea Snake

4084 ***Hydrophis belcheri***
Squamata - Elapidae
e Belcher's Sea Snake
d Walo Walo-Ruderschlange

4085 ***Hydrophis bituberculatus***
Squamata - Elapidae
e Peters's Sea Snake

4086 ***Hydrophis brookei***
Squamata - Elapidae
e Brooke's Sea Snake

4087 ***Hydrophis caerulescens***
Squamata - Elapidae
e Dark Blue-banded Sea Snake; Many-toothed Sea Snake; Merrem's Sea Snake
d Blaugebänderte Ruderschlange

4088 ***Hydrophis cantoris***
Squamata - Elapidae
e Günther's Sea Snake

4089 ***Hydrophis coggeri***
Squamata - Elapidae
e Coggers Sea Snake

4090 ***Hydrophis cyanocinctus***
Squamata - Elapidae
e Annulated Sea Snake; Chittul (ISC); Banded Sea Snake

4091 ***Hydrophis czeblukovi***
Squamata - Elapidae
e Czeblukov's Sea Snake

4092 ***Hydrophis elegans***
Squamata - Elapidae
e Elegant Sea Snake

4093 ***Hydrophis fasciatus***
Squamata - Elapidae
e Striped Sea Snake; Banded Small-headed Sea Snake
d Gebänderte Ruderschlange

4094 ***Hydrophis geometricus***
Squamata - Elapidae
e Geometric Sea Snake

4095 ***Hydrophis gracilis***
Squamata - Elapidae
e Smallhead Sea Snake

4096 ***Hydrophis inornatus***
Squamata - Elapidae
e Blue-grey Sea Snake

4097 ***Hydrophis klossi***
Squamata - Elapidae
e Kloss's Sea Snake

4098 ***Hydrophis lamberti***
Squamata - Elapidae
e Lambert's Sea Snake

4099 ***Hydrophis lapemoides***
Squamata - Elapidae
e Persian Gulf Sea Snake

4100 ***Hydrophis macdowelli***
Squamata - Elapidae
e McDowell's Sea Snake

4101 ***Hydrophis mamillaris***
Squamata - Elapidae
e Bombay Sea Snake; Broad-banded Sea Snake

4102 ***Hydrophis melanocephalus***
Squamata - Elapidae
e Black-headed Sea Snake

4103 ***Hydrophis melanosoma***
Squamata - Elapidae
e Black-striped Sea Snake

4104 ***Hydrophis obscurus***
Squamata - Elapidae
e Russel's Sea Snake; Estuarine Sea Snake (ISC)

4105 ***Hydrophis ornatus***
Squamata - Elapidae
e Reef Sea Snake; Gray's Sea Snake

4106 ***Hydrophis ornatus maresinensis***
Squamata - Elapidae
e Ryukyu Ornate Sea Snake

4107 ***Hydrophis pacificus***
Squamata - Elapidae
e Pacific Sea Snake

4108 ***Hydrophis semperi***
Squamata - Elapidae
e Garman's Sea Snake; Lake Taal Snake
d Sempersche Ruderschlange; Taal-Seeschlange; Philippinen-Seeschlange; Duhol

4109 ***Hydrophis spiralis***
Squamata - Elapidae
e Yellow Sea Snake (ISC); Narrow-banded Sea Snake
d Ruderschlangen

4110 ***Hydrophis stricticolis***
Squamata - Elapidae
e Collared Sea Snake

4111 ***Hydrophis torquatus***
Squamata - Elapidae
e West Coast Black-headed Sea Snake; Günther's Sea Snake

4112 ***Hydrophis vorisi***
Squamata - Elapidae
e Kharin's Sea Snake

4113 ***Hydrops***
Squamata - Colubridae
e Amazon Water Snakes; Mud Snakes
d Amazonas-Nattern

4114 ***Hydrops martii***
Squamata - Colubridae
e Amazon Water Snake; Coral Mud Snake
d Amazonas-Natter

4115 ***Hydrops triangularis***
Squamata - Colubridae
e Triangle Water Snake; Common Mud Snake

4116 ***Hydrosaurus***
Squamata - Agamidae
e Sailfin Lizards; Water Lizards; Sail-tailed Lizards; Sailing Lizards; Sailfin Dragons; Sailtail Dragons; Crested Water Dragons
d Segelechsen
f Hydrosaures

4117 ***Hydrosaurus amboinensis***
Squamata - Agamidae
e Sailfin Lizard; Common Sailfin Lizard; Soa-Soa
d Segelechse; Soa-Soa
f Hydrosaure d'Amboine; Porte-crête

4118 ***Hydrosaurus pustulatus***
Squamata - Agamidae
e Philippine Sailfin Lizard; Sailfin Water Lizard; Water Lizard; Philippine Sailfin
d Philippinische Segelechse
f Hydrosaure des Philippines

4119 ***Hydrosaurus weberi***
Squamata - Agamidae
e Weber's Sailfin Lizard; Weber's Sailing Lizard; Philippine Water Lizard; Indonesian Water Lizard; Emerald Water Dragon; Emerald Water Lizard; Philippine Water

Dragon; Indonesian Water Dragon; Weber's Sailfin; Moluccas Sailfin
d Webers Segelechse
f Hydrosaure de Weber

4120 ***Hylagama***
Squamata - Agamidae
e Wood Agamas
d Waldagamen

4121 ***Hylagama borneensis***
Squamata - Agamidae
e Wood Agama
d Waldagame

4122 ***Hypnale***
Squamata - Viperidae
e Humpnose Vipers; Hump-nosed Vipers
d Ceylon-Nasenottern

4123 ***Hypnale hypnale***
Squamata - Viperidae
e Indian Humpnose Viper; Hump-nosed Moccasin; Hump-nosed Pit Viper (ISC); Merrem's Hump-nosed Viper; Humpnose Viper
d Indische Nasenotter; Indische Höckernasengrubenotter

4124 ***Hypnale nepa***
Squamata - Viperidae
e Sri Lanka Humpnose Viper; Laurenti's Hump-nosed Viper
d Ceylon-Nasenottern

4125 ***Hypnale walli***
Squamata - Viperidae
e Wall's Humpnose Viper

4126 ***Hypoptophis***
Squamata - Colubridae
e African Bighead Snakes

4127 ***Hypoptophis wilsoni***
Squamata - Colubridae
e African Bighead Snake

4128 ***Hypsiglena***
Squamata - Colubridae
e Night Snakes; Fangless Night Snakes (NA)
d Nachtschlangen; Nachtnattern
f Couleuvres nocturnes

4129 ***Hypsiglena gularis***
Squamata - Colubridae
e Partida Norte Island Night Snake

4130 ***Hypsiglena tanzeri***
Squamata - Colubridae
e Tanzer's Night Snake

4131 ***Hypsiglena torquata***
Squamata - Colubridae
e Night Snake
d Nachtschlange; Nachtnatter
f Couleuvre nocturne

4132 ***Hypsiglena torquata baueri***
Squamata - Colubridae
e Cedros Island Night Snake

4133 ***Hypsiglena torquata chlorophaea***
Squamata - Colubridae
e Sonoran Night Snake

4134 ***Hypsiglena torquata deserticola***
Squamata - Colubridae
e Desert Night Snake

4135 ***Hypsiglena torquata jani***
Squamata - Colubridae
e Texas Night Snake

4136 ***Hypsiglena torquata klauberi***
Squamata - Colubridae
e San Diego Night Snake

4137 ***Hypsiglena torquata loreala***
Squamata - Colubridae
e Mesa Verde Night Snake

4138 ***Hypsiglena torquata martinensis***
Squamata - Colubridae
e San Martin Island Night Snake

4139 ***Hypsiglena torquata nuchalata***
Squamata - Colubridae
e California Night Snake

4140 ***Hypsiglena torquata ochrorhyncha***
Squamata - Colubridae
e Spotted Night Snake
d Gepunktete Bachschlange

4141 ***Hypsilurus***
Squamata - Agamidae
e Anglehead Forest Dragons; Australian Forest Dragons; Angle-headed Forest Dragons; Forest Dragons
d Winkelkopfagamen
f Hypsilures

4142 ***Hypsilurus binotatus***
Squamata - Agamidae
e Two-marked Forest Dragon

4143 ***Hypsilurus boydii***
Squamata - Agamidae
e Boyd's Forest Dragon; Boyd's Rainforest Dragon
d Boyds Agame; Boyds Winkelkopfagame
f Hypsilure du Boyd

4144 ***Hypsilurus bruijnii***
Squamata - Agamidae
e Bruijn's Forest Dragon

4145 ***Hypsilurus dilophus***
Squamata - Agamidae
e Indonesian Forest Dragon

4146 ***Hypsilurus geelvinkianus***
Squamata - Agamidae
e New Guinea Forest Dragon

4147 ***Hypsilurus godeffroyi***
Squamata - Agamidae
e Northern Forest Dragon; Godeffroy's Forest Dragon
d Godeffroys Winkelkopfagame
f Hypsilure de Godeffroy

4148 ***Hypsilurus modestus***
Squamata - Agamidae
e Modest Forest Dragon

4149 ***Hypsilurus nigrilabris***
Squamata - Agamidae
e Black-lipped Forest Dragon

4150 ***Hypsilurus papuensis***
Squamata - Agamidae
e Papua Forest Dragon
d Papua-Winkelkopfagame
f Hypsilure de Papua

4151 ***Hypsilurus schoedei***
Squamata - Agamidae
e Vogt's Forest Dragon

4152 ***Hypsilurus spinipes***
Squamata - Agamidae
e Southern Forest Dragon (ANZ); Southern Anglehead Dragon; Southern Angle-headed Dragon (ANZ); Rain Forest Dragon

4153 ***Hypsirhynchus***
Squamata - Colubridae
e Hispaniola Cat-eyed Snakes

4154 ***Hypsirhynchus ferox***
Squamata - Colubridae
e Hispaniola Cat-eyed Snake; Common Hog-nosed Racer

I

4155 ***Ialtris***
Squamata - Colubridae
e Fanged Snakes

4156 ***Ialtris agyrtes***
Squamata - Colubridae
e Barreras Fanged Snake; Southern Desert Racer

4157 ***Ialtris dorsalis***
Squamata - Colubridae
e Brown Fanged Snake; W-headed Racer

4158 ***Ialtris parishi***
Squamata - Colubridae
e Parish's Fanged Snake; Tiburon Banded Racer

4159 ***Ichnotropis***
Squamata - Lacertidae
e Rough-scaled Lizards
d Rauschuppige Echsen

4160 ***Ichnotropis bivittatus***
Squamata - Lacertidae
e Angola Rough-scaled Lizard

4161 ***Ichnotropis capensis***
Squamata - Lacertidae
e Cape Rough-scaled Lizard; Cape Rough-scaled Sand Lizard (CSA)
d Rauschuppige Kap-Eidechse

4162 ***Ichnotropis grandiceps***
Squamata - Lacertidae
e Caprivi Rough-scaled Lizard

4163 ***Ichnotropis microlepidota***
Squamata - Lacertidae
e Marx's Rough-scaled Lizard

4164 ***Ichnotropis squamulosa***
Squamata - Lacertidae
e Common Rough-scaled Lizard; Mozambique Rough-scaled Sand Lizard (CSA); Tropical Rough-scaled Sand Lizard (CSA)

4165 ***Iguana***
Squamata - Iguanaidae
e Common Iguanas
d Leguane; Grüne Leguane

4166 ***Iguana delicatissima***
Squamata - Iguanaidae
e West Indian Iguana; Lesser Antillean Green Iguana; Antillean Iguana; Lesser Antillean Iguana (WI)
d Grüner Inselleguan
f Iguane des Antilles; Iguane tuberculeux

4167 ***Iguana iguana***
Squamata - Iguanaidae
e Green Iguana; Common Iguana; Green Mexican Iguana
d Grüner Leguan
f Iguane verte; Iguane vrai; Iguane commun
i Iguana verde; Iguana dei tubercoli; Iguana comune; Iguana

4168 **Iguanaidae**
Squamata
e Iguanids
d Leguane
f Iguanes; Iguanidés

4169 ***Iguanognathus***
Squamata - Colubridae
e Spatula-tooth Snakes

4170 ***Iguanognathus werneri***
Squamata - Colubridae
e Spatula-tooth Snake

4171 ***Imantodes***
Squamata - Colubridae
e Central American Tree Snakes; Blunt-headed Tree Snakes; Tree Snakes
d Riesennattern

4172 ***Imantodes cenchoa***
Squamata - Colubridae
e Blunthead Tree Snake; Blunt-headed Tree Snake; Common Blunt-headed Tree Snake
d Riesennatter

4173 ***Imantodes gemmistratus***
Squamata - Colubridae
e Central American Tree Snake; Spotted Blunt-headed Tree Snake

4174 ***Imantodes inornatus***
Squamata - Colubridae
e Western Tree Snake; Brown Blunt-headed Tree Snake; Yellow Blunt-headed Tree Snake

4175 ***Imantodes lentiferus***
Squamata - Colubridae
e Amazon Basin Tree Snake; Amazon Tree Snake

4176 ***Imantodes phantasma***
Squamata - Colubridae
e Phantasma Tree Snake; Fantastic Blunt-headed Tree Snake

4177 ***Imantodes tenuissimus***
Squamata - Colubridae
e Yucatan Blunthead Snake; Yucatan Blunt-headed Tree Snake

4178 ***Incaspis***
Squamata - Colubridae
e Donoso-Barros's Snakes

4179 ***Incaspis cercostsropha***
Squamata - Colubridae
e Donoso-Barros's Snake

4180 ***Indotestudo***
Testudines - Testudinidae
e Asian Tortoises
d Asiatische Landschildkröten

4181 ***Indotestudo elongata***
Testudines - Testudinidae
e Elongate Tortoise; Red-nosed Tortoise; Elongated Tortoise
d Gelbkopflandschildkröte; Gelbkopfschildkröte
f Tortue à tête jaune
i Testuggine a testa gialla

4182 ***Indotestudo forstenii***
Testudines - Testudinidae
e Travancore Tortoise; Celebes Tortoise
d Celebes-Landschildkröte; Travancore-Landschildkröte
f Tortue des Célèbes; Tortue de Travancore
i Testuggine di Forsten; Testuggine di Travancore

4183 ***Internatus***
Squamata - Colubridae
e Slug Snakes; Vine Snakes

4184 ***Internatus laevis***
Squamata - Colubridae
e Smooth Slug Snake; Smooth Vine Snake

4185 ***Internatus malaccanus***
Squamata - Colubridae
e Malayan Slug Snake; Malayan Vine Snake

4186 ***Iphisa***
Squamata - Teiidae
e Iphisas; Glossy Shade Lizards

4187 ***Iphisa elegans***
Squamata - Teiidae
e Common Iphisa; Glossy Shade Lizard

4188 ***Isopachys***
Squamata - Scincidae
e Isopachys

4189 ***Isopachys anguinoides***
Squamata - Scincidae
e Heyer's Isopachys

4190 ***Isopachys borealis***
Squamata - Scincidae
e Lang's Isopachys

4191 ***Isopachys hyldenstolpei***
Squamata - Scincidae
e Gyldenstolpe's Isopachys

4192 ***Isopachys roulei***
Squamata - Scincidae
e Roule's Isopachys

4193 ***Ithycyphus***
Squamata - Colubridae
e Forest Night Snakes

4194 ***Ithycyphus goudoti***
Squamata - Colubridae
e Forest Night Snake

4195 ***Ithycyphus miniatus***
Squamata - Colubridae
e Tiny Night Snake

J

4196 ***Janetascincus***
Squamata - Scincidae
e Janet's Skinks

4197 ***Janetascincus braueri***
Squamata - Scincidae
e Brauer's Skink

4198 ***Janetascincus veseyfitzgeraldi***
Squamata - Scincidae
e Vesey's Skink

4199 ***Japalura***
Squamata - Agamidae
e Mountain Lizards; Japalures; Japalura Lizards (ISC); Mountain Agamas
d Bergagamen
f Japalures

4200 ***Japalura andersoniana***
Squamata - Agamidae
e Anderson's Japalure
d Andersons Bergagame
f Japalure d'Anderson

4201 ***Japalura brevipes***
Squamata - Agamidae
e Short-legged Japalure

4202 ***Japalura chapaensis***
Squamata - Agamidae
e Japalure

4203 ***Japalura dymondi***
Squamata - Agamidae
e Dymond's Japalure
d Dymonds Bergagame
f Japalure de Dymond

4204 ***Japalura fasciata***
Squamata - Agamidae
e Banded Japalure
d Gebänderte Bergagame
f Japalure rayé

4205 ***Japalura flavipes***
Squamata - Agamidae
e Szechwan Japalure

4206 ***Japalura grahami***
Squamata - Agamidae
e Graham's Japalure

4207 ***Japalura hamptoni***
Squamata - Agamidae
e Hampton's Japalure
d Hamptons Bergagame
f Japalure de Hampton

4208 ***Japalura kaulbacki***
Squamata - Agamidae
e Smith's Japalure

4209 ***Japalura kumaonensis***
Squamata - Agamidae
e Kumaon Japalure
d Kumaonische Bergagame
f Japalure de Kumaon

4210 ***Japalura major***
Squamata - Agamidae
e Himalayan Japalure

4211 ***Japalura makii***
Squamata - Agamidae
e Ota's Japalure

4212 ***Japalura mitsukuruii***
Squamata - Agamidae

e Botel Tobago Japalure

4213 ***Japalura nasuta***
Squamata - Agamidae
e De Jong's Japalure

4214 ***Japalura ornata***
Squamata - Agamidae
e Ornate Japalure

4215 ***Japalura planidorsata***
Squamata - Agamidae
e Khasi Japalure

4216 ***Japalura polygonata***
Squamata - Agamidae
e Ryukyu Japalure

4217 ***Japalura polygonata ishigakiensis***
Squamata - Agamidae
e Okinawan Tree Lizard; Sakishima Tree Lizard

4218 ***Japalura sagittifera***
Squamata - Agamidae
e Burmese Japalure

4219 ***Japalura splendida***
Squamata - Agamidae
e Splendid Japalure
d Prachtvolle Bergagame
f Japalure splendide

4220 ***Japalura swinhonis***
Squamata - Agamidae
e Taiwan Japalure

4221 ***Japalura tricarinata***
Squamata - Agamidae
e Cloud Forest Japalure

4222 ***Japalura varcoae***
Squamata - Agamidae
e Chinese Japalure

4223 ***Japalura variegata***
Squamata - Agamidae
e Variegated Japalure
d Veränderliche Bergagame
f Japalure varié

4224 ***Japalura yunnanensis***
Squamata - Agamidae
e Yunnan Japalure

K

4225 ***Kachuga***
Testudines - Emydidae
e Indian Roofed Turtles; Roofed Turtles; Roof Turtles
d Dachschildkröten
f Kachugas

4226 ***Kachuga dhongoka***
Testudines - Emydidae
e Three-striped Roofed Turtle; Dhongoka Turtle; Dhoor (ISC); Three-striped Roofed Terrapin (ISC); Three-striped Roof Turtle; Dhomgoka Roof Turtle
d Dhongoka-Dachschildkröte
f Kachuga à front rayé

4227 ***Kachuga kachuga***
Testudines - Emydidae
e Red-crowned Roofed Turtle; Dura Turtle; Red-crowned Roofed Terrapin (ISC); Bengal Roofed Turtle (ISC); Bengal Roof Turtle; Red-crowned Roof Turtle
d Bengalische Dachschildkröte; Kachuga-Dachschildkröte
f Kachuga à front rouge
i Kachuga fronte rossa

4228 ***Kachuga smithii***
Testudines - Emydidae
e Brown-roofed Turtle; Brown Roofed River Turtle; Brown River Turtle; Brown Roofed Terrapin (ISC); Chapant (ISC); Brown Roof Turtle
d Smiths Dachschildkröte; Smith-Dachschildkröte
f Kachuga de Smith
i Kachuga di Smith

4229 ***Kachuga sylhetensis***
Testudines - Emydidae
e Assam Roofed Turtle; Assam Roofed Terrapin (ISC); Assam Saw-back Turtle; Assam Roof Turtle
d Assam-Dachschildkröte
f Kachuga de l'Assam

4230 ***Kachuga tecta***
Testudines - Emydidae
e Indian Roofed Turtle; Indian Sawback Turtle; Indian Sawback (ISC); Roofed Terrapin (ISC); North Indian Roofed Turtle; Indian Roof Turtle; Indian Tent Turtle; Indian Turtle
d Indische Dachschildkröte
f Kachuga à dos en toit; Tortue à toit de l'Inde
i Tartaruga dal dorso a tetto

4231 ***Kachuga tentoria***
Testudines - Emydidae
e Indian Tent Turtle; Deccan Sawback (ISC); Indian Tent Terrapin (ISC); South Indian Roofed Turtle; Indian Sawback Turtle
f Kachuga carénée
i Tartaruga a tetto indiana

4232 ***Kachuga trivittata***
Testudines - Emydidae
e Burmese Roofed Turtle; Burmese Roof Turtle
d Dreistreifendachschildkröte
f Kachuga de Birmanie; Tortue de Birmanie

4233 ***Kentropyx***
Squamata - Teiidae
e Kentropyx; Whiptails
d Kieltejus

4234 ***Kentropyx altamazonicus***
Squamata - Teiidae
e Amazon Kentropyx; Cocha Whiptail

4235 ***Kentropyx borckiana***
Squamata - Teiidae
e Guyana Kentropyx

4236 ***Kentropyx calcaratus***
Squamata - Teiidae
e Spix's Kentropyx
d Afterspornechse

4237 ***Kentropyx paulensis***
Squamata - Teiidae
e Boettger's Kentropyx

4238 ***Kentropyx striatus***
Squamata - Teiidae
e Striped Kentropyx
d Kielteju

4239 ***Kentropyx vanzoi***
Squamata - Teiidae
e Gallagher's Kentropyx

4240 ***Kentropyx viridistriga***
Squamata - Teiidae
e Green Kentropyx

4241 ***Kentropyx williamsoni***
Squamata - Teiidae
e Williamson's Kentropyx

4242 ***Kerilia***
Squamata - Elapidae
e Jerdon's Sea Snakes

4243 ***Kerilia jerdoni***
Squamata - Elapidae
e Jerdon's Sea Snake

4244 ***Kinixys***
Testudines - Testudinidae
e Hingeback Tortoises
d Gelenkschildkröten
f Kinyx

4245 ***Kinixys belliana***
Testudines - Testudinidae
e Bell's Hingeback Tortoise; Bell's Hinged Tortoise(CSA); Bell's Eastern Hinged Tortoise; Hinged Tortoise (CSA)
d Glattrandgelenkschildkröte
f Kinyx de Bell
i Cinisside di Bell; Testuggine di Bell

4246 ***Kinixys erosa***
Testudines - Testudinidae
e Serrated Hingeback Tortoise; Serrated Hinged-back Tortoise ; Eroded Hingeback Tortoise; Rosy Hingeback; Schweigger's Hingeback Tortoise
d Stachelrandgelenkschildkröte
f Kinixys rongée
i Cinisside denticolata; Testuggine serrata

4247 ***Kinixys homeana***
Testudines - Testudinidae
e Home's Hingeback Tortoise; Home's Hinge-backed Tortoise; Home's Hinged Tortoise
d Stutzgelenkschildkröte
f Kinyx de Home
i Cinisside di Home; Testuggine di Home

4248 ***Kinixys lobatsiana***
Testudines - Testudinidae
e Lobatsa Hingeback Tortoise; Lobatsa Hinged Tortoise

4249 ***Kinixys natalensis***
Testudines - Testudinidae
e Natal Hingeback Tortoise; Natal Hinged Tortoise (CSA); Natal Hinge-backed Tortoise
d Natal-Gelenkschildkröte

4250 ***Kinixys spekii***
Testudines - Testudinidae
e Speke's Hingeback Tortoise; Savanna Hingeback

4251 **Kinosternidae**
Testudines
e Mud and Musk Turtles; American Mud and Musk Turtles; Musk Turtles
d Schlammschildkröten; Moschusschildkröten; Klappschildkröten; Klappbrust- und Moschusschildkröten
f Tortues bourbeuses et musquées; Kinosternidés
i Tartarughe muschiate; Tartarughe del fango

4252 ***Kinosternon***
Testudines - Kinosternidae
e American Mud Turtles; Mud Turtles; Musk Turtles
d Klappschildkröten; Moschusschildkröten
f Tortues bourbeuses; Kinosternes; Cinosternes

4253 ***Kinosternon acutum***
Testudines - Kinosternidae
e Tabasco Mud Turtle
d Tabasco-Klappschildkröte; Spitzschnauzenklappschildkröte
f Cinosterne de Tabasco

4254 ***Kinosternon alamosae***
Testudines - Kinosternidae
e Alamos Mud Turtle; American Musk Turtle
f Cinosterne d'Alamos

4255 ***Kinosternon angustipons***
Testudines - Kinosternidae
e Narrow-bridged Mud Turtle; Central American Mud Turtle; Caribbean Mud Turtle
f Cinosterne à pont étroit

4256 ***Kinosternon arizonense***
Testudines - Kinosternidae
e Arizoona Mud Turtle

4257 ***Kinosternon baurii***
Testudines - Kinosternidae
e Striped Mud Turtle
d Dreistreifenklappschildkröte; Streifenklappschildkröte
f Tortue bourbeuse rayé; Kinosterne de Baur; Cinosterne de Baur
i Tartaruga del fango striata

4258 ***Kinosternon bauri bauri***
Testudines - Kinosternidae
e Key Mud Turtle

4259 ***Kinosternon carinatum***
Testudines - Kinosternidae
e Razorback Musk Turtle; Razor-backed Musk Turtle; Mississippi Musk Turtle; Keel-backed Musk Turtle; Razor-backed Mud Turtle
d Dachmoschusschildkröte; Gekielte Moschusschildkröte
f Tortue musquée carénée; Cinosterne cariné
i Tartaruga del muschio carenata

4260 ***Kinosternon chimalhuaca***
Testudines - Kinosternidae
e Jalisco Mud Turtle

4261 ***Kinosternon creaseri***
Testudines - Kinosternidae
e Creaser's Mud Turtle; Yucatan Mud Turtle
d Yucatan-Klappschildkröte
f Cinosterne du Yucatan
i Tartaruga del fango di Creaser

4262 ***Kinosternon depressum***
Testudines - Kinosternidae
e Flat Musk Turtle; Flattened Musk Turtle
f Tortue musquée déprimée;

Cinosterne aplati
i Tartaruga del muschio appiattita

4263 ***Kinosternon dunni***
Testudines - Kinosternidae
e Dunn's Mud Turtle; Colombian Mud Turtle
d Dunns Klappschildkröte
f Cinosterne de Colombie

4264 ***Kinosternon flavescens***
Testudines - Kinosternidae
e Yellow Mud Turtle
d Gelbliche Klappschildkröte
f Tortue bourbeuse jaune; Cinosterne jaune
i Tartaruga del fango gialla

4265 ***Kinosternon flavescens arizonense***
Testudines - Kinosternidae
e Southwestern Mud Turtle

4266 ***Kinosternon flavescens flavescens***
Testudines - Kinosternidae
e Yellow Mud Turtle

4267 ***Kinosternon flavescens spooneri***
Testudines - Kinosternidae
e Illinois Mud Turtle

4268 ***Kinosternon herrerai***
Testudines - Kinosternidae
e Herrera's Mud Turtle
d Herreras Klappschildkröte
f Cinosterne d'Herrera

4269 ***Kinosternon hirtipes***
Testudines - Kinosternidae
e Mexican Roughfoot Mud Turtle; Mexican Mud Turtle; Rough-footed Mud Turtle; Mexican Rough-footed Mud Turtle; Big Bend Mud Turtle
d Raufußklappschildkröte
f Tortue bourbeuse du Mexique; Cinosterne du Mexique central
i Tartaruga del fango dai piedi rugosi

4270 ***Kinosternon hirtipes chapalensis***
Testudines - Kinosternidae
e Lake Chapala Mud Turtle
f Cinosterne duLac Chapala

4271 ***Kinosternon hirtipes hirtipes***
Testudines - Kinosternidae
e Mexican Mud Turtle
f Cinosterne de Mexique

4272 ***Kinosternon hirtipes magdalense***
Testudines - Kinosternidae
e San Juanico Mud Turtle
f Cinosterne de San Juanico

4273 ***Kinosternon hirtipes megacephalum***
Testudines - Kinosternidae
e Viesca Mud Turtle
f Cinosterne de Viesca

4274 ***Kinosternon hirtipes murrayi***
Testudines - Kinosternidae
e Big Bend Mud Turtle; Mexican Plateau Mud Turtle
f Cinosterne du plateau mexicain

4275 ***Kinosternon hirtipestarascene***
Testudines - Kinosternidae
e Patzuaro Mud Turtle
f Cinosterne de Patzcuaro

4276 ***Kinosternon integrum***
Testudines - Kinosternidae
e Mexican Mud Turtle; Le Conte's Mud Turtle
d Mexiko-Klappschildkröte
f Cinosterne du Mexique occidental
i Testuggine del fango messicana

4277 ***Kinosternon leucostomum***
Testudines - Kinosternidae
e White-lipped Mud Turtle

d Weißmaulklappschildkröte
f Cinosterne à bouche blanche
i Tartaruga del fango dalla bocca bianca

4278 ***Kinosternon leucostomum leucostomum***
Testudines - Kinosternidae
e Northern White-lipped Mud Turtle
f Cinosterne à bouche blanche du Nord

4279 ***Kinosternon leucostomum postinguinale***
Testudines - Kinosternidae
e Southern White-lipped Mud Turtle
f Cinosterne à bouche blanche du Sud

4280 ***Kinosternon minor***
Testudines - Kinosternidae
e Loggerhead Musk Turtle
d Kleine Moschusschildkröte
f Tortue musquée mineure; Cinosterne à grosse tête
i Tartaruga del muschio capogrosso

4281 ***Kinosternon minor minor***
Testudines - Kinosternidae
e Loggerhead Musk Turtle
f Cinostserne à grosse tête commune

4282 ***Kinosternon minor peltifer***
Testudines - Kinosternidae
e Stripeneck Musk Turtle
f Cinosterne à cou rayé

4283 ***Kinosternon oaxacae***
Testudines - Kinosternidae
e Oaxacan Mud Turtle; Oaxaca Mud Turtle
f Cinosterne d'Oaxaca

4284 ***Kinosternon odoratum***
Testudines - Kinosternidae
e Common Musk Turtle; Stinkin' Jenny (NA); Musk Turtle; Stinkin' Jim (NA); Stinkpot (NA)
d Gewöhnliche Moschusschildkröte; Moschusschildkröte; Stinkender Jim
f Tortue musquée odorante; Sternothère odorant; Tortue musquée (Qué); Cinosterne odorant
i Tartaruga del muschio comune; Tartaruga muschiata comune

4285 ***Kinosternon scorpioides***
Testudines - Kinosternidae
e Scorpion Mud Turtle
d Skorpionklappschildkröte
f Cinosterne scorpioïde; Tortue-scorpion
i Tartaruga scorpione

4286 ***Kinosternon scorpioides abaxillare***
Testudines - Kinosternidae
e Chiapas Mud Turtle
f Cinostserne de Chiapas

4287 ***Kinosternon scorpioides albogulare***
Testudines - Kinosternidae
e White-throated Mud Turtle
f Cinosterne à gorge blanche

4288 ***Kinosternon scorpioides carajensis***
Testudines - Kinosternidae
e Carajas Mud Turtle
f Cinosterne de Carajas

4289 ***Kinosternon scorpioides cruentatum***
Testudines - Kinosternidae
e Redhead Mud Turtle
f Cinossterne ensanglanté

4290 ***Kinosternon scorpioides scorpioides***
Testudines - Kinosternidae
e Amazon Mud Turtle
f Cinosterne de l'Amazone

4291 ***Kinosternon scorpioides seriei***
Testudines - Kinosternidae
e Argentine Mud Turtle

f Cinosterne d'Argentine

4292 ***Kinosternon sonoriense***
Testudines - Kinosternidae
e Sonoran Mud Turtle; Sonora Mud Turtle
d Sonora-Klappschildkröte
f Tortue bourbeuse du Sonora; Cinosterne de Sonora

4293 ***Kinosternon sonoriense longifemorale***
Testudines - Kinosternidae
e Sonoyta Mud Turtle
f Cinosterne de Sonoyta

4294 ***Kinosternon sonoriense sonoriense***
Testudines - Kinosternidae
e Sonoran Mud Turtle; Common Sonora Mud Turtle; Sonora Mud Turtle
f Cinosterne du Sonora commune

4295 ***Kinosternon subrubrum***
Testudines - Kinosternidae
e Mud Turtle (NA); Common Mud Turtle; Eastern Mud Turtle
d Pennsylvania-Klappschildkröte
f Tortue bourbeuse des marais; Tortue des marais; Cinosterne rougeâtre
i Tartaruga del fango comune

4296 ***Kinosternon subrubrum hippocrepis***
Testudines - Kinosternidae
e Mississippi Mud Turtle; Mississippi Musk Turtle
f Cinosterne rougeâtre du Mississippi

4297 ***Kinosternon subrubrum steindachneri***
Testudines - Kinosternidae
e Florida Mud Turtle
f Cinosterne rougeâtre de Floride

4298 ***Kinosternon subrubrum subrubrum***
Testudines - Kinosternidae
e Eastern Mud Turtle
f Cinosterne rougeâtre commune

4299 ***Klauberina***
Squamata - Xantusiidae
e Island Night Lizards
d Insel-Nachtechsen
f Lézards nocturnes des îles

4300 ***Klauberina riversiana***
Squamata - Xantusiidae
e Island Night Lizard
d Insel-Nachtechse
f Lézard nocturne des îles

4301 ***Klauberina riversiana reticulata***
Squamata - Xantusiidae
e San Clemente Night Lizar

4302 ***Klauberina riversiana riversiana***
Squamata - Xantusiidae
e San Nicolas Night Lizard

4303 ***Kleiometopon***
Squamata - Colubridae
e Booliati Snakes

4304 ***Kleiometopon booliati***
Squamata - Colubridae
e Booliati Snake

4305 ***Kolpophis***
Squamata - Elapidae
e Bighead Sea Snakes

4306 ***Kolpophis annandalei***
Squamata - Elapidae
e Bighead Sea Snake

L

4307 ***Lacerta***
Squamata - Lacertidae
e Common Eurasian Lizards; Mediterranean Lizards (CSA); Lizards
d Echte Eidechsen; Halsbandeidechsen; Eidechsen
f Lézards
i Lucertole

4308 ***Lacerta agilis***
Squamata - Lacertidae
e Sand Lizard
d Zauneidechse; Flinke Eidechse; Graue Eidechse; Gewöhnliche Eidechse; Gemeine Eidechse; Kleinäugige Eidechse
f Lézard des souches; Lézard agile
i Lucertola dei ceppi; Lucertola agile; Lucertola delle sterpi; Lucertola degli arbusti

4309 ***Lacerta andreanskyi***
Squamata - Lacertidae
e Andreansky's Lizard; Wall Lizard
d Atlas-Bergeidechse
f Lézard du Haut Atlas; Lézard d'Andreansky

4310 ***Lacerta armeniaca***
Squamata - Lacertidae
e Armenian Lizard
d Armenische Felseneidechse; Armenische Eidechse
f Lézard arménien; Lézard d'Arménie
i Lucertola armena

4311 ***Lacerta australis***
Squamata - Lacertidae
e Southern Rock Lizard

4312 ***Lacerta bedriagae***
Squamata - Lacertidae
e Bédriaga's Rock Lizard
d Tyrrhenische Gebirgseidechse; Bedriagas Gebirgseidechse
f Lézard de montagne corse; Lézard de Bédriaga; Lézard montagnard corse
i Lucertola di Bedriaga

4313 ***Lacerta bilineata***
Squamata - Lacertidae
e Western Green Lizard
d Westliche Smaragdeidechse
f Lézard vert
i Ramarro occidentale; Ramarro

4314 ***Lacerta bithynica***
Squamata - Lacertidae
e Méhely's Lizard
f Lézard de Bithynie
i Lucertola bitinica

4315 ***Lacerta brandti***
Squamata - Lacertidae
e Persian Lizard
d Persische Eidechse
f Lézard de Perse; Lézard nain
i Lucertola di Brandt

4316 ***Lacerta cappadocica***
Squamata - Lacertidae
e Anatolian Lizard; Cappadocian Lizard
d Kappadokische Eidechse
f Lézard d'Anatolie; Lézard de Cappadoce
i Lucertola anatolica; Lucertola cappadocica

4317 ***Lacerta caucasica***
Squamata - Lacertidae
e Caucasian Lizard

d Kaukasus-Eidechse; Kaukasische Eidechse; Kaukasus-Gebirgseidechse
f Lézard du Caucase
i Lucertola caucasica

4318 ***Lacerta chlorogaster***
Squamata - Lacertidae
e Greenbelly Lizard; Green-bellied Lizard
d Grünbaucheidechse
f Lézard à ventre vert
i Lucertola chlorogaster

4319 ***Lacerta clarkorum***
Squamata - Lacertidae
e Clark's Lizard
d Clarks Eidechse; Clarks Feldeidechse
f Lézard de Clark
i Lucertola di Clark

4320 ***Lacerta cyanura***
Squamata - Lacertidae
e Omman's Lizard

4321 ***Lacerta dahli***
Squamata - Lacertidae
e Dahl's Lizard
d Dahls Eidechse; Dahls Feldeidechse
f Lézard de Dahl
i Lucertola di Dahl

4322 ***Lacerta danfordi***
Squamata - Lacertidae
e Danford's Lizard
d Danford-Eidechse; Danfords Gebirgseidechse; Anatolische Eidechse
f Lézard de Danford
i Lucertola di Danford

4323 ***Lacerta defilippii***
Squamata - Lacertidae
e Elburs Lizard
d Elburs-Eidechse
f Lézard d'Elburs

4324 ***Lacerta derjugini***
Squamata - Lacertidae
e Derjugin's Lizard
d Artwinner-Eidechse; Derjugins Eidechse
f Lézard de Derjugin
i Lucertola di Derjugin

4325 ***Lacerta echinata***
Squamata - Lacertidae
e Cope's Spinytail Lizard
d Stachelschwanzeidechse

4326 ***Lacerta fraasii***
Squamata - Lacertidae
e Fraas's Lizard
d Fraas-Eidechse
f Lézard de Fraas

4327 ***Lacerta graeca***
Squamata - Lacertidae
e Greek Rock Lizard
d Griechische Felseidechse; Griechische Spitzkopfeidechse
f Lézard grec
i Lucertola greca

4328 ***Lacerta horvathi***
Squamata - Lacertidae
e Horvath's Rock Lizard
d Kroatische Gebirgseidechse
f Lézard de Horvath
i Lucertola di Horvath; Lucertola croata

4329 ***Lacerta jayakari***
Squamata - Lacertidae
e Jayakar Lizard

4330 ***Lacerta laevis***
Squamata - Lacertidae
e Lebanon Lizard; Smooth Lizard
d Syrische Eidechse; Libanon-Eidechse
f Lézard du Liban; Lézard lisse
i Lucertola liscia

4331 ***Lacerta lepida***
Squamata - Lacertidae
e Jewelled Lizard; European Eyed Lizard; Eyed Lizard; Ocellated Lizard; Spanish Lizard
d Perleidechse
f Lézard ocellé; Rassade; Lambert; Lézard ocellé d'Europe
i Lucertola ocellata; Lucertola occhiata

4332 ***Lacerta mixta***
Squamata - Lacertidae
e Adjar Lizard
d Bastardeidechse; Adsharische Eidechse
f Lézard d'Adjarie; Lézard bâtard
i Lucertola bastarda

4333 ***Lacerta monticola***
Squamata - Lacertidae
e Iberian Mountain Lizard
d Iberische Gebirgseidechse
f Lézard de montagne pyrénéen; Lézard montagnard pyrénéen; Lézard montagnard
i Lucertola montana

4334 ***Lacerta mosorensis***
Squamata - Lacertidae
e Mosor Rock Lizard
d Mosor-Eidechse; Mosor-Gebirgseidechse
f Lézard de Mosor
i Lucertola di Mosor

4335 ***Lacerta oxycephala***
Squamata - Lacertidae
e Sharp-snouted Rock Lizard; Long-headed Lizard
d Dalmatinische Spitzkopfeidechse; Dalmatinische Spitzköpfige Eidechse; Spitzkopfeidechse
f Lézard à tête pointue; Lézard oxycéphale
i Lucertola ossicefala

4336 ***Lacerta parva***
Squamata - Lacertidae
e Dwarf Lizard; North African Eyed Lizard
d Zwergeidechse; Zwergzauneidechse
f Lézard d'Asie Mineure; Lézard nain
i Lucertola parva

4337 ***Lacerta parvula***
Squamata - Lacertidae
e Redbelly Lizard; Red-bellied Lizard
d Rotbauchfelseneidechse; Rotbaucheidechse
f Lézard à ventre rouge
i Lucertola parvola

4338 ***Lacerta pater***
Squamata - Lacertidae
e Lataste's Lizard; North African Eyed Lizard
d Berbereidechse
f Lézard ocellé d'Afrique du Nord

4339 ***Lacerta portschinskii***
Squamata - Lacertidae
e River Kura Lizard
d Portschinskis Eidechse; Kuriner Eidechse; Kuriner Feldeidechse
f Lézard de Portschinski
i Lucertola di Portschinski

4340 ***Lacerta praticola***
Squamata - Lacertidae
e Meadow Lizard; Pontic Lizard
d Wieseneidechse; Kaukasische Wieseneidechse
f Lézard de prairie; Lézard des prairies
i Lucertola dei prati

4341 ***Lacerta princeps***
Squamata - Lacertidae
e Zagrosian Lizard
d Zagros-Eidechse
f Lézard de Zagros; Lézard de Blanford

i Lucertola principe

4342 ***Lacerta raddei***
Squamata - Lacertidae
e Azerbaijan Lizard
d Aserbaidshanische Eidechse; Aserbaidshanische Feldeidechse
f Lézard d'Azerbaïdjan
i Lucertola di Radde

4343 ***Lacerta rostombekovi***
Squamata - Lacertidae
e Rostombekov's Lizard
d Rostombekows Eidechse; Rostombekows Feldeidechse
f Lézard de Rostombekov
i Lucertola di Rostombekov

4344 ***Lacerta rudis***
Squamata - Lacertidae
e Spinytail Lizard; Spiny-tailed Lizard
d Kielschwanz-Felseneidechse; Georgische Eidechse; Kielschwanzfelseidechse; Grusinische Eidechse
f Lézard à queue épineuse

4345 ***Lacerta rupicola***
Squamata - Lacertidae
e Zoutspansberg Rock Lizard; Zoutspansberg Lizard

4346 ***Lacerta schreiberi***
Squamata - Lacertidae
e Iberian Emerald Lizard; Schreiber's Lizard; Schreiber's Green Lizard
d Iberische Smaragdeidechse; Schreibers Smaragdeidechse
f Lézard vert d'Espagne; Lézard de Schreiber
i Ramarro di Schreiber; Ramarro iberico; Lucertola di Schreiber

4347 ***Lacerta strigata***
Squamata - Lacertidae
e Caucasus Emerald Lizard; Five-streaked Lizard; Caspian Green Lizard
d Streifeneidechse; Streifensmaragdeidechse; Kaspische Smaragdeidechse
f Lézard vert de la Caspienne
i Ramarro caspico

4348 ***Lacerta trilineata***
Squamata - Lacertidae
e Balkan Emerald Lizard; Three-lined Lizard; Giant Green Lizard; Balkan Green Lizard; Three-lined Emerald Lizard
d Riesensmaragdeidechse; Gelbkehlige Smaragdeidechse
f Lézard vert à trois lignes
i Ramarro trilineato; Ramarro gigante

4349 ***Lacerta unisexualis***
Squamata - Lacertidae
e Whitebelly; White-bellied Lizard
d Weißbaucheidechse; Weißbauchfeldeidechse
f Lézard à ventre blanc; Lézard unisexué
i Lucertola unisessuale

4350 ***Lacerta uzzelli***
Squamata - Lacertidae
e Uzzell's Lizard
d Uzzells Eidechse; Uzzells Feldeidechse
f Lézard d'Uzell
i Lucertola di Uzzell

4351 ***Lacerta valentini***
Squamata - Lacertidae
e Valentin's Lizard
d Valentins Eidechse; Vlentins Feldeidechse
f Lézard de Valentin
i Lucertola di Valentin

4352 ***Lacerta viridis***
Squamata - Lacertidae
e Emerald Lizard; Green Lizard; Green Lacerta Lizard

d Smaragdeidechse
f Lézard vert
i Ramarro; Ramarro europeo; Lucertola ramarro; Ramàrr; Màrr

4353 ***Lacerta vivipara***
Squamata - Lacertidae
e Lacertid Lizard; Common Viviparous Lizard; Common Lizard; Scaly Lizard; Common English Lizard; Swift Lizard; Viviparous Lizard
d Waldeidechse; Bergeidechse; Schwarze Waldechse; Safranbäuchige Eidechse; Mooreidechse; Gebäreidechse
f Lézard vivipare
i Lucertola vivipara

4354 ***Lacerta vivipara***
Squamata - Lacertidae
e Common Lizard
d Bergeidechse

4355 ***Lacerta vivipara pannonica***
Squamata - Lacertidae
d Pannonische Bergeidechse

4356 ***Lacertaspis***
Squamata - Scincidae
e Lidless Skinks

4357 ***Lacertaspis reichenowi***
Squamata - Scincidae
e Reichenow's Lidless Skink; Reichenow's Snake-eyed Skink

4358 **Lacertidae**
Squamata
e Lacertids; Long-tailed Lizards; Old World Lizards; True Lizards; Sand Lizards
d Halsbandeidechsen; Landeidechsen; Echte Eidechsen; Lacerten
f Lézards typiques; Lacertidés; Lézards vrais

4359 ***Lacertus***
Squamata - Scincidae
e Eared Skinks

4360 ***Lacertus auratus***
Squamata - Scincidae
e Eared Skink

4361 ***Lachesis***
Squamata - Viperidae
e Bushmasters; Surucucus
d Buschmeister; Surucucus
f Surucucus

4362 ***Lachesis melanocephala***
Squamata - Viperidae
e Black-headed Bushmaster

4363 ***Lachesis muta***
Squamata - Viperidae
e South American Bushmaster; Surucucu; Bushmaster
d Buschmeister; Surucucu
f Surucucu; Maître de la brousse; Grage grands carreaux (Ants)

4364 ***Lachesis muta muta***
Squamata - Viperidae
e Amazonian Bushmaster; Amazon Bushmaster
d Buschmeister

4365 ***Lachesis stenophrys***
Squamata - Viperidae
e Central American Bushmaster
d Mittelamerikanischer Buschmeister

4366 ***Laemanctus***
Squamata - Iguanaidae
e Casquehead Iguanas; Casque-headed Lizards
d Helmkopfbasilisken; Kronenbasilisken
f Laimanctes

4367 ***Laemanctus longipes***
Squamata - Iguanaidae
e Eastern Casquehead Iguana; Conehead Lizard; Common Iguana; Casque-headed Iguana
f Laimancte longipède

4368 ***Laemanctus serratus***
Squamata - Iguanaidae
e Serrated Casquehead Iguana; Cone-headed Lizard (NA); Serrated Conehead Lizard; Casque-headed Lizard; Serrated Casque-headed Iguana
d Helmkopfleguan; Helmkopfbasilisk

4369 ***Lamprolepis***
Squamata - Scincidae
e Emerald Skinks
d Baumskinks

4370 ***Lamprolepis nieuwenhuisi***
Squamata - Scincidae
e Bieuwenhuis's Skink
d Nieuwenhuisens Baumskink

4371 ***Lamprolepis smaragdina***
Squamata - Scincidae
e Emerald Skink; Emerald Tree Skink; Green Tree Skink
d Smaragdskink
f Léiolopisme émeraudin

4372 ***Lamprolepis vyneri***
Squamata - Scincidae
e Shelford's Skink

4373 ***Lampropeltis***
Squamata - Colubridae
e King Snakes; Milk Snakes; New World King Snakes
d Königsnattern; Milchschlangen
f Couleuvres royales; Couleuvres du roi; Serpents du roi; Serpents-roi

4374 ***Lampropeltis alterna***
Squamata - Colubridae
e Gray-banded King Snake; Davis's Mountain King Snake; Grey-banded King Snake
d Graugebänderte Königsnatter; Alterna-Königsnatter
f Serpent-roi gris à bandes
i Serpente reale a bande grigie; Serpente reale grigio

4375 ***Lampropeltis alterna alterna***
Squamata - Colubridae
e Gray-banded King Snake

4376 ***Lampropeltis alterna blairi***
Squamata - Colubridae
e Blair's King Snake

4377 ***Lampropeltis calligaster***
Squamata - Colubridae
e Prairie King Snake; Mole Snake (NA); Evans's King Snake; Yellow-bellied King Snake
f Couleuvre royal des prairies
i Serpente reale di prateria

4378 ***Lampropeltis calligaster calligaster***
Squamata - Colubridae
e Prairie King Snake

4379 ***Lampropeltis calligaster occipitolineata***
Squamata - Colubridae
e South Florida Mole King Snake; Gray-banded King Snake (NA); South Florida Mole Snake; Grey-banded King Snake

4380 ***Lampropeltis calligaster rhombomaculata***
Squamata - Colubridae
e Mole King Snake; Brown King Snake ; Mole Snake

4381 ***Lampropeltis catalinensis***
Squamata - Colubridae
e Santa Catalina Island King Snake

4382 ***Lampropeltis getula***
Squamata - Colubridae
e Common King Snake; Chain Snake
d Kettenkönigsnatter; Kettennatter; Königsnatter
f Couleuvre royale commune; Couleuvre du roi; Serpent du roi; Serpent-roi
i Serpente reale ; Serpente reale comune

4383 ***Lampropeltis getula brooksi***
Squamata - Colubridae
e Brooks Kingsnake

4384 ***Lampropeltis getula californiae***
Squamata - Colubridae
e California King Snake; Californian King Snake; Cali Kingsnake (NA)
d Kalifornische Königsnatter; Kalifornische Kettennatter
f Serpent-roi de Californie
i Serpente reale californiano; Serpente reale della California

4385 ***Lampropeltis getula floridana***
Squamata - Colubridae
e Florida King Snake
d Florida-Kettennatter
f Serpent-roi de Floride
i Serpente reale della Florida

4386 ***Lampropeltis getula getula***
Squamata - Colubridae
e Eastern King Snake; Chain Snake (NA); King Snake (NA); Chain King Snake
d Östliche Kettennatter
i Serpente reale dell'est

4387 ***Lampropeltis getula goini***
Squamata - Colubridae
e Goin's Kingsnake

4388 ***Lampropeltis getula holbrooki***
Squamata - Colubridae
e Speckled King Snake; Holbrook's King Snake
d Gesprenkelte Königsnatter; Gesprenkelte Kettennatter
i Serpente reale screziato

4389 ***Lampropeltis getula niger***
Squamata - Colubridae
e Black King Snake; Eastern Black King Snake
d Schwarze Kettennatter
i Serpente reale nero

4390 ***Lampropeltis getula nigrita***
Squamata - Colubridae
e Black Desert King Snake; Mexican Black King Snake; Western Black King Snake
d Schwarze Wüstenkettennatter
f Serpent-roi noir du Mexique; Serpent-roi noir
i Serpente reale nero messicano

4391 ***Lampropeltis getula splendida***
Squamata - Colubridae
e Desert King Snake; Sonoran King Snake
d Texas-Königsnatter; Wüstenkettennatter
i Serpente reale del deserto di Sonora

4392 ***Lampropeltis getula sticticeps***
Squamata - Colubridae
e Outer Banks King Snake

4393 ***Lampropeltis herrerae***
Squamata - Colubridae
e Todos Santos Island King Snake

4394 ***Lampropeltis mexicana***
Squamata - Colubridae
e Mexican King Snake; Variable King Snake; Grey-banded King Snake
d Transpecos-Königsnatter; Mexikanische Königsnatter
f Couleuvre royale mexicaine;

Serpent-roi noir du Mexique
i Serpente reale messicano

4395 ***Lampropeltis mexicana blairi***
Squamata - Colubridae
e Milk Snake
d Transpecos-Königsnatter

4396 ***Lampropeltis mexicana greeri***
Squamata - Colubridae
e Greer's King Snake
d Greers Königsnatter
i Serpente reale di Greer

4397 ***Lampropeltis mexicana mexicana***
Squamata - Colubridae
e San Luis Potosi King Snake; Mexican King Snake
d Mexikanische Königsnatter; Transpecos-Kettennatter
i Serpente reale di San Luis Potosi; Serpente reale messicano

4398 ***Lampropeltis mexicana thayeri***
Squamata - Colubridae
e Nuevo Leon King Snake; Thayer's Variable King Snake; Thayer's King Snake; Variable King Snake
i Serpente reale del Nuevo Leon

4399 ***Lampropeltis pyromelana***
Squamata - Colubridae
e Sonoran Mountain King Snake; Sonora Mountain King Snake; Arizona Coral King Snake; Huachuca Mountain King Snake
d Arizona-Königsnatter; Arizona-Natter; Bergkönigsnatter
f Couleuvre royale du Sonora
i Serpente reale di Sonora; Serpente reale delle montagne

4400 ***Lampropeltis pyromelana infralabialis***
Squamata - Colubridae
e Utah Mountain King Snake
d Utah-Königsnatter

4401 ***Lampropeltis pyromelana knoblochi***
Squamata - Colubridae
e Chihuahua Mountain King Snake; Tarahumara Mountain King Snake

4402 ***Lampropeltis pyromelana pyromelana***
Squamata - Colubridae
e Arizona Mountain King Snake; Arizona King Snake (NA); Banded King Snake; Sonoran Mountain King Snake
d Arizona-Königsnatter
i Serpente reale dell'Arizona

4403 ***Lampropeltis pyromelana woodini***
Squamata - Colubridae
e Huachuca Mountain King Snake

4404 ***Lampropeltis ruthveni***
Squamata - Colubridae
e Ruthven's King Snake
d Ruthvens Königsnatter; Königsnatter
i Serpente reale di Ruthven

4405 ***Lampropeltis triangulum***
Squamata - Colubridae
e Milk Snake; American Milk Snake; Tropical Milk Snake
d Rote Königsnatter; Milchschlange; Dreiecksnatter
f Couleuvre royale tachetée; Couleuvre tachetée (Qué); Faux serpent corail; Serpent du lait
i Serpente del latte

4406 ***Lampropeltis triangulum abnorma***
Squamata - Colubridae
e Guatemalan Milk Snake

4407 ***Lampropeltis triangulum amaura***
Squamata - Colubridae
e Louisiana Milk Snake

4408 ***Lampropeltis triangulum andesiana***
Squamata - Colubridae

e Andean Milk Snake

4409 ***Lampropeltis triangulum annulata***
Squamata - Colubridae
e Mexican Milk Snake; House Snake; Milk Adder
d Mexikanische Milchschlange; Mexikanische Königsnatter

4410 ***Lampropeltis triangulum arcifera***
Squamata - Colubridae
e Jalisco Milk Snake

4411 ***Lampropeltis triangulum blanchardi***
Squamata - Colubridae
e Blanchard's Milk Snake

4412 ***Lampropeltis triangulum celaenops***
Squamata - Colubridae
e New Mexico Milk Snake; Big Bend Milk Snake

4413 ***Lampropeltis triangulum conanti***
Squamata - Colubridae
e Conant's Milk Snake
d Conants Dreiecknatter

4414 ***Lampropeltis triangulum dixoni***
Squamata - Colubridae
e Dixon's Milk Snake

4415 ***Lampropeltis triangulum elapsoides***
Squamata - Colubridae
e Scarlet King Snake
d Rote Königsnatter

4416 ***Lampropeltis triangulum gaigeae***
Squamata - Colubridae
e Black Milk Snake

4417 ***Lampropeltis triangulum gentilis***
Squamata - Colubridae
e Central Plains Milk Snake
d Central-Plains-Milchschlange; Ebenen-Dreiecksnatter

4418 ***Lampropeltis triangulum hondurensis***
Squamata - Colubridae
e Honduran Milk Snake
d Honduras-Königsnatter; Honduras-Milchschlange; Honduras-Dreiecksnatter

4419 ***Lampropeltis triangulum micropholis***
Squamata - Colubridae
e Ecuadorian Milk Snake; Ecuadoran Milk Snake

4420 ***Lampropeltis triangulum multistriata***
Squamata - Colubridae
e Pale Milk Snake

4421 ***Lampropeltis triangulum nelsoni***
Squamata - Colubridae
e Nelson's Milk Snake

4422 ***Lampropeltis triangulum oligozona***
Squamata - Colubridae
e Pacific Central American Milk Snake

4423 ***Lampropeltis triangulum polyzona***
Squamata - Colubridae
e Atlantic Central American Milk Snake; Central American Milk Snake

4424 ***Lampropeltis triangulum sinaloae***
Squamata - Colubridae
e Sinaloa Milk Snake; Sinaloan Milk Snake
d Sinaloa-Königsnatter; Rote Königsnatter; Sinaloa-Milchschlange; Königsnatter
i Serpente del latte sinaloan

4425 ***Lampropeltis triangulum smithi***
Squamata - Colubridae
e Smith's Milk Snake

4426 ***Lampropeltis triangulum stuarti***
Squamata - Colubridae
e Stuart's Milk Snake

4427 ***Lampropeltis triangulum syspila***
Squamata - Colubridae
e Red Milk Snake

4428 ***Lampropeltis triangulum taylori***
Squamata - Colubridae
e Utah Milk Snake

4429 ***Lampropeltis triangulum triangulum***
Squamata - Colubridae
e Eastern Milk Snake; Common Milk Snake (NA); Milk Snake (NA); Southern Milk Snake (NA)
d Östliche Dreiecksnatter

4430 ***Lampropeltis trianuglum campbelli***
Squamata - Colubridae
e Pueblan Milk Snake; Milk Snake
d Milchschlange; Königsnatter
f Faux serpent corail; Serpent du lait

4431 ***Lampropeltis zonata***
Squamata - Colubridae
e California Mountain King Snake; Coast Range Coral Snake; California King Snake
d Bergkönigsnatter; Korallenkönigsnatter
f Couleuvre royale de Californie
i Serpente reale di montagna

4432 ***Lampropeltis zonata agalma***
Squamata - Colubridae
e San Pedro King Snake; Baja California Mountain King Snake; San Pedro Mountain King Snake
d San Pedro Königsnatter

4433 ***Lampropeltis zonata herrerae***
Squamata - Colubridae
e Todos Santos Island Milk Snake

4434 ***Lampropeltis zonata multicinctata***
Squamata - Colubridae
e Sierra Mountain King Snake; Coral King Snake

4435 ***Lampropeltis zonata multifasciata***
Squamata - Colubridae
e Coast Mountain King Snake; Coastal Mountain King Snake

4436 ***Lampropeltis zonata parvirubra***
Squamata - Colubridae
e San Bernardino Mountain King Snake

4437 ***Lampropeltis zonata pulchra***
Squamata - Colubridae
e San Diego Mountain King Snake
d Kalifornische Bergkönigsnatter

4438 ***Lampropeltis zonata zonata***
Squamata - Colubridae
e St. Helena Mountain King Snake
d Korallenkönigsnatter

4439 ***Lamprophis***
Squamata - Colubridae
e House Snakes
d Hausschlangen; Südafrikanische Hausschlangen
f Serpents des maisons

4440 ***Lamprophis aurora***
Squamata - Colubridae
e Aurora House Snake; Aurora Snake (CSA); Night Snake
d Aurora-Hausschlange; Aurora-Natter

4441 ***Lamprophis fiskii***
Squamata - Colubridae
e Fisk's House Snake; Fisk's Snake
d Fisk-Hausschlange

4442 ***Lamprophis fuliginosus***
Squamata - Colubridae

e Brown House Snake; Common House Snake (CSA); Common African House Snake; African House Snake
d Braune Schlange; Hausschlange; Afrikanische Hausschlange; Braune Hausschlange
f Serpent des maisons; Couleuvre commune d'Afrique; Serpent africain des maisons
i Serpente bruno delle case

4443 ***Lamprophis fuscus***
Squamata - Colubridae
e Yellowbelly House Snake; Yellow-bellied House Snake; Yellow-lipped Snake; Yellow-bellied Snake (CSA)
d Gelbbauchhausschlange

4444 ***Lamprophis geometricus***
Squamata - Colubridae
e Seychelles House Snake

4445 ***Lamprophis guttatus***
Squamata - Colubridae
e Spotted House Snake
d Olivfarbene Hausschlange

4446 ***Lamprophis inornatus***
Squamata - Colubridae
e Olive House Snake (CSA); Black House Snake; Brown Snake
d Schwarze Hausschlange; Olivfarbene Hausschlange

4447 ***Lamprophis lineatus***
Squamata - Colubridae
e African House Snake; Striped House Snake; Lined House Snake
d Gestreifte Hausschlange; Afrikanische Hausnatter; Gewöhnliche Hausschlange; Westafrikanische Hausschlange
f Boaédon à quatre raies

4448 ***Lamprophis maculatus***
Squamata - Colubridae
e Dotted House Snake

4449 ***Lamprophis swazicus***
Squamata - Colubridae
e Swazi House Snake; Swaziland House Snake; Swazi Rock Snake (CSA); Eastern House Snake (CSA)
d Swasiland-Hausschlange

4450 ***Lamprophis virgatus***
Squamata - Colubridae
e Hallowell's House Snake

4451 ***Lampropholis***
Squamata - Scincidae
e Water Skinks; Metallic Skinks

4452 ***Lampropholis amicula***
Squamata - Scincidae
e Brown Water Skink

4453 ***Lampropholis basiliscus***
Squamata - Scincidae
e Basilisk Water Skink

4454 ***Lampropholis caligula***
Squamata - Scincidae
e Grey Water Skink

4455 ***Lampropholis czechuci***
Squamata - Scincidae
e Czech Water Skink

4456 ***Lampropholis delicata***
Squamata - Scincidae
e Delicate Water Skink; Metallic Skink; Rainbow Skink; Delicate Skink (ANZ)

4457 ***Lampropholis guichenoti***
Squamata - Scincidae
e Guichenot's Water Skink; Guichenot's Skink; Guichenot's Grass Skink; Garden Skink (ANZ)
d Guichenots Skink

f Scinque de Guichenot

4458 ***Lampropholis mirabilis***
Squamata - Scincidae
e Olive Water Skink

4459 ***Langaha***
Squamata - Colubridae
e Leafnose Snakes; Leaf-nosed Snakes; Madagascan Twig Snakes
d Blattnasennattern; Madagassische Baumschlangen

4460 ***Langaha alluaudi***
Squamata - Colubridae
e Southern Leafnose Snake; Southern Leafnose
i Cresto di gallo

4461 ***Langaha madagascariensis***
Squamata - Colubridae
e Madagascar Leafnose Vine Snake
d Madagassische Blasennatter
i Cresto di gallo del Madagascar

4462 ***Langaha pseudoalluaudi***
Squamata - Colubridae
e Northern Leafnose Snake; Northern Leafnose

4463 ***Lankascincus***
Squamata - Scincidae
e Tree Skinks

4464 ***Lankascincus deignani***
Squamata - Scincidae
e Deignan's Tree Skink

4465 ***Lankascincus deraniyagalae***
Squamata - Scincidae
e Deraniyagalae's Tree Skink

4466 ***Lankascincus fallax***
Squamata - Scincidae
e Peters's Tree Skink

4467 ***Lankascincus gansi***
Squamata - Scincidae
e Gans's Tree Skink; Carl Gans's Skink

4468 ***Lankascincus taprobanensis***
Squamata - Scincidae
e Ceylon Tree Skink
d Ceylon-Waldskink

4469 ***Lankascincus taylori***
Squamata - Scincidae
e Taylor's Tree Skink

4470 ***Lanthanotus***
Squamata - Varanidae
e Earless Monitors; Earless Lizards
d Taubwarane
f Varans sans oreilles

4471 ***Lanthanotus borneensis***
Squamata - Varanidae
e Borneo Earless Monitor; Bornean Earless Monitor; Bornean Earless Lizard
d Borneo-Taubwaran; Taubwaran
f Varan sans oreilles

4472 ***Lapemis***
Squamata - Elapidae
e Shaw's Sea Snakes
d Plump-Seeschlangen

4473 ***Lapemis curtus***
Squamata - Elapidae
e Shaw's Sea Snake (CSI); Short Sea Snake; Hardwicke's Sea Snake
d Plump-Seeschlange

4474 ***Latastia***
Squamata - Lacertidae
e Longtail Lizards; Long-tailed Lacertas; Long-tailed Lizards
d Langschwanzeidechsen

4475 ***Latastia bosacai***
Squamata - Lacertidae
e Eritrea Longtail Lizard

4476 ***Latastia carinata***
Squamata - Lacertidae
e Peters's Longtail Lizard

4477 ***Latastia johnstoni***
Squamata - Lacertidae
e Nyasa Longtail Lizard; Nyasaland Long-tailed Lizard

4478 ***Latastia longicaudata***
Squamata - Lacertidae
e Southern Longtail Lizard; Southern Long-tailed Lizard

4479 ***Latastia siebenrocki***
Squamata - Lacertidae
e Siebenrock's Longtail Lizard

4480 ***Latastia taulori***
Squamata - Lacertidae
e Taylor's Longtail Lizard

4481 ***Laticauda***
Squamata - Elapidae
e Banded Sea Snakes; Flat-tailed Sea Snakes; Sea Kraits; Amphibious Sea Snakes
d Plattschwänze; Plattschwanzseeschlangen
f Serpents à queue plate
i Serpenti di mare

4482 ***Laticauda colubrina***
Squamata - Elapidae
e Banded Sea Snake; Yellow-lipped Sea Snake; Ringed Sea Snake; Schneider's Sea Snake; Giant Banded Sea Krait; Yellow-lipped Sea Krait
d Natternplattschwanz
i Serpente di mare

4483 ***Laticauda crockeri***
Squamata - Elapidae
e Crocker's Sea Snake

4484 ***Laticauda laticaudata***
Squamata - Elapidae
e Black-banded Sea Snake; Large-scaled Sea Snake; Linnaeus's Sea Krait; Banded Amphibious Sea Snake
d Gewöhnlicher Plattschwanz; Plattschwanzseeschlange
f Serpent à queue plate

4485 ***Laticauda schistorhynchus***
Squamata - Elapidae
e Flat-tail Sea Snake

4486 ***Laticauda semifasciata***
Squamata - Elapidae
e Chinese Sea Snake; Erabu Sea Snake
d Halbgebänderter Plattschwanz

4487 ***Leiocephalus***
Squamata - Iguanaidae
e Curlytail Lizards; Crested Keeled Lizards; Curly-tailed Lizards; Curlytails
d Ringelschwanzleguane; Rollschwanzleguane; Maskenleguane; Glattkopfleguane

4488 ***Leiocephalus barahonensis***
Squamata - Iguanaidae
e Barahona Curlytail Lizard

4489 ***Leiocephalus carinatus***
Squamata - Iguanaidae
e Northern Curlytail Lizard; Northern Curly-tailed Lizard; Curly-tailed Lizard; Bahaman Crested Lizard; Rough-scaled Lizard; Crested Keeled Lizard; Carinate Curly-tailed Lizard; Saw-scaled Curlytail (WI)
d Strandhündchen; Rollschwanzleguan; Ringelschwanzleguan; Maskenleguan

f Lézard à queue recourbée

4490 ***Leiocephalus cubensis***
Squamata - Iguanaidae
e Cuban Curlytail Lizard

4491 ***Leiocephalus eremitus***
Squamata - Iguanaidae
e Navassa Curlytail Lizard

4492 ***Leiocephalus greenwayi***
Squamata - Iguanaidae
e Plana Cay Curlytail Lizard; East Plana Curlytail (WI)

4493 ***Leiocephalus herminieri***
Squamata - Iguanaidae
e Martinique Curlytail Lizard; Martinique Curlytail (WI)

4494 ***Leiocephalus inaguae***
Squamata - Iguanaidae
e Inagua Curlytail Lizard; Inagua Curlytail (WI)

4495 ***Leiocephalus loxogrammus***
Squamata - Iguanaidae
e Rum Cay Curlytail Lizard; San Salvador Curlytail (WI)

4496 ***Leiocephalus lunatus***
Squamata - Iguanaidae
e Santo Domingo Curlytail Lizard; Santo Domingo Curly-tailed Lizard

4497 ***Leiocephalus macropus***
Squamata - Iguanaidae
e Monte Verde Curlytail Lizard; Cuban Side-blotched Curlytail

4498 ***Leiocephalus melanochlorus***
Squamata - Iguanaidae
e Jeremie Curlytail Lizard

4499 ***Leiocephalus onaneyi***
Squamata - Iguanaidae
e Sierra Curlytail Lizard

4500 ***Leiocephalus personatus***
Squamata - Iguanaidae
e Haitian Curlytail Lizard
d Bunter Haiti-Maskenleguan; Gewöhlicher Maskenleguan; Maskenleguan; Bunter Maskenleguan
f Lézard à queue frisée; Lézard à queue courbée; Curlytail

4501 ***Leiocephalus pratensis***
Squamata - Iguanaidae
e Atalaye Curlytail Lizard

4502 ***Leiocephalus psammodromus***
Squamata - Iguanaidae
e Bastion Cay Curlytail Lizard; Turks and Caicos Curlytail (WI)

4503 ***Leiocephalus punctatus***
Squamata - Iguanaidae
e Spotted Curlytail Lizard; Crooked-Acklins Curlytail (WI)

4504 ***Leiocephalus raviceps***
Squamata - Iguanaidae
e Mountain Curlytail Lizard; Pallid Curlytail

4505 ***Leiocephalus rhutidira***
Squamata - Iguanaidae
e Lapierre Curlytail Lizard

4506 ***Leiocephalus schreibersii***
Squamata - Iguanaidae
e Red-sided Curlytail Lizard; Red-sided Curly-tailed Lizard; Schreiber's Curly-tailed Lizard
d Schreibers Glattkopfleguan; Maskenleguan; Großer Maskenleguan; Haiti-Glattkopfleguan; Glattkopfleguan

4507 ***Leiocephalus semilineatus***
Squamata - Iguanaidae

e Thomazeau Curlytail Lizard

4508 ***Leiocephalus stictigaster***
Squamata - Iguanaidae
e Cabo Corrientes Curlytail Lizard

4509 ***Leiocephalus vinculum***
Squamata - Iguanaidae
e Cochran's Curlytail Lizard

4510 ***Leioheterodon***
Squamata - Colubridae
e Madagascar Brown Snakes; Madagascar Hognose Snakes; Madagascar Hognoses
d Hakennattern; Madagaskar-Hakennattern; Madagaskar-Nattern

4511 ***Leioheterodon geayi***
Squamata - Colubridae
e Madagascar Dune Snake

4512 ***Leioheterodon madagascariensis***
Squamata - Colubridae
e Madagascar Menarana Snake; Giant Madagascar Hognose Snake; Madagascar Giant Hognose Snake
d Madagaskar-Hakennatter; Madagaskar-Natter

4513 ***Leioheterodon modestus***
Squamata - Colubridae
e Madagascar Brown Snake; Brown Hognose Snake

4514 ***Leiolepis***
Squamata - Agamidae
e Butterfly Lizards; Granular-scaled Lizards; Smooth-scaled Agamas; Butterfly Agamas
d Schmetterlingsagamen
f Léiolépides

4515 ***Leiolepis belliana***
Squamata - Agamidae
e Butterfly Lizard; Butterfly Agama; Common Butterfly Lizard
d Prachtschmetterlingsagame; Schmetterlingsagame
f Agame-papillon

4516 ***Leiolepis belliana rubritaeniata***
Squamata - Agamidae
e Sand Lizard

4517 ***Leiolepis guttata***
Squamata - Agamidae
e Spotted Butterfly Lizard
d Fleckige Schmetterlingsagame
f Léiolépide à gouttelettes

4518 ***Leiolepis peguensis***
Squamata - Agamidae
e Burmese Butterfly Lizard

4519 ***Leiolepis reevesii***
Squamata - Agamidae
e Reeves's Butterfly Lizard

4520 ***Leiolepis triploida***
Squamata - Agamidae
e Thai Butterfly Lizard

4521 ***Leiolopisma***
Squamata - Scincidae
e Ground Skinks; Window-eyed Skinks; Leiolopismas
d Leilopismaskinke
f Léiolopismes

4522 ***Leiolopisma acrinasum***
Squamata - Scincidae
e Hardy's Ground Skink

4523 ***Leiolopisma alazon***
Squamata - Scincidae
e Lauan Ground Skink

4524 ***Leiolopisma austrocaledonicum***
Squamata - Scincidae

e Caledonian Ground Skink

4525 ***Leiolopisma cyclodina***
Squamata - Scincidae
e Common Ground Skink

4526 ***Leiolopisma eulepis***
Squamata - Scincidae
e Fitzinger's Ground Skink

4527 ***Leiolopisma fallai***
Squamata - Scincidae
e McCann's Ground Skink

4528 ***Leiolopisma gracilicorpus***
Squamata - Scincidae
e Graceful Ground Skink

4529 ***Leiolopisma grande***
Squamata - Scincidae
e Large Ground Skink; Large Leiolopisma

4530 ***Leiolopisma hombronia***
Squamata - Scincidae
e Girard's Ground Skink

4531 ***Leiolopisma infrapunctatum***
Squamata - Scincidae
e Dotted Ground Skink

4532 ***Leiolopisma lineoocellatum***
Squamata - Scincidae
e Eyespot Ground Skink

4533 ***Leiolopisma lioscincus***
Squamata - Scincidae
e Bocage's Ground Skink

4534 ***Leiolopisma lygosomella***
Squamata - Scincidae
e Island Ground Skink

4535 ***Leiolopisma moco***
Squamata - Scincidae
e Moco Ground Skink
f Léiolopisme moco

4536 ***Leiolopisma nigriplantare***
Squamata - Scincidae
e Dark Ground Skink

4537 ***Leiolopisma nigrofasciatum***
Squamata - Scincidae
e Black-banded Ground Skink

4538 ***Leiolopisma oligosoma***
Squamata - Scincidae
e Longtail Ground Skink

4539 ***Leiolopisma orocryptum***
Squamata - Scincidae
e Hutchinson's Ground Skink

4540 ***Leiolopisma pseudemoi***
Squamata - Scincidae
e Fuhn's Ground Skink

4541 ***Leiolopisma rawlinsoni***
Squamata - Scincidae
e Rawlinson's Ground Skink

4542 ***Leiolopisma sauroscincus***
Squamata - Scincidae
e Peters's Ground Skink

4543 ***Leiolopisma smithi***
Squamata - Scincidae
e Smith's Ground Skink; Smith's Leiolopisma

4544 ***Leiolopisma steindachneri***
Squamata - Scincidae
e Steindachner's Ground Skink; Steindachner's Window-eyed Skink

4545 ***Leiolopisma striatum***
Squamata - Scincidae

e Striped Ground Skink

4546 ***Leiolopisma suteri***
Squamata - Scincidae
e Suter's Ground Skink; Suter's Window-eyed Skink

4547 ***Leiolopisma telfairi***
Squamata - Scincidae
e Telford's Ground Skink; Round Island Skink
d Telfords Skink
f Scinque de l'île Ronde; Léiolopisme de Telford; Scinque de Telfair

4548 ***Leiolopisma tricolor***
Squamata - Scincidae
e Three-colour Ground Skink

4549 ***Leiolopisma trilloni***
Squamata - Scincidae
e Ineich's Ground Skink

4550 ***Leiolopisma tropidoscincus***
Squamata - Scincidae
e Tropical Ground Skink

4551 ***Leiolopisma variabile***
Squamata - Scincidae
e Variable Ground Skink

4552 ***Leiolopisma zelandicum***
Squamata - Scincidae
e Zelandic Ground Skink

4553 ***Leiopython***
Squamata - Boidae
e White-lipped Pythons
d Weißlippenpythone
f Pythons de la Nouvelle Guinée

4554 ***Leiopython albertisii***
Squamata - Boidae
e White-lipped Python; D'Albertis's Python; Purple-brown Python
d Weißlippenpython
f Python de la Nouvelle Guinée

4555 ***Leiosaurus***
Squamata - Iguanaidae
e Southern Anoles; Matuastos

4556 ***Leiosaurus bardensis***
Squamata - Iguanaidae
e Southern Anole; Matuasto

4557 ***Leiosaurus bellii***
Squamata - Iguanaidae
e Bell's Anole; Bell's Matuasto

4558 ***Lepidoblepharis***
Squamata - Gekkonidae
e Scaly-eyed Geckos; Dwarf Geckos

4559 ***Lepidoblepharis buchwaldi***
Squamata - Gekkonidae
e Buchwald's Scaly-eyed Gecko

4560 ***Lepidoblepharis colombianus***
Squamata - Gekkonidae
e Colombian Scaly-eyed Gecko

4561 ***Lepidoblepharis duolepis***
Squamata - Gekkonidae
e Antioquia Scaly-eyed Gecko

4562 ***Lepidoblepharis festae***
Squamata - Gekkonidae
e Amazon Scaly-eyed Gecko; Brown Dwarf Gecko

4563 ***Lepidoblepharis grandis***
Squamata - Gekkonidae
e Grand Scaly-eyed Gecko; Greater Dwarf Gecko

4564 ***Lepidoblepharis heyerorum***
Squamata - Gekkonidae
e Vanzolini's Scaly-eyed Gecko;

Amazon Dwarf Gecko

4565 ***Lepidoblepharis intermedius***
Squamata - Gekkonidae
e Lowland Scaly-eyed Gecko

4566 ***Lepidoblepharis microlepis***
Squamata - Gekkonidae
e Tiny Scaly-eyed Gecko

4567 ***Lepidoblepharis miyata***
Squamata - Gekkonidae
e Miyata Scaly-eyed Gecko

4568 ***Lepidoblepharis oxycephalus***
Squamata - Gekkonidae
e Sharp Scaly-eyed Gecko

4569 ***Lepidoblepharis peraccae***
Squamata - Gekkonidae
e Peracca's Scaly-eyed Gecko

4570 ***Lepidoblepharis ruthveni***
Squamata - Gekkonidae
e Ruthven's Scaly-eyed Gecko

4571 ***Lepidoblepharis sanctaemartae***
Squamata - Gekkonidae
e Panama Scaly-eyed Gecko

4572 ***Lepidoblepharis williamsi***
Squamata - Gekkonidae
e Williams's Scaly-eyed Gecko

4573 ***Lepidoblepharis xanthostigma***
Squamata - Gekkonidae
e Costa Rica Scaly-eyed Gecko

4574 ***Lepidochelys***
Testudines - Cheloniidae
e Ridley Turtles; Ridleys; Ridley Sea Turtles
d Bastardschildkröten
f Tortues bâtardes

4575 ***Lepidochelys kempii***
Testudines - Cheloniidae
e Kemp's Ridley Turtle; Kemp Turtle; Bastard Turtle; Kemp's Loggerhead; Kemp's Ridley; Atlantic Ridley Sea Turtle; Atlantic Olive Ridley; Atlantic Ridley; Gulf Ridley Turtle; Kemp's Ridley Sea Turtle
d Atlantik-Bastardschildkröte; Kemps Bastardschildkröte; Bastardschildkröte; Atlantische Bastardschildkröte; Kemps Schildkröte
f Ridley de Kemp; Tortue bâtarde; Tortue mexicaine; Chelonée de Kemp; Tortue d'Espagne; Tortue de Kemp; Lépidochélyde
i Tartaruga di Kemp; Tartaruga Kemp; Tartaruga bastarda

4576 ***Lepidochelys olivacea***
Testudines - Cheloniidae
e Olive Ridley Turtle; Ridley; Olive Ridley Sea Turtle (NA); Pacific Ridley (ANZ); Olive Ridley; Ridley Sea Turtle; Olive Loggerhead; Pacific Olive Ridley
d Pazifische Bastardschildkröte; Bastardschildkröte; Olive Bastardschildkröte
f Ridley olivacée; Tortue bâtarde; Chelonée olivâtre; Tortue olivacée
i Tartaruga olivacea

4577 ***Lepidodactylus***
Squamata - Gekkonidae
e Scaly-toed Geckos
d Schuppenfingergeckos; Schuppenfinger

4578 ***Lepidodactylus aurelineatus***
Squamata - Gekkonidae
e Golden Scaly-toed Gecko

4579 ***Lepidodactylus balioburius***
Squamata - Gekkonidae
e Batan Scaly-toed Gecko

4580 ***Lepidodactylus browni***
Squamata - Gekkonidae
e Brown's Scaly-toed Gecko

4581 ***Lepidodactylus christiani***
Squamata - Gekkonidae
e Christian Scaly-toed Gecko

4582 ***Lepidodactylus euaensis***
Squamata - Gekkonidae
e Eua Scaly-toed Gecko

4583 ***Lepidodactylus flaviocularis***
Squamata - Gekkonidae
e Yellow-eyed Scaly-toed Gecko

4584 ***Lepidodactylus gardineri***
Squamata - Gekkonidae
e Rotuman Scaly-toed Gecko

4585 ***Lepidodactylus guppyi***
Squamata - Gekkonidae
e Solomon Scaly-toed Gecko

4586 ***Lepidodactylus herreri***
Squamata - Gekkonidae
e Negros Scaly-toed Gecko

4587 ***Lepidodactylus listeri***
Squamata - Gekkonidae
e Christmas Scaly-toed Gecko

4588 ***Lepidodactylus lugubris***
Squamata - Gekkonidae
e Mourning Gecko; Sad Gecko; Morning Gecko
d Schuppenfinger; Schuppenfingergecko; Jungferngecko
f Veuf

4589 ***Lepidodactylus magnus***
Squamata - Gekkonidae
e Mountain Scaly-toed Gecko

4590 ***Lepidodactylus manni***
Squamata - Gekkonidae
e Fiji Scaly-toed Gecko

4591 ***Lepidodactylus mutahi***
Squamata - Gekkonidae
e Bougainville's Scaly-toed Gecko

4592 ***Lepidodactylus novaeguineae***
Squamata - Gekkonidae
e New Guinea Scaly-toed Gecko

4593 ***Lepidodactylus oorti***
Squamata - Gekkonidae
e Indonesia Scaly-toed Gecko

4594 ***Lepidodactylus orientalis***
Squamata - Gekkonidae
e Oriental Scaly-toed Gecko

4595 ***Lepidodactylus planicaudus***
Squamata - Gekkonidae
e Mindanao Scaly-toed Gecko

4596 ***Lepidodactylus pulcher***
Squamata - Gekkonidae
e Wild Scaly-toed Gecko

4597 ***Lepidodactylus pumilus***
Squamata - Gekkonidae
e Torres Scaly-toed Gecko

4598 ***Lepidodactylus pusillus***
Squamata - Gekkonidae
e Cope's Scaly-toed Gecko

4599 ***Lepidodactylus shebae***
Squamata - Gekkonidae
e Guadalcanal Scaly-toed Gecko

4600 ***Lepidodactylus woodfordi***
Squamata - Gekkonidae
e Woodford's Scaly-toed Gecko

4601 ***Lepidodactylus yami***
Squamata - Gekkonidae
e Lanyu Scaly-toed Gecko

4602 ***Lepidodactyus ranauensis***
Squamata - Gekkonidae
e Sabah Scaly-toed Gecko

4603 ***Lepidophyma***
Squamata - Xantusiidae
e Tropical Night Lizards
d Mittelamerikanische Nachtechsen

4604 ***Lepidophyma alvarezi***
Squamata - Xantusiidae
e Alvarez del Toro's Night Lizard

4605 ***Lepidophyma chicoaensis***
Squamata - Xantusiidae
e Sumidero Night Lizard

4606 ***Lepidophyma dontomasi***
Squamata - Xantusiidae
e MacDougall's Night Lizard

4607 ***Lepidophyma flavimaculatum***
Squamata - Xantusiidae
e Yellow-spotted Night Lizard; Tropical Night Lizard; Yellow-spotted Tropical Night Lizard
d Krokodilnachtechsen

4608 ***Lepidophyma gaigae***
Squamata - Xantusiidae
e Gaige's Night Lizard; Gaige's Tropical Night Lizard

4609 ***Lepidophyma lipetzi***
Squamata - Xantusiidae
e Lipetz's Night Lizard

4610 ***Lepidophyma mayae***
Squamata - Xantusiidae
e Maya Night Lizard

4611 ***Lepidophyma micropholis***
Squamata - Xantusiidae
e Cave Night Lizard; Cave Tropical Night Lizard
Squamata - Xantusiidae
e Japlan Night Lizard

4613 ***Lepidophyma pajapanensis***
Squamata - Xantusiidae
e Pajapan Night Lizard; Pajapan Tropical Night Lizard

4614 ***Lepidophyma radula***
Squamata - Xantusiidae
e Yautepec Night Lizard

4615 ***Lepidophyma sawini***
Squamata - Xantusiidae
e Sawin's Night Lizard

4616 ***Lepidophyma smithii***
Squamata - Xantusiidae
e Smith's Night Lizard; Smith's Tropical Night Lizard

4617 ***Lepidophyma sylvaticum***
Squamata - Xantusiidae
e Madrean Night Lizard; Madrean Tropical Night Lizard

4618 ***Lepidophyma tarascae***
Squamata - Xantusiidae
e Tarascan Night Lizard

4619 ***Lepidophyma tuxtlae***
Squamata - Xantusiidae
e Tuxtla Night Lizard; Tuxtla Tropical Night Lizard

4620 ***Leposoma***
Squamata - Gymnophthalmidae
e Skink Tegus; Spectacled Lizards

4621 ***Leposoma***
Squamata - Teiidae

e Leposomas

4622 ***Leposoma annectans***
Squamata - Gymnophthalmidae
e Ruibal's Brown Tegu

4623 ***Leposoma guianense***
Squamata - Teiidae
e Guyana Leposoma

4624 ***Leposoma hexalepis***
Squamata - Gymnophthalmidae
e Six-scaled Tegu

4625 ***Leposoma joanna***
Squamata - Teiidae
e Joanna Leposoma

4626 ***Leposoma parietale***
Squamata - Teiidae
e Cope's Leposoma; Common Root Lizard

4627 ***Leposoma percarinatum***
Squamata - Gymnophthalmidae
e Müller's Tegu; Keeled Microteiid

4628 ***Leposoma rugiceps***
Squamata - Teiidae
e Red Leposoma

4629 ***Leposoma scincoides***
Squamata - Gymnophthalmidae
e Skink Tegu

4630 ***Leposoma southi***
Squamata - Teiidae
e Southern Leposoma; Northern Spectacled Lizard

4631 ***Leposternon***
Squamata - Amphisbaenidae
e South American Worm Lizards

4632 ***Leposternon affine***
Squamata - Amphisbaenidae
e Paraguay Worm Lizard

4633 ***Leposternon boettgeri***
Squamata - Amphisbaenidae
e Boettger's Worm Lizard

4634 ***Leposternon borelli***
Squamata - Amphisbaenidae
e Borelli's Worm Lizard

4635 ***Leposternon boulengeri***
Squamata - Amphisbaenidae
e Boulanger's Worm Lizard

4636 ***Leposternon carcani***
Squamata - Amphisbaenidae
e Luque Worm Lizard

4637 ***Leposternon carcani***
Squamata - Amphisbaenidae
e Tebicuari Worm Lizard

4638 ***Leposternon crassum***
Squamata - Amphisbaenidae
e Thick Worm Lizard

4639 ***Leposternon guentheri***
Squamata - Amphisbaenidae
e Günther's Worm Lizard

4640 ***Leposternon infraorbitale***
Squamata - Amphisbaenidae
e Berthold's Worm Lizard

4641 ***Leposternon laticeps***
Squamata - Amphisbaenidae
e Matto Grosso Worm Lizard

4642 ***Leposternon latifrontale***
Squamata - Amphisbaenidae
e Asuncion Worm Lizard

4643 ***Leposternon microcephalum***
Squamata - Amphisbaenidae
e Smallhead Worm Lizard
d Kleinköpfige Doppelschleiche
Squamata - Amphisbaenidae
e Duméril's Worm Lizard

4645 ***Leposternon onychocephalum***
Squamata - Amphisbaenidae
e Yellow-headed Worm Lizard

4646 ***Leposternon petersi***
Squamata - Amphisbaenidae
e Peters's Worm Lizard

4647 ***Leposternon pfefferi***
Squamata - Amphisbaenidae
e Pfeffer's Worm Lizard

4648 ***Leposternon phacaena***
Squamata - Amphisbaenidae
e Buenos Aires Worm Lizard

4649 ***Leposternon polystegoides***
Squamata - Amphisbaenidae
e Lake Papary Worm Lizard

4650 ***Leposternon polystegum***
Squamata - Amphisbaenidae
e Bahia Worm Lizard

4651 ***Leposternon rostratum***
Squamata - Amphisbaenidae
e Beaked Worm Lizard

4652 ***Leposternon scutigerum***
Squamata - Amphisbaenidae
e Shielded Worm Lizard

4653 ***Leposternon sinuosum***
Squamata - Amphisbaenidae
e Winding Worm Lizard

4654 ***Leposternon strauchi***
Squamata - Amphisbaenidae
e Strauch's Worm Lizard

4655 ***Leposternon wuchereri***
Squamata - Amphisbaenidae
e Wucherer's Worm Lizard

4656 ***Leptodeira***
Squamata - Colubridae
e Cat-eyed Snakes; Annulated Snakes
d Katzenaugennattern
f Couleuvres oeil-de-chat

4657 ***Leptodeira annulata***
Squamata - Colubridae
e Banded Cat-eyed Snake; Cat-eyed Snake; Annulated Snake; Common Cat-eyed Snake
d Namamen-Natter; Katzenaugennatter; Bananennatter
f Serpent coco (Ants)

4658 ***Leptodeira annulata annulata***
Squamata - Colubridae
e Common Cat-eyed Snake

4659 ***Leptodeira bakeri***
Squamata - Colubridae
e Baker's Cat-eyed Snake

4660 ***Leptodeira frenata***
Squamata - Colubridae
e Rainforest Cat-eyed Snake; Yucatan Cat-eyed Snake

4661 ***Leptodeira maculata***
Squamata - Colubridae
e Southwestern Cat-eyed Snake

4662 ***Leptodeira nigrofasciata***
Squamata - Colubridae
e Black-banded Cat-eyed Snake; Banded Cat-eyed Snake

4663 ***Leptodeira punctata***
Squamata - Colubridae

e Western Cat-eyed Snake

4664 ***Leptodeira septentrionalis***
Squamata - Colubridae
e Cat-eyed Snake; Northern Cat-eyed Snake; Annulated Snake; Ringed Snake; Spotted Cat-eyed Snake
d Gebänderte Katzenaugennatter
f Couleuvre oeil-de-chat

4665 ***Leptodeira septentrionalis septentrionalis***
Squamata - Colubridae
e Northern Cat-eyed Snake

4666 ***Leptodeira splendida***
Squamata - Colubridae
e Splendid Cat-eyed Snake

4667 ***Leptodrymus***
Squamata - Colubridae
e Striped Lowland Snakes

4668 ***Leptodrymus pulcherrimus***
Squamata - Colubridae
e Striped Lowland Snake; Green-headed Racer

4669 ***Leptomicrurus***
Squamata - Elapidae
e Thread Coral Snakes; Slender Coral Snakes
d Schlanke Korallenottern; Schlankkorallenottern; Korallenschlangen
f Serpents corail

4670 ***Leptomicrurus collaris***
Squamata - Elapidae
e Schlegel's Thread Coral Snake

4671 ***Leptomicrurus narduccii***
Squamata - Elapidae
e Jan's Thread Coral Snake; Schlegel's Thread Coral Snake

4672 ***Leptomicrurus scutiventris***
Squamata - Elapidae
e Cope's Thread Coral Snake

4673 ***Leptophis***
Squamata - Colubridae
e Parrot Snakes
d Schlanknattern; Papageienschlangen; Dünnschlangen
i Serpenti papagallo

4674 ***Leptophis ahaetulla***
Squamata - Colubridae
e Parrot Snake; Common Parrot Snake; Greater Green Parrot Snake
d Brasilianische Schlanknatter; Brasilianische Dünnschlange

4675 ***Leptophis ahaetulla nigromarginatus***
Squamata - Colubridae
e Black-skinned Parrot Snake

4676 ***Leptophis cupreus***
Squamata - Colubridae
e Copper Parrot Snake; Brown Parrot Snake

4677 ***Leptophis depressirostris***
Squamata - Colubridae
e Cope's Parrot Snake; Lesser Green Parrot Snake; Satiny Parrot Snake

4678 ***Leptophis diplotropis***
Squamata - Colubridae
e Pacific Coast Parrot Snake; Pacific Parrot Snake

4679 ***Leptophis mexicanus***
Squamata - Colubridae
e Mexican Parrot Snake; Bronze-backed Parrot Snake
d Mexikanische Dünnschlange; Papageienschlange

4680 ***Leptophis modestus***
Squamata - Colubridae
e Cloud Forest Parrot Snake

4681 ***Leptophis nebulosus***
Squamata - Colubridae
e Oliver's Parrot Snake; Striped Parrot Snake

4682 ***Leptophis riveti***
Squamata - Colubridae
e Despax's Parrot Snake; Panamanian Parrot Snake; Turquoise Parrot Snake

4683 ***Leptosiaphos***
Squamata - Scincidae
e Five-toed Skinks

4684 ***Leptosiaphos aloysiisabaudiae***
Squamata - Scincidae
e Uganda Five-toed Skink; Uganda Lidless Skink

4685 ***Leptosiaphos amieti***
Squamata - Scincidae
e Cameroon Five-toed Skink; Cameroon Lidless Skink

4686 ***Leptosiaphos blochmanni***
Squamata - Scincidae
e Zaire Five-toed Skink; Zaire Lidless Skink

4687 ***Leptosiaphos dewittei***
Squamata - Scincidae
e Dewitte's Five-toed Skink

4688 ***Leptosiaphos fuhni***
Squamata - Scincidae
e Fuhn's Five-toed Skink; Fuhn's Lidless Skink

4689 ***Leptosiaphos gemmiventris***
Squamata - Scincidae
e Sjosstedt's Five-toed Skink

4690 ***Leptosiaphos graueri***
Squamata - Scincidae
e Rwanda Five-toed Skink

4691 ***Leptosiaphos hackarsi***
Squamata - Scincidae
e Hackars's Five-toed Skink

4692 ***Leptosiaphos hylophilus***
Squamata - Scincidae
e Laurent's Five-toed Skink

4693 ***Leptosiaphos ianthwoxantha***
Squamata - Scincidae
e Böhme's Five-toed Skink

4694 ***Leptosiaphos kilemensis***
Squamata - Scincidae
e Kilimanjaro Five-toed Skink

4695 ***Leptosiaphos lepesmei***
Squamata - Scincidae
e Angel's Five-toed Skink

4696 ***Leptosiaphos luberoensis***
Squamata - Scincidae
e Witte's Five-toed Skink

4697 ***Leptosiaphos meleagris***
Squamata - Scincidae
e Ruwenzori Four-toed Skink

4698 ***Leptosiaphos pauliana***
Squamata - Scincidae
e Pauliana's Five-toed Skink

4699 ***Leptosiaphos rhodurus***
Squamata - Scincidae
e Red Five-toed Skink

4700 ***Leptosiaphos vigintiserierum***
Squamata - Scincidae
e African Five-toed Skink

4701 **Leptotyphlopidae**
Squamata
e Blind Snakes; Slender Blind Snakes (NA); Thread Snakes (CSA); Worm Snakes (NA); Leptotyphlopids
d Blindschlangen; Schlankblindschlangen; Engmundschlangen
f Leptotyphlopidés; Serpents aveugles
i Leptotiflopidi

4702 ***Leptotyphlops***
Squamata - Leptotyphlopidae
e Slender Blind Snakes; Worm Snakes; Blind Snakes; Thread Snakes
d Blindschlangen; Schlankblindschlangen
f Serpents aveugles; Serpents minutes; Leptotyphlops

4703 ***Leptotyphlops affinis***
Squamata - Leptotyphlopidae
e Venezuela Blind Snake

4704 ***Leptotyphlops albifrons***
Squamata - Leptotyphlopidae
e Wagler's Blind Snake; Wagler's Worm Snake; White-fronted Thread Snake

4705 ***Leptotyphlops albipunctatus***
Squamata - Leptotyphlopidae
e Tucuman Blind Snake

4706 ***Leptotyphlops amazonicus***
Squamata - Leptotyphlopidae
e Amazon Blind Snake

4707 ***Leptotyphlops anthracinus***
Squamata - Leptotyphlopidae
e Bailey's Blind Snake

4708 ***Leptotyphlops asbolepis***
Squamata - Leptotyphlopidae
e Martin Garcia Tree Snake

4709 ***Leptotyphlops australis***
Squamata - Leptotyphlopidae
e Freiberg's Blind Snake

4710 ***Leptotyphlops bicolor***
Squamata - Leptotyphlopidae
e Two-coloured Blind Snake

4711 ***Leptotyphlops bilineatus***
Squamata - Leptotyphlopidae
e Two-lined Blind Snake; Lesser Antillean Thread Snake (WI)

4712 ***Leptotyphlops blanfordi***
Squamata - Leptotyphlopidae
e Ethiopia Blind Snake; Ethiopian Blind Snake; Sind Thread Snake

4713 ***Leptotyphlops borrichianus***
Squamata - Leptotyphlopidae
e Degerbol's Blind Snake

4714 ***Leptotyphlops boulengeri***
Squamata - Leptotyphlopidae
e Manda Flesh-pink Blind Snake; Manda Flesh-pink Worm Snake

4715 ***Leptotyphlops braccianii***
Squamata - Leptotyphlopidae
e Eritrea Blind Snake

4716 ***Leptotyphlops brasiliensis***
Squamata - Leptotyphlopidae
e Brazilian Blind Snake

4717 ***Leptotyphlops bressoni***
Squamata - Leptotyphlopidae
e Michoacan Slender Blind Snake

4718 ***Leptotyphlops brevicaudus***
Squamata - Leptotyphlopidae
e Shorthead Blind Snake

4719 ***Leptotyphlops brevissimus***
Squamata - Leptotyphlopidae
e Caqueta Blind Snake

4720 ***Leptotyphlops burii***
Squamata - Leptotyphlopidae
e Arabian Blind Snake

4721 ***Leptotyphlops cairi***
Squamata - Leptotyphlopidae
e Cairo Blind Snake
d Ägyptische Schlankblindschlange
f Leptotyphlops du Caire

4722 ***Leptotyphlops calypso***
Squamata - Leptotyphlopidae
e Samana Thread Snake

4723 ***Leptotyphlops collaris***
Squamata - Leptotyphlopidae
e Collared Blind Snake

4724 ***Leptotyphlops columbi***
Squamata - Leptotyphlopidae
e San Salvador Blind Snake; Bahamian Thread Snake (WI)

4725 ***Leptotyphlops conjunctus***
Squamata - Leptotyphlopidae
e Cape Blind Snake; Cape Thread Snake (CSA); Jan's Black Worm Snake (CSA); Jan's All-black Worm Snake; Jan's All-black Thread Snake; Jan's Black Thread Snake; Cape Worm Snake (CSA)

4726 ***Leptotyphlops cupinensis***
Squamata - Leptotyphlopidae
e Mato Grosso Blind Snake

4727 ***Leptotyphlops debilis***
Squamata - Leptotyphlopidae
e West African Blind Snake

4728 ***Leptotyphlops diaplocius***
Squamata - Leptotyphlopidae
e Common Peru Blind Snake; Amazonian Thread Snake

4729 ***Leptotyphlops dimidiatus***
Squamata - Leptotyphlopidae
e Dainty Blind Snake

4730 ***Leptotyphlops dissimilis***
Squamata - Leptotyphlopidae
e Sudan Blind Snake

4731 ***Leptotyphlops distanti***
Squamata - Leptotyphlopidae
e Distant's Blind Snake; Distant's Thread Snake (CSA); Distant's Snake; Transvaal Worm Snake (CSA); Transvaal Thread Snake

4732 ***Leptotyphlops dugandi***
Squamata - Leptotyphlopidae
e Dugand's Blind Snake

4733 ***Leptotyphlops dulcis***
Squamata - Leptotyphlopidae
e Texas Blind Snake; Plains Blind Snake
d Texas-Schlankblindschlange
f Serpent aveugle du Texas; Leptotyphlops du Texas

4734 ***Leptotyphlops dulcis dissectus***
Squamata - Leptotyphlopidae
e New Mexico Blind Snake; New Mexican Blind Snake; New Mexico Blind Snake

4735 ***Leptotyphlops dulcis dulcis***
Squamata - Leptotyphlopidae
e Plains Blind Snake; Texas Blind Snake

4736 ***Leptotyphlops erythraeus***
Squamata - Leptotyphlopidae
e Red Blind Snake

4737 ***Leptotyphlops filiformis***
Squamata - Leptotyphlopidae
e Socotra Island Blind Snake

4738 ***Leptotyphlops fitzingeri***
Squamata - Leptotyphlopidae
e Fitzinger's Blind Snake

4739 ***Leptotyphlops gestri***
Squamata - Leptotyphlopidae
e Central African Blind Snake

4740 ***Leptotyphlops goudoti***
Squamata - Leptotyphlopidae
e Black Blind Snake; Common Worm Snake
d Mittelamerikanische Blindschlange

4741 ***Leptotyphlops gracilior***
Squamata - Leptotyphlopidae
e Slender Blind Snake; Slender Thread Snake (CSA)

4742 ***Leptotyphlops guayaquilensis***
Squamata - Leptotyphlopidae
e Guyaquila Blind Snake

4743 ***Leptotyphlops hamulirostris***
Squamata - Leptotyphlopidae
e Nikolsky's Blind Snake

4744 ***Leptotyphlops humilis***
Squamata - Leptotyphlopidae
e Western Blind Snake; Western Thread Snake; Trans-Pecos Blind Snake; Western Slender Blind Snake
d Mexikanische Schlankblindschlange; Westamerikanische Blindschlange
f Serpent aveugle de l'Ouest; Leptotyphlops du Mexique

4745 ***Leptotyphlops humilis cahuilae***
Squamata - Leptotyphlopidae
e Desert Blind Snake; Desert Thread Snake

4746 ***Leptotyphlops humilis humilis***
Squamata - Leptotyphlopidae
e Southwestern Blind Snake; Southwestsern Thread Snake

4747 ***Leptotyphlops humilis segregus***
Squamata - Leptotyphlopidae
e Trans-Pecos Blind Snake; TransPecos Thread Snake

4748 ***Leptotyphlops humilis utahensis***
Squamata - Leptotyphlopidae
e Utah Blind Snake; Utah Thread Snake

4749 ***Leptotyphlops joshuai***
Squamata - Leptotyphlopidae
e Joshua's Blind Snake

4750 ***Leptotyphlops koppesi***
Squamata - Leptotyphlopidae
e Amaral's Blind Snake

4751 ***Leptotyphlops labialis***
Squamata - Leptotyphlopidae
e Damara Blind Snake; Damara Thread Snake (CSA)

4752 ***Leptotyphlops leptipileptus***
Squamata - Leptotyphlopidae
e Haitian Border Thread Snake

4753 ***Leptotyphlops longicaudus***
Squamata - Leptotyphlopidae
e Long-tail Blind Snake; Long-tailed Thread Snake (CSA); Long-tailed Flesh-pink Thread Snake; Long-tailed Flesh-pink Worm Snake; Long-tailed Worm Snake (CSA)

4754 ***Leptotyphlops macrolepis***
Squamata - Leptotyphlopidae
e Big-scaled Blind Snake; Large-scaled Worm Snake; Big-eyed Thread Snake

4755 ***Leptotyphlops macrorhynchus***
Squamata - Leptotyphlopidae
e Beaked Blind Snake; Long-nosed Worm Snake; Beaked Thread Snake; Worm Snake
d Langnasige Schlankblindschlange; Hakenschnabelblindschlange; Nordafrikanische Schlankblindschlange
f Leptotyphlops oncirostre; Leptotyphlops macrorhynque
i Leptotiflope macrorinco

4756 ***Leptotyphlops macrurus***
Squamata - Leptotyphlopidae
e Boulenger's Blind Snake

4757 ***Leptotyphlops maximus***
Squamata - Leptotyphlopidae
e Giant Blind Snake

4758 ***Leptotyphlops melanotermus***
Squamata - Leptotyphlopidae
e Latin American Blind Snake

4759 ***Leptotyphlops melanurus***
Squamata - Leptotyphlopidae
e Dark Blind Snake

4760 ***Leptotyphlops munoi***
Squamata - Leptotyphlopidae
e Rio Grande del Sul Blind Snake

4761 ***Leptotyphlops nasalis***
Squamata - Leptotyphlopidae
e Taylor's Blind Snake; Managuan Worm Snake

4762 ***Leptotyphlops nasirostris***
Squamata - Leptotyphlopidae
e Bignose Blind Snake

4763 ***Leptotyphlops natatrix***
Squamata - Leptotyphlopidae
e Gambia Blind Snake

4764 ***Leptotyphlops nicefori***
Squamata - Leptotyphlopidae
e Santander Blind Snake

4765 ***Leptotyphlops nigricans***
Squamata - Leptotyphlopidae
e Schlegel's Blind Snake; Black Thread Snake (CSA); Black Blind Snake; Cape Worm Snake; Black Worm Snake (CSA)

4766 ***Leptotyphlops nursii***
Squamata - Leptotyphlopidae
e Nurse's Blind Snake

4767 ***Leptotyphlops occidentalis***
Squamata - Leptotyphlopidae
e Occident Blind Snake; Western Thread Snake (CSA)

4768 ***Leptotyphlops peruvianus***
Squamata - Leptotyphlopidae
e Peru Blind Snake
f Leptotyphlops de Phillips; Leptotyphlops d'Israël

4769 ***Leptotyphlops phillipsi***
Squamata - Leptotyphlopidae
e Phillips's Blind Snake
d Israelische Schlankblindschlange

4770 ***Leptotyphlops pyrites***
Squamata - Leptotyphlopidae
e Thomas's Blind Snake; Barahona Thread Snake

4771 ***Leptotyphlops reticulatus***
Squamata - Leptotyphlopidae
e Reticulate Blind Snake

4772 ***Leptotyphlops rostratus***
Squamata - Leptotyphlopidae
e Bocage's Blind Snake

4773 ***Leptotyphlops rubrolineatus***
Squamata - Leptotyphlopidae
e Red-lined Blind Snake

4774 ***Leptotyphlops rufidorsus***
Squamata - Leptotyphlopidae
e Rose Blind Snake

4775 ***Leptotyphlops salgueiroi***
Squamata - Leptotyphlopidae
e Espirito Santo Blind Snake

4776 ***Leptotyphlops scutifrons***
Squamata - Leptotyphlopidae
e Shielded Blind Snake; Peters's Thread Snake (CSA); Peters's Black Thread Snake; Peters's Black Worm Snake (CSA); Glossy Worm Snake; Glossy Thread Snake; Scaly-fronted Worm Snake (CSA)

4777 ***Leptotyphlops scutifrons merkeri***
Squamata - Leptotyphlopidae
e Merker's Worm Snake

4778 ***Leptotyphlops septemstriatus***
Squamata - Leptotyphlopidae
e Seven-striped Blind Snake

4779 ***Leptotyphlops signatus***
Squamata - Leptotyphlopidae
e South American Blind Snake; Peruvian Thread Snake

4780 ***Leptotyphlops subcrotillus***
Squamata - Leptotyphlopidae
e Klauber's Blind Snake

4781 ***Leptotyphlops sundevalli***
Squamata - Leptotyphlopidae
e Sundevall Blind Snake; Sundevall's Worm Snake
d Sundevall-Schlankblindschlange

4782 ***Leptotyphlops teaguei***
Squamata - Leptotyphlopidae
e Northern Blind Snake

4783 ***Leptotyphlops telloi***
Squamata - Leptotyphlopidae
e Tello's Blind Snake; Tello's Thread Snake (CSA)

4784 ***Leptotyphlops tenellus***
Squamata - Leptotyphlopidae
e Guyana Blind Snake; Neotropical Thread Snake (WI)

4785 ***Leptotyphlops tesselatus***
Squamata - Leptotyphlopidae
e Tschudi's Blind Snake

4786 ***Leptotyphlops tricolor***
Squamata - Leptotyphlopidae
e Three-coloured Blind Snake

4787 ***Leptotyphlops undecimstriatus***
Squamata - Leptotyphlopidae
e Eleven-striped Blind Snake

4788 ***Leptotyphlops unguirostris***
Squamata - Leptotyphlopidae
e Southern Blind Snake

4789 ***Leptotyphlops variabilis***
Squamata - Leptotyphlopidae
e Variable Blind Snake

4790 ***Leptotyphlops weyrauchi***
Squamata - Leptotyphlopidae
e Argentine Blind Snake

4791 ***Leptotyphlops wilsoni***
Squamata - Leptotyphlopidae
e Wilson's Blind Snake

4792 ***Lepturophis***
Squamata - Colubridae
e Slender Wolf Snakes

4793 ***Lepturophis borneensis***
Squamata - Colubridae
e Slender Wolf Snake

4794 ***Lerista***
Squamata - Scincidae
e Leristas; Rhodonas
f Léristes

4795 ***Lerista aericeps***
Squamata - Scincidae
e Sandridge Lerista

4796 ***Lerista allanae***
Squamata - Scincidae
e Allan's Lerista

4797 ***Lerista allochira***
Squamata - Scincidae
e Gorge Lerista

4798 ***Lerista ameles***
Squamata - Scincidae
e Woodland Lerista

4799 ***Lerista apoda***
Squamata - Scincidae
e Legless Lerista

4800 ***Lerista bipes***
Squamata - Scincidae
e Two-legged Lerista; Two-toed Skink

4801 ***Lerista borealis***
Squamata - Scincidae
e Boreal Lerista

4802 ***Lerista bougainvillii***
Squamata - Scincidae
e Bougainville's Lerista; Bougainville's Skink
f Scinque de Bougainville

4803 ***Lerista carpentariae***
Squamata - Scincidae
e Carpentaria Lerista

4804 ***Lerista chalybura***
Squamata - Scincidae
e Olive-grey Lerista

4805 ***Lerista christinae***
Squamata - Scincidae
e Christina's Lerista

4806 ***Lerista cinerea***
Squamata - Scincidae
e Charters Towers Lerista

4807 ***Lerista connivens***
Squamata - Scincidae
e Eyelid Lerista

4808 ***Lerista desertorium***
Squamata - Scincidae
e Desert Lerista

4809 ***Lerista distinguenda***
Squamata - Scincidae
e Distinguished Lerista

4810 ***Lerista dorsalis***
Squamata - Scincidae
e Shiny Lerista

4811 ***Lerista elegans***
Squamata - Scincidae
e Elegant Lerista

4812 ***Lerista flammicauda***
Squamata - Scincidae
e Redtail Lerista

4813 ***Lerista fragilis***
Squamata - Scincidae
e Fragile Lerista

4814 ***Lerista frosti***
Squamata - Scincidae
e Frost's Lerista

4815 ***Lerista gascoynensis***
Squamata - Scincidae
e Gascoyne River Lerista

4816 ***Lerista gerrardi***
Squamata - Scincidae
e Gerrard's Lerista

4817 ***Lerista greeri***
Squamata - Scincidae
e Greer's Lerista

4818 ***Lerista griffini***
Squamata - Scincidae
e Griffin's Lerista

4819 ***Lerista haroldi***
Squamata - Scincidae
e Harold's Lerista

4820 ***Lerista humphriesi***
Squamata - Scincidae
e Humphrey's Lerista

4821 ***Lerista ips***
Squamata - Scincidae
e Pink Lerista

4822 ***Lerista kalumburu***
Squamata - Scincidae
e Napier Broome Bay Lerista

4823 ***Lerista karlschmidti***
Squamata - Scincidae
e Karl's Lerista

4824 ***Lerista kennedyensis***
Squamata - Scincidae
e Kendrick's Lerista

4825 ***Lerista labialis***
Squamata - Scincidae
e Interior Lerista; Southern Sandslider

4826 ***Lerista lineata***
Squamata - Scincidae
e Lined Lerista; Lineated Lerista
d Australischer Skink
f Lériste à quatre raies

4827 ***Lerista lineopunctata***
Squamata - Scincidae
e Six-lined Lerista
f Scinque linéo-ponctué

4828 ***Lerista macropisthopus***
Squamata - Scincidae
e Purple Lerista

4829 ***Lerista microtis***
Squamata - Scincidae
e Small-eared Lerista

4830 ***Lerista muelleri***
Squamata - Scincidae
e Müller's Lerista; Müller's Skink (ANZ)
d Müllers Waldskink

4831 ***Lerista neanderi***
Squamata - Scincidae
e Opthalmia Range Lerista

4832 ***Lerista nicholsii***
Squamata - Scincidae
e Nichols's Lerista

4833 ***Lerista onsloviana***
Squamata - Scincidae
e Exmouth Gulf Lerista

4834 ***Lerista orientalis***
Squamata - Scincidae
e Bronze Lerista

4835 ***Lerista picturata***
Squamata - Scincidae
e Fry's Lerista

4836 ***Lerista planiventralis***
Squamata - Scincidae
e Two-fingered Lerista

4837 ***Lerista praefrontalis***
Squamata - Scincidae
e King Hall Island Lerista

4838 ***Lerista praepedita***
Squamata - Scincidae
e Spotted Lerista

4839 ***Lerista punctatovittata***
Squamata - Scincidae
e Western Lerista; Spotted Burrowing Skink (ANZ)

4840 ***Lerista separanda***
Squamata - Scincidae
e Dampier Land Lerista

4841 ***Lerista simillima***
Squamata - Scincidae
e Pindan Lerista

4842 ***Lerista stictopleura***
Squamata - Scincidae
e Mount Augustus Lerista

4843 ***Lerista storri***
Squamata - Scincidae
e Storr's Lerista

4844 ***Lerista stylis***
Squamata - Scincidae
e Carpentaria Lerista

4845 ***Lerista taeniata***
Squamata - Scincidae
e Speckled Lerista

4846 ***Lerista terdigitata***
Squamata - Scincidae
e Island Lerista

4847 ***Lerista uniduo***
Squamata - Scincidae
e Shrub Lerista

4848 ***Lerista varia***
Squamata - Scincidae
e Variable Lerista

4849 ***Lerista vermicularis***
Squamata - Scincidae
e Great Sandy Desert Lerista

4850 ***Lerista vittata***
Squamata - Scincidae
e Brown-flecked Lerista

4851 ***Lerista walkeri***
Squamata - Scincidae
e Walker's Lerista

4852 ***Lerista wilkinsi***
Squamata - Scincidae
e Wilkins's Lerista

4853 ***Lerista xanthura***
Squamata - Scincidae
e Buff Lerista

4854 ***Lialis***
Squamata - Gekkonidae
e Snake Lizards
d Flossenfüße
f Lialis

4855 ***Lialis burtonis***
Squamata - Gekkonidae
e Burton's Snake Lizard; Sharp-snouted Snake Lizard
d Spitzkopfflossenfuß
f Lialis de Burton

4856 ***Lialis jicari***
Squamata - Gekkonidae
e Papua Snake Lizard; New Guinea Snake Lizard; Jicar's Snake Lizard; Jicar's Scalefoot
d Neuguinea-Flossenfuß
f Lialis de la Nouvelle Guinée

4857 ***Liasis***
Squamata - Boidae
e Rock Pythons; Australian Water Pythons; Water Pythons
d Australische Felsenpythone

4858 ***Liasis fuscus***
Squamata - Boidae
e Brown Python; Water Python; New Guinea Orange-bellied Black Water Python; New Guinea Python; Brown Water Python
d Brauner Wasserpython; Wasserpython

4859 ***Liasis mackloti***
Squamata - Boidae
e Macklot's Python; Spotted Python; Indonesian Water Python
d Macklots Python; Neuguinea-Wasserpython

4860 ***Liasis mackloti dunni***
Squamata - Boidae
e Dunn's Python

4861 ***Liasis mackloti mackloti***
Squamata - Boidae
e Macklot's Python

4862 ***Liasis mackloti savuensis***
Squamata - Boidae
e White-eyed Python; Savu Python; Sawu Python

4863 ***Liasis olivaceus***
Squamata - Boidae
e Olive Python (ANZ); Olive-brown Python; Papuan Python
d Olivpython; Olivfarbiger Python

4864 ***Lichanura***
Squamata - Boidae
e Rosy Boas
d Rosenboas
f Boas roses

4865 ***Lichanura trivirgata***
Squamata - Boidae
e Rosy Boa; Mexican Rosy Boa
d Dreistreifenrosenboa
f Boa rose; Boa à trois bandes; Boa ligné
i Boa roseo

4866 ***Lichanura trivirgata gracia***
Squamata - Boidae
e Desert Rosy Boa
d Dreistreifenrosenboa
f Boa rose du désert
i Boa roseo del deserto

4867 ***Lichanura trivirgata roseofusca***
Squamata - Boidae
e Coastal Rosy Boa (NA); California Boa (NA); Rosy Boa
d Echte Rosenboa; Rosenboa
i Boa roseo della costa

4868 ***Lichanura trivirgata saslowi***
Squamata - Boidae
e Baja California Rosy Boa
d Baja-Rosenboa
i Boa roseo della Baja California

4869 ***Lichanura trivirgata trivirgata***
Squamata - Boidae
e Mexican Rosy Boa
d Rosenboa; Dreistreifenrosenboa
i Boa roseo messicano

4870 ***Limnophis***
Squamata - Colubridae
e Striped Swamp Snakes; Tropical Green Snakes; Bicoloured Swamp Snakes
d Sumpfnattern

4871 ***Limnophis bicolor***
Squamata - Colubridae
e Striped Swamp Snake; Tropical Green Snake; Bicoloured Swamp Snake
d Zweifarbige Sumpfnatter

4872 ***Liochlorophis***
Squamata - Colubridae
e Smooth Green Snakes

4873 ***Liochlorophis vernalis***
Squamata - Colubridae
e Smooth Green Snake
d Glatte Grasnatter
f Serpent vert lisse; Couleuvre verte
i Serpente verde liscio

4874 ***Liochlorophis vernalis blanchardi***
Squamata - Colubridae
e Western Smooth Green Snake

4875 ***Liochlorophis vernalis vernalis***
Squamata - Colubridae
e Eastern Smooth Green Snake

4876 ***Lioheterophis***
Squamata - Colubridae
e Ihering's Snakes

4877 ***Lioheterophis iheringi***
Squamata - Colubridae
e Ihering's Snake

4878 ***Liolaemus***
Squamata - Iguanaidae
e Tree Iguanas; Smooth-throated Lizards
d Erdleguane

4879 ***Liolaemus alticolor***
Squamata - Iguanaidae
e Brilliant Tree Iguana

4880 ***Liolaemus andinus***
Squamata - Iguanaidae
e Andes Tree Iguana

4881 ***Liolaemus annectens***
Squamata - Iguanaidae
e United Tree Iguana

4882 ***Liolaemus archeforus***
Squamata - Iguanaidae
e Main Tree Iguana

4883 ***Liolaemus austromendocinus***
Squamata - Iguanaidae
e Austromendocino Tree Iguana; Lagarto de las piedras (NA)

4884 ***Liolaemus bibroni***
Squamata - Iguanaidae
e Bibron's Tree Iguana

4885 ***Liolaemus bisignatus***
Squamata - Iguanaidae
e Philippi's Tree Iguana

4886 ***Liolaemus bitaeniatus***
Squamata - Iguanaidae
e Striped Tree Iguana

4887 ***Liolaemus boulengeri***
Squamata - Iguanaidae
e Boulenger's Tree Iguana

4888 ***Liolaemus buergeri***
Squamata - Iguanaidae
e Buerger's Tree Iguana

4889 ***Liolaemus capillitas***
Squamata - Iguanaidae
e Hulse's Tree Iguana

4890 ***Liolaemus ceii***
Squamata - Iguanaidae
e Cei's Iguana

4891 ***Liolaemus chacoensis***
Squamata - Iguanaidae
e Chaco Tree Iguana
Squamata - Iguanaidae
e Chilean Tree Iguana

4893 ***Liolaemus coerules***
Squamata - Iguanaidae
e Blue Tree Iguana

4894 ***Liolaemus constanzae***
Squamata - Iguanaidae
e Constanze's Tree Iguana

4895 ***Liolaemus copiapensis***
Squamata - Iguanaidae
e Müller's Tree Iguana

4896 ***Liolaemus cranwelli***
Squamata - Iguanaidae
e Cranwell's Tree Iguana

4897 ***Liolaemus curicensis***
Squamata - Iguanaidae
e Curicen Tree Iguana

4898 ***Liolaemus curis***
Squamata - Iguanaidae
e Nunez Tree Iguana

4899 ***Liolaemus cyanogaster***
Squamata - Iguanaidae
e Cyan Tree Iguana

4900 ***Liolaemus darwinii***
Squamata - Iguanaidae
e Darwin's Tree Iguana; Darwins Lagartiga (NA)

4901 ***Liolaemus donosobarrosi***
Squamata - Iguanaidae
e Donoso-Barros's Tree Iguana

4902 ***Liolaemus donosoi***
Squamata - Iguanaidae
e Karen's Tree Iguana

4903 ***Liolaemus dorbignyi***
Squamata - Iguanaidae
e D'Orbigny's Tree Iguana

4904 ***Liolaemus duellmani***
Squamata - Iguanaidae
e Duellman's Tree Iguana

4905 ***Liolaemus elegans***
Squamata - Iguanaidae
e Elegant Tree Iguana

4906 ***Liolaemus eleodori***
Squamata - Iguanaidae
e Eleodor's Tree Iguana

4907 ***Liolaemus elongtus***
Squamata - Iguanaidae
e Elongate Tree Iguana

4908 ***Liolaemus fabianus***
Squamata - Iguanaidae
e Yanez's Tree Iguana

4909 ***Liolaemus fitzgeraldi***
Squamata - Iguanaidae
e Fitzgerald's Tree Iguana

4910 ***Liolaemus fitzingerii***
Squamata - Iguanaidae
e Fitzinger's Tree Iguana

4911 ***Liolaemus forsteri***
Squamata - Iguanaidae
e Forster's Tree Iguana

4912 ***Liolaemus fuscus***
Squamata - Iguanaidae
e Brown Tree Iguana

4913 ***Liolaemus goetschi***
Squamata - Iguanaidae
e Goetsch's Tree Iguana

4914 ***Liolaemus gracilis***
Squamata - Iguanaidae
e Graceful Tree Iguana

4915 ***Liolaemus gravenhorstii***
Squamata - Iguanaidae
e Gravenhorst's Tree Iguana; Gravenhorst's Larartija (NA)

4916 ***Liolaemus griseus***
Squamata - Iguanaidae
e Grey Tree Iguana

4917 ***Liolaemus hellmichi***
Squamata - Iguanaidae
e Hellmich's Tree Iguana

4918 ***Liolaemus jamesi***
Squamata - Iguanaidae
e James's Tree Iguana

4919 ***Liolaemus kingii***
Squamata - Iguanaidae
e King's Tree Iguana

4920 ***Liolaemus kriegi***
Squamata - Iguanaidae
e Krieg's Tree Iguana

4921 ***Liolaemus kuhlmanni***
Squamata - Iguanaidae
e Kuhlmnan's Tree Iguana

4922 ***Liolaemus lemniscutus***
Squamata - Iguanaidae
e Wreath Tree Iguana

4923 ***Liolaemus leopardinus***
Squamata - Iguanaidae
e Leopard Tree Iguana

4924 ***Liolaemus lineomaculatus***
Squamata - Iguanaidae
e Decorated Tree Iguana

4925 ***Liolaemus lorenzmuelleri***
Squamata - Iguanaidae
e Lorenz's Tree Iguana

4926 ***Liolaemus lutzae***
Squamata - Iguanaidae
e Lutz's Tree Iguana

4927 ***Liolaemus magellanicus***
Squamata - Iguanaidae
e Magellan's Tree Iguana; Magellan's Smooth-throated Lizard
d Magellans Erdleguan; Erdleguan; Magellan-Erdleguan

4928 ***Liolaemus mocquardi***
Squamata - Iguanaidae
e Mocquard's Tree Iguana

4929 ***Liolaemus modestus***
Squamata - Iguanaidae
e Modest Tree Iguana

4930 ***Liolaemus montanus***
Squamata - Iguanaidae
e Mountain Tree Iguana

4931 ***Liolaemus monticola***
Squamata - Iguanaidae
e Peak Tree Iguana

4932 ***Liolaemus multicolor***
Squamata - Iguanaidae
e Many-coloured Tree Iguana

4933 ***Liolaemus multiformis***
Squamata - Iguanaidae
e Peru Tree Iguana; Peruvian Smooth-throated Lizard

4934 ***Liolaemus multimaculatus***
Squamata - Iguanaidae
e Many-spotted Tree Iguana

4935 ***Liolaemus nigriceps***
Squamata - Iguanaidae
e Black Tree Iguana

4936 ***Liolaemus nigromaculatus***
Squamata - Iguanaidae
e Black-spotted Tree Iguana

4937 ***Liolaemus nigroviridis***
Squamata - Iguanaidae
e Black-green Tree Iguana

4938 ***Liolaemus nitidus***
Squamata - Iguanaidae
e Shining Tree Iguana

4939 ***Liolaemus occipitalis***
Squamata - Iguanaidae
e Skull Tree Iguana

4940 ***Liolaemus ornatus***
Squamata - Iguanaidae
e Ortez's Tree Iguana

4941 ***Liolaemus ortizii***
Squamata - Iguanaidae
e Ornate Tree Iguana

4942 ***Liolaemus patherinus***
Squamata - Iguanaidae
e Panther Tree Iguana

4943 ***Liolaemus paulinae***
Squamata - Iguanaidae
e Paulina Tree Iguana

4944 ***Liolaemus pictus***
Squamata - Iguanaidae
e Painted Tree Iguana

4945 ***Liolaemus platei***
Squamata - Iguanaidae
e Braided Tree Iguana

4946 ***Liolaemus pulcher***
Squamata - Iguanaidae
e Beautiful Tree Iguana

4947 ***Liolaemus rabinoi***
Squamata - Iguanaidae
e Rabino's Tree Iguana

4948 ***Liolaemus robertmertensi***
Squamata - Iguanaidae
e Robert's Tree Iguana

4949 ***Liolaemus rothi***
Squamata - Iguanaidae
e Roth's Tree Iguana

4950 ***Liolaemus ruibali***
Squamata - Iguanaidae
e Ruitbal's Tree Iguana

4951 ***Liolaemus ruizleali***
Squamata - Iguanaidae
e Ruizleal Tree Iguana

4952 ***Liolaemus sanjuanensis***
Squamata - Iguanaidae
e San Juan Tree Iguana

4953 ***Liolaemus scapularis***
Squamata - Iguanaidae
e Shoulder Tree Iguana

4954 ***Liolaemus schmidti***
Squamata - Iguanaidae
e Schmidt's Tree Iguana

4955 ***Liolaemus schroederi***
Squamata - Iguanaidae
e Schroeder's Tree Iguana

4956 ***Liolaemus signifer***
Squamata - Iguanaidae
e Zodiac Tree Iguana

4957 ***Liolaemus simonsi***
Squamata - Iguanaidae
e Simons's Tree Iguana

4958 ***Liolaemus tachae***
Squamata - Iguanaidae
e Spotted Tree Iguana

4959 ***Liolaemus tenuis***
Squamata - Iguanaidae
e Thin Tree Iguana

4960 ***Liolaemus uspallatensis***
Squamata - Iguanaidae
e Macola's Tree Iguana

4961 ***Liolaemus variegatus***
Squamata - Iguanaidae
e Variegated Tree Iguana

4962 ***Liolaemus walkeri***
Squamata - Iguanaidae
e Walker's Tree Iguana

4963 ***Liolaemus wiegmanni***
Squamata - Iguanaidae
e Wiegmann's Tree Iguana; Lagartija de la arena (NA)

4964 ***Liolaemus zapallarensis***
Squamata - Iguanaidae
e Zapallaren Tree Iguana

4965 ***Liopeltis***
Squamata - Colubridae
e Reed Snakes; Smooth Snakes

4966 ***Liopeltis calamaria***
Squamata - Colubridae
e Calamaria Reed Snake

4967 ***Liopeltis frenatus***
Squamata - Colubridae
e Günther's Reed Snake; Günther's Smooth Snake; Reed Snake

4968 ***Liopeltis tricolor***
Squamata - Colubridae
e Malayan Reed Snake; Schlegel's Smooth Snake

4969 ***Liophidium***
Squamata - Colubridae
e Bright Snakes

4970 ***Liophidium mayottensis***
Squamata - Colubridae
e Peters's Bright Snake

4971 ***Liophidium rhodogaster***
Squamata - Colubridae
e Gold-collared Bright Snake

4972 ***Liophidium trilineatum***
Squamata - Colubridae
e Madagascar Bright Snake

4973 ***Liophis***
Squamata - Colubridae
e Amazon Ground Snakes; Swamp Snakes; Ground Snakes (WI)

4974 ***Liophis albiceps***
Squamata - Colubridae
e Pale Ground Snake

4975 ***Liophis albiventris***
Squamata - Colubridae

e Jan's Ground Snake

4976 ***Liophis almadensis***
Squamata - Colubridae
e Almaden's Ground Snake

4977 ***Liophis amarali***
Squamata - Colubridae
e Amaral's Ground Snake

4978 ***Liophis amazonicus***
Squamata - Colubridae
e Amazon Ground Snake

4979 ***Liophis anomolus***
Squamata - Colubridae
e Strange Ground Snake
d Parana-Goldbauchnatter; Goldbauchnatter

4980 ***Liophis atraventer***
Squamata - Colubridae
e Dixon's Ground Snake

4981 ***Liophis bimaculatus***
Squamata - Colubridae
e Two-spotted Ground Snake

4982 ***Liophis biridis***
Squamata - Colubridae
e Green Ground Snake

4983 ***Liophis brazili***
Squamata - Colubridae
e Brazilian Ground Snake

4984 ***Liophis breviceps***
Squamata - Colubridae
e Short Ground Snake

4985 ***Liophis breviceps***
Squamata - Colubridae
e Amazon Swamp Snake

4986 ***Liophis canaima***
Squamata - Colubridae
e Roze's Ground Snake

4987 ***Liophis carajascinsis***
Squamata - Colubridae
e Carajascin Ground Snake

4988 ***Liophis chrysostoma***
Squamata - Colubridae
e Cope's Ground Snake

4989 ***Liophis cobella***
Squamata - Colubridae
e South American Ground Snake

4990 ***Liophis cobella dyticus***
Squamata - Colubridae
e South American Swamp Snake

4991 ***Liophis coralliventris***
Squamata - Colubridae
e Coral Ground Snake; Boulenger's Tropical Snake

4992 ***Liophis cursor***
Squamata - Colubridae
e Lacepède's Ground Snake; Martinique Ground Snake (WI)

4993 ***Liophis dilepis***
Squamata - Colubridae
e Lema's Ground Snake

4994 ***Liophis epinephalus***
Squamata - Colubridae
e Night Ground Snake; Fire-bellied Snake

4995 ***Liophis festae***
Squamata - Colubridae
e Drab Ground Snake

4996 ***Liophis flavifrenatus***
Squamata - Colubridae
e Fronted Ground Snake

4997 ***Liophis fraseri***
Squamata - Colubridae
e Fraser's Ground Snake

4998 ***Liophis frenata***
Squamata - Colubridae
e Common Ground Snake

4999 ***Liophis ingeri***
Squamata - Colubridae
e Inger's Ground Snake

5000 ***Liophis jaegeri***
Squamata - Colubridae
e Jaeger's Ground Snake

5001 ***Liophis juliae***
Squamata - Colubridae
e Julia's Ground Snake; Leeward Ground Snake (WI)

5002 ***Liophis lineatus***
Squamata - Colubridae
e Lined Ground Snake; Lined Tropical Snake; Striped Ground Snake

5003 ***Liophis longiventris***
Squamata - Colubridae
e Long Ground Snake

5004 ***Liophis maryellenae***
Squamata - Colubridae
e Maryellen's Ground Snake

5005 ***Liophis melanotus***
Squamata - Colubridae
e Shaw's Dark Ground Snake; Grenada Ground Snake (WI)

5006 ***Liophis miliaris***
Squamata - Colubridae
e Military Ground Snake

5007 ***Liophis miliaris chrysostomus***
Squamata - Colubridae
e White-lipped Swamp Snake

5008 ***Liophis mossoroensis***
Squamata - Colubridae
e Mossoro Ground Snake

5009 ***Liophis ornatus***
Squamata - Colubridae
e Ornate Ground Snake; St. Lucia Ground Snake (WI)

5010 ***Liophis paucidens***
Squamata - Colubridae
e Hoge's Ground Snake

5011 ***Liophis perfuscus***
Squamata - Colubridae
e Tan Ground Snake; Barbados Ground Snake (WI)

5012 ***Liophis poecilogyrus***
Squamata - Colubridae
e Wied's Ground Snake
d Buntnatter; Argentinische Buntnatter

5013 ***Liophis problematicus***
Squamata - Colubridae
e Problem Ground Snake

5014 ***Liophis pseudocobella***
Squamata - Colubridae
e Peracca's Ground Snake

5015 ***Liophis purpurans***
Squamata - Colubridae
e Purple Ground Snake

5016 ***Liophis reginae***
Squamata - Colubridae
e Royal Ground Snake

5017 ***Liophis reginae semilineatus***
Squamata - Colubridae
e Common Swamp Snake

5018 ***Liophis sagittifer***
Squamata - Colubridae
e Arrow Ground Snake

5019 ***Liophis taeniurus***
Squamata - Colubridae
e Thin Ground Snake

5020 ***Liophis trebbaui***
Squamata - Colubridae
e Trebbau's Ground Snake

5021 ***Liophis triscalis***
Squamata - Colubridae
e Three-scaled Ground Snake

5022 ***Liophis typhlus***
Squamata - Colubridae
e Blind Ground Snake

5023 ***Liophis typhlus typhlus***
Squamata - Colubridae
e Velvety Swamp Snake

5024 ***Liophis vanzolinii***
Squamata - Colubridae
e Vanzolini's Ground Snake

5025 ***Liophis williamsi***
Squamata - Colubridae
e Williams's Ground Snake

5026 ***Liophis zweifeli***
Squamata - Colubridae
e Zweifel's Ground Snake

5027 ***Liopholidophis***
Squamata - Colubridae
e Madagascar Water Snakes
d Kurznattern

5028 ***Liopholidophis grandidieri***
Squamata - Colubridae
e Grandidier's Water Snake

5029 ***Liopholidophis lateralis***
Squamata - Colubridae
e Lateral Water Snake

5030 ***Liopholidophis pinguis***
Squamata - Colubridae
e Madagascar Water Snake

5031 ***Liopholidophis pseudolateralis***
Squamata - Colubridae
e False-striped Water Snake

5032 ***Liopholidophis sexlineatus***
Squamata - Colubridae
e Six-lined Water Snake

5033 ***Liopholidophis stumpffi***
Squamata - Colubridae
e Yellow-striped Water Snake

5034 ***Liopholidophis thielei***
Squamata - Colubridae
e Thiele's Water Snake

5035 ***Liotyphlops***
Squamata - Anomalepididae
e Lesser Blind Snakes; Lesser Worm Snakes; Slender Blind Snakes

5036 ***Liotyphlops albirostris***
Squamata - Anomalepididae
e Whitenose Blind Snake; Whitenose Worm Snake

5037 ***Liotyphlops anops***
Squamata - Anomalepididae
e Cope's Blind Snake; Cope's Worm Snake

5038 ***Liotyphlops metae***
Squamata - Anomalepididae
e Meta Blind Snake; Meta Worm Snake

5039 ***Liotyphlops schubarti***
Squamata - Anomalepididae
e Sao Paolo Blind Snake; Sao Paolo Worm Snake

5040 ***Liotyphlops teretzii***
Squamata - Anomalepididae
e Ternetz's Blind Snake; Ternetz's Worm Snake; Ternetz's Lesser Blind Snake

5041 ***Liotyphlops wilderi***
Squamata - Anomalepididae
e Wilder's Blind Snake; Wilder's Worm Snake

5042 ***Lipinia***
Squamata - Scincidae
e Lipinias

5043 ***Lipinia auriculata***
Squamata - Scincidae
e Taylor's Lipinia

5044 ***Lipinia cheesmanae***
Squamata - Scincidae
e Cheesman's Lipinia

5045 ***Lipinia leptosoma***
Squamata - Scincidae
e Slender Lipinia; Slender Skink

5046 ***Lipinia longiceps***
Squamata - Scincidae
e Long Lipinia

5047 ***Lipinia macrotympanum***
Squamata - Scincidae
e Big-eared Lipinia

5048 ***Lipinia miangensis***
Squamata - Scincidae
e Werner's Lipinia

5049 ***Lipinia miotis***
Squamata - Scincidae
e Boulenger's Lipinia

5050 ***Lipinia noctua***
Squamata - Scincidae
e Moth Lipinia; Moth Skink; Brown Skink

5051 ***Lipinia pulchella***
Squamata - Scincidae
e Beautiful Lipinia

5052 ***Lipinia pulchra***
Squamata - Scincidae
e Small-legged Lipinia

5053 ***Lipinia quadrivittata***
Squamata - Scincidae
e Four-striped Lipinia

5054 ***Lipinia rabori***
Squamata - Scincidae
e Rabor's Lipinia

5055 ***Lipinia relicta***
Squamata - Scincidae
e Vinciguerra's Lipinia

5056 ***Lipinia rouxi***
Squamata - Scincidae
e Roux's Lipinia

5057 ***Lipinia semperi***
Squamata - Scincidae
e Semper's Lipinia

5058 ***Lipinia subvittata***
Squamata - Scincidae
e Striped Lipinia

5059 ***Lipinia venemai***
Squamata - Scincidae
e Brongersma's Lipinia

5060 ***Lipinia vittigera***
Squamata - Scincidae
e Banded Lipinia

5061 ***Lipinia vulcania***
Squamata - Scincidae
e Vulcan Lipinia

5062 ***Lipinia zamboangensis***
Squamata - Scincidae
e Zamboang Lipinia

5063 ***Lissemys***
Testudines - Trionychidae
e Indian Flapshell Turtles; Indian Mud Turtles
d Indische Klappenweichschildkröten

5064 ***Lissemys punctata***
Testudines - Trionychidae
e Indian Flapshell Turtle; Indian Flap-shelled Turtle; Indian Mud Turtle (ISC); Indian Flapshell
d Indische Klappenweichschildkröte; Klappenweichschildkröte
f Trionyx à clapets de l'Inde; Tortue de l'Inde
i Tartaruga alata indiana

5065 ***Lissemys punctata andersoni***
Testudines - Trionychidae
f Trionyx à clapets de l'Inde du Nord

5066 ***Lissemys punctata granosa***
Testudines - Trionychidae
e Peninsular Mud Turtle (ISC); Peninsular Flapshell (ISC); Southern Flap-shelled Turtle (ISC)

5067 ***Lissemys punctata punctata***
Testudines - Trionychidae
e Indo-Gangetic Mud Turtle (ISC); Indo-Gangetic Flapshell (ISC); North Indian Flap-shelled Turtle (ISC)
f Trionyx à clapets de l'Inde du Sud

5068 ***Lissemys scutata***
Testudines - Trionychidae
e Burmese Flapshell Turtle; Burmese Flapshelled Turtle
d Burma-Klappenweichschildkröte
f Trionyx à clapets de Birmanie

5069 ***Lobulia***
Squamata - Scincidae
e Lobulias; Eared Skinks
d Neuguinea-Skinke

5070 ***Lobulia brongersmai***
Squamata - Scincidae
e Brongersma's Lobulia; Brongersma's Eared Skink

5071 ***Lobulia elegans***
Squamata - Scincidae
e Elegant Lobulia

5072 ***Lobulia phaeodes***
Squamata - Scincidae
e New Guinea Lobulia

5073 ***Lobulia stanleyana***
Squamata - Scincidae
e Stanley's Lobulia

5074 ***Lophocalotes***
Squamata - Agamidae
e Crested Lizards
d Kammeidechsen
f Lophocalotes

5075 ***Lophocalotes ludekingi***
Squamata - Agamidae
e Crested Lizard
d Kammeidechse
f Lophocalote

5076 ***Lophognathus***
Squamata - Agamidae
e Water Dragons; Australian Water Dragons
d Australische Wasserdrachen

5077 ***Lophognathus gilberti***
Squamata - Agamidae
e Gilbert's Water Dragon; Gilbert's Lashtail; Gilbert's Dragon (ANZ)
d Gilberts Wasserdrachen

5078 ***Lophognathus longirostris***
Squamata - Agamidae
e Australian Water Dragon; Long-nosed Water Dragon (ANZ); Long-snouted Lashtail (ANZ)

5079 ***Lophognathus temporalis***
Squamata - Agamidae
e Papua Water Dragon; Swamplands Lashtail (ANZ); Northern Water Dragon (ANZ); Southeast Asian Water Dragon

5080 ***Loveridgea***
Squamata - Amphisbaenidae
e Tanzania Worm Lizards

5081 ***Loveridgea ionidesi***
Squamata - Amphisbaenidae
e Liwale Worm Lizard

5082 ***Loveridgea phylophiniens***
Squamata - Amphisbaenidae
e Udjitiji Worm Lizard; Ujiji Worm Lizard

5083 ***Loveridgelaps***
Squamata - Elapidae
e Orange-banded Snakes

5084 ***Loveridgelaps elapsoides***
Squamata - Elapidae
e Orange-banded Snake

5085 ***Loxocemus***
Squamata - Boidae
e Mexican Burrowing Pythons; Mexican Pythons
d Spitzkopfphythone
f Pythons mexicains
i Serpenti scavatore del Messico

5086 ***Loxocemus bicolor***
Squamata - Boidae
e Mexican Burrowing Python; Neotropical Sunbeam Snake; Mexican Python; Mexican Burrowing Snake
d Spitzkopfphython
i Pitone del Nuevo Monde; Serpente scavatore del Messico

5087 ***Lubuya***
Squamata - Scincidae
e Iven's Skinks

5088 ***Lubuya ivensi***
Squamata - Scincidae
e Iven's Skink

5089 ***Luperosaurus***
Squamata - Gekkonidae
e Wolf Lizards; Wolf Geckos

5090 ***Luperosaurus brooksii***
Squamata - Gekkonidae
e Brooks's Wolf Lizard; Brooks's Wolf Gecko

5091 ***Luperosaurus browni***
Squamata - Gekkonidae
e Brown's Wolf Lizard; Brown's Wolf Gecko

5092 ***Luperosaurus cunningii***
Squamata - Gekkonidae
e Philippine Wolf Lizard; Philippine Wolf Gecko

5093 ***Luperosaurus joloensis***
Squamata - Gekkonidae
e Taylor's Wolf Lizard; Taylor's Wolf Gecko

5094 ***Luperosaurus macgregori***
Squamata - Gekkonidae
e MacGregor's Wolf Lizard; MacGregor's Wolf Gecko

5095 ***Luperosaurus palawanensis***
Squamata - Gekkonidae
e Palawan Wolf Lizard; Palawan Wolf Gecko

5096 ***Lycodon***
Squamata - Colubridae
e Wolf Snakes; Asian Wolf Snakes; Lycodontine Snakes
d Wolfszahnnattern
f Serpents-loup

5097 ***Lycodon albofuscus***
Squamata - Colubridae
e Brown Wolf Snake

5098 ***Lycodon alcalai***
Squamata - Colubridae
e Alcala's Wolf Snake

5099 ***Lycodon aulicus***
Squamata - Colubridae
e Indian Wolf Snake; House Snake; Indian Lycodontine Snake; Common Wolf Snake (ISC); Wolf Snake (ISC)
d Gewöhnliche Wolfszahnnatter

5100 ***Lycodon bibonius***
Squamata - Colubridae
e Ota's Wolf Snake

5101 ***Lycodon butleri***
Squamata - Colubridae
e Butler's Wolf Snake

5102 ***Lycodon chrysoprateros***
Squamata - Colubridae
e Ross's Wolf Snake

5103 ***Lycodon dumerili***
Squamata - Colubridae
e Duméril's Wolf Snake

5104 ***Lycodon fasciatus***
Squamata - Colubridae
e Banded Wolf Snake; Anderson's Wolf Snake; Yellow-banded Wolf Snake (ISC)

5105 ***Lycodon flavomaculatus***
Squamata - Colubridae
e Yellow-spotted Wolf Snake
d Gelbgepunkte Wolfszahnnatter

5106 ***Lycodon florensis***
Squamata - Colubridae
e Flores Wolf Snake

5107 ***Lycodon jara***
Squamata - Colubridae
e Twin-spotted Wolf Snake

5108 ***Lycodon kindui***
Squamata - Colubridae
e Burmese Wolf Snake

5109 ***Lycodon laoensis***
Squamata - Colubridae
e Laotian Wolf Snake; Indo-Chinese Wolf Snake

5110 ***Lycodon mackinnoni***
Squamata - Colubridae
e Mackinnon's Wolf Snake

5111 ***Lycodon muelleri***
Squamata - Colubridae
e Java Wolf Snake

5112 ***Lycodon osmanhilli***
Squamata - Colubridae
e Colombo Wolf Snake

5113 ***Lycodon paucifasciatus***
Squamata - Colubridae
e Rendahl's Wolf Snake

5114 ***Lycodon ruhstrati***
Squamata - Colubridae
e Formosa Wolf Snake

5115 ***Lycodon ruhstrati multifasciatus***
Squamata - Colubridae
e White-banded Wolf Snake; Formosa Wolf Snake

5116 ***Lycodon solivagus***
Squamata - Colubridae
e Common Wolf Snake

5117 ***Lycodon stormi***
Squamata - Colubridae
e Sulawesi Wolf Snake

5118 ***Lycodon striatus***
Squamata - Colubridae
e Northern Wolf Snake; Shaw's Wolf Snake (ISC)

5119 ***Lycodon subcinctus***
Squamata - Colubridae
e Malayan Banded Wolf Snake; Banded Wolf Snake

5120 ***Lycodon tesselatus***
Squamata - Colubridae
e Manila Wolf Snake

5121 ***Lycodon tiwari***
Squamata - Colubridae
e Andaman Wolf Snake

5122 ***Lycodon travancoricus***
Squamata - Colubridae
e Travancore Wolf Snake

5123 ***Lycodonomorphus***
Squamata - Colubridae
e African Water Snakes; Water Snakes (CSA)
d Sumpfnattern

5124 ***Lycodonomorphus bicolor***
Squamata - Colubridae
e Tanganyika Water Snake; Tanganyika White-bellied Water Snake; Fish-eating Water Snake

5125 ***Lycodonomorphus laevissimus***
Squamata - Colubridae
e Duskbelly Water Snake; Dusky-bellied Water Snake (CSA); Vatermeyer's Black Snake; Black Water Snake (CSA)

5126 ***Lycodonomorphus leleupi***
Squamata - Colubridae
e Mulanje Water Snake; Leleup's Water Snake (CSA)

5127 ***Lycodonomorphus rufulus***
Squamata - Colubridae
e Lichtenstein's Water Snake; Common Water Snake (CSA); Brown Water Snake (CSA); Olive-brown Water Snake; Black Water Snake
d Braune Sumpfnatter

5128 ***Lycodonomorphus sybtaeniatus***
Squamata - Colubridae
e Lined Water Snake

5129 ***Lycodonomorphus whytii***
Squamata - Colubridae
e Whyte's Water Snake (CSA); White-lipped Water Snake (CSA); Dark-bellied Water Snake

5130 ***Lycodryas***
Squamata - Colubridae

e Banded Tree Snakes
d Kurznattern

5131 ***Lycodryas arctifasciatus***
Squamata - Colubridae
e Banded Tree Snake

5132 ***Lycodryas betsileanus***
Squamata - Colubridae
e Betsileo Tree Snake

5133 ***Lycodryas gaimardi***
Squamata - Colubridae
e Rusty Tree Snake

5134 ***Lycodryas guentheri***
Squamata - Colubridae
e Günther's Tree Snake

5135 ***Lycodryas inornatus***
Squamata - Colubridae
e Dull Tree Snake

5136 ***Lycodryas maculatus***
Squamata - Colubridae
e St. Johann's Tree Snake

5137 ***Lycodryas sanctijohannis***
Squamata - Colubridae
e Orange Tree Snake

5138 ***Lycodryas seychellensis***
Squamata - Colubridae
e Seychelles Tree Snake

5139 ***Lycodryas variabilis***
Squamata - Colubridae
e Orange Tree Snake

5140 ***Lycognathophis***
Squamata - Colubridae
e Seychelles Wolf Snakes

5141 ***Lycognathophis seychellensis***
Squamata - Colubridae
e Seychelles Wolf Snake

5142 ***Lycophidion***
Squamata - Colubridae
e African Wolf Snakes; Wolf Snakes (CSA)
d Wolfsnattern; Afrikanische Wolfsnattern

5143 ***Lycophidion capense***
Squamata - Colubridae
e Cape Wolf Snake; Common Wolf Snake (CSA); Wolf Snake (CSA)
d Kap-Wolfsnatter

5144 ***Lycophidion capense jacksoni***
Squamata - Colubridae
e Jackson's Wolf Snake

5145 ***Lycophidion depressirostris***
Squamata - Colubridae
e Flat-snouted Wolf Snake

5146 ***Lycophidion irroratum***
Squamata - Colubridae
e Leach's Wolf Snake

5147 ***Lycophidion laterale***
Squamata - Colubridae
e Flat Wolf Snake

5148 ***Lycophidion meleagre***
Squamata - Colubridae
e Boulenger's Wolf Snake

5149 ***Lycophidion ornatum***
Squamata - Colubridae
e Ornate Wolf Snake

5150 ***Lycophidion polylepis***
Squamata - Colubridae
e Many-scaled Wolf Snake

5151 ***Lycophidion semiannule***
Squamata - Colubridae
e Mozambique Wolf Snake; Eastern Wolf Snake (CSA); Black-speckled Wolf Snake; Northern Wolf Snake; Steel Snake
d Östliche Wolfsnatter

5152 ***Lycophidion uzungwense***
Squamata - Colubridae
e Red-snouted Wolf Snake

5153 ***Lycophidion variegatum***
Squamata - Colubridae
e Variegated Wolf Snake
d Bunte Wolfsnatter

5154 ***Lygisaurus***
Squamata - Scincidae
e Four-fingered Skinks

5155 ***Lygisaurus aeratus***
Squamata - Scincidae
e Queensland Four-fingered Skink

5156 ***Lygisaurus burnetti***
Squamata - Scincidae
e Snake-eyed Four-fingered Skink; Snake-eyed Carlia

5157 ***Lygisaurus foliorum***
Squamata - Scincidae
e De Vis's Four-fingered Skink

5158 ***Lygisaurus laevis***
Squamata - Scincidae
e Oudeman's Four-fingered Skink

5159 ***Lygisaurus mcfarlani***
Squamata - Scincidae
e McFarlan's Four-fingered Skink

5160 ***Lygisaurus novaeguineae***
Squamata - Scincidae
e New Guinea Four-fingered Skink

5161 ***Lygisaurus rococo***
Squamata - Scincidae
e Rococo Four-fingered Skink

5162 ***Lygisaurus sesbrauna***
Squamata - Scincidae
e Cape York Four-fingered Skink

5163 ***Lygisaurus tanneri***
Squamata - Scincidae
e Tanner's Four-fingered Skink

5164 ***Lygisaurus timlowi***
Squamata - Scincidae
e Low's Four-fingered Skink

5165 ***Lygodactylus***
Squamata - Gekkonidae
e Dwarf Geckos
d Zerggeckos; Haftschwanzgeckos; Haftschwanzgeckonen
f Lygodactyles

5166 ***Lygodactylus angolensis***
Squamata - Gekkonidae
e Angola Dwarf Gecko

5167 ***Lygodactylus angularis***
Squamata - Gekkonidae
e Angled Dwarf Gecko; Yellowhead Daygecko

5168 ***Lygodactylus arnoulti***
Squamata - Gekkonidae
e Pasteur's Dwarf Gecko

5169 ***Lygodactylus bernardi***
Squamata - Gekkonidae
e Fitzsimons's Dwarf Gecko; Bernard's Dwarf Gecko (CSA); Inyangani Dwarf Gecko (CSA)

5170 ***Lygodactylus blanci***
Squamata - Gekkonidae
e Blanc's Dwarf Gecko

5171 ***Lygodactylus bradfieldi***
Squamata - Gekkonidae
e Bradfield's Dwarf Gecko; Kalahari Dwarf Gecko

5172 ***Lygodactylus capensis***
Squamata - Gekkonidae
e Cape Dwarf Gecko
d Kap-Zwerggecko; Gemeiner Zwerggecko
f Gecko du Cap

5173 ***Lygodactylus capensis capensis***
Squamata - Gekkonidae
e Common Dwarf Gecko (CSA)

5174 ***Lygodactylus chobiensis***
Squamata - Gekkonidae
e Okavango Dwarf Gecko; Chobe Dwarf Gecko (CSA)
d Haftschwanzgecko

5175 ***Lygodactylus conradti***
Squamata - Gekkonidae
e Matschie's Dwarf Gecko

5176 ***Lygodactylus conrani***
Squamata - Gekkonidae
e Cameroon Dwarf Gecko

5177 ***Lygodactylus decaryi***
Squamata - Gekkonidae
e Angel's Dwarf Gecko

5178 ***Lygodactylus depressus***
Squamata - Gekkonidae
e Zaire Dwarf Gecko

5179 ***Lygodactylus expectatus***
Squamata - Gekkonidae
e Ambilobe Dwarf Gecko; Hispaniolan Desert Gecko

5180 ***Lygodactylus fischeri***
Squamata - Gekkonidae
e Fischer's Dwarf Gecko

5181 ***Lygodactylus grandisonae***
Squamata - Gekkonidae
e Kenya Dwarf Gecko

5182 ***Lygodactylus gravis***
Squamata - Gekkonidae
e Usambara Dwarf Gecko

5183 ***Lygodactylus guibei***
Squamata - Gekkonidae
e Western Dwarf Gecko

5184 ***Lygodactylus gutturalis***
Squamata - Gekkonidae
e Uganda Dwarf Gecko

5185 ***Lygodactylus heterurus***
Squamata - Gekkonidae
e Boettger's Dwarf Gecko

5186 ***Lygodactylus inexpectatus***
Squamata - Gekkonidae
e Dar-es-Salaam Dwarf Gecko

5187 ***Lygodactylus insularis***
Squamata - Gekkonidae
e Insular Dwarf Gecko

5188 ***Lygodactylus keniensis***
Squamata - Gekkonidae
e Parker's Dwarf Gecko

5189 ***Lygodactylus klemmeri***
Squamata - Gekkonidae
e Klemmer's Dwarf Gecko; Neon Day Gecko

5190 ***Lygodactylus klugei***
Squamata - Gekkonidae
e Kluge's Dwarf Gecko

5191 ***Lygodactylus lawrencei***
Squamata - Gekkonidae
e Lawrence's Dwarf Gecko

5192 ***Lygodactylus luteopicturatus***
Squamata - Gekkonidae
e Zanzibar Dwarf Gecko

5193 ***Lygodactylus madagascariensis***
Squamata - Gekkonidae
e Madagascar Dwarf Gecko

5194 ***Lygodactylus manni***
Squamata - Gekkonidae
e Mann's Dwarf Gecko

5195 ***Lygodactylus methueni***
Squamata - Gekkonidae
e Woodbrush Dwarf Gecko; Methuen's Dwarf Gecko (CSA)

5196 ***Lygodactylus miops***
Squamata - Gekkonidae
e Gunther's Dwarf Gecko

5197 ***Lygodactylus montanus***
Squamata - Gekkonidae
e Mount Ivohibe Gecko

5198 ***Lygodactylus ocellatus***
Squamata - Gekkonidae
e Ocellated Dwarf Gecko; Spotted Dwarf Gecko (CSA)

5199 ***Lygodactylus ornatus***
Squamata - Gekkonidae
e Ornate Dwarf Gecko

5200 ***Lygodactylus picturatus***
Squamata - Gekkonidae
e Painted Dwarf Gecko; Yellow-headed Dwarf Gecko; White-headed Dwarf Gecko
d Gelbkopfhaftschwanzgecko; Gelbkopfzwerggecko

5201 ***Lygodactylus pictus***
Squamata - Gekkonidae
e Robust Dwarf Gecko

5202 ***Lygodactylus rarus***
Squamata - Gekkonidae
e Thin Dwarf Gecko

5203 ***Lygodactylus rex***
Squamata - Gekkonidae
e King Dwarf Gecko

5204 ***Lygodactylus scheffleri***
Squamata - Gekkonidae
e Scheffler's Dwarf Gecko

5205 ***Lygodactylus scorteccii***
Squamata - Gekkonidae
e Scortecci's Dwarf Gecko

5206 ***Lygodactylus septemtuberculatus***
Squamata - Gekkonidae
e Seven-bump Dwarf Gecko

5207 ***Lygodactylus somalicus***
Squamata - Gekkonidae
e Somali Dwarf Gecko

5208 ***Lygodactylus stevensoni***
Squamata - Gekkonidae
e Stevenson's Dwarf Gecko

5209 ***Lygodactylus thomensis***
Squamata - Gekkonidae
e Annobon Dwarf Gecko; Thomé Dwarf Gecko
d Thomé-Zwerggecko

5210 ***Lygodactylus tolampyae***
Squamata - Gekkonidae
e Grandidier's Dwarf Gecko

5211 ***Lygodactylus tuberosus***
Squamata - Gekkonidae

e Bumpy Dwarf Gecko

5212 ***Lygodactylus verticillatus***
Squamata - Gekkonidae
e Mocquard's Dwarf Gecko

5213 ***Lygodactylus wetzeli***
Squamata - Gekkonidae
e South American Dwarf Gecko

5214 ***Lygophis***
Squamata - Colubridae
e Tropical Snakes

5215 ***Lygophis flavifrenatus***
Squamata - Colubridae
e Cope's Tropical Snake

5216 ***Lygophis paucidens***
Squamata - Colubridae
e Goias Tropical Snake

5217 ***Lygophis tristriatus***
Squamata - Colubridae
e Three-striped Tropical Snake

5218 ***Lygosoma***
Squamata - Scincidae
e Writhing Skinks; Slender Skinks; Supple Skinks; Lygosomic Skinks; Sand Skinks
d Prachtskinke; Schlankskinke
f Ligosomes

5219 ***Lygosoma afer***
Squamata - Scincidae
e Peters's Writhing Skink; Mozambique Writhing Skink (CSA)
d Mozambique-Schlankskink; Moçambique-Schlankskink; Mozambik Schlankskink

5220 ***Lygosoma albopunctatum***
Squamata - Scincidae
e White-dotted Writhing Skink

5221 ***Lygosoma angeli***
Squamata - Scincidae
e Angel's Writhing Skink; Angel's Slender Skink

5222 ***Lygosoma anguininum***
Squamata - Scincidae
e Burmese Writhing Skink; Burmese Slender Skink
d Burma-Schlankskink

5223 ***Lygosoma bamfyldei***
Squamata - Scincidae
e Bamfylde's Writhing Skink

5224 ***Lygosoma bowringii***
Squamata - Scincidae
e Bowring's Writhing Skink; Bowring's Slender Skink

5225 ***Lygosoma brevicaudus***
Squamata - Scincidae
e Greer's Writhing Skink

5226 ***Lygosoma corpulentum***
Squamata - Scincidae
e Fat Skink

5227 ***Lygosoma fernandi***
Squamata - Scincidae
e Fire Skink; Fernando Po Fire Skink; Fernando Po Skink; Red-and-black Skink
d Fernand-Schlankskink; Fernand-Prachtskink
f Scinque à flancs rouges; Scinque rouge
i Scinco did fuoco

5228 ***Lygosoma frontoparietale***
Squamata - Scincidae
e Taylor's Writhing Skink

5229 ***Lygosoma grandisonianum***
Squamata - Scincidae

e Lanza's Writhing Skink

5230 ***Lygosoma guentheri***
Squamata - Scincidae
e Günther's Writhing Skink; Günther's Slender Skink

5231 ***Lygosoma guineense***
Squamata - Scincidae
e Guinea Writhing Skink; Guinea Slender Skink

5232 ***Lygosoma haroldyoungi***
Squamata - Scincidae
e Harold's Writhing Skink; Harold Young's Slender Skink

5233 ***Lygosoma herberti***
Squamata - Scincidae
e Herbert's Writhing Skink

5234 ***Lygosoma isodactylum***
Squamata - Scincidae
e Siamese Writhing Skink; Siamese Slender Skink
d Siam-Schlankskink

5235 ***Lygosoma koratense***
Squamata - Scincidae
e Korat Writhing Skink
d Khorat-Skink

5236 ***Lygosoma laeviceps***
Squamata - Scincidae
e Common Writhing Skink

5237 ***Lygosoma lanceolatum***
Squamata - Scincidae
e Broadley's Writhing Skink

5238 ***Lygosoma lineatum***
Squamata - Scincidae
e Lined Writhing Skink; Writhing Skink; Lined Slender Skink
f Skinque à bandes

5239 ***Lygosoma lineolatum***
Squamata - Scincidae
e Striped Writhing Skink; Striped Slender Skink

5240 ***Lygosoma mabuiiforme***
Squamata - Scincidae
e Mabuya-like Skink

5241 ***Lygosoma miopus***
Squamata - Scincidae
e African Horn Skink

5242 ***Lygosoma mocquardi***
Squamata - Scincidae
e Mocqauard's Writhing Skink

5243 ***Lygosoma opisthorhodum***
Squamata - Scincidae
e Sumatra Writhing Skink; Sumatran Slender Skink
d Sumatra-Schlankskink

5244 ***Lygosoma paedocarinatum***
Squamata - Scincidae
e Abyssinia Writhing Skink; Abyssinian Slender Skink

5245 ***Lygosoma pembanum***
Squamata - Scincidae
e Pemba Island Writhing Skink

5246 ***Lygosoma popae***
Squamata - Scincidae
e Pope's Writhing Skink; Pope's Lygosoma

5247 ***Lygosoma productum***
Squamata - Scincidae
e Boulenger's Writhing Skink

5248 ***Lygosoma punctatum***
Squamata - Scincidae
e Dotted Writhing Skink; Snake Skink (ISC); Dotted Slender Skink

5249 ***Lygosoma quadrupes***
Squamata - Scincidae
e Linnaeus's Writhing Skik

5250 ***Lygosoma simonettai***
Squamata - Scincidae
e Simonetta's Writhing Skink

5251 ***Lygosoma singhum***
Squamata - Scincidae
e Singh Writhing Skink

5252 ***Lygosoma somalicum***
Squamata - Scincidae
e Somali Writhing Skink; Somali Slender Skink

5253 ***Lygosoma sundevalli***
Squamata - Scincidae
e Sundevall's Writhing Skink; African Snake Skink
d Sundevall-Prachtskink; Sundevall-Schlankskink; Sundevalls Schlankskink
f Lygosome de Sundevall

5254 ***Lygosoma tanae***
Squamata - Scincidae
e Loveridge's Writhing Skink

5255 ***Lygosoma vinciguerra***
Squamata - Scincidae
e Vinciguerra's Writhing Skink; Vinciguerra's Slender Skink

5256 ***Lygosoma vosmaeri***
Squamata - Scincidae
e Vosmer's Writhing Skink; Vosmer's Slender Skink

5257 ***Lyriocephalus***
Squamata - Agamidae
e Lyrehead Lizards; Lyre-headed Lizards
d Lyrakopfagamen; Leierkopfagamen
f Lyriocéphales

5258 ***Lyriocephalus scutatus***
Squamata - Agamidae
e Lyrehead Lizard; Lyre-headed Lizard; Ceylonese Ballnose
d Lyrakopfagame; Stumpfnase; Leieragame; Leierkopfagame; Lyrakopf
f Lyriocéphale commun; Lyriocéphale perlé

5259 ***Lystrophis***
Squamata - Colubridae
e South American Hognose Snakes; South American Hognoses
d Südamerikanische Hakennattern

5260 ***Lystrophis dorbignyi***
Squamata - Colubridae
e South American Hognose Snake

5261 ***Lystrophis histricus***
Squamata - Colubridae
e Jan's Hognose Snake

5262 ***Lystrophis semicinctus***
Squamata - Colubridae
e Ringed Hognose Snake; Tri-coloured Hognose Snake

5263 ***Lytorhynchus***
Squamata - Colubridae
e Leafnose Snakes; Awl-headers; Awl-heded Snakes
d Schnauzennattern

5264 ***Lytorhynchus diadema***
Squamata - Colubridae
e Crowned Leafnose Snake; Diadem Sand Snake; Common Leaf-nosed Snake
d Gekrönte Schnauzennatter
f Couleuvre diadème; Lytorhynque diadème; Couleuvre fouisseuse à diadème

i Colubro diadema

5265 ***Lytorhynchus gaddi***
Squamata - Colubridae
e Diamond Snake

5266 ***Lytorhynchus gasperetti***
Squamata - Colubridae
e Keviton's Leafnose Snake

5267 ***Lytorhynchus kennedyi***
Squamata - Colubridae
e Kennedy's Leafnose Snake

5268 ***Lytorhynchus maynardi***
Squamata - Colubridae
e Maynard's Leafnose Snake; Maynard's Awl-headed Snake

5269 ***Lytorhynchus paradoxus***
Squamata - Colubridae
e Sind Leafnose Snake; Sind Awl-headed Snake

5270 ***Lytorhynchus ridgewayi***
Squamata - Colubridae
e Derfshi Snake; Afghan Awl-headed Snake

M

5271 ***Mabuya***
Squamata - Scincidae
e Mabuyas; Typical Skinks (CSA); Skinks
d Mabujen; Mabuyen
f Mabuyas; Mabuias

5272 ***Mabuya acutilabris***
Squamata - Scincidae
e Sharp-lipped Mabuya; Wedge-snouted Skink (CSA)
d Keilschnauzenmabuya

5273 ***Mabuya affinis***
Squamata - Scincidae
e Senegal Mabuya

5274 ***Mabuya albilabris***
Squamata - Scincidae
e Guinea Mabuya

5275 ***Mabuya andamanensis***
Squamata - Scincidae
e Andaman Mabuya
d Andaman-Mabuje; Andaman-Mabuya

5276 ***Mabuya arjara***
Squamata - Scincidae
e Arajara Mabuya

5277 ***Mabuya aurata***
Squamata - Scincidae
e Golden Grass Mabuya; Golden Grass Skink
d Östlicher Goldskink; Goldmabuje; Goldstreifenskink; Goldene Mabuje; Goldene Mabuya; Goldfarbige Mabuya; Goldfarbige Mabuje; Goldmabuya
f Mabuya doré
i Mabuia dorata

5278 ***Mabuya aureopunctata***
Squamata - Scincidae
e Gold-spotted Mabuya

5279 ***Mabuya bayonii***
Squamata - Scincidae
e Bayon's Mabuya; Bayon's Skink

5280 ***Mabuya beddomii***
Squamata - Scincidae
e Beddome's Mabuya; Beddome's Skink

5281 ***Mabuya bensoni***
Squamata - Scincidae
e Benson's Mabuya; Benson's Skink
d Bensons Skink

5282 ***Mabuya betsileana***
Squamata - Scincidae
e Betsileo Mabuya; Madagascan Mabuya

5283 ***Mabuya bibroni***
Squamata - Scincidae
e Bibron's Mabuya; Sand Skink (ISC); Bibron's Skink
d Bibrons Skink
f Mabuya de Bibron

5284 ***Mabuya binotata***
Squamata - Scincidae
e Bocage's Mabuya; Ovambo Tree Skink (CSA)

5285 ***Mabuya bistriata***
Squamata - Scincidae
e Two-striped Mabuya; Two-lined Skink

5286 ***Mabuya boettgeri***
Squamata - Scincidae
e Boettger's Mabuya; Boettger's Skink
d Boettgers Skink

5287 ***Mabuya boulengeri***
Squamata - Scincidae
e Boulenger's Mabuya; Boulenger's Skink (CSA)
d Boulengers Skink

5288 ***Mabuya brauni***
Squamata - Scincidae
e Braun's Mabuya

5289 ***Mabuya brevicollis***
Squamata - Scincidae
e Sudan Mabuya; Short-necked Skink

5290 ***Mabuya breviparietalis***
Squamata - Scincidae
e Chabanaud's Mabuya

5291 ***Mabuya buttneri***
Squamata - Scincidae
e Buttner's Mabuya

5292 ***Mabuya caissara***
Squamata - Scincidae
e Caissara Mabuya

5293 ***Mabuya capensis***
Squamata - Scincidae
e Cape Three-striped Mabuya; Cape Skink (CSA); Cape Three-striped Skink (CSA); Three-lined Skink (CSA)
d Kap-Mabuya
f Scinque du Cap

5294 ***Mabuya carinata***
Squamata - Scincidae
e Keeled Indian Mabuya; Common Skink (ISC); Brahminy Skink (ISC)
d Glanzskink; Kielmabuje; Kielmabuya

5295 ***Mabuya carvalhoi***
Squamata - Scincidae
e Carvalho's Mabuya

5296 ***Mabuya chimbana***
Squamata - Scincidae
e Chimban Mabuya; Chimba Skink (CSA)

5297 ***Mabuya clivicola***
Squamata - Scincidae
e Inger's Mabuya

5298 ***Mabuya croizati***
Squamata - Scincidae
e Horton's Mabuya

5299 ***Mabuya cunningi***
Squamata - Scincidae
e Cunning's Mabuya

5300 ***Mabuya dissimilis***
Squamata - Scincidae
e Striped Grass Mabuya; Striped Grass Skink (ISC)

5301 ***Mabuya dorsivittata***
Squamata - Scincidae
e Paraguay Mabuya

5302 ***Mabuya elegans***
Squamata - Scincidae
e Elegant Mabuya

5303 ***Mabuya engelei***
Squamata - Scincidae
e Engel's Mabuya

5304 ***Mabuya ferrarai***
Squamata - Scincidae
e Ferrara's Mabuya

5305 ***Mabuya ficta***
Squamata - Scincidae
e Rebouch's Mabuya

5306 ***Mabuya frenata***
Squamata - Scincidae
e Cope's Mabuya

5307 ***Mabuya gravenhorstii***
Squamata - Scincidae
e Gravenhorst's Mabuya

5308 ***Mabuya guaporicola***
Squamata - Scincidae
e Dunn's Mabuya

5309 ***Mabuya heathi***
Squamata - Scincidae
e Brazilian Mabuya

5310 ***Mabuya hemmingi***
Squamata - Scincidae
e Somali Mabuya

5311 ***Mabuya hildenbrandtii***
Squamata - Scincidae
e Hildenbrandt's Mabuya

5312 ***Mabuya hoeschi***
Squamata - Scincidae
e Hoesch's Mabuya; Hoesch's Skink (CSA)

5313 ***Mabuya homalocephala***
Squamata - Scincidae
e Red-Sided Skink; Smooth-headed Skink (CSA); Cape Speckled Skink

5314 ***Mabuya homalocephala depressa***
Squamata - Scincidae
e Black-flanked Skink (CSA)

5315 ***Mabuya indeprensa***
Squamata - Scincidae
e Brown's Mabuya

5316 ***Mabuya innovata***
Squamata - Scincidae
e Blanford's Mabuya

5317 ***Mabuya irregularis***
Squamata - Scincidae
e Alpine Meadow Mabuya; Alpine Meadow Skink

5318 ***Mabuya lacertiformis***
Squamata - Scincidae
e Rock Mabuya; Bronze Rock Skink (CSA); Lacertiform Rock Skink (CSA)

5319 ***Mabuya lineolata***
Squamata - Scincidae
e Lined Mabuya

5320 ***Mabuya longicauda***
Squamata - Scincidae
e Longtail Mabuya; Long-tailed Skink

5321 ***Mabuya macleani***
Squamata - Scincidae
e Carrot Rock Skink

5322 ***Mabuya macrorhyncha***
Squamata - Scincidae
e Hoge's Mabuya

5323 ***Mabuya macularia***
Squamata - Scincidae
e Bronze Mabuya; Little Skink (ISC); Bronze Grass Skink; Little Indian Skink
d Gefleckte Mabuje; Fleckenmabuje; Gefleckte Mabuya; Fleckenmabuya

5324 ***Mabuya maculata***
Squamata - Scincidae
e Spotted Mabuya

5325 ***Mabuya maculilabris***
Squamata - Scincidae
e Speckle-lipped Mabuya; Speckle-lipped Skink (CSA)

5326 ***Mabuya madagascariensis***
Squamata - Scincidae
e Malagasy Mabuya

5327 ***Mabuya megalura***
Squamata - Scincidae
e Grass-top Mabuya; Grass-top Skink (CSA)

5328 ***Mabuya multifasciata***
Squamata - Scincidae
e East Indian Brown Mabuya; East Indian Brown-sided Skink; Indonesian Skink
d Vielstreifenmabuje; Vielstreifenmabuya
f Mabuya multibande

5329 ***Mabuya multifasciata balinensis***
Squamata - Scincidae
e Bali Skink

5330 ***Mabuya nagarjuni***
Squamata - Scincidae
e Sharma's Mabuya

5331 ***Mabuya nigropalmata***
Squamata - Scincidae
e Black Mabuya; Black-handed Skink

5332 ***Mabuya novemcarinata***
Squamata - Scincidae
e Andersson's Mabuya

5333 ***Mabuya occidentalis***
Squamata - Scincidae
e Western Three-striped Mabuya; Western Three-striped Skink (CSA)

5334 ***Mabuya perrotetii***
Squamata - Scincidae
e Teita Mabuya; Perrotet's Skink; Perrotet's Mabuya; Red-sided Skink
f Mabuya de Perrotet

5335 ***Mabuya polytropis***
Squamata - Scincidae
e Tropical Mabuya

5336 ***Mabuya praesigne***
Squamata - Scincidae
e Early Mabuya

5337 ***Mabuya quadrivittata***
Squamata - Scincidae
e Beautiful Mabuya

5338 ***Mabuya quinquetaeniata***
Squamata - Scincidae
e Rainbow Mabuya; Five-lined Skink (CSA); Rainbow Skink (CSA); Bean Skink
d Blauschwanzskink; Fünfstreifenmabuje; Blauschwanzmabuje; Fünfstreifenmabuya; Blauschwanzmabuya
f Scinque à queue bleue d'Afrique

5339 ***Mabuya quinquetaeniata margaritifer***
Squamata - Scincidae
e Blue-tailed Kopje Skink (CSA)

5340 ***Mabuya rodenburgi***
Squamata - Scincidae
e Rodenburg's Mabuya

5341 ***Mabuya rudis***
Squamata - Scincidae
e Rough Mabuya

5342 ***Mabuya seychellensis***
Squamata - Scincidae
e Seychelles Mabuya
f Mabuya des Seychelles

5343 ***Mabuya spilogaster***
Squamata - Scincidae
e Spiny Mabuya; Kalahari Tree Skink (CSA)

5344 ***Mabuya stanjorgeri***
Squamata - Scincidae
e Stanjorger's Mabuya

5345 ***Mabuya striata***
Squamata - Scincidae
e African Striped Mabuya; Striped Mabuya; Striped Skink (CSA); Common Striped Skink (CSA); African Striped Skink; Eastern Striped Skink (CSA)
d Gestreifte Mabuje; Streifenskink; Hausskink; Zweistreifenskink
f Scinque répandu

5346 ***Mabuya striata punctatissimus***
Squamata - Scincidae
e Striped Skink

5347 ***Mabuya striata striata***
Squamata - Scincidae
e Striped Skink

5348 ***Mabuya sulcata***
Squamata - Scincidae
e Kopje Mabuya; Western Rock Skink (CSA)

5349 ***Mabuya trivittata***
Squamata - Scincidae
e Three-banded Mabuya

5350 ***Mabuya tytleri***
Squamata - Scincidae
e Tytler's Mabuya

5351 ***Mabuya unimarginata***
Squamata - Scincidae
e Central American Mabuya; Central American Skink

5352 ***Mabuya vaillanti***
Squamata - Scincidae
e Vaillant's Mabuya
d Vaillant-Skink

5353 ***Mabuya varia***
Squamata - Scincidae
e Savanna Variable Mabuya; Variable Skink (CSA); Savannah Variable Mabuya

5354 ***Mabuya variegata***
Squamata - Scincidae
e Variegated Mabuya; Variegated Skink (CSA)

5355 ***Mabuya variegata punctulata***
Squamata - Scincidae
e Long-lobed Skink

5356 ***Mabuya vato***
Squamata - Scincidae
e Boulder Mabuya

5357 ***Mabuya vittata***
Squamata - Scincidae
e Bridled Mabuya; Bridled Skink
d Streifenmabuje; Streifenmabuya
f Mabuya à bandes; Mabuya d'Olivier
i Mabuia vittata

5358 ***Mabuya wrightii***
Squamata - Scincidae
e Wright's Mabuya

5359 ***Macrelaps***
Squamata - Colubridae
e Natal Black Snakes
d Natal-Schwarzschlangen

5360 ***Macrelaps microlepidotus***
Squamata - Colubridae
e Natal Black Snake
d Natal-Schwarzschlange

5361 ***Macrocalamus***
Squamata - Colubridae
e Mountain Reed Snakes

5362 ***Macrocalamus jasoni***
Squamata - Colubridae
e Jason's Mountain Reed Snake

5363 ***Macrocalamus lateralis***
Squamata - Colubridae
e Malayan Mountain Reed Snake

5364 ***Macrocalamus tweediei***
Squamata - Colubridae
e Tweedie's Mountain Reed Snake

5365 ***Macroclemys***
Testudines - Chelydridae
e Alligator Snapping Turtles; Alligator Snappers
d Geierschildkröten
f Macrolémydes; Tortues-alligator

5366 ***Macroclemys temminckii***
Testudines - Chelydridae
e Alligator Snapping Turtle; Alligator Snapper; Alligator Turtle (NA)
d Geierschildkröte
f Tortue-alligator; Macrolémyde de Temminck; Tortue vorace
i Tartaruga alligatore

5367 ***Macrolepis***
Squamata - Cordylidae
e Snake-Lizards

5368 ***Macrolepis miopropus***
Squamata - Cordylidae
e Northern Rhodesia Snake-Lizard

5369 ***Macrophistodon***
Squamata - Colubridae
e Varied Keelbacks; Keelbacks

5370 ***Macrophistodon flaviceps***
Squamata - Colubridae
e Orangeneck Keelback; Malayan Keelback; Orange-necked Keelback

5371 ***Macrophistodon plumbicolor***
Squamata - Colubridae
e Lead Keelback; Common Indian Whip Snake; Green Keelback (ISC)

5372 ***Macrophistodon rhodomelas***
Squamata - Colubridae
e Blueneck Keelback; Boie's Keelback; Blue-necked Keelback

5373 ***Macrophistodon rudis***
Squamata - Colubridae
e Red Keelback

5374 ***Macropholidus***
Squamata - Teiidae
e Macropholiduses

5375 ***Macropholidus ruthveni***
Squamata - Teiidae
e Ruthven's Macropholidus

5376 ***Macroprotodon***
Squamata - Colubridae
e False Smooth Snakes
d Kaputzennattern
f Couleuvres à capuchon

5377 ***Macroprotodon cucullatus***
Squamata - Colubridae
e False Smooth Snake; Hooded Smooth Snake; False Hooded Snake; Cowled Snake; Hooded Snake
d Kaputzennatter
f Couleuvre à capuchon
i Colubro dal cappucio; Colubro cucullato

5378 ***Macroprotodon cucullatus textilis***
Squamata - Colubridae
f Couleuvre à capuchon commune

Squamata - Scincidae
e Cape Verde Giant Skinks
d Kapverdische Riesenskinke
f Macroscinques

5380 ***Macroscincus cocteaui***
Squamata - Scincidae
e Cape Verde Giant Skink
d Kapverdischer Riesenskink
f Macroscinque de Cocteau

5381 ***Madagascarophis***
Squamata - Colubridae
e Moila Snakes

5382 ***Madagascarophis colubrina***
Squamata - Colubridae
e Madagascar Common Snake

5383 ***Malaclemys***
Testudines - Emydidae
e Diamondback Terrapins; Diamond-backed Terrapins
d Diamantschildkröten
f Tortues diamantines; Tortues diamantées; Tortues d'eau saumâtre; Malaclemmydes

5384 ***Malaclemys terrapin***
Testudines - Emydidae
e Diamondback Terrapin; Diamond-backed Terrapins
d Diamantschildkröte
f Tortue diamantine; Tortue diamantée; Tortue à dos diamanté; Malaclemmyde; Tortue d'eau saumâtre; Malaclemmyde concentrique
i Testuggine dal dorso di diamante

5385 ***Malaclemys terrapin centrata***
Testudines - Emydidae
e Carolina Diamondback Terrapin; Diamondback Terrapin (NA); Carolina Diamond-backed Terrapin

5386 ***Malaclemys terrapin littoralis***
Testudines - Emydidae
e Texas Diamondback Terrapin; Texas Diamond-backed Terrapin

5387 ***Malaclemys terrapin macrospilota***
Testudines - Emydidae
e Ornate Diamondback Terrapin; Ornate Diamond-backed Terrapin

5388 ***Malaclemys terrapin pileata***
Testudines - Emydidae
e Mississippi Diamondback Terrapin; Mississippi Diamond-backed Terrapin

5389 ***Malaclemys terrapin rhizophorarum***
Testudines - Emydidae
e Mangrove Diamondback Terrapin; Mangrove Diamond-backed Terrapin

5390 ***Malaclemys terrapin tequesta***
Testudines - Emydidae
e Florida East Coast Terrapin

5391 ***Malaclemys terrapin terrapin***
Testudines - Emydidae
e Northern Diamondback Terrapin; Northern Diamond-backed Terrapin

5392 ***Malacochersus***
Testudines - Testudinidae
e Pancake Tortoises
d Spaltenschildkröten
f Tortues de Tornier
i Malacochersi

5393 ***Malacochersus tornieri***
Testudines - Testudinidae
e African Pancake Tortoise; Pancake Tortoise; Soft-shelled Tortoise
d Ostafrikanische Spaltenschildkröte; Spaltenschildkröte
f Tortue de Tornier
i Malacocherso; Testuggine foccacia africana

5394 ***Malayemys***
Testudines - Emydidae
e Malayan Snail-eating Turtles
d Malayen-Sumpfschildkröten
f Tortues malayses; Malayémides

5395 ***Malayemys subtrijuga***
Testudines - Emydidae
e Malayan Snail-eating Turtle; Snail-eating Turtle
d Malayen-Sumpfschildkröte
f Tortue malays à trois carènes; Malayémide à trois arêtes; Tortue malayse mangeuse d'escargots
i Testuggine d'acqua malese; Tartaruga malese

5396 ***Malpolon***
Squamata - Colubridae
e Montpellier Snakes
d Eidechsennattern
f Couleuvres de Montpellier

5397 ***Malpolon moilensis***
Squamata - Colubridae
e Moila Snake
d Arabische Eidechsennatter; Moila-Natter
f Couleuvre de Moila

5398 ***Malpolon monspessulanus***
Squamata - Colubridae
e Montpellier Snake
d Eidechsennatter; Europäische Eidechsennatter; Rieseneidechsennatter
f Couleuvre de Montpellier; Couleuvre maillée; Colubre; Ser
i Colubro lacertino; Colubro di Montpellier; Malpolo di Montpellier

5399 ***Manolepis***
Squamata - Colubridae
e Ridgehead Snakes

5400 ***Manolepis putnami***
Squamata - Colubridae
e Ridgehead Snake

5401 ***Manouria***
Testudines - Testudinidae
e Indo-Chinese Tortoises
d Hinterindische Landschildkröten

5402 ***Manouria emys***
Testudines - Testudinidae
e Asian Brown Tortoise; Eastern Hill Tortoise (ISC); Yellow-and-black Giant Tortoise (ISC); Assam Land Tortoise; Burmese Brown Tortoise; Six-footed Tortoise; Brown Tortoise; Mountain Tortoise
d Braune Landschildkröte
f Tortue brune de Birmanie
i Tartaruga bruna birmana

5403 ***Manouria impressa***
Testudines - Testudinidae
e Impressed Tortoise
d Hinterindische Landschildkröte
f Tortue imprimée
i Testuggine impressa

5404 ***Masticophis***
Squamata - Colubridae
e Whip Snakes; Coachwhips; Australian Whip Snakes
d Peitschennattern; Zornnattern; Kutscherpeitschennattern
f Serpents-fouet; Couleuvres à fouet

5405 ***Masticophis anthonyi***
Squamata - Colubridae
e Clarion Island Whip Snake

5406 ***Masticophis aurigulus***
Squamata - Colubridae
e Baja California Striped Whip Snake; Cape Whip Snake (NA)

5407 ***Masticophis barbouri***
Squamata - Colubridae

e Espiritu Santo Island Striped Whip Snake

5408 ***Masticophis bilineatus***
Squamata - Colubridae
e Sonoran Whip Snake; Sonora Whip Snake; Sonora Racer
d Sonora-Zornnatter
f Serpent-fouet du Sonora

5409 ***Masticophis bilineatus bilineatus***
Squamata - Colubridae
e Sonoran Mountain Whip Snake; Sonoran Whip Snake

5410 ***Masticophis bilineatus lineolatus***
Squamata - Colubridae
e Ajo Mountain Whip Snake

5411 ***Masticophis flagellum***
Squamata - Colubridae
e Coachwhip; Coachwhip Snake; San Joaquin Whip Snake
d Kutscherpeitschennatter; Peitschenschlange; Peitschennatter
f Serpent-fouet flagellé

5412 ***Masticophis flagellum cingulum***
Squamata - Colubridae
e Sonoran Coachwhip

5413 ***Masticophis flagellum flagellum***
Squamata - Colubridae
e Eastern Coachwhip
d Östliche Kutscherpeitschennatter

5414 ***Masticophis flagellum fuliginosus***
Squamata - Colubridae
e Baja California Coachwhip; Baja Coachwhip

5415 ***Masticophis flagellum lineatulus***
Squamata - Colubridae
e Lined Coachwhip

5416 ***Masticophis flagellum piceus***
Squamata - Colubridae
e Red Coachwhip; Red Racer
d Rote Kutscherpeitschennatter

5417 ***Masticophis flagellum ruddocki***
Squamata - Colubridae
e San Joaquin Coachwhip

5418 ***Masticophis flagellum testaceus***
Squamata - Colubridae
e Western Coachwhip; Central Coachwhip
d Westliche Kutscherpeitschennatter

5419 ***Masticophis lateralis***
Squamata - Colubridae
e Striped Racer (NA); Chaparral Snake; California Whip Snake; California Striped Whip Snake; California Striped Racer
d Streifenzornnatter
f Serpent-fouet agile

5420 ***Masticophis lateralis euryxanthus***
Squamata - Colubridae
e Alameda Striped Racer; Alameda Whip Snake
d Alameda-Peitschenschlange

5421 ***Masticophis lateralis lateralis***
Squamata - Colubridae
e Chaparral Whip Snake; California Striped Whip Snake; California Striped Racer

5422 ***Masticophis mentovarius***
Squamata - Colubridae
e Neotropical Whip Snake

5423 ***Masticophis pulchericeps***
Squamata - Colubridae
e Cope's Whip Snake

5424 ***Masticophis schotti***
Squamata - Colubridae
e Schott's Whip Snake

5425 ***Masticophis schotti ruthvenii***
Squamata - Colubridae
e Ruthven's Whip Snake

5426 ***Masticophis schotti schotti***
Squamata - Colubridae
e Schott's Whip Snake

5427 ***Masticophis slevini***
Squamata - Colubridae
e San Esteban Island Whip Snake

5428 ***Masticophis stigodryas***
Squamata - Colubridae
e Amaral's Whip Snake

5429 ***Masticophis taeniatus***
Squamata - Colubridae
e Striped Whip Snake; Striped Racer
d Bandnatter
f Serpent-fouet ligné

5430 ***Masticophis taeniatus girardi***
Squamata - Colubridae
e Central Texas Whip Snake

5431 ***Masticophis taeniatus taeniatus***
Squamata - Colubridae
e Desert Striped Whip Snake; Desert Striped Racer

5432 ***Mastigodryas***
Squamata - Colubridae
e Tropical Racers; Racers
d Panthernattern

5433 ***Mastigodryas amarali***
Squamata - Colubridae
e Amaral's Tropical Racer

5434 ***Mastigodryas bifossatus***
Squamata - Colubridae
e Rio Tropical Racer
d Panthernatter

5435 ***Mastigodryas boddaerti***
Squamata - Colubridae
e Boddaert's Tropical Racer

5436 ***Mastigodryas boddaerti boddaerti***
Squamata - Colubridae
e Tan Racer

5437 ***Mastigodryas bruesi***
Squamata - Colubridae
e Barbour's Tropical Racer; Windward Tree Racer (WI)

5438 ***Mastigodryas danieli***
Squamata - Colubridae
e Daniel's Tropical Racer

5439 ***Mastigodryas heathii***
Squamata - Colubridae
e Heath's Tropical Racer

5440 ***Mastigodryas melanolomus***
Squamata - Colubridae
e Dark Tropical Racer; Salmon-bellied Racer

5441 ***Mastigodryas pleei***
Squamata - Colubridae
e Plee's Tropical Racer

5442 ***Mastigodryas pulchriceps***
Squamata - Colubridae
e Cope's Tropical Racer

5443 ***Mastigodryas sanguiventris***
Squamata - Colubridae
e Taylor's Tropical Racer

5444 ***Maticora***
Squamata - Elapidae
e Malaysian Coral Snakes; Long-glanded Coral Snakes
d Bauchdrüsenottern

5445 ***Maticora beddomei***
Squamata - Elapidae
e Beddome's Malaysian Coral Snake

5446 ***Maticora bivirgata***
Squamata - Elapidae
e Blue Malaysian Coral Snake
d Rotbäuchige Bauchdrüsenotter; Indoaustralische Bauchdrüsenotter; Bauchdrüsenotter; Blaue Korallenotter

5447 ***Maticora intestinalis***
Squamata - Elapidae
e Banded Malaysian Coral Snake
d Rotschwanzbauchdrüsenotter

5448 ***Maticora maculiceps***
Squamata - Elapidae
e Günther's Malaysian Coral Snake

5449 ***Maticora nigrescens***
Squamata - Elapidae
e Black Malaysian Coral Snake

5450 ***Mauremys***
Testudines - Emydidae
e Stripeneck Turtles; Eurasian Freshwater Turtles; Pond Turtles
d Eurasische Wasserschildkröten; Eurasiatische Wasserschildkröten
f Émydes

5451 ***Mauremys annamensis***
Testudines - Emydidae
e Annam Leaf Turtle; Annam Pond Turtle; Vietnamese Leaf Turtle
d Annam-Sumpfschildkröte; Annam-Schildkröte
f Émyde de l'Annam
i Tartaruga foglia del Vietnam

5452 ***Mauremys caspica***
Testudines - Emydidae
e Caspian Turtle; Stripe-necked Terrapin; Caspian Terrapin; Stripe-necked Pond Turtle; Caspian Pond Turtle
d Kaspische Bachschildkröte; Kaspische Wasserschildkröte; Kaspische Sumpfschildkröte
f Émyde caspienne; Cistude
i Mauremide caspica; Clemmide

5453 ***Mauremys caspica caspica***
Testudines - Emydidae
e Eastern Caspian Turtle
f Émyde caspienne orientale

5454 ***Mauremys caspica cretica***
Testudines - Emydidae
e Cretan Caspian Turtle
f Émyde de Crète

5455 ***Mauremys caspica rivulata***
Testudines - Emydidae
e Western Caspian Turtle
f Émyde caspienne occidentale

5456 ***Mauremys iversoni***
Testudines - Emydidae
e Fujian Pond Turtle

5457 ***Mauremys japonica***
Testudines - Emydidae
e Japanese Pond Turtle; Japanese Turtle; Japanese Water Tortoise
d Japanische Wasserschildkröte
f Émyde du Japon
i Testuggine del Giappone

5458 ***Mauremys leprosa***
Testudines - Emydidae
e Mediterranean Turtle; Mediterranean Terrapin; Spanish Terrapin; Spanish

Turtle; Leprous Terrapin; Stripe-necked Terrapin; Spanish Pond Turtle; Mediterranean Pond Turtle; Leprous Pond Turtle

d Maurische Sumpfschildkröte; Mittelmeer-Sumpfschildkröte; Spanische Wasserschildkröte; Spanische Sumpfschildkröte; Nordafrikanische Sumpfschildkröte

f Émyde lépreuse; Tortue d'Espagne; Clemmyde lépreuse; Cistude de Maurétanie

i Clemmide iberico; Mauremide leprosa; Mauremide occidentale; Clemmide; Testuggine palustre leprosa spagnola

5459 ***Mauremys mutica***

Testudines - Emydidae

e Yellow Pond Turtle; Asian Yellow Pond Turtle

d Dreikielwasserschildkröte

f Émyde mutique

i Testuggine palustre asiatica gialla

5460 ***Mediodactylus***

Squamata - Gekkonidae

e Middle-toed Geckos

d Nacktfingergeckos

5461 ***Mediodactylus amictopholis***

Squamata - Gekkonidae

e Israeli Middle-toed Gecko

5462 ***Mediodactylus brevipes***

Squamata - Gekkonidae

e Iranian Middle-toed Gecko

5463 ***Mediodactylus heterocercus***

Squamata - Gekkonidae

e Blanford's Middle-toed Gecko

5464 ***Mediodactylus kotschyi***

Squamata - Gekkonidae

e Kotchy's Gecko; Naked-fingered Gecko; Crimean Gecko; European Naked-fingered Gecko

d Europäischer Nacktfingergecko; Ägäischer Bogenfingergecko; Ägäischer Nacktfingergecko; Ägäischer Nacktfinger

f Gymnodactyle de Kotchyi; Cyrtopodion de Kotschy; Cyrtopodion de la Crimée; Gecko à doigts nus; Cyrtopodion

i Geco di Kotschy; Gimnodattilo

5465 ***Mediodactylus russowii***

Squamata - Gekkonidae

e Caucasus Middle-toed Gecko; Caucasian Gecko

d Transkaspischer Bogenfingergecko; Transkaspischer Nacktfingergecko

f Cyrtopodion transcaspien

i Geco di Russow

5466 ***Mediodactylus sagittifer***

Squamata - Gekkonidae

e Nikolsky's Middle-toed Gecko; Jaz Maurian Bent-toed Gecko

5467 ***Mediodactylus spinicauda***

Squamata - Gekkonidae

e Thorny Middle-toed Gecko

5468 ***Mehelya***

Squamata - Colubridae

e File Snakes

d Feilennattern; Feilenschlangen

5469 ***Mehelya capensis***

Squamata - Colubridae

e Cape File Snake; File Snake (CSA); Three-cornered Snake

d Gewöhnliche Feilennatter; Feilennatter; Kap-Feilennatter

5470 ***Mehelya crossi***

Squamata - Colubridae

e African File Snake

5471 ***Mehelya egbensis***

Squamata - Colubridae

e Dunger's File Snake

5472 ***Mehelya guirali***
Squamata - Colubridae
e Mocquard's File Snake

5473 ***Mehelya nyassae***
Squamata - Colubridae
e Nyasa File Snake; Black File Snake (CSA); Northern File Snake (CSA); Northern Three-cornered Snake
d Schwarze Feilenschlange

5474 ***Mehelya poensis***
Squamata - Colubridae
e Western Forest File Snake; Purple File Snake

5475 ***Mehelya riggenbachi***
Squamata - Colubridae
e Riggenbach's File Snake

5476 ***Mehelya stenophtalmus***
Squamata - Colubridae
e Small-eyed File Snake

5477 ***Mehelya vernayi***
Squamata - Colubridae
e Angola File Snake
d Angola-Feilennatter

5478 ***Meizodon***
Squamata - Colubridae
e Crowned Snakes; African Smooth Snakes (CSA)
d Buschschlangen

5479 ***Meizodon coronatus***
Squamata - Colubridae
e Western Crowned Snake

5480 ***Meizodon semiornatus***
Squamata - Colubridae
e Semiornate Snake

5481 ***Melanochelys***
Testudines - Emydidae
e Indian Black Turtles
d Vorderindische Erdschildkröten; Dreikielerdschildkröten

5482 ***Melanochelys tricarinata***
Testudines - Emydidae
e Tricarinate Hill Turtle; Eastern Hill Terapin (ISC); Three-keeled Tortoise; Terrestial Turtle; Burmese Swamp Turtle; Burmese Peacock Turtle; Three-keeled Asian Turtle
d Dreikielschildkröte; Dreikielerdschildkröte
f Tortue à trois carènes; Émyde indienne à trois carènes; Tortue de Birmanie; Tortue tricarénée

5483 ***Melanochelys trijuga***
Testudines - Emydidae
e Indian Black Turtle; Ceylon Terrapin; Asiatic Geoemyda; Indian Pond Terrapin (ISC); Black Turtle (ISC); Madras Pond Tortoise
d Schwarzbaucherdschildkröte
f Géoémyde d'Asie; Émyde noire à trois bandes; Émyde noire à trois arrètes

5484 ***Melanochelys trijuga coronata***
Testudines - Emydidae
e Cochin Black Turtle
f Émyde noire de Cochin

5485 ***Melanochelys trijuga edeniana***
Testudines - Emydidae
e Burma Black Turtle
f Émyde noire de Birmanie

5486 ***Melanochelys trijuga indopeninsularis***
Testudines - Emydidae
e Bengal lack Turtle
f Émyde noire du Bengale

5487 ***Melanochelys trijuga parkeri***
Testudines - Emydidae
e Parker's Black Turtle
f Émyde noire de Parker

5488 ***Melanochelys trijuga thermalis***
Testudines - Emydidae
e Sri Lanka Black Turtle
f Émyde noire du Sri Lanka

5489 ***Melanochelys trijuga trijuga***
Testudines - Emydidae
e Peninsular Black Turtle
f Éyde noire péninsulaire

5490 ***Melanochelys trijuga wiroti***
Testudines - Emydidae
e Indian Snail-eating Turtle
f Émyde noire de Thaïlande

5491 ***Melanophidium***
Squamata - Uropeltidae
e Black Earth Snakes

5492 ***Melanophidium bilineatum***
Squamata - Uropeltidae
e Two-lined Black Earth Snake

5493 ***Melanophidium punctatum***
Squamata - Uropeltidae
e Beddome's Black Earth Snake

5494 ***Melanophidium wynaudensis***
Squamata - Uropeltidae
e Indian Black Earth Snake

5495 ***Melanoseps***
Squamata - Scincidae
e Limbless Skinks

5496 ***Melanoseps ater***
Squamata - Scincidae
e Longtail Limbless Skink; Eastern Legless Skink; Long-tailed Limbless Skink

5497 ***Melanoseps longicauda***
Squamata - Scincidae
e Pangani Black Limbless Skink

5498 ***Melanoseps occidentalis***
Squamata - Scincidae
e Western Limbless Skink; Western Legless Skink

5499 ***Melanosuchus***
Crocodylia - Alligatoridae
e Black Caimans
d Mohrenkaimane
f Caïmans noirs

5500 ***Melanosuchus niger***
Crocodylia - Alligatoridae
e Black Caiman; Large Caiman; Jacare-assu
d Mohrenkaiman
f Caïman noir; Caïman blanc; Caïman noir de Guyane
i Caimano nero

5501 ***Menetia***
Squamata - Scincidae
e Menetias

5502 ***Menetia alanae***
Squamata - Scincidae
e Coastal Menetia

5503 ***Menetia amaura***
Squamata - Scincidae
e Shark Bay Menetia

5504 ***Menetia concinna***
Squamata - Scincidae
e Silver Menetia

5505 ***Menetia greyii***
Squamata - Scincidae
e Grey-brown Menetia; Grey's Menetia

5506 ***Menetia koshlandae***
Squamata - Scincidae
e Koshland Menetia

5507 ***Menetia maini***
Squamata - Scincidae
e Metallic Menetia

5508 ***Menetia sadlieri***
Squamata - Scincidae
e Sadlier's Menetia

5509 ***Menetia surda***
Squamata - Scincidae
e Inland Menetia

5510 ***Meroles***
Squamata - Lacertidae
e South African Sand Lizards; Desert Lizards (CSA); Sand Lizards
d Scharr-Eidechsen

5511 ***Meroles ctenodactylus***
Squamata - Lacertidae
e Smith's Sand Lizard; Smith's Desert Lizard (CSA)

5512 ***Meroles cuneirostris***
Squamata - Lacertidae
e Wedge-snouted Sand Lizard; Wedge-snouted Desert Lizard (CSA); Wedge-snouted Lizard
d Wüsteneidechse; Scharr-Eidechse

5513 ***Meroles knoxii***
Squamata - Lacertidae
e Knox's Ocellated Sand Lizard; Knox's Desert Lizard (CSA)
d Knox-Scharreidechse

5514 ***Meroles micropholidotus***
Squamata - Lacertidae
e Small-scaled Sand Lizard; Small-scaled Desert Lizard (CSA)

5515 ***Meroles reticulatus***
Squamata - Lacertidae
e Reticulate Sand Lizard; Reticulated Desert Lizard (CSA)

5516 ***Meroles suborbitalis***
Squamata - Lacertidae
e Spotted Sand Lizard; Spotted Desert Lizard (CSA)

5517 ***Mesalina***
Squamata - Lacertidae
e Sand Lizards; Mesalinas

5518 ***Mesalina adramitana***
Squamata - Lacertidae
e Hadramawt Sand Lizard; Arabian Mesalina

5519 ***Mesalina ayunensis***
Squamata - Lacertidae
e Arnold's Sand Lizard

5520 ***Mesalina guttulata***
Squamata - Lacertidae
e Small-spotted Sand Lizard; Small-spotted Lizard; Long-tailed Desert Lacerta
f Lézard à goutteletes; Érémias à goutelettes

5521 ***Mesalina olivieri***
Squamata - Lacertidae
e Oliviers Sand Lizard; Olivier's Small Lizard
f Érémias d'Olivier

5522 ***Mesalina pasteuri***
Squamata - Lacertidae
e Pasteur's Lizard; Pasteur's Small Lizard
f Érémias de Pasteur

5523 ***Mesalina regeli***
Squamata - Lacertidae
e Tajik Racerunner

5524 ***Mesalina rubropunctata***
Squamata - Lacertidae
e Red-spotted Lizard; Red-spotted Small Lizard
f Érémias à points rouges

5525 ***Mesaspis***
Squamata - Anguidae
e Mexican Alligator Lizards
d Hochland-Krokodilschleichen

5526 ***Mesaspis antauges***
Squamata - Anguidae
e Mount Orizaba Alligator Lizard

5527 ***Mesaspis gadovii***
Squamata - Anguidae
e Gadow's Alligator Lizard

5528 ***Mesaspis juarezi***
Squamata - Anguidae
e Sierra Juarez Alligator Lizard

5529 ***Mesaspis monticolus***
Squamata - Anguidae
e Cope's Alligator Lizard

5530 ***Mesaspis moreletii***
Squamata - Anguidae
e Morelet's Alligator Lizard; Highland Alligator Lizard

5531 ***Mesaspis viridiflava***
Squamata - Anguidae
e Dwarf Alligator Lizard

5532 ***Mesobaena***
Squamata - Amphisbaenidae
e Inirida Worm Lizards; Dwarf Amphisbaenas

5533 ***Mesobaena huebneri***
Squamata - Amphisbaenidae
e Inirida Worm Lizard; Dwarf Amphisbaena

5534 ***Micreablepharus***
Squamata - Teiidae
e Tinyfoot Teiids

5535 ***Micreablepharus dunni***
Squamata - Teiidae
e Dunn's Tinyfoot Teiid

5536 ***Micreablepharus maximiliani***
Squamata - Teiidae
e Maximilian's Tinyfoot Teiid

5537 ***Micrelaps***
Squamata - Colubridae
e Two-headed Snakes

5538 ***Micrelaps bicoloratus***
Squamata - Colubridae
e Kenya Two-headed Snake

5539 ***Micrelaps boettgeri***
Squamata - Colubridae
e Boettger's Two-headed Snake

5540 ***Micrelaps muelleri***
Squamata - Colubridae
e Müller's Two-headed Snake
d Erdviper; Müllers Erdviper
f Vipère de Mueller

5541 ***Micrelaps vaillanti***
Squamata - Colubridae
e Somali Two-headed Snake

5542 ***Microgecko***
Squamata - Gekkonidae
e Tiny Geckos

5543 ***Microgecko helenae***
Squamata - Gekkonidae
e Tiny Helena Gecko

5544 ***Microgecko latifi***
Squamata - Gekkonidae

e Tiny Zagros Gecko

5545 ***Microgecko persicus***
Squamata - Gekkonidae
e Tiny Persia Gecko

5546 ***Microlophus***
Squamata - Iguanaidae
e Pacific Iguanas

5547 ***Microlophus atacamensis***
Squamata - Iguanaidae
e Atacamen Pacific Iguana; Donoso's Lava Lizard

5548 ***Microlophus delalonis***
Squamata - Iguanaidae
e Hood Island Lava Lizard

5549 ***Microlophus grayi***
Squamata - Iguanaidae
e Gray's Pacific Iguana; Gray's Lava Lizard

5550 ***Microlophus occipitalis***
Squamata - Iguanaidae
e Knobbed Pacific Iguana; Brush Lava Lizard

5551 ***Microlophus pacificus***
Squamata - Iguanaidae
e Comon Pacific Iguana

5552 ***Microlophus peruvianus***
Squamata - Iguanaidae
e Peru Pacific Iguana; Peru Lava Lizard

5553 ***Microlophus quadrivittatus***
Squamata - Iguanaidae
e Four-banded Pacific Iguana

5554 ***Microlophus stolzmanni***
Squamata - Iguanaidae
e Stolzmann's Pacific Iguana

5555 ***Microlophus tarapacensis***
Squamata - Iguanaidae
e Tarapaca Pacific Iguana

5556 ***Microlophus theresiae***
Squamata - Iguanaidae
e Theresia's Pacific Iguana

5557 ***Microlophus theresioides***
Squamata - Iguanaidae
e Small Pacific Iguana

5558 ***Microlophus thoracicus***
Squamata - Iguanaidae
e Tschudi Pacific Iguana

5559 ***Microlophus tigris***
Squamata - Iguanaidae
e Tiger Pacific Iguana

5560 ***Microphistodon***
Squamata - Colubridae
e Flathead Snakes

5561 ***Microphistodon ochraceus***
Squamata - Colubridae
e Flathead Snake

5562 ***Micropochis***
Squamata - Elapidae
e Pacific Coral Snakes

5563 ***Micropochis ikaheka***
Squamata - Elapidae
e Pacific Coral Snake; New Guinea Small-eyed Snake

5564 ***Microscalabotes***
Squamata - Gekkonidae
e Tiny Scaled Geckos

5565 ***Microscalabotes bivittis***
Squamata - Gekkonidae
e Tiny Scaled Gecko

5566 ***Micruroides***
Squamata - Elapidae
e Western Coral Snakes (NA); Arizona Coral Snakes; Sonoran Coral Snakes
d Arizona-Korallenottern; Arizona-Korallenschlangen; Korallenschlangen
f Serpents-corail de l'Arizona

5567 ***Micruroides euryxanthus***
Squamata - Elapidae
e Western Coral Snake; Sonoran Coral Snake; Sonora Coral Snake; Arizona Coral Snake
d Arizona-Korallenschlange; Arizona-Korallenotter
f Serpent-corail de l'Arizona

5568 ***Micruroides euryxanthus euryxanthus***
Squamata - Elapidae
e Arizona Coral Snake
d Arizona-Korallenotter; Arizona-Korallenschlange
f Serpent-corail de l'Arizona

5569 ***Micrurus***
Squamata - Elapidae
e Common Coral Snakes; Coral Snakes; Eastern Coral Snakes; American Coral Snakes
d Echte Korallenottern; Korallenschlangen; Echte Korallenschlangen
f Serpents-corail

5570 ***Micrurus alleni***
Squamata - Elapidae
e Allen's Coral Snake; Black-headed Coral Snake

5571 ***Micrurus ancoralis***
Squamata - Elapidae
e Anchor Coral Snake

5572 ***Micrurus annelatus***
Squamata - Elapidae
e Annellated Coral Snake

5573 ***Micrurus anomalis***
Squamata - Elapidae
e Regal Coral Snake

5574 ***Micrurus averyi***
Squamata - Elapidae
e Black-headed Coral Snake; Avery's Coral Snake

5575 ***Micrurus bernardi***
Squamata - Elapidae
e Blotched Coral Snake; Bernard's Coral Snake

5576 ***Micrurus bocourti***
Squamata - Elapidae
e Ecuador Coral Snake; Bocourt's Coral Snake; San Gil Coral Snake

5577 ***Micrurus bogerti***
Squamata - Elapidae
e Bogert's Coral Snake

5578 ***Micrurus browni***
Squamata - Elapidae
e Brown's Coral Snake; Sierra Madre Coral Snake

5579 ***Micrurus catamayensis***
Squamata - Elapidae
e Catamayo Coral Snake

5580 ***Micrurus clarki***
Squamata - Elapidae
e Clark's Coral Snake

5581 ***Micrurus collaris***
Squamata - Elapidae
e Guyana Blackback Coral Snake
d Schwarzkopfkorallenschlange; Schwarzkopfkorallenotter
f Serpent-corail

5582 ***Micrurus corallinus***
Squamata - Elapidae
e Painted Coral Snake
d Gewöhnliche Korallenschlange
f Serpent-corail commun

5583 ***Micrurus decoratus***
Squamata - Elapidae
e Brazilian Coral Snake; Common Snake; Cobra Common Snake

5584 ***Micrurus diastema***
Squamata - Elapidae
e Variable Coral Snake

5585 ***Micrurus dissoleucus***
Squamata - Elapidae
e Pygmy Coral Snake; Dunn's Coral Snake

5586 ***Micrurus distans***
Squamata - Elapidae
e West Mexican Coral Snake; Broad-banded Coral Snake

5587 ***Micrurus dumerili***
Squamata - Elapidae
e Duméril's Coral Snake; Rough-tailed Coral Snake

5588 ***Micrurus elegans***
Squamata - Elapidae
e Elegant Coral Snake

5589 ***Micrurus ephippifer***
Squamata - Elapidae
e Oaxacan Coral Snake

5590 ***Micrurus filiformis***
Squamata - Elapidae
e Slender Coral Snake; Thread-form Coral Snake

5591 ***Micrurus frontalis***
Squamata - Elapidae
e Coral Snake (NA); Arizona Coral Snake; Southern Coral Snake; Coral Cobra
d Kobra-Korallenschlange; Kobra-Korallennatter; Brasilianische Korallenotter

5592 ***Micrurus frontifasciatus***
Squamata - Elapidae
e Bolivian Coral Snake

5593 ***Micrurus fulvius***
Squamata - Elapidae
e Eastern Coral Snake; Florida Coral Snake (NA); Thunder Snake (NA); Harlequin Snake; Bead Snake; Harlequin Coral Snake
d Harlekin-Korallennatter; Harlekin-Korallenschlange
f Serpent-corail de l'Est; Serpent-arlequin

5594 ***Micrurus fulvius fulvius***
Squamata - Elapidae
e Eastern Coral Snake

5595 ***Micrurus hemprichii***
Squamata - Elapidae
e Hemprich's Coral Snake
f Serpent corail

5596 ***Micrurus hemprichii ortonii***
Squamata - Elapidae
e Orange-ringed Coral Snake

5597 ***Micrurus hippocrepis***
Squamata - Elapidae
e Maya Coral Snake; Horseshoe Coral Snake

5598 ***Micrurus ibiboboca***
Squamata - Elapidae
e Caatinga Coral Snake; Merrem's Coral Snake

5599 ***Micrurus isozonus***
Squamata - Elapidae
e Venezuela Coral Snake; Amazonian Coral Snake

5600 ***Micrurus karlschmidti***
Squamata - Elapidae
e Pygmy Blackback Coral Snake

5601 ***Micrurus langsdorffi***
Squamata - Elapidae
e Langsdorff's Coral Snake

5602 ***Micrurus laticollaris***
Squamata - Elapidae
e Balsan Coral Snake

5603 ***Micrurus latifasciatus***
Squamata - Elapidae
e Broad-ringed Coral Snake; Long-banded Coral Snake

5604 ***Micrurus lemniscatus***
Squamata - Elapidae
e South American Coral Snake; Large Coral Snake; Tropical Coral Snake
d Ibiboca

5605 ***Micrurus limbatus***
Squamata - Elapidae
e Tuxtlan Coral Snake

5606 ***Micrurus margaritiferus***
Squamata - Elapidae
e Speckled Coral Snake; Roze's Coral Snake

5607 ***Micrurus mertensi***
Squamata - Elapidae
e Mertens's Coral Snake

5608 ***Micrurus mipartitus***
Squamata - Elapidae
e Redtail Coral Snake; Black-ringed Coral Snake
d Costa-Rica-Korallenschlange

5609 ***Micrurus multifasciatus***
Squamata - Elapidae
e Many-banded Coral Snake

5610 ***Micrurus multiscutatus***
Squamata - Elapidae
e Cauca Coral Snake

5611 ***Micrurus narduccii***
Squamata - Elapidae
e Andean Blackback Coral Snake; Andean Black-backed Coral Snake

5612 ***Micrurus nebularis***
Squamata - Elapidae
e Cloud Forest Coral Snake

5613 ***Micrurus nigricinctus***
Squamata - Elapidae
e Central American Coral Snake; Central America Coral Snake; Black-banded Coral Snake

5614 ***Micrurus peruvianus***
Squamata - Elapidae
e Peru Coral Snake; Peruvian Coral Snake

5615 ***Micrurus petersi***
Squamata - Elapidae
e Peters's Coral Snake

5616 ***Micrurus proximans***
Squamata - Elapidae
e Nayarit Coral Snake

5617 ***Micrurus psyches***
Squamata - Elapidae
e Carib Coral Snake; Small Coral Snake
f Serpent corail

5618 ***Micrurus putamayensis***
Squamata - Elapidae
e Putamayo Coral Snake; Putamayan Coral Snake; Sooty Coral Snake

5619 ***Micrurus rondonianus***
Squamata - Elapidae
e Rondo Coral Snake

5620 ***Micrurus ruatanus***
Squamata - Elapidae
e Roatan Coral Snake

5621 ***Micrurus sangilensis***
Squamata - Elapidae
e Santander Coral Snake

5622 ***Micrurus spixii***
Squamata - Elapidae
e Amazon Coral Snake; Wagler's Coral Snake; Spix's Coral Snake; Amazonian Coral Snake
d Riesenkorallenotter

5623 ***Micrurus spurelli***
Squamata - Elapidae
e Colombian Coral Snake; Spurell's Coral Snake

5624 ***Micrurus steindachneri***
Squamata - Elapidae
e Steindachner's Coral Snake

5625 ***Micrurus stewarti***
Squamata - Elapidae
e Panamanian Coral Snake

5626 ***Micrurus stuarti***
Squamata - Elapidae
e Stuart's Coral Snake

5627 ***Micrurus surinamensis***
Squamata - Elapidae
e Aquatic Coral Snake; Surinam Coral Snake
f Serpent corail

5628 ***Micrurus tener***
Squamata - Elapidae
e Texas Coral Snake
d Texas-Korallenschlange

5629 ***Micrurus tener tener***
Squamata - Elapidae
e Texas Coral Snake

5630 ***Micrurus tschudii***
Squamata - Elapidae
e Desert Coral Snake; Bolivian Coral Snake; Tschudi's Coral Snake

5631 ***Mictopholis***
Squamata - Agamidae
e Annandale's Dragons

5632 ***Mictopholis austeniana***
Squamata - Agamidae
e Annandale's Dragon

5633 ***Millotisaurus***
Squamata - Gekkonidae
e Marvellous Geckos

5634 ***Millotisaurus mirabilis***
Squamata - Gekkonidae
e Marvellous Gecko

5635 ***Mimophis***
Squamata - Colubridae
e Big-eyed Snakes

5636 ***Mimophis mahfalensis***
Squamata - Colubridae
e Big-eyed Snake; Common Big-eyed Snake

5637 ***Moloch***
Squamata - Agamidae
e Thorny Devils; Molochs; Thorny

Devil Lizards
d Dornteufel; Wüstenteufel; Dornteufelchen
f Molochs

5638 ***Moloch horridus***
Squamata - Agamidae
e Thorny Devil; Moloch; Thorny Devil Lizard
d Dornteufel; Wüstenteufel; Moloch
f Diable épineux; Moloch d'Australie; Moloch épineux; Moloch hérissé; Diable cornu; Moloch

5639 ***Monopeltis***
Squamata - Amphisbaenidae
e Wedge-snouted Worm Lizards; Spade-snouted Worm Lizards (CSA)

5640 ***Monopeltis adercae***
Squamata - Amphisbaenidae
e Lualuba Worm Lizard

5641 ***Monopeltis anchieate***
Squamata - Amphisbaenidae
e Angola Worm Lizard; Angolan Spade-snouted Worm Lizard (CSA); Anchieta's Wedge-snouted Amphisbaenian (CSA)

5642 ***Monopeltis capensis***
Squamata - Amphisbaenidae
e Cape Wedge-snouted Worm Lizard; Cape Spade-snouted Worm Lizard (CSA); Cape Wedge-snouted Amphisbaenian (CSA); Cape Worm Lizard (CSA)
d Afrikanische Doppelschleiche

5643 ***Monopeltis decosteri***
Squamata - Amphisbaenidae
e Delagoa Bay Worm Lizard

5644 ***Monopeltis galeata***
Squamata - Amphisbaenidae
e Helmeted Worm Lizard

5645 ***Monopeltis guentheri***
Squamata - Amphisbaenidae
e Western Congo Worm Lizard

5646 ***Monopeltis habenichti***
Squamata - Amphisbaenidae
e Mahinga Worm Lizard

5647 ***Monopeltis jugularis***
Squamata - Amphisbaenidae
e Gaboon Worm Lizard

5648 ***Monopeltis kabindae***
Squamata - Amphisbaenidae
e Kabinda Worm Lizard

5649 ***Monopeltis leonhardi***
Squamata - Amphisbaenidae
e Kalahari Worm Lizard; Leonhard's Spade-snouted Worm Lizard (CSA); Kalahari Wedge-snouted Amphisbaenian (CSA)

5650 ***Monopeltis mauricei***
Squamata - Amphisbaenidae
e Maurice Worm Lizard; Slender Wedge-snouted Amphisbaenian (CSA)

5651 ***Monopeltis ocularis***
Squamata - Amphisbaenidae
e Bechuanaland Worm Lizard

5652 ***Monopeltis remaclei***
Squamata - Amphisbaenidae
e De Witte's Worm Lizard

5653 ***Monopeltis scalper***
Squamata - Amphisbaenidae
e Carved Worm Lizard

5654 ***Monopeltis schoutedeni***
Squamata - Amphisbaenidae
e Middle Congo Worm Lizard

5655 ***Monopeltis sphenorhynchus***
Squamata - Amphisbaenidae
e Slender Wedge-snouted Worm Lizard; Slender Spade-snouted Worm Lizard (CSA); Slender Wedge-snouted Amphisbaenian (CSA); Wedge-snouted Worm Lizard (CSA)

5656 ***Monopeltis vanderysti***
Squamata - Amphisbaenidae
e Vanderyst Worm Lizard

5657 ***Monopeltis zambezensis***
Squamata - Amphisbaenidae
e Zambezi Spade-snouted Worm Lizard

5658 ***Morelia***
Squamata - Boidae
e Carpet Pythons; Diamond Pythons; Australian Diamond Pythons
d Rautenpythone; Teppichschlangen; Teppichpythone; Teppichpythonschlangen; Rautenschlangen

5659 ***Morelia amethistina***
Squamata - Boidae
e Scrub Python; Amethystine Python (ANZ); Queensland Rock Python; Merauke Scrub Python
d Amethystpython; Amethystschlange
f Liasis améthyste; Python des rochers
i Pitone ametistino

5660 ***Morelia boeleni***
Squamata - Boidae
e Boelen's Python; Boelen's Water Python
d Boelens Python; Schwarzer Python

5661 ***Morelia bredli***
Squamata - Boidae
e Blotched Python; Centralian Python
d Inland-Teppichpython
i Morelia tappeto centroaustralico

5662 ***Morelia carinata***
Squamata - Boidae
e Rough-scaled Python; Rough-scaled Carpet Python
d Rauschuppiger Python; Rauschuppige Teppichschlange; Felsenpython

5663 ***Morelia oenpelliensis***
Squamata - Boidae
e Oenpelli Python; South African Water Python; Oenpelli Rock Python
d Oenpelli-Python

5664 ***Morelia spilota***
Squamata - Boidae
e Carpet Python; Diamond Python; Australian Python
d Teppichpython; Rautenpython; Australischer Python; Teppichschlange; Rautenschlange
f Python tapis; Serpent diamant
i Morelia tappeto; Pitone diamantino; Pitone australiano

5665 ***Morelia spilota cheynei***
Squamata - Boidae
e Jungle Carpet Python
d Dschungel-Rautenpython

5666 ***Morelia spilota imbricata***
Squamata - Boidae
e Southwestern Carpet Python(ANZ)

5667 ***Morelia spilota mcdowelli***
Squamata - Boidae
e Coastal Carpet Python
d Küstenrautenpython

5668 ***Morelia spilota spilota***
Squamata - Boidae
e Inland Carpet Python; Diamond Python
d Rautenpython; Diamantpython

5669 ***Morelia spilota variegata***
Squamata - Boidae
e Australian Carpet Python; Northwestern Carpet Python (ANZ); Carpet Python
d Australischer Teppichpython; Teppichpython
f Python tapis

5670 ***Morelia viridis***
Squamata - Boidae
e Green Tree Python; Green Python (ANZ); Aru Green Python
d Grüner Baumpython; Chondropython
f Python vert; Python vert d'Indonésie
i Morelia verde

5671 ***Morenia***
Testudines - Emydidae
e Eyed Turtles
d Pfauenaugensumpfschildkröten; Pfauenaugesumpfschildkröten
f Émydes à ocelles

5672 ***Morenia ocellata***
Testudines - Emydidae
e Burmese Eyed Turtle; Burmese Swamp Turtle
d Hinterindische Pfauenaugensumpfschildkröte ; Hinterindische Pfauenaugesumpfschildkröte
f Émyde à ocelles de Birmanie
i Tartaruga ocellata birmana

5673 ***Morenia petersi***
Testudines - Emydidae
e Indian Eyed Turtle
d Indische Pfauenaugensumpfschildkröte; Vorderindische Pfauenaugensumpfschildkröte; Indische Pfauenaugesumpfschildkröte; Vorderindische Pfauenaugesumpfschildkröte
f Émyde à ocelles de Bengale

5674 ***Morethia***
Squamata - Scincidae
e Morethias

5675 ***Morethia adalaidensis***
Squamata - Scincidae
e Adelaide Morethia; Samphire Skink (ANZ)

5676 ***Morethia boulengeri***
Squamata - Scincidae
e Boulenger's Morethia; Boulenger's Skink (ANZ)

5677 ***Morethia butleri***
Squamata - Scincidae
e Butler's Morethia

5678 ***Morethia lineoocellata***
Squamata - Scincidae
e Line-eyed Morethia; Line-eyed Skink; Eyed Pete
f Scinque rayé-et-ocellé

5679 ***Morethia obscura***
Squamata - Scincidae
e Obscure Morethia; Obscure Skink (ANZ)

5680 ***Morethia ruficauda***
Squamata - Scincidae
e Redtail Morethia

5681 ***Morethia storri***
Squamata - Scincidae
e Storr's Morethia

N

5688 *Nactus*
Squamata - Gekkonidae
e Sea Geckos

5689 *Nactus coindermirensis*
Squamata - Gekkonidae
e Mauritius Accidental Gecko

5690 *Nactus galgajuga*
Squamata - Gekkonidae
e Trevathan Range Gecko

5691 *Nactus pelagicus*
Squamata - Gekkonidae
e Pelagic Gecko; Pelagic Rock Gecko

5692 *Nactus serpeninsula*
Squamata - Gekkonidae
e Serpent Island Gecko
d Serpent-Insel-Gecko

5693 *Nactus vankampeni*
Squamata - Gekkonidae
e Vankampen's Gecko

5694 *Naja*
Squamata - Elapidae
e Cobras; True Cobras
d Kobras; Echte Kobras; Königshutschlangen; Hutschlangen
f Cobras

5695 *Naja annulifera*
Squamata - Elapidae
e Snouted Cobra

5696 *Naja atra*
Squamata - Elapidae
e Chinese Cobra
d Chinesische Kobra; Indochinesische Kobra
i Cobra cinese

5697 *Naja haje*
Squamata - Elapidae
e Egyptian Cobra; Asp; South African Cobra; Banded Cobra
d Uräusschlange
f Cobra égyptien; Cobra d'Égypte; Cobra d'Afrique du Nord

5698 *Naja kaouthia*
Squamata - Elapidae
e Monocled Cobra; Thai Monoculate Cobra; Monocellate Cobra (ISC)
d Monokelkobra; Thai-Kobra
i Cobra monocolo

5699 *Naja katiensis*
Squamata - Elapidae
e Mali Cobra; West African Brown Spitting Cobra
d Mali-Kobra

5700 *Naja melanoleuca*
Squamata - Elapidae
e Forest Cobra (CSA); Black-and-white Cobra; Black-lipped Cobra
d Schwarzweiße Kobra; Schwarzweiße Hutschlange
f Cobra noir-et-blanc; Cobra de forêt; Cobra des forêts

5701 *Naja mossambica*
Squamata - Elapidae
e Mozambique Spitting Cobra; M'fezi (CSA); Mossambic Cobra
d Moçambique-Speikobra; Mozambique-Speikobra; Mozambik-Speikobra

5702 ***Naja naja***
Squamata - Elapidae
e Indian Cobra; Biocellate Cobra (ISC); Asiatic Cobra; Indian Spectacled Cobra; Common Cobra (ISC); Asian Cobra; Spectacled Cobra (ISC)
d Brillenschlange; Kobra
f Cobra à lunettes; Cobra indien; Naja; Cobra d'Afrique du Nord

5703 ***Naja naja naja***
Squamata - Elapidae
d Vorderindische Kobra

5704 ***Naja naja polyocellata***
Squamata - Elapidae
d Gepunktete Ceylon-Kobra

5705 ***Naja nigricollis***
Squamata - Elapidae
e Black-necked Cobra; Blackneck Spitting Cobra; Black-necked Spitting Cobra; Pink Spitting Cobra; Red Spitting Cobra
d Afrikanische Speikobra; Speikobra; Schwarzhalskobra
f Cobra à cou noir; Cobra cracheur; Cobra

5706 ***Naja nigricollis nigricincta***
Squamata - Elapidae
e Western Barred Spitting Cobra; Zebra Snake; Zebra Spitting Cobra
d Speikobra

5707 ***Naja nigricollis nigricollis***
Squamata - Elapidae
e Black-necked Spitting Cobra
d Speikobra

5708 ***Naja nigricollis woodi***
Squamata - Elapidae
e Black Spitting Cobra; South-western Spitting Cobra
d Speikobra

5709 ***Naja nivea***
Squamata - Elapidae
e Cape Cobra; Cape Yellow Cobra; Yellow Cobra
d Kap-Kobra
f Cobra du Cap

5710 ***Naja nubiae***
Squamata - Elapidae
e Nubian Spitting Cobra
d Nübische Speikobra

5711 ***Naja oxiana***
Squamata - Elapidae
e Central Asia Cobra; Central Asian Cobra; Trans-Caspian Cobra; Oxus Cobra; Black Cobra (ISC); Acellate Cobra (ISC)
d Mittelasiatische Kobra; Russische Kobra
f Naja

5712 ***Naja pallida***
Squamata - Elapidae
e African Cobra; Red Spitting Cobra
d Rote Speikobra
i Cobra sputatore rosso

5713 ***Naja philippinensis***
Squamata - Elapidae
e Philippine Cobra
d Philippinen-Kobra

5714 ***Naja samarensis***
Squamata - Elapidae
e Peters's Cobra
d Samar Kobra

5715 ***Naja siamensis***
Squamata - Elapidae
e Indo-Chinese Spitting Cobra
d Siamesische Kobra; Thai-Speikobra

5716 ***Naja sputatrix***
Squamata - Elapidae

e Indonesian Cobra; Indonesian Spitting Cobra
d Javanische Speikobra; Indonesische Speikobra; Malayische Kobra; Sunda-Speikobra

5717 ***Naja sumatrana***
Squamata - Elapidae
e Golden Spitting Cobra; Equatorial Spitting Cobra; Sumatran Cobra
d Goldene Speikobra; Speikobra

5718 ***Nannoscincus***
Squamata - Scincidae
e Elf Skinks

5719 ***Nannoscincus gracilis***
Squamata - Scincidae
e Slender Elf Skink

5720 ***Nannoscincus greeri***
Squamata - Scincidae
e Greer's Elf Skink

5721 ***Nannoscincus maccoyi***
Squamata - Scincidae
e McCoy's Elf Skink; McCoy's Skink (ANZ)

5722 ***Nannoscincus mariae***
Squamata - Scincidae
e Maria's Elf Skink

5723 ***Nannoscincus rankini***
Squamata - Scincidae
e Rankin's Elf Skink

5724 ***Nannoscincus slevini***
Squamata - Scincidae
e Slevin's Elf Skink

5725 ***Narudasia***
Squamata - Gekkonidae
e Festive Geckos

5726 ***Narudasia festiva***
Squamata - Gekkonidae
e Festive Gecko

5727 ***Natator***
Testudines - Cheloniidae
e Flatback Turtles; Flatback Sea Turtles
d Australische Flachrückenschildkröten
f Chélonées à dos plats

5728 ***Natator depressa***
Testudines - Cheloniidae
e Flatback Turtle; Flatback Sea Turtle; Flatback Green Turtle
d Australische Suppenschildkröte; Australische Flachrückenschildkröte
f Chélonée à dos plat; Chélone à dos plat
i Tartaruga franca del Pacifico

5729 ***Natriceteres***
Squamata - Colubridae
e Marsh Snakes
d Sumpfnattern

5730 ***Natriceteres fuliginoides***
Squamata - Colubridae
e Marsh Snake

5731 ***Natriceteres olivacea***
Squamata - Colubridae
e Olive Marsh Snake (CSA); Banded Marsh Snake
d Olivfarbige Sumpfnatter; Afrikanische Olive Sumpfnatter

5732 ***Natriceteres variegata***
Squamata - Colubridae
e Forest Marsh Snake

5733 ***Natrix***
Squamata - Colubridae
e Grass Snakes; Eurasian Water Snakes; Common Water Snakes; Water Snakes; Ringed Snakes

d Kielrückennattern; Schwimmnattern
i Natrice

5734 ***Natrix maura***
Squamata - Colubridae
e Viperine Grass Snake; Viperine Snake; Viperine Water Snake
d Vipernatter
f Couleuvre vipérine
i Natrice viperina; Biscia viperina

5735 ***Natrix megalocephala***
Squamata - Colubridae
e Large-headed Grass Snake; Large-headed Water Snake; Bighead European Grass Snake
d Großkopfrngelnatter
f Couleuvre à grosse tête; Couleuvre mégalocéphale
i Natrice megacefala

5736 ***Natrix natrix***
Squamata - Colubridae
e Grass Snake; European Grass Snake; Common Grass Snake; Ringed Snake
d Ringelnatter; Wassernatter; Schwimmnatter; Gemeine Natter
f Couleuvre à collier
i Natrice dal collare; Biscia dal collare; Natrice; Biscia viperina; Biscia; Biscia d'acqua europea; Bèssa; Serpe bottaia; Biscia d'acqua dal collare

5737 ***Natrix natrix astreptohora***
Squamata - Colubridae
e Green-collared Grass Snake; Grass Snake; Ringed Snake
d Ringelnatter; Spanische Ringelnatter

5738 ***Natrix natrix helvetica***
Squamata - Colubridae
e European Grass Snake
d Barrenringelnatter

5739 ***Natrix natrix natrix***
Squamata - Colubridae
e Common Grass Snake
d Gewöhnliche Ringelnatter

5740 ***Natrix natrix persa***
Squamata - Colubridae
e Balkan Grass Snake
d Gestreifte Ringelnatter; Streifenringelnatter

5741 ***Natrix natrix schweizeri***
Squamata - Colubridae
e European Grass Snake
d Ringelnatter; Milos-Ringelnatter

5742 ***Natrix natrix scutata***
Squamata - Colubridae
e Crimean Grass Snake
d Ringelnatter

5743 ***Natrix natrix scutata***
Squamata - Colubridae
d Russische Ringelnatter

5744 ***Natrix natrix sicula***
Squamata - Colubridae
d Sizilianische Ringelnatter

5745 ***Natrix tessellata***
Squamata - Colubridae
e Dice Snake; Tessellated Snake; Tessellated Water Snake; Diced Snake
d Würfelnatter; Gefleckte Wassernatter
f Couleuvre tesselée; Serpent nageur
i Natrice tesselata; Biscia tesselata; Bèssa

5746 ***Natrix tessellata heinrothi***
Squamata - Colubridae
d Heinroths Würfelnatter

5747 ***Natrix tessellata tessellata***
Squamata - Colubridae
d Gewöhnliche Würfelnatter

5748 ***Naultinus***
Squamata - Gekkonidae
e Tree Geckos
d Baumgeckos; Neuseeländerische Baumgeckos; Grüne Baumgeckos

5749 ***Naultinus elegans***
Squamata - Gekkonidae
e Green Tree Gecko
d Grüner Baumgecko; Neuseeländerischer Baumgecko
f Gecko vert des arbres

5750 ***Naultinus gemmeus***
Squamata - Gekkonidae
e South Island Tree Gecko
d Juwelengecko

5751 ***Naultinus grayii***
Squamata - Gekkonidae
e Gray's Tree Gecko

5752 ***Naultinus manukanus***
Squamata - Gekkonidae
e Northern Tree Gecko

5753 ***Naultinus poecilochlorus***
Squamata - Gekkonidae
e Central Tree Gecko

5754 ***Naultinus rudis***
Squamata - Gekkonidae
e Natural Tree Gecko

5755 ***Naultinus stellatus***
Squamata - Gekkonidae
e Starry Tree Gecko
d Sterngecko

5756 ***Naultinus tuberculatus***
Squamata - Gekkonidae
e Warty Tree Gecko

5757 ***Neoseps***
Squamata - Scincidae
e Sand Skinks; American Burrowing Skinks; Florida Burrowing Skinks; Florida Worm Lizards; Florida Sand Skinks (NA)
d Florida-Sandskinke
f Scinques de Reynolds; Néoseps

5758 ***Neoseps reynoldsi***
Squamata - Scincidae
e American Sand Skink; Florida Sand Skink; Small Sand Skink; Sand Skink
d Florida-Sandskink
f Scinque de Reynolds; Néoseps de Reynolds

5759 ***Nephrurus***
Squamata - Gekkonidae
e Knobtail Geckos; Kidney-tailed Geckos
d Keulenschwanzgeckos; Knopfschwanzgeckos

5760 ***Nephrurus asper***
Squamata - Gekkonidae
e Rough Knobtail Gecko
d Keulenschwanzgecko

5761 ***Nephrurus deleani***
Squamata - Gekkonidae
e Acacia Knobtail Gecko

5762 ***Nephrurus laevis***
Squamata - Gekkonidae
e Smooth Knobtail Gecko
d Glatter Keulenschwanzgecko

5763 ***Nephrurus laevissimus***
Squamata - Gekkonidae
e Spinifex Knobtail Gecko; Pale Smooth Knob-tailed Gecko

5764 ***Nephrurus milii***
Squamata - Gekkonidae
e Knobtail Gecko; Thick-tailed Gecko (ANZ)

d Knopfschwanzgecko; Dickschwänziger Rindengecko

5765 ***Nephrurus sphyrurus***
Squamata - Gekkonidae
e Ogilby's Knobtail Gecko

5766 ***Nephrurus stellatus***
Squamata - Gekkonidae
e Starry Knobtail Gecko

5767 ***Nephrurus vertebratus***
Squamata - Gekkonidae
e Storr's Knobtail Gecko

5768 ***Nephrurus wheeleri***
Squamata - Gekkonidae
e Wheeler's Knobtail Gecko

5769 ***Nerodia***
Squamata - Colubridae
e Water Snakes; American Water Snakes; Water and Salt Marsh Snakes; North American Water Snakes
d Amerikanische Wassernattern; Wassernattern; Nordamerikanische Wassernattern; Amerikanische Schwimmnattern
f Couleuvres d'eau

5770 ***Nerodia clarkii***
Squamata - Colubridae
e Saltmarsh Snake; Ring Snake

5771 ***Nerodia clarkii clarkii***
Squamata - Colubridae
e Saltmarsh Snake; Ring Snake; Gulf Salt Marsh Snake

5772 ***Nerodia clarkii compressicauda***
Squamata - Colubridae
e Mangrove Salt Marsh Snake; Flat-tailed Water Snake; Saltwater Moccasin (NA)

5773 ***Nerodia clarkii taeniata***
Squamata - Colubridae
e Atlantic Salt Marsh Snake; Gulf Saltmarsh Snake

5774 ***Nerodia cyclopion***
Squamata - Colubridae
e Mississippi Green Water Snake; Green Water Snake; Cyclops Water Snake
d Grüne Schwimmnatter; Grüne Wasserschlange
f Couleuvre d'eau verte

5775 ***Nerodia erythrogaster***
Squamata - Colubridae
e Plain-bellied Water Snake; Red-bellied Water Snake; Plainbelly Water Snake; Red Water Snake
d Rotbäuchige Schwimmnatter; Rotbauchwassernatter
f Couleuvre d'eau à ventre unicolore; Couleuvre à ventre rouge

5776 ***Nerodia erythrogaster erythrogaster***
Squamata - Colubridae
e Redbelly Water Snake; Red-bellied Water Snake; Copper-bellied Moccasin (NA)

5777 ***Nerodia erythrogaster flavigaster***
Squamata - Colubridae
e Yellowbelly Water Snake; Yellow-bellied Water Snake

5778 ***Nerodia erythrogaster neglecta***
Squamata - Colubridae
e Copperbelly Water Snake; Copper-bellied Water Snake

5779 ***Nerodia erythrogaster transversa***
Squamata - Colubridae
e Blotched Water Snake
d Sattelfleckwassernatter

5780 ***Nerodia fasciata***
Squamata - Colubridae

e Southern Water Snake; Banded Watersnake; Saltmarsh Snake
d Südliche Schwimmnatter; Gebänderte Wassernatter
f Couleuvre d'eau du Sud

5781 ***Nerodia fasciata confluens***
Squamata - Colubridae
e Broad-banded Water Snake
d Gebänderte Wassernatter

5782 ***Nerodia fasciata fasciata***
Squamata - Colubridae
e Banded Water Snake
d Gebänderte Wassernatter

5783 ***Nerodia fasciata pictiventris***
Squamata - Colubridae
e Florida Water Snake
d Gebänderte Wassernatter

5784 ***Nerodia floridana***
Squamata - Colubridae
e Florida Green Water Snake

5785 ***Nerodia harteri***
Squamata - Colubridae
e Brazos Water Snake; Harter's Water Snake; Brazos River Water Snake
d Harters Schwimmnatter; Harters Wassernatter
f Couleuvre d'eau de Harter

5786 ***Nerodia paucimaculata***
Squamata - Colubridae
e Concho Water Snake

5787 ***Nerodia rhombifer***
Squamata - Colubridae
e Diamondback Water Snake; Diamondback (NA); Diamond-backed Water Snake
d Rautenschwimmnatter; Diamantwassernatter; Rautenwassernatter
f Couleuvre d'eu diamantine

5788 ***Nerodia rhombifer rhombifer***
Squamata - Colubridae
e Diamondback Water Snake; Northern Diamondback Water Snake; Northern Diamond-backed Water Snake
d Diamantwassernatter

5789 ***Nerodia sipedon***
Squamata - Colubridae
e Northern Water Snake; Common Water Snake
d Siegelringnatter; Siegelringelnatter
f Couleuvre d'eau du Nord; Couleuvre d'eau (Qué)

5790 ***Nerodia sipedon insularum***
Squamata - Colubridae
e Lake Erie Water Snake

5791 ***Nerodia sipedon pleuralis***
Squamata - Colubridae
e Midland Water Snake

5792 ***Nerodia sipedon sipedon***
Squamata - Colubridae
e Northern Water Snake; Banded Water Snake; Common Water Snake
d Siegelringnatter
f Couleuvre d'eau

5793 ***Nerodia sipedon williamengelsi***
Squamata - Colubridae
e Carolina Water Snake

5794 ***Nerodia taxispilota***
Squamata - Colubridae
e Brown Water Snake; Water Pilot
d Braune Schwimmnatter; Braune Wassernatter; Braune Wasserschlange
f Couleuvre d'eu brune; Couleuvre jaune

5795 ***Nerodia valida***
Squamata - Colubridae
e Kennicott's Water Snake; Pacific

Water Snake (NA); Pacific Garter Snake; Mexican West Coast Watersnake
d Mexikanische Westküsten-Wassernatter

5796 ***Nessia***
Squamata - Scincidae
e Nessias; Ceylon Skinks
d Ceylon-Skinke; Glattechsen
f Nessies

5797 ***Nessia bipes***
Squamata - Scincidae
e Two-legged Nessia

5798 ***Nessia burtonii***
Squamata - Scincidae
e Burton's Nessia
d Burtons Glattechse
f Nessie de Burton

5799 ***Nessia deraniyagalai***
Squamata - Scincidae
e Deraniyagla's Nessia

5800 ***Nessia didactyla***
Squamata - Scincidae
e Two-toed Nessia
d Zweizehenfaultier; Zweizehen Ceylon-Skink; Ceylon-Skink

5801 ***Nessia hickanala***
Squamata - Scincidae
e Hickanala Nessia

5802 ***Nessia layardi***
Squamata - Scincidae
e Layard's Nessia

5803 ***Nessia monodactyla***
Squamata - Scincidae
e One-toed Nessia

5804 ***Nessia sarasinorum***
Squamata - Scincidae
e Müller's Nessia

5805 ***Neusticurus***
Squamata - Gymnophthalmidae
e Neusticuruses; Stream Lizards
d Wassertejus

5806 ***Neusticurus apodemus***
Squamata - Gymnophthalmidae
e Uzzell's Neusticurus; Water Teiid

5807 ***Neusticurus bicarinatus***
Squamata - Gymnophthalmidae
e Two-faced Neusticurus
d Wasserteju; Eigentlicher Wasserteju

5808 ***Neusticurus cochranae***
Squamata - Gymnophthalmidae
e Cochran's Neusticurus
d Cochrans Wasserteju

5809 ***Neusticurus eopleopus***
Squamata - Gymnophthalmidae
e Cope's Neusticurus; Common Stream Lizard

5810 ***Neusticurus medemi***
Squamata - Gymnophthalmidae
e Medem's Neusticurus

5811 ***Neusticurus racenisi***
Squamata - Gymnophthalmidae
e Roze's Neusticurus

5812 ***Neusticurus rudis***
Squamata - Gymnophthalmidae
e Red Neusticurus

5813 ***Neusticurus strangulatus***
Squamata - Gymnophthalmidae
e Big-scaled Neusticurus

5814 ***Neusticurus tatei***
Squamata - Gymnophthalmidae
e Tate's Neusticurus

5815 ***Nilssonia***
Testudines - Trionychidae
e Burmese Peacock Softshells
d Burma-Weichschildkröten

5816 ***Nilssonia formosa***
Testudines - Trionychidae
e Burmese Peacock Softshell; Burmese Soft-shelled Turtle; Burmese Softshell (ISC); Burmese Peacock Soft-shelled Turtle; Burmese Peacock Softshell Turtle; Burmese Softshell Turtle
d Burma-Weichschildkröte
f Trionyx de Birmanie

5817 ***Ninia***
Squamata - Colubridae
e Common Coffee Snakes; Coffee Snakes
d Kaffeenattern

5818 ***Ninia atrata***
Squamata - Colubridae
e Hallowell's Coffee Snake; Southern Coffee Snake

5819 ***Ninia celata***
Squamata - Colubridae
e Rare Coffee Snake

5820 ***Ninia diademata***
Squamata - Colubridae
e Ringneck Coffee Snake; Ring-necked Coffee Snake

5821 ***Ninia espinali***
Squamata - Colubridae
e Espinal's Coffee Snake

5822 ***Ninia hudsoni***
Squamata - Colubridae
e Hudson's Coffee Snake; Amazon Coffee Snake

5823 ***Ninia maculata***
Squamata - Colubridae
e Spotted Coffee Snake; Checkered Coffee Snake

5824 ***Ninia oxynota***
Squamata - Colubridae
e Werner's Coffee Snake

5825 ***Ninia psephota***
Squamata - Colubridae
e Cope's Coffee Snake

5826 ***Ninia sebae***
Squamata - Colubridae
e Redback Coffee Snake; Red Coffee Snake

5827 ***Notechis***
Squamata - Elapidae
e Black Tiger Snakes; Tiger Snakes; Australian Tiger Snakes
d Tigerottern
f Serpents tigrés
i Serpenti tigre

5828 ***Notechis ater***
Squamata - Elapidae
e Black Tiger Snake (ANZ); Western Tiger Snake; Peninsula Tiger Snake; Chapell Island Tiger Snake
d Schwarze Tigerotter
f Serpent tigré noir
i Serpente tigre nero

5829 ***Notechis scutatus***
Squamata - Elapidae
e Mainland Tiger Snake (ANZ); Eastern Tiger Snake; Common Tiger Snake; Tiger Snake (ANZ); Australian Tiger Snake
d Östliche Tigerotter; Tigerotter

f Serpent tigré

5830 ***Nothobachia***
Squamata - Teiidae
e Nothobachias

5831 ***Nothobachia ablephara***
Squamata - Teiidae
e Nothobachia

5832 ***Nothopsis***
Squamata - Colubridae
e Rough Coffee Snakes

5833 ***Nothopsis rugosus***
Squamata - Colubridae
e Rough Coffee Snake; Rough Forest Snake

5834 ***Notochelys***
Testudines - Emydidae
e Malayan Flatshell Turtles
d Plattrückenschildkröten

5835 ***Notochelys platynota***
Testudines - Emydidae
e Malayan Flatshell Turtle; Malayan Flat-shelled Turtle; Purple-bellied Turtle; Spotted-shelled Turtle
d Plattrückenschildkröte
f Tortue-boîte à dos plat

5836 ***Notoscincus***
Squamata - Scincidae
e Snake-eyed Skinks

5837 ***Notoscincus butleri***
Squamata - Scincidae
e Butler's Snake-eyed Skink

5838 ***Notoscincus ornatus***
Squamata - Scincidae
e Ornate Snake-eyed Skink

5839 ***Nucras***
Squamata - Lacertidae
e Tiger Lizards; Sandveld Lizards (CSA); Blunt-headed Lacertids
d Nucras-Eidechsen; Nukras-Eidechsen

5840 ***Nucras boulengeri***
Squamata - Lacertidae
e Uganda Savanna Lizard; Savanna Lizard; Savannah Lizard; Uganda Savannah Lizard; Boulenger's Scrub Lizard

5841 ***Nucras caesicaudata***
Squamata - Lacertidae
e Bluetail Scrub Lizard; Blue-tailed Sandveld Lizard (CSA); Blue-tailed Scrub Lizard (CSA)

5842 ***Nucras emini***
Squamata - Lacertidae
e Boulenger's Sandveld Lizard

5843 ***Nucras intertexta***
Squamata - Lacertidae
e Spotted Sandveld Lizard; Spotted Scrub Lizard (CSA)
d Stumpfkopfeidechse

5844 ***Nucras lalandii***
Squamata - Lacertidae
e Delalandes Sandveld Lizard; Delalande's Lizard; Lalande's Sandveld Lizard (CSA)
d Delalandes Eidechse
f Lézard de Delalande

5845 ***Nucras scalaris***
Squamata - Lacertidae
e Scaled Sandveld Lizard

5846 ***Nucras taeniolata***
Squamata - Lacertidae
e Striped Scrub Lizard
f Scinque à queue cuivrée

5847 ***Nucras taeniolata ornata***
Squamata - Lacertidae
e Ornate Sandveld Lizard (CSA)

5848 ***Nucras tesselata***
Squamata - Lacertidae
e Tiger Lizard; Striped Sandveld Lizard (CSA); Banded Sand Lizard; Chequered Lizard
d Westliche Scharreidechse
f Lézard marqueté

O

5849 ***Ocadia***
Testudines - Emydidae
e Chinese Stripeneck Turtles
d Streifenschildkröten; Chinesische Streifenschildkröten

5850 ***Ocadia glyphistoma***
Testudines - Emydidae
e Stripeneck Turtle

5851 ***Ocadia philippeni***
Testudines - Emydidae
e Philippen's Stripeneck Turtle

5852 ***Ocadia sinensis***
Testudines - Emydidae
e Chinese Stripeneck Turtle; Chinese Stripe-necked Turtle; Green-headed Turtle; Chinese Golden Thread Turtle
d Chinesische Streifenschildkröte
f Émyde commune à coup rayé
i Testuggine cinese a collo striato

5853 ***Oedura***
Squamata - Gekkonidae
e Velvet Geckos; Fat-tailed Geckos
d Samtgeckos; Fettschwanzgeckos

5854 ***Oedura castelnaui***
Squamata - Gekkonidae
e Northern Velvet Gecko
d Fettschwanzgecko; Samtgecko

5855 ***Oedura coggeri***
Squamata - Gekkonidae
e Northern Spotted Velvet Gecko
d Coggers Fettschwanzgecko

5856 ***Oedura filicipoda***
Squamata - Gekkonidae
e Mitchell Velvet Gecko

5857 ***Oedura gemmata***
Squamata - Gekkonidae
e Bejewelled Velvet Gecko

5858 ***Oedura gracilis***
Squamata - Gekkonidae
e Graceful Velvet Gecko

5859 ***Oedura lesueurii***
Squamata - Gekkonidae
e Lesueur's Velvet Gecko

5860 ***Oedura marmorata***
Squamata - Gekkonidae
e Marbled Velvet Gecko; Fat-tailed Gecko; Marbled Fat-tailed Gecko
d Gefleckter Samtgecko; Gefleckter Fettschwanzgecko

5861 ***Oedura monilis***
Squamata - Gekkonidae
e Ocellated Velvet Gecko
d Augenfleckengecko; Fettschwanzgecko; Samtgecko

5862 ***Oedura obscura***
Squamata - Gekkonidae
e Obscure Velvet Gecko

5863 ***Oedura reticulata***
Squamata - Gekkonidae
e Reticulate Velvet Gecko; Reticulated Velvet Gecko (ANZ)

5864 ***Oedura rhombifer***
Squamata - Gekkonidae
e Zigzag Velvet Gecko

5865 ***Oedura robusta***
Squamata - Gekkonidae
e Robust Velvet Gecko

5866 ***Oedura tryoni***
Squamata - Gekkonidae
e Southern Spotted Velvet Gecko; Tryon's Velvet Gecko; Velvet Gecko

5867 ***Ogmodon***
Squamata - Elapidae
e Fiji Cobras; Fiji Cobra Snakes
d Fidschi-Giftnattern; Fidschi-Schlangen

5868 ***Ogmodon vitianus***
Squamata - Elapidae
e Fiji Cobra
d Fidschi-Giftnatter; Fidschi-Schlange

5869 ***Oligodon***
Squamata - Colubridae
e Kukri Snakes
d Kukrinattern

5870 ***Oligodon affinis***
Squamata - Colubridae
e Kukri Snake

5871 ***Oligodon albocinctus***
Squamata - Colubridae
e Light-barred Kukri Snake; Ladderback (ISC); Ladderback Kukri Snake

5872 ***Oligodon annamensis***
Squamata - Colubridae
e Leviton's Kukri Snake

5873 ***Oligodon annulifer***
Squamata - Colubridae
e Ringed Kukri Snake

5874 ***Oligodon arnensis***
Squamata - Colubridae
e Russet Kukri Snake; Common Kukri Snake (ISC); Banded Kukri Snake

5875 ***Oligodon barroni***
Squamata - Colubridae
e Barron's Kukri Snake; Smith's Kukri Snake

5876 ***Oligodon bellus***
Squamata - Colubridae
e Pretty Kukri Snake

5877 ***Oligodon bitorquata***
Squamata - Colubridae
e Shorthead Kukri Snake

5878 ***Oligodon bitorquatus***
Squamata - Colubridae
e Boie's Kukri Snake

5879 ***Oligodon calamarius***
Squamata - Colubridae
e Sri Lanka Kukri Snake

5880 ***Oligodon catenata***
Squamata - Colubridae
e Assam Kukri Snake

5881 ***Oligodon chinensis***
Squamata - Colubridae
e Chinese Kukri Snake

5882 ***Oligodon cinereus***
Squamata - Colubridae
e Günther's Kukri Snake

5883 ***Oligodon cruentatus***
Squamata - Colubridae
e Pegu Kukri Snake

5884 ***Oligodon cyclurus***
Squamata - Colubridae
e Cantor's Kukri Snake

5885 ***Oligodon dorsalis***
Squamata - Colubridae
e Gray's Kukri Snake

5886 ***Oligodon durkheimi***
Squamata - Colubridae
e Durkheim's Kukri Snake

5887 ***Oligodon erythrogaster***
Squamata - Colubridae
e Nagarkot Kukri Snake; Coral-bellied Kukri Snake

5888 ***Oligodon erythrorachis***
Squamata - Colubridae
e Namsang Kukri Snake

5889 ***Oligodon everetti***
Squamata - Colubridae
e Everett's Kukri Snake

5890 ***Oligodon forbesi***
Squamata - Colubridae
e Forbes's Kukri Snake

5891 ***Oligodon formosanus***
Squamata - Colubridae
e Formosa Kukri Snake; Formosan Kukri Snake

5892 ***Oligodon inornatus***
Squamata - Colubridae
e Inornate Kukri Snake

5893 ***Oligodon jacobi***
Squamata - Colubridae
e Jacob's Kukri Snake

5894 ***Oligodon joynsoni***
Squamata - Colubridae
e Grey Kukri Snake

5895 ***Oligodon juglandifer***
Squamata - Colubridae
e Walnut Kukri Snake; Large-spotted Kukri Snake

5896 ***Oligodon kunmigeosis***
Squamata - Colubridae
e Kuo's Kukri Snake

5897 ***Oligodon lacroixi***
Squamata - Colubridae
e Lacroix Kukri Snake

5898 ***Oligodon macrurus***
Squamata - Colubridae
e Angel's Kukri Snake

5899 ***Oligodon maculatus***
Squamata - Colubridae
e Spotted Kukri Snake

5900 ***Oligodon mcdougalli***
Squamata - Colubridae
e McDougall's Kukri Snake

5901 ***Oligodon melaneus***
Squamata - Colubridae
e Bluebelly Kukri Snake; Blue-bellied Kukri Snake

5902 ***Oligodon melanozonatus***
Squamata - Colubridae
e Abor Hills Kukri Snake

5903 ***Oligodon meyerinkii***
Squamata - Colubridae
e Sulu Kukri Snake

5904 ***Oligodon modestum***
Squamata - Colubridae
e Luzon Kukri Snake

5905 ***Oligodon mouhoti***
Squamata - Colubridae
e Cambodian Kukri Snake; Boulenger's Kukri Snake

5906 ***Oligodon multizonatus***
Squamata - Colubridae
e Luding Kukri Snake

5907 ***Oligodon ningshaanensis***
Squamata - Colubridae
e Ningshaan Kukri Snake

5908 ***Oligodon notospilus***
Squamata - Colubridae
e Asian Kukri Snake

5909 ***Oligodon ocellatus***
Squamata - Colubridae
e Ocellated Kukri Snake

5910 ***Oligodon octolineatus***
Squamata - Colubridae
e Grace's Kukri Snake; Striped Kukri Snake

5911 ***Oligodon ornatus***
Squamata - Colubridae
e Ornate Kukri Snake

5912 ***Oligodon purpurescens***
Squamata - Colubridae
e Brown Kukri Snake; Himalayo-Chinese Kukri Snake; Schlegel's Brown Kukri Snake
d Braune Kukrinatter

5913 ***Oligodon signata***
Squamata - Colubridae
e Barred Kukri Snake

5914 ***Oligodon splendidus***
Squamata - Colubridae
e Splendid Kukri Snake

5915 ***Oligodon subcarinatus***
Squamata - Colubridae
e Beautiful Kukri Snake

5916 ***Oligodon sublineatus***
Squamata - Colubridae
e Duméril's Kukri Snake

5917 ***Oligodon taeniatus***
Squamata - Colubridae
e Striped Kukri Snake

5918 ***Oligodon taeniolatus***
Squamata - Colubridae
e Loos Snake; Streaked Kukri Snake; Variegated Kukri Snake (ISC); Russell's Kukri Snake (ISC); Common Kukri Snake
d Gestreifte Kukrinatter

5919 ***Oligodon theobaldi***
Squamata - Colubridae
e Theobald's Kukri Snake

5920 ***Oligodon travancoricus***
Squamata - Colubridae
e Travancore Kukri Snake

5921 ***Oligodon trilineatus***
Squamata - Colubridae
e Three-lined Kukri Snake

5922 ***Oligodon venustus***
Squamata - Colubridae
e Jerdon's Kukri Snake

5923 ***Oligodon vertebralis***
Squamata - Colubridae
e Vertebral Kukri Snake

5924 ***Oligodon waandersi***
Squamata - Colubridae
e Bleck's Kukri Snake

5925 ***Ommateremias***
Squamata - Lacertidae
e Steppe Racerunners; Steppe Lizards

5926 ***Ommateremias arguta***
Squamata - Lacertidae
e Steppe Racerunner; Racerunner; Stepperunner; Arguta
d Bunter Wüstenrenner; Steppeneidechse; Buckelnase
f Érérmias élégant; Érémias variable
i Eremia steppica

5927 ***Ommateremias aria***
Squamata - Lacertidae
e Aria Racerunner; Aria Lizard; Afghanistan Racerunner; Afghan Racerunner
d Afghanistan-Wüstenrenner

5928 ***Ommateremias nigrocellata***
Squamata - Lacertidae
e Black-spotted Racerunner; Black-ocellated Racerunner
d Schwarzäugiger Wüstenrenner

5929 ***Opheodrys***
Squamata - Colubridae
e Green Snakes; North American Green Snakes; Rough Green Snakes
d Grasnattern
f Serpents verts

5930 ***Opheodrys aestivus***
Squamata - Colubridae
e Rough Green Snake; Small Kukri Snake; Keeled Green Snake
d Rauhe Grasnatter
f Serpent vert rugueux
i Serpente verde della pelle ruvida

5931 ***Opheodrys aestivus aestivus***
Squamata - Colubridae
e Eastern Rough Green Snake; Northern Rough Green Snake
f Serpent vert rugueux de l'Est

5932 ***Opheodrys aestivus carinatus***
Squamata - Colubridae
e Florida Rough Green Snake

Squamata - Colubridae
e Western Rough Green Snake
f Serpent vert rugueux de l'Ouest

5934 ***Ophidiocephalus***
Squamata - Gekkonidae
e Snakehead Lizards

5935 ***Ophidiocephalus taeniatus***
Squamata - Gekkonidae
e Snakehead Lizard

5936 ***Ophiodes***
Squamata - Anguidae
e South American Worm Lizards; Worm Lizards
d Schlangenschleichen
f Ophiodes

5937 ***Ophiodes intermedius***
Squamata - Anguidae
e Middle Worm Lizard

5938 ***Ophiodes striatus***
Squamata - Anguidae
e Striped Worm Lizard; Cariococca
f Ophiode strié

5939 ***Ophiodes vertebralis***
Squamata - Anguidae
e Jointed Worm Lizard

5940 ***Ophiodes yacupoi***
Squamata - Anguidae
e Yacupoi Worm Lizard

5941 ***Ophiomorus***
Squamata - Scincidae
e Greater Snake Skinks; Sand Skinks; Ophiomores
d Schlangenskinke
f Ophiomores
i Offiomoro punteggiato

5942 ***Ophiomorus blanfordi***
Squamata - Scincidae
e Blanford's Snake Skink
d Blanfords Schlangenskink
f Ophiomore de Blanford

5943 ***Ophiomorus brevipes***
Squamata - Scincidae
e Short-legged Snake Skink
d Kurzfußschlangenskink

5944 ***Ophiomorus chernovi***
Squamata - Scincidae
e Chernov's Snake Skink

5945 ***Ophiomorus latasti***
Squamata - Scincidae
e Lataste's Snake Skink; Lataste's Sand Skink

5946 ***Ophiomorus nuchalis***
Squamata - Scincidae
e Nilson's Snake Skink

5947 ***Ophiomorus persicus***
Squamata - Scincidae
e Persia Snake Skink; Persian Sand Skink
d Persischer Schlangenskink
f Ophiomore persan

5948 ***Ophiomorus punctatissimus***
Squamata - Scincidae
e Greek Snake Skink; Spotted Greek Legless Skink; Speckled Snake Skink; Spotted Skink; Legless Skink; Speckled Sand Skink
d Gesprenkelter Schlangenskink
f Ophiomore ponctué; Scinque serpentiforme

5949 ***Ophiomorus streeti***
Squamata - Scincidae
e Street's Snake Skink; Street's Skink

5950 ***Ophiomorus tridactylus***
Squamata - Scincidae
e Three-toed Snake Skink; Sandfish (ISC); Indian Sandswimmer (ISC); Three-toed Sand Skink
d Dreizehiger Schlangenskink

5951 ***Ophiophagus***
Squamata - Elapidae
e King Cobras; Hamadryads; Snake-eating Cobras
d Königskobras
f Cobras

5952 ***Ophiophagus hannah***
Squamata - Elapidae
e King Cobra; Hamadryad
d Königskobra; Riesenhutschlange; Hamadryad
f Cobra hannah; Cobra royal
i Cobra reale

5953 ***Ophioscincus***
Squamata - Scincidae
e Lesser Snake Skinks
d Schlangenskinke

5954 ***Ophioscincus coolooensis***
Squamata - Scincidae
e Rainforest Snake Skink

5955 ***Ophioscincus ophioscincus***
Squamata - Scincidae
e Snake Skink

5956 ***Ophioscincus truncatus***
Squamata - Scincidae
e Burrowing Snake Skink

5957 ***Ophisaurus***
Squamata - Anguidae
e Glass Lizards; Slender Glass Lizards
d Glasschleichen; Panzerschleichen
f Ophisaures

5958 ***Ophisaurus apodus***
Squamata - Anguidae
e Armoured Glass Lizard; Sheltopusik; Sheltopusic; Glass Snake; Armoured Glass Snake; Balkan Glass Snake; Glass Lizard; European Glass Lizard; European Glass Snake; European Legless Lizard
d Seitenfalter; Panzerschleiche; Scheltopusik
f Sheltopusik; Pseudope de Pallas; Serpent-lézard sans pattes; Ophisaure
i Ofisauro apodo; Pseudopo; Pseudopo europeo; Ofisauro europeo

5959 ***Ophisaurus attenuatus***
Squamata - Anguidae
e Slender Glass Lizard; Slender Glass Snake
d Schlanke Glasschleiche
f Ophisaure élancé; Ophisaure svelte

5960 ***Ophisaurus attenuatus attenuatus***
Squamata - Anguidae
e Western Slender Glass Lizard; Western Slender Glass Snake

5961 ***Ophisaurus attenuatus longicaudus***
Squamata - Anguidae
e Eastern Slender Glass Lizard; Eastern Slender Glass Snake

5962 ***Ophisaurus buettikofferi***
Squamata - Anguidae
e Buettikofer's Glass Lizard; Buettikofer's Glass Snake
d Kalamantan-Schleiche; Borneo-Schleiche

5963 ***Ophisaurus ceroni***
Squamata - Anguidae
e Ceron's Glass Lizard; Ceron's Glass Snake

5964 ***Ophisaurus compressus***
Squamata - Anguidae
e Island Glass Lizard; Island Glass Snake
d Küstenglasschleiche
f Ophisaure compact

5965 ***Ophisaurus gracilis***
Squamata - Anguidae
e Burma Glass Lizard; Burma Glass Snake; Burman Glass Lizard; Burmese Glass Snake (ISC)
d Burma-Schleiche

5966 ***Ophisaurus hainanus***
Squamata - Anguidae
e Hainan Glass Lizard; Hainan Glass Snake

5967 ***Ophisaurus harti***
Squamata - Anguidae
e Hart's Glass Lizard; Hart's Glass Snake
d Südchinesische Schleiche

5968 ***Ophisaurus incomptus***
Squamata - Anguidae
e Plain-neck Glass Lizard; Plain-neck Glass Snake

5969 ***Ophisaurus koellikeri***
Squamata - Anguidae
e Koelliker's Glass Lizard; Koelliker's Glass Snake; Moroccan Glass Lizard
d Türkis-Panzerschleiche; Marokko-Schleiche
f Orvet du Maroc

5970 ***Ophisaurus mimicus***
Squamata - Anguidae
e Mimic Glass Lizard; Mimic Glass Snake

5971 ***Ophisaurus sokolovi***
Squamata - Anguidae
e Sokolov's Glass Lizard; Sokolov's Glass Snake; Sokolov's Glass Lizard

5972 ***Ophisaurus ventralis***
Squamata - Anguidae
e Eastern Glass Lizard; Eastern Glass Snake
d Östliche Glasschleiche
f Ophisaure de l'Est; Ophisaure ventrale

5973 ***Ophisaurus wegneri***
Squamata - Anguidae
e Wegner's Glass Lizard; Wegner's Glass Snake; Wegner's Glass Lizard
d Sumatra-Schleiche

5974 ***Ophisops***
Squamata - Lacertidae
e Snake-eyed Lizards; Snake-eyed Lacertas; Snake-eyed Lacertids
d Schlangenaugen; Schlangenaugeeidechsen; Schlangenaugeneidechsen
f Ophisops; Lézards à l'oeil de serpent

5975 ***Ophisops elbaensis***
Squamata - Lacertidae
e Elbaen Snake-eyed Lizard

5976 ***Ophisops elegans***
Squamata - Lacertidae
e Snake-eyed Lizard; Ménétries's Lizard
d Europäische Schlangenaugeeidechse; Europäisches Schlangenauge; Schlangenauge
f Ophisops élégant; Lézard à l'oeil de serpent
i Ofisope elegante

5977 ***Ophisthoplus***
Squamata - Colubridae
e Rio Grande Snakes

5978 ***Ophisthoplus degener***
Squamata - Colubridae
e Rio Grande Snake

5979 ***Ophisthotropis***
Squamata - Colubridae
e Mountain Keelbacks
d Bergwassernattern

5980 ***Ophisthotropis alcali***
Squamata - Colubridae
e Gary's Mountain Keelback

5981 ***Ophisthotropis andersoni***
Squamata - Colubridae
e Anderson's Mountain Keelback

5982 ***Ophisthotropis balteatus***
Squamata - Colubridae
e Hainan Mountain Keelback

5983 ***Ophisthotropis guangxiensis***
Squamata - Colubridae
e Guanxi Mountain Keelback

5984 ***Ophisthotropis jacobi***
Squamata - Colubridae
e Chapa Mountain Keelback

5985 ***Ophisthotropis kuatunensis***
Squamata - Colubridae
e Chinese Mountain Keelback

5986 ***Ophisthotropis lateralis***
Squamata - Colubridae
e Tonkin Mountain Keelback

5987 ***Ophisthotropis latouchii***
Squamata - Colubridae
e Sichuan Mountain Keelback

5988 ***Ophisthotropis maxwelli***
Squamata - Colubridae
e Maxwell's Mountain Keelback

5989 ***Ophisthotropis praemaxillaris***
Squamata - Colubridae
e Angel's Mountain Keelback

5990 ***Ophisthotropis rugos***
Squamata - Colubridae
e Sumatra Mountain Keelback

5991 ***Ophisthotropis spenceri***
Squamata - Colubridae
e Smith's Mountain Keelback

5992 ***Ophisthotropis typica***
Squamata - Colubridae
e Olive Mountain Keelback

5993 ***Ophryacus***
Squamata - Viperidae
e Mexican Horned Vipers; Mexican Horned Pit Viper
d Mexikanische Hornlanzenottern

5994 ***Ophryacus undulatus***
Squamata - Viperidae
e Mexican Horned Viper; Mexican Horned Pit Viper
d Mexikanische Hornlanzenotter

5995 ***Ophryoessoides***
Squamata - Iguanaidae
e Keeled Iguanas
d Festlandglattkopfleguane

5996 ***Ophryoessoides aculeatus***
Squamata - Iguanaidae
e O'Shaughnessy's Keeled Iguana

5997 ***Ophryoessoides caducus***
Squamata - Iguanaidae
e Cope's Lava Lizard

5998 ***Ophryoessoides erythrogaster***
Squamata - Iguanaidae
e Redbelly Lava Lizard

5999 ***Ophryoessoides iridescens***
Squamata - Iguanaidae
e Iridescent Lava Lizard

6000 ***Ophryoessoides scapularis***
Squamata - Iguanaidae
e Boulenger's Lava Lizard

6001 ***Ophryoessoides tricristatus***
Squamata - Iguanaidae
e Duméril's Lava Lizard

6002 ***Opipeuter***
Squamata - Teiidae
e River Teiids

6003 ***Opipeuter xestus***
Squamata - Teiidae
e River Teiid

6004 ***Oplurus***
Squamata - Iguanaidae
e Madagascar Swifts
d Madagaskar-Leguane

6005 ***Oplurus cuvieri***
Squamata - Iguanaidae
e Cuvier's Madagascar Swift; Cuvier's Madagascar Tree Swift

6006 ***Oplurus cyclurus***
Squamata - Iguanaidae
e Merrem's Madagascar Swift; Madagascar Tree Swift
d Riesen Madagaskar-Leguan; Madagaskar-Leguan; Kugelechse

6007 ***Oplurus fierinensis***
Squamata - Iguanaidae
e Anzamala Madagascar Swift
d Blauer Madagaskar-Leguan

6008 ***Oplurus grandidieri***
Squamata - Iguanaidae
e Grandidier's Madagascar Swift

6009 ***Oplurus quadrimaculatus***
Squamata - Iguanaidae
e Duméril's Madagascar Swift

6010 ***Oplurus saxicola***
Squamata - Iguanaidae
e Marked Madagascar Swift

6011 ***Oreocalamus***
Squamata - Colubridae
e Kalimantan Burrowing Snakes

6012 ***Oreocalamus hanitschi***
Squamata - Colubridae
e Kalimantan Burrowing Snake

6013 ***Oriocalotes***
Squamata - Agamidae
e Smith's Agamas

6014 ***Oriocalotes paulus***
Squamata - Agamidae
e Smith's Agama

6015 ***Orlitia***
Testudines - Emydidae
e Malaysia Giant Turtle; Malaysian Giant Turtles; Malayan Giant Turtles
d Borneo-Flussschildkröten
f Émydes géantes de Bornéo

6016 ***Orlitia borneensis***
Testudines - Emydidae
e Malaysia Giant Turtle; Malaysian Giant Turtle; Bornean River Turtle; Malayan Giant Turtle
d Borneo-Flussschildkröte
f Émyde géante de Bornéo
i Tartaruga gigante delle Malesia

6017 ***Osteolaemus***
Crocodylia - Crocodylidae
e African Dwarf Crocodiles; Dwarf Crocodiles
d Stumpfkrokodile
f Crocodiles nains
i Osteolemi

6018 ***Osteolaemus tetraspis***
Crocodylia - Crocodylidae
e African Dwarf Crocodile; Dwarf Crocodile; West African Dwarf Crocodile
d Stumpfkrokodil
f Crocodile des marais; Crocodile cuirassée; Crocodile à front large; Crocodile noir à front large; Crodile d'Osborn; Crocodile nain
i Osteolemo

6019 ***Otocryptis***
Squamata - Agamidae
e Hidden-eared Agamas; Ground Agamas
f Otocryptes

6020 ***Otocryptis beddomi***
Squamata - Agamidae
e South India Agama

6021 ***Otocryptis wiegmanni***
Squamata - Agamidae
e Wiegmann's Agama; Sri Lankan Kangaroo Lizard
f Otocrypte de Wiegmann; Otocrypte à deux bandes

6022 ***Ovophis***
Squamata - Viperidae
e Asian Pit Vipers

6023 ***Ovophis chaseni***
Squamata - Viperidae
e Mount Kinabalu Pit Viper

6024 ***Ovophis convictus***
Squamata - Viperidae
e Stoliczka's Pit Viper

6025 ***Ovophis monticola***
Squamata - Viperidae
e Chinese Mountain Pit Viper
d Berglanzenotter

6026 ***Ovophis okinavensis***
Squamata - Viperidae
e Ryukyu Island Pit Viper; Himehabu
d Okinawa-Habusschlange

6027 ***Ovophis tokarensis***
Squamata - Viperidae
e Tokara Habu Pit Viper

6028 ***Oxybelis***
Squamata - Colubridae
e Vine Snakes; Pike-headed Tree Snakes; American Vine Sn akes
d Spitznattern
f Serpents des vignes mexicains; Serpents-liane

6029 ***Oxybelis aeneus***
Squamata - Colubridae
e Brown Vine Snake; Mexican Vine Snake; Pike-headed Tree Snake; American Vine Snake; Tropical Vine Snake
d Erzspitznatter; Erdspitzschlange; Mexikanische Spitznatter
f Serpent des vignes mexicain; Serpent-liane

6030 ***Oxybelis argenteus***
Squamata - Colubridae
e Daudin's Vine Snake
f Serpent-liane

6031 ***Oxybelis brevirostris***
Squamata - Colubridae
e Cope's Vine Snake; Short-nosed Vine Snake

6032 ***Oxybelis fulgidus***
Squamata - Colubridae
e Green Vine Snake; Green Whip Snake
d Glanzspitznatter
f Serpent-liane

6033 ***Oxybelis wilsoni***
Squamata - Colubridae
e Yellow Vine Snake

6034 ***Oxyrhabdium***
Squamata - Colubridae
e Shrub Snakes; Neotropical Vine Snakes

6035 ***Oxyrhabdium leporinum***
Squamata - Colubridae
e Günther's Philippine Shrub Snake

6036 ***Oxyrhabdium modestum***
Squamata - Colubridae
e Philippine Shrub Snake

6037 ***Oxyrhopus***
Squamata - Colubridae
e False Coral Snakes; Calico Snakes
d Mondnattern

6038 ***Oxyrhopus clathratus***
Squamata - Colubridae
e Duméril's False Coral Snake

6039 ***Oxyrhopus doliatus***
Squamata - Colubridae
e Bibron's False Coral Snake

6040 ***Oxyrhopus fitzingeri***
Squamata - Colubridae
e Fitzinger's False Coral Snake

6041 ***Oxyrhopus formosus***
Squamata - Colubridae
e Formosa False Coral Snake; Yellow-headed Calico Snake

6042 ***Oxyrhopus leucomelas***
Squamata - Colubridae
e Werner's False Coral Snake

6043 ***Oxyrhopus marcapatae***
Squamata - Colubridae

e Boulenger's False Coral Snake

6044 ***Oxyrhopus melanogenys***
Squamata - Colubridae
e Tschudi's False Coral Snake

6045 ***Oxyrhopus melanogenys melanogenys***
Squamata - Colubridae
e Black-headed Calico Snake

6046 ***Oxyrhopus petola***
Squamata - Colubridae
e Calico False Coral Snake; Forest Flame Snake; White-bellied False Coral Snake

6047 ***Oxyrhopus petola digitalis***
Squamata - Colubridae
e Banded Calico Snake

6048 ***Oxyrhopus rhombifer***
Squamata - Colubridae
e Amazon False Coral Snake

6049 ***Oxyrhopus trigeminus***
Squamata - Colubridae
e Brazilian False Coral Snake; South American Coral Mimic; Brazilian Calico Snake
d Mondnatter

6050 ***Oxyrhopus venezuelanus***
Squamata - Colubridae
e Amazon False Coral Snake

6051 ***Oxyuranus***
Squamata - Elapidae
e Taipans
d Taipane

6052 ***Oxyuranus microlepidotus***
Squamata - Elapidae
e Fierce Snake; Inland Taipan
d Schreckensotter; Inland Taipan
i Taipan; Taipan australiano

6053 ***Oxyuranus scutellatus***
Squamata - Elapidae
e Taipan; Coastal Taipan
d Taipan; Östlicher Taipan
i Taipan; Serpente bruno australiano

P

6054 ***Pachydactylus***
Squamata - Gekkonidae
e Thick-toed Geckos; Gecko (CSA)
d Dickfingergeckos
f Pachydactyles

6055 ***Pachydactylus austeni***
Squamata - Gekkonidae
e Austen Thick-toed Gecko; Austen's Gecko (CSA)

6056 ***Pachydactylus bibronii***
Squamata - Gekkonidae
e Bibron's Thick-toed Gecko; Bibron's Gecko (CSA); Bibron's Clawless Gecko; Bibron's Thick-toed Rock Gecko; Bibron Gecko
d Bibrons Dickfingergecko
f Pachydactyle de Bibron; Gecko de Bibron

6057 ***Pachydactylus bicolor***
Squamata - Gekkonidae
e Two-coloured Thick-toed Gecko; Velvety Gecko (CSA)

6058 ***Pachydactylus capensis***
Squamata - Gekkonidae
e Cape Thick-toed Gecko; Cape Gecko (CSA); Cape Ground Gecko (CSA); Cape Thick-toed Ground Gecko

6059 ***Pachydactylus capensis affinis***
Squamata - Gekkonidae
e Transvaal Gecko; Transvaal Thick-toed Gecko

6060 ***Pachydactylus capensis capensis***
Squamata - Gekkonidae
e Cape Gecko; Cape Thick-toed Gecko (CSA)

6061 ***Pachydactylus capensis oshauyghnessyi***
Squamata - Gekkonidae
e O'Shaughnessy's Gecko

6062 ***Pachydactylus capensis vansoni***
Squamata - Gekkonidae
e Van Son's Gecko; Van Son's Thick-toed Gecko (CSA)

6063 ***Pachydactylus caraculicus***
Squamata - Gekkonidae
e Angola Thick-toed Gecko; Angolan Banded Gecko (CSA)

6064 ***Pachydactylus fasciatus***
Squamata - Gekkonidae
e Striped Thick-toed Gecko; Banded Gecko (CSA)

6065 ***Pachydactylus formosus***
Squamata - Gekkonidae
e Smith's Thick-toed Gecko

6066 ***Pachydactylus gaiasensis***
Squamata - Gekkonidae
e Brandberg Thick-toed Gecko; Brandberg Gecko (CSA)

6067 ***Pachydactylus geitje***
Squamata - Gekkonidae
e Craddock Thick-toed Gecko; Ocellated Gecko (CSA)

6068 ***Pachydactylus haackei***
Squamata - Gekkonidae
e Haacke's Thick-toed Gecko

6069 ***Pachydactylus kochii***
Squamata – Gekkonidae
e Cape Cross Thick-toed Gecko; Koch's Gecko (CSA)

6070 ***Pachydactylus labialis***
Squamata - Gekkonidae
e Calvinia Thick-toed Gecko; Western Cape Gecko (CSA)

6071 ***Pachydactylus laevigatus***
Squamata - Gekkonidae
e Fischer's Thick-toed Gecko; Button-scaled Gecko (CSA)

6072 ***Pachydactylus maculatus***
Squamata - Gekkonidae
e Spotted Thick-toed Gecko; Spotted Gecko (CSA); Puff-adder Lizard
d Gefleckter Dickfingergecko

6073 ***Pachydactylus mariquensis***
Squamata - Gekkonidae
e Ceres Thick-toed Gecko; Marico Gecko (CSA)

6074 ***Pachydactylus namaquensis***
Squamata - Gekkonidae
e Namaqua Thick-toed Gecko; Namaqua Gecko (CSA)

6075 ***Pachydactylus oculatus***
Squamata - Gekkonidae
e Inland Thick-toed Gecko; Golden Spotted Gecko (CSA)

6076 ***Pachydactylus oreophilus***
Squamata - Gekkonidae
e African Thick-toed Gecko; Kaokoveld Rock Gecko (CSA)

6077 ***Pachydactylus punctatus***
Squamata - Gekkonidae
e Pointed Thick-toed Gecko; Speckled Gecko (CSA); Spotted Thick-toed Gecko; Spotted Ground Gecko (CSA); Spotted Thick-toed Ground Gecko; Speckled Thick-toed Gecko (CSA)

6078 ***Pachydactylus rugosus***
Squamata - Gekkonidae
e Wrinkled Thick-toed Gecko; Rough-scaled Gecko (CSA)

6079 ***Pachydactylus sansteyni***
Squamata - Gekkonidae
e Coastal Thick-toed Gecko; Sansteyn's Gecko (CSA)

6080 ***Pachydactylus scutatus***
Squamata - Gekkonidae
e Shielded Thick-toed Gecko; Large-scaled Gecko (CSA)

6081 ***Pachydactylus serval***
Squamata - Gekkonidae
e Werner's Thick-toed Gecko; Western Spotted Gecko (CSA)

6082 ***Pachydactylus tetensis***
Squamata - Gekkonidae
e Zambezi Thick-toed Gecko; Tete Gecko (CSA); Zambezi Rock Gecko; Zambezi Thick-toed Rock Gecko

6083 ***Pachydactylus tigrinus***
Squamata - Gekkonidae
e Tiger Thick-toed Gecko; Tiger Gecko (CSA); Tiger Rock Gecko (CSA); Tigroid Thick-toed Gecko (CSA); Tiger Thick-toed Ground Gecko

6084 ***Pachydactylus tsodiloensis***
Squamata - Gekkonidae
e Tsodilo Thick-toed Gecko; Tsodilo Gecko (CSA)
d Tsodilo-Dickfingergecko

6085 ***Pachydactylus tuberculosus***
Squamata - Gekkonidae
e Warty Thick-toed Gecko

6086 ***Pachydactylus weberi***
Squamata - Gekkonidae
e Weber's Thick-toed Gecko;

Weber's Gecko (CSA)

6087 ***Palea***
Testudines - Trionychidae
e Wattleneck Turtles; Wattleneck Softshells; Wattle-necked Softshells
d Nackendornenweichschildkröten

6088 ***Palea steindachneri***
Testudines - Trionychidae
e Wattleneck Turtle; Wattleneck Softshell; Wattle-necked Soft-shelled Turtle (NA); Wattle-necked Softshell Turtle; Spiny-necked Softshell Turtle; Wattle-necked Softshells
d Nackendornenweichschildkröte
f Trionyx à tubercules

6089 ***Paleosuchus***
Crocodylia - Alligatoridae
e Smooth-fronted Caimans; Dwarf Caimans
d Glattstirnkaimane
f Paléosuches; Caïmans à front lisse; Caïmans cuirassés

6090 ***Paleosuchus palpebrosus***
Crocodylia - Alligatoridae
e Dwarf Caiman; Cuvier's Dwarf Caiman; Cuvier's Smooth-fronted Caiman; Musky Caiman
d Brauen-Glattstirnkaiman
f Caïman à paupières osseuses
i Caimano nano

6091 ***Paleosuchus trigonotus***
Crocodylia - Alligatoridae
e Smooth-fronted Caiman; Schneider's Dwarf Caiman; Schneider's Smooth-fronted Caiman
d Keilkopf-Glattstirnkaiman
f Caïman à front lisse; Caïman gris; Caiman de Schneider

6092 ***Palmatogecko***
Squamata - Gekkonidae
e Namib Geckos; Web-footed Geckos (CSA)
d Namib-Geckos; Wüstengeckos; Sandgeckos
f Geckos du désert

6093 ***Palmatogecko rangei***
Squamata - Gekkonidae
e Namib Sand Gecko; Namib Web-footed Gecko; Web-footed Gecko (CSA)
d Namib-Gecko; Sandgecko; Wüstengecko; Schwimmfußgecko
f Gecko du désert; Gecko palmé des sables; Gecko palmé namibien; Gecko palmé de Namibie

6094 ***Palmatogecko vanzyli***
Squamata - Gekkonidae
e Namib Desert Gecko; Kaoko Web-footed Gecko

6095 ***Pamelaescincus***
Squamata - Scincidae
e Gardiner's Skinks

6096 ***Pamelaescincus gardinieri***
Squamata - Scincidae
e Gardinier's Skink

6097 ***Panaspis***
Squamata - Scincidae
e Lidless Skinks; Snake-eyed Skinks (CSA); African Lidless Skinks

6098 ***Panaspis africana***
Squamata - Scincidae
e Guinea Lidless Skink

6099 ***Panaspis annobonensis***
Squamata - Scincidae
e Annobon Lidless Skink

6100 ***Panaspis breviceps***
Squamata - Scincidae
e Peters's Lidless Skink

6101 ***Panaspis cabindaee***
Squamata - Scincidae
e Cabinda Lidless Skink

6102 ***Panaspis nimbaensis***
Squamata - Scincidae
e Ruwenzori Lidless Skink

6103 ***Panaspis rohdei***
Squamata - Scincidae
e Gaboon Lidless Skink; Snake-eyed Skink

6104 ***Panaspis togoensis***
Squamata - Scincidae
e Togo Lidless Skink

6105 ***Panaspis wahlbergi***
Squamata - Scincidae
e Savanna Lidless Skink; Wahlberg's Snake-eyed Skink (CSA); Savannah Lidless Skink

6106 ***Pantodactylus***
Squamata - Teiidae
e Many-fingered Teiids

6107 ***Pantodactylus quadrilineatus***
Squamata - Teiidae
e Lined Many-fingered Teiid

6108 ***Pantodactylus schreibersii***
Squamata - Teiidae
e Schreibers's Many-fingered Teiid

6109 ***Papuascincus***
Squamata - Scincidae
e Emo Skinks

6110 ***Papuascincus buergersi***
Squamata - Scincidae
e Bürger's Emo Skink

6111 ***Papuascincus morocana***
Squamata - Scincidae
e Morocco Emo Skink

6112 ***Paracontias***
Squamata - Scincidae
e Stone Skinks

6113 ***Paracontias brocchii***
Squamata - Scincidae
e Stone Skink

6114 ***Paracontias hildenbrandti***
Squamata - Scincidae
e Hildenbrandt's Skink

6115 ***Paracontias holomelas***
Squamata - Scincidae
e Anzahamaru Skink

6116 ***Paracontias milloti***
Squamata - Scincidae
e Nosy Mamoko Skink

6117 ***Paracontias rothschildi***
Squamata - Scincidae
e Rothschild's Skink

6118 ***Paradelma***
Squamata - Gekkonidae
e Queensland Legless Lizards

6119 ***Paradelma orientalis***
Squamata - Gekkonidae
e Queensland Legless Lizard

6120 ***Paragehyra***
Squamata - Gekkonidae
e Angel's Petite Geckos

6121 ***Paragehyra petiti***
Squamata - Gekkonidae
e Angel's Petite Gecko

6122 ***Parahydrophis***
Squamata - Elapidae
e Roux's Sea Snakes

6123 ***Parahydrophis mertoni***
Squamata - Elapidae
e Roux's Sea Snake

6124 ***Paranaja***
Squamata - Elapidae
e Many-banded Snakes

6125 ***Paranaja multifasciata***
Squamata - Elapidae
e Many-banded Snake; Burrowing Cobra

6126 ***Parapistocalamus***
Squamata - Elapidae
e Hediger's Snakes

6127 ***Parapistocalamus hedigeri***
Squamata - Elapidae
e Hediger's Snake

6128 ***Parapostolepis***
Squamata - Colubridae
e Shingled Water Snakes

6129 ***Parapostolepis polylepis***
Squamata - Colubridae
e Shingled Water Snake

6130 ***Pararhabdophis***
Squamata - Colubridae
e Vietnam Water Snakes

6131 ***Pararhabdophis chapaensis***
Squamata - Colubridae
e Vietnam Water Snake

6132 ***Pararhadinea***
Squamata - Colubridae
e Madagascar Burrowing Snakes

6133 ***Pararhadinea melanogaster***
Squamata - Colubridae
e Madagascar Burrowing Snake

6134 ***Pareas***
Squamata - Colubridae
e Asian Slug Snakes; Slug-eating Snakes; Slug Snakes
d Asiatische Schneckennattern

6135 ***Pareas boulengeri***
Squamata - Colubridae
e Boulenger's Slug Snake

6136 ***Pareas carinatus***
Squamata - Colubridae
e Keeled Slug Snake

6137 ***Pareas chinensis***
Squamata - Colubridae
e Chinese Slug Snake

6138 ***Pareas formosensis***
Squamata - Colubridae
e Formosa Slug Snake

6139 ***Pareas hamptoni***
Squamata - Colubridae
e Hampton's Slug Snake

6140 ***Pareas komaii***
Squamata - Colubridae
e Taiwan Slug Snake

6141 ***Pareas macularius***
Squamata - Colubridae
e Spotted Slug Snake

6142 ***Pareas margaritophorus***
Squamata - Colubridae
e Mountain Slug Snake; White-spotted Slug Snake
d Weißfleckenschneckennatten

6143 ***Pareas monticola***
Squamata - Colubridae
e Common Slug Snake; Cantor's Slug Snake; Snail Snake

6144 ***Pareas stanleyi***
Squamata - Colubridae
e Stanley's Slug Snake

6145 ***Pareas vertebralis***
Squamata - Colubridae
e Vertebral Slug Snake

6146 ***Paroedura***
Squamata - Gekkonidae
e Madagascar Ground Geckos

6147 ***Paroedura androyensis***
Squamata - Gekkonidae
e Grandidier's Madagascar Ground Gecko

6148 ***Paroedura bastardi***
Squamata - Gekkonidae
e Mocquard's Madagascar Ground Gecko

6149 ***Paroedura gracilis***
Squamata - Gekkonidae
e Graceful Madagascar Ground Gecko

6150 ***Paroedura guibeae***
Squamata - Gekkonidae
e SouthernMadagascar Ground Gecko

6151 ***Paroedura homalorhinus***
Squamata - Gekkonidae
e Northern Madagascar Ground Gecko

6152 ***Paroedura oviceps***
Squamata - Gekkonidae
e Nosy Be Ground Gecko

6153 ***Paroedura pictus***
Squamata - Gekkonidae
e Madagascar Ground Gecko; Madagascar Big-headed Gecko; Ocelot Gecko
d Madagaskar-Großkopfgecko
f Gecko-panthère; Gecko à grosse tête; Gecko peint de Madagascar

6154 ***Paroedura sanctijohannis***
Squamata - Gekkonidae
e Comoro Ground Gecko

6155 ***Paroedura stumpffi***
Squamata - Gekkonidae
e Stumpff's Madagascar Ground Gecko

6156 ***Pedioplanis***
Squamata - Lacertidae
e Sand Lizards
d Wüstenrenner

6157 ***Pedioplanis benguelensis***
Squamata - Lacertidae
e Bocage's Sand Lizard

6158 ***Pedioplanis breviceps***
Squamata - Lacertidae
e Sternfeld's Sand Lizard; Short-headed Sand Lizard (CSA)
d Kurzkopfwüstenrenner

6159 ***Pedioplanis burchelli***
Squamata - Lacertidae
e Burchell's Sand Lizard

6160 ***Pedioplanis gaerdesi***
Squamata - Lacertidae
e Mayer's Sand Lizard

6161 ***Pedioplanis husabensis***
Squamata - Lacertidae
e Husab Sand Lizard

6162 ***Pedioplanis inornata***
Squamata - Lacertidae

e Plain Sand Lizard

6163 ***Pedioplanis laticeps***
Squamata - Lacertidae
e Cape Sand Lizard

6164 ***Pedioplanis lineoocellata***
Squamata - Lacertidae
e Spotted Sand Lizard; Ocellated Sand Lizard
f Lézard linéo-ocellé

6165 ***Pedioplanis namaquensis***
Squamata - Lacertidae
e Namaqua Sand Lizard
d Namaqua-Wüstenrenner
f Lézard namaquensis

6166 ***Pedioplanis rubens***
Squamata - Lacertidae
e Ruben's Sand Lizard

6167 ***Pedioplanis undata***
Squamata - Lacertidae
e Western Sand Lizard

6168 ***Pelamis***
Squamata - Elapidae
e Pelagic Snakes; Pelagic Sea Snakes; Yellow-bellied Sea Snakes
d Plättchenseeschlangen

6169 ***Pelamis platurus***
Squamata - Elapidae
e Pelagic Snake; Pelagic Sea Snake; Yellow-bellied Sea Snake; Yellowbelly Sea Snake; Parti-coloured Sea Snake; Flat-tail (ISC); Black-and-yellow Sea Snake (ISC)
d Plättchenseeschlange; Hochseeschlange

6170 ***Pelochelys***
Testudines - Trionychidae
e Asian Giant Softshells
d Riesenweichschildkröten

6171 ***Pelochelys bibroni***
Testudines - Trionychidae
e Asian Giant Softshell; Asian Giant Soft-shelled Turtle; Asian Giant Softshell Turtle; New Guinea Giant Softshell Turtle
d Riesenweichschildkröte
f Trionyx géant

6172 ***Pelochelys cantorii***
Testudines - Trionychidae
e Asian Giant Softshell Turtle; Cantor's Giant Softshell Turtle; Antipa Turtle; Blunt-headed Giant Softshell Turtle

6173 ***Pelodiscus***
Testudines - Trionychidae
e Chinese Softshells; Chinese Soft-shelled Turtles
d Chinesische Weichschildkröten

6174 ***Pelodiscus sinensis***
Testudines - Trionychidae
e Chinese Softshell Turtle; Chinese Soft-shelled Turtle; Spiny Softshell Turtle; Chinese Softshell
d Chinesische Weichschildkröte
f Trionyx de Chine
i Tartaruga a guscio molle cinese

6175 ***Pelomedusa***
Testudines - Pelomedusidae
e Helmeted Turtles; Helmeted Terrrapins
d Starrbrustpelomedusenschildkröten
f Péloméduses
i Pelomeduse; Pelomeduse africane

6176 ***Pelomedusa subrufa***
Testudines - Pelomedusidae
e Helmeted Turtle; African Helmeted Turtle; Helmeted Terrapin (CSA); Marsh Terrapin (CSA); Common African Helmeted Turtle; African Hingeless Terrapin; Cape Terrapin

(CSA)
d Starrbrustschildkröte; Weißbrustpelomedusenschildkröte; Starrbrustpelomedusenschildkröte
f Tortue aquatique; Péloméduse roussâtre; Pélusios rougâtre
i Pelomedusa; Tartaruga elmetto

6177 Pelomedusidae
Testudines
e Side-necked Turtles; Afro-American Side-necked Turtles; Hidden-neck Turtles; River Turtles
d Pelomedusenschildkröten
f Pélomédusidés
i Tartarughe con il collo di lato; Pelomedusidi

6178 *Peltocephalus*
Testudines - Pelomedusidae
e Bighead Amazon River Turtles; Big-headed River Turtles
d Duméril-Schienenschildkröten
f Peltocéphales

6179 *Peltocephalus dumerilianus*
Testudines - Pelomedusidae
e Bighead Amazon River Turtle; Big-headed Amazon River Turtle; Big-headed Side-necked Turtle; Big-headed River Turtle
d Duméril-Schienenschildkröte
f Peltocéphale; Podocnémide de Duméril
i Tartaruga capogrosso

6180 *Pelusios*
Testudines - Pelomedusidae
e African Hinged Terrapins; Hinged Terrapins (CSA)
d Klappbrustpelomedusen
f Péluses
i Tartarughe scatola africane

6181 *Pelusios adansonii*
Testudines - Pelomedusidae
e Adanson's Mud Turtle; Adanson's Turtle; Adanson's Terrapin; Sahelian Terrapin; Adanson's Hinged Terrrapin
d Weißbrustpelomedusenschildkröte; Weißbrustpelomeduse
f Péluse d'Adanson; Pélusios d'Adanson

6182 *Pelusios bechuanicus*
Testudines - Pelomedusidae
e Okavango Mud Turtle; Okavanga Hinged Terrapin (CSA); Okavanga Terrapin; Fitzsimons's Terrapin
f Péluse de l'Okavango

6183 *Pelusios broadleyi*
Testudines - Pelomedusidae
e Turkana Mud Turtle; Lake Turkana Mud Turtle; Broadley's Mud Terrapin; Kenya Mud Terrapin
f Péluse de Broadley

6184 *Pelusios carinatus*
Testudines - Pelomedusidae
e African Keeled Mud Turtle; Laurent's Terrapin; Keel-backed Terrapin
f Péluse à dos caréné
i Tartaruga del fango africana carenata

6185 *Pelusios castaneus*
Testudines - Pelomedusidae
e West African Mud Turtle; African Mud Turtle; Chestnut Hinged Terrapin (CSA); Schweigger's Terrapin; Chestnut Terrapin; Brown Hinged Terrapin; Swamp Hinged Terrapin
f Péluse de Schweigger; Pseudémyde
i Tartaruga del fango africana

6186 *Pelusios castanoides*
Testudines - Pelomedusidae
e Yellowbelly Mud Turtle; Eastern Hinged Terrpin (CSA); Yellow-bellied Mud Turtle; Bianconi's

Terrapin; East African Yellow-bellied Mud Turtle
i Tartaruga del fango gialla

6187 ***Pelusios chapini***
Testudines - Pelomedusidae
e Central African Mud Turtle; Centrtal African Giant Terrapin; Chapin's Terrapin

6188 ***Pelusios gabonsensis***
Testudines - Pelomedusidae
e African Forest Turtle; Gabon Turtle; African Forest Terrapin; Gabon Terrapin; African Black Terrapin; Gabon Hinged Terrapin
d Rückstreifenpelomedusenschildkröte; Rückenstreifenpelomeduse
f Péluse du Gabon; Pélusios noir

6189 ***Pelusios nanus***
Testudines - Pelomedusidae
e African Dwarf Mud Turtle; African Dwarf Terrapin
f Péluse naine

6190 ***Pelusios niger***
Testudines - Pelomedusidae
e West African Black Turtle; West African Black Forest Turtle; West African Black Forest Terrapin; Black Terrapin; Hooked-beak Terrapin; African Black Terrapin; Black Hinged Terrapin
d Schwarze Pelomedusenschildkröte; Schwarze Pelomeduse
f Péluse à bec crochu; Pélusios noir
i Tartaruga di foresta africana nera

6191 ***Pelusios rhodesianus***
Testudines - Pelomedusidae
e Variable Mud Turtle; Mashona Hinged Terrapin(CSA); Hewitt's Terrapin

6192 ***Pelusios seychellensis***
Testudines - Pelomedusidae
e Seychelles Mud Turtle; Siebenrock's Terrapin; Seychelles Black Terrapin

6193 ***Pelusios sinuatus***
Testudines - Pelomedusidae
e East African Serrated Mud Turtle; Serrated Turtle; Serrated Hinged Terrapin (CSA); Hinged Terrapin (CSA); Smith's Terrapin
d Gezähnelte Pelomedusenschildkröte; Gezähnelte Pelomeduse
f Péluse à bord sineux
i Pelusio a bordo sinuato

6194 ***Pelusios subniger***
Testudines - Pelomedusidae
e East African Black Mud Turtle; Pan Hinged Terrapin (CSA); East African Side-necked Turtle
d Dunkle Pelomedusenschildkröte; Dunkle Pelomeduse
f Péluse noirâtre
i Pelusio nerastro

6195 ***Pelusios upembae***
Testudines - Pelomedusidae
e Upemba Mud Turtle; Upemba Basin Terrapin; Broadley's Terrapin

6196 ***Pelusios williamsi***
Testudines - Pelomedusidae
e Williams's Mud Turtle; Williams's African Mud Turtle; Lake Victoria Mud Turtle; Williams's Terrapin
f Péluse duLac Victoria

6197 ***Perochirus***
Squamata - Gekkonidae
e Tropical Geckos

6198 ***Perochirus ateles***
Squamata - Gekkonidae
e Duméril's Tropical Gecko

6199 ***Perochirus guentheri***
Squamata - Gekkonidae

e Günther'sTropical Gecko

6200 ***Perochirus scutellatus***
Squamata - Gekkonidae
e Shielded Tropical Gecko

6201 ***Petrosaurus***
Squamata - Iguanaidae
e Rock Lizards; California Rock Lizards
d Felsensleguane

6202 ***Petrosaurus mearnsi***
Squamata - Iguanaidae
e Banded Rock Lizard
d Gebänderter Felsenleguan
f Streptosaure de Mearns

6203 ***Petrosaurus mearnsi mearnsi***
Squamata - Iguanaidae
e Mearns's Rock Lizard

6204 ***Petrosaurus thalassinus***
Squamata - Iguanaidae
e Baja California Rock Lizard

6205 ***Petrosaurus thalassinus repens***
Squamata - Iguanaidae
e Short-nosed Rock Lizard; Shortnose Rock Lizard

6206 ***Petrosaurus thalassinus thalassinus***
Squamata - Iguanaidae
e San Lucan Rock Lizard

6207 ***Phelsuma***
Squamata - Gekkonidae
e Day Geckos; Madagascar Green Geckos; Madagascar Day Geckos
d Taggeckos; Madagassische Taggeckos; Phelsumen
f Phelsumes; Geckos diurnes

6208 ***Phelsuma abbotti***
Squamata - Gekkonidae
e Aldabra Day Gecko; Abbott's Gecko
d Taggecko; Aldabra-Taggecko; Olivgrüner Taggecko; Seychellen-Taggecko
f Phelsumes des Seychelles

6209 ***Phelsuma andamenensis***
Squamata - Gekkonidae
e Andaman Day Gecko; Andaman Emerald Gecko (ISC); Emerald Gecko (ISC)
d Andamanen-Taggecko

6210 ***Phelsuma astriata***
Squamata - Gekkonidae
e Seychelles Day Gecko; Day Gecko
d Kleiner Seychellen-Taggecko

6211 ***Phelsuma barbouri***
Squamata - Gekkonidae
e Barbour's Day Gecko
d Barbour's Taggecko; Gebirgstaggecko

6212 ***Phelsuma befotakensis***
Squamata - Gekkonidae
e Northwestern Day Gecko

6213 ***Phelsuma borbonica***
Squamata - Gekkonidae
e Reunion Day Gecko

6214 ***Phelsuma cepediana***
Squamata - Gekkonidae
e Bluetail Day Gecko; Madagascar Gecko; Blue-tailed Day Gecko
d Blauschwanztaggecko; Blauschwanzphelsume
f Phelsuma

6215 ***Phelsuma chekei***
Squamata - Gekkonidae
e Northern Day Gecko
f Gecko diurne du Nord

6216 ***Phelsuma comorensis***
Squamata - Gekkonidae
e Comoro Day Gecko
d Komoren-Taggecko

6217 ***Phelsuma dubia***
Squamata - Gekkonidae
e Zanzibar Day Gecko

6218 ***Phelsuma flavigularis***
Squamata - Gekkonidae
e Yellow-throated Day Gecko
d Gelbkehliger Taggecko

6219 ***Phelsuma gigas***
Squamata - Gekkonidae
e Rodrigues's Day Gecko

6220 ***Phelsuma guentheri***
Squamata - Gekkonidae
e Round Island Day Gecko
d Günthers Taggecko

6221 ***Phelsuma guimbeaui***
Squamata - Gekkonidae
e Mauritius Day Gecko; Orange-spotted Day Gecko (NA); Day Gecko
d Guimbeau-Taggecko; Guimbeaus Taggecko

6222 ***Phelsuma guimbeaui guimbeaui***
Squamata - Gekkonidae
e Guimbeau's Day Gecko

6223 ***Phelsuma guttata***
Squamata - Gekkonidae
e Speckled Day Gecko

6224 ***Phelsuma laticauda***
Squamata - Gekkonidae
e Broad-tailed Day Gecko; Madagascar Gecko
d Goldstaubtaggecko; Flachschwanztaggecko

6225 ***Phelsuma laticauda laticauda***
Squamata - Gekkonidae
e Gold-dust Day Gecko

6226 ***Phelsuma lineata***
Squamata - Gekkonidae
e Striped Day Gecko; Lined Day Gecko
d Streifentaggecko
f Gecko diurne à ligne; Gecko diurne ligné; Gecko diurne à bande

6227 ***Phelsuma madagascariensis***
Squamata - Gekkonidae
e Madagascar Day Gecko; Madagascar Gecko; Giant Day Gecko; Grandis Day Gecko
d Großer Madagaskar-Taggecko; Madagassischer Taggecko; Madagassischer Großer Taggecko; Großer Taggecko
f Phelsume de Madagascar; Gecko diurne géant

6228 ***Phelsuma madagascariensis grandis***
Squamata - Gekkonidae
e Madagascar Giant Day Gecko
f Gecko diurne géant de Madagascar

6229 ***Phelsuma minuthi***
Squamata - Gekkonidae
e Borner's Day Gecko

6230 ***Phelsuma modesta***
Squamata - Gekkonidae
e Modest Day Gecko
f Gecko diurne modeste

6231 ***Phelsuma mutabilis***
Squamata - Gekkonidae
e Thicktail Day Gecko
d Veränderlicher Taggecko
f Gecko diurne à queue fine

6232 ***Phelsuma nigristriata***
Squamata - Gekkonidae

e Island Day Gecko
d Schwarzstreifentaggecko

6233 ***Phelsuma ocellata***
Squamata - Gekkonidae
e Ocellated Day Gecko; Namaqua Day Gecko (CSA)

6234 ***Phelsuma ornata***
Squamata - Gekkonidae
e Ornate Day Gecko; Day Gecko
f Gecko diurne orné de Mauritius

6235 ***Phelsuma parkeri***
Squamata - Gekkonidae
e Parker's Day Gecko

6236 ***Phelsuma quadriocellata***
Squamata - Gekkonidae
e Peacock Day Gecko
d Augenflecktaggecko; Pfauenaugentaggecko; Pfauenaugetaggecko
f Phelsume à quatre ocelles

6237 ***Phelsuma robertmertensi***
Squamata - Gekkonidae
e Mertens's Day Gecko
d Mertens-Taggecko

6238 ***Phelsuma serraticauda***
Squamata - Gekkonidae
e Serrated Day Gecko
d Sägeschwanztaggecko

6239 ***Phelsuma standingi***
Squamata - Gekkonidae
e Standing's Day Gecko
d Standing-Taggecko

6240 ***Phelsuma sundbergi***
Squamata - Gekkonidae
e Seychelles Giant Day Gecko
d Großer Seychellen-Taggecko

6241 ***Phelsuma sundbergi ladiguensis***
Squamata - Gekkonidae
d La-Digue-Taggecko

6242 ***Phelsuma sundbergi sundbergi***
Squamata - Gekkonidae
e Seychelles Giant Day Gecko
d Großer Seychellen-Taggecko

6243 ***Phelsuma trilieanta***
Squamata - Gekkonidae
e Three-lined Day Gecko

6244 ***Phelsuma v-nigra***
Squamata - Gekkonidae
e Indian Day Gecko
d Komoren-Zwergtaggecko

6245 ***Phenacosaurus***
Squamata - Iguanaidae
e Andes Anoles; False Anoles
d Anden-Anolis

6246 ***Phenacosaurus heterodermus***
Squamata - Iguanaidae
e Flat Andes Anole

6247 ***Phenacosaurus neblinae***
Squamata - Iguanaidae
e Myers's Andes Anole

6248 ***Phenacosaurus nicefori***
Squamata - Iguanaidae
e Nicéforo's Andes Anole

6249 ***Phenacosaurus orcesi***
Squamata - Iguanaidae
e Orces's Andes Anole

6250 ***Phenacosaurus richteri***
Squamata - Iguanaidae
e Richter's Andes Anole

6251 ***Philochortus***
Squamata - Lacertidae
e Orangetail Lizards; Shield-backed Ground Lizards

6252 ***Philochortus hardeggeri***
Squamata - Lacertidae
e Hardegger's Orangetail Lizard

6253 ***Philochortus intermedia***
Squamata - Lacertidae
e Southern Orangetail Lizard; Southern Shield-backed Ground Lizard

6254 ***Philochortus lhotei***
Squamata - Lacertidae
e Lhote Orangetail Lizard

6255 ***Philochortus neumanni***
Squamata - Lacertidae
e Neumann's Orangetail Lizard; Neumann's Lizard

6256 ***Philochortus phillipsii***
Squamata - Lacertidae
e Phillips's Orangetail Lizard

6257 ***Philochortus spinalis***
Squamata - Lacertidae
e Eritrea Orangetail Lizard

6258 ***Philochortus zolli***
Squamata - Lacertidae
e Scortecci's Orangetail Lizard

6259 ***Philodryas***
Squamata - Colubridae
e Green Racers; Cobra-cipos; Palm Snakes
d Grüne Strauchnattern

6260 ***Philodryas aestivus***
Squamata - Colubridae
e Brazilian Green Racer

6261 ***Philodryas arnaldoi***
Squamata - Colubridae
e Arnaldo's Green Racer

6262 ***Philodryas baroni***
Squamata - Colubridae
e Baron's Green Racer
d Langnasenstrauchnatter; Argentinische Langnasenstrauchnatter

6263 ***Philodryas burmeisteri***
Squamata - Colubridae
e Burmeister's Green Racer

6264 ***Philodryas carbonelli***
Squamata - Colubridae
e Roze's Green Racer

6265 ***Philodryas chamissonis***
Squamata - Colubridae
e Chilean Green Racer

6266 ***Philodryas elegans***
Squamata - Colubridae
e Elegant Green Racer

6267 ***Philodryas inca***
Squamata - Colubridae
e Tropical Green Racer

6268 ***Philodryas mattogrossensis***
Squamata - Colubridae
e Miranda Green Racer

6269 ***Philodryas nattereri***
Squamata - Colubridae
e Paraguay Green Racer

6270 ***Philodryas olfersii***
Squamata - Colubridae
e Lichtenstein's Green Racer
d Grüne Strauchnatter

6271 ***Philodryas olfersii herbeus***
Squamata - Colubridae
e Green-bellied Palm Snake

6272 ***Philodryas oligolepis***
Squamata - Colubridae
e Gomes's Green Racer

6273 ***Philodryas patagoniensis***
Squamata - Colubridae
e Patagonia Green Racer

6274 ***Philodryas psammophideus***
Squamata - Colubridae
e Günther's Green Racer

6275 ***Philodryas simonsi***
Squamata - Colubridae
e Simon's Green Racer

6276 ***Philodryas tachymenoides***
Squamata - Colubridae
e Schmidt's Green Racer

6277 ***Philodryas varius***
Squamata - Colubridae
e Jan's Green Racer

6278 ***Philodryas viridissimus***
Squamata - Colubridae
e Common Green Racer; Green Palm Snake

6279 ***Philothamnus***
Squamata - Colubridae
e African Green Snakes; Green Bush Snakes; Bush Snakes
d Grüne Buschschlangen; Grüne Buschnattern; Grünnattern

6280 ***Philothamnus angolensis***
Squamata - Colubridae
e Angola Green Snake; Western Green Snake (CSA)

6281 ***Philothamnus battersbyi***
Squamata - Colubridae
e Battersby's Green Snake

6282 ***Philothamnus bequaerti***
Squamata - Colubridae
e Bequaert's Green Snake

6283 ***Philothamnus carinatus***
Squamata - Colubridae
e Thirteen-scaled Green Snake

6284 ***Philothamnus dorsalis***
Squamata - Colubridae
e Striped Green Snake; Striped Wood Snake

6285 ***Philothamnus girardi***
Squamata - Colubridae
e Girard's Green Snake

6286 ***Philothamnus heterodermus***
Squamata - Colubridae
e Variable Green Snake; Thirteen-scaled Snake

6287 ***Philothamnus heterolepidotus***
Squamata - Colubridae
e Slender Green Snake

6288 ***Philothamnus hoplogaster***
Squamata - Colubridae
e Southeastern Green Snake (CSA); Green Water Snake (CSA); Common Green Water Snake (CSA); Common Green Bush Snake

6289 ***Philothamnus hughesi***
Squamata - Colubridae
e Hughes's Green Snake

6290 ***Philothamnus irregularis***
Squamata - Colubridae
e Western Green Snake (CSA); Northern Green Bush Snake;

Northern Green Snake (CSA)
d Grüne Sumpfschlange

6291 ***Philothamnus macrops***
Squamata - Colubridae
e Usambara Green Snake

6292 ***Philothamnus natalensis***
Squamata - Colubridae
e East Natal Green Snake; Natal Green Water Snake; Natal Green Bush Snake

6293 ***Philothamnus natalensis natalensis***
Squamata - Colubridae
e Natal Green Snake (CSA)

6294 ***Philothamnus natalensis occidentalis***
Squamata - Colubridae
e Natal Green Snake (CSA)

6295 ***Philothamnus nitidus***
Squamata - Colubridae
e Green Bush Snake

6296 ***Philothamnus ornatus***
Squamata - Colubridae
e Ornate Green Snake

6297 ***Philothamnus punctatus***
Squamata - Colubridae
e Spotted Green Snake; Green Bush Snake

6298 ***Philothamnus semivariegatus***
Squamata - Colubridae
e Variegated Green Snake; Spotted Bush Snake (CSA); Green Bush Snake; Variegated Bush Snake (CSA); Bush Snake (CSA)
d Gefleckte Buschschlange

6299 ***Phimophis***
Squamata - Colubridae
e Pampas Snakes
d Schnauzennattern

6300 ***Phimophis guerini***
Squamata - Colubridae
e Argentine Pampas Snake

6301 ***Phimophis guianensis***
Squamata - Colubridae
e Troschel's Pampas Snake; Southern Hook-nosed Snake

6302 ***Phimophis iglesias***
Squamata - Colubridae
e Gomes's Pampas Snake

6303 ***Phimophis vittatus***
Squamata - Colubridae
e Banded Pampas Snake

6304 ***Phoboscincus***
Squamata - Scincidae
e Garnier's Skinks

6305 ***Phoboscincus garnieri***
Squamata - Scincidae
e Garnier's Skink

6306 ***Pholidobolus***
Squamata - Teiidae
e Pholidoboluses

6307 ***Pholidobolus affinis***
Squamata - Teiidae
e Peters's Pholidobolus

6308 ***Pholidobolus annectens***
Squamata - Teiidae
e Parker's Pholidobolus

6309 ***Pholidobolus anomalus***
Squamata - Teiidae
e Strange Pholidobolus

6310 ***Pholidobolus macbrydei***
Squamata - Teiidae
e MacBryde's Pholidobolus

6311 ***Pholidobolus montium***
Squamata - Teiidae
e Mountain Pholidobolus
d Zwergteju; Gebirgszwergteju

6312 ***Pholidobolus prefrontalis***
Squamata - Teiidae
e Montanucci's Pholidobolus

6313 ***Phoxophrys***
Squamata - Agamidae
e Eyebrow Lizards; Eyebrowed Lizards
d Borneo-Bergagamen

6314 ***Phoxophrys borneensis***
Squamata - Agamidae
e Sabah's Eyebrow Lizard

6315 ***Phoxophrys cephalum***
Squamata - Agamidae
e Mocquard's Eyebrow Lizard

6316 ***Phoxophrys nigrilabris***
Squamata - Agamidae
e Black-lipped Eyebrow Lizard

6317 ***Phoxophrys spiniceps***
Squamata - Agamidae
e Sarawak Eyebrow Lizard
d Borneo-Bergagame

6318 ***Phoxophrys tuberculata***
Squamata - Agamidae
e Hubrecht's Eyebrow Lizard

6319 ***Phrynocephalus***
Squamata - Agamidae
e Toadhead Agamas; Toad Agamas (ISC); Toad-headed Agamas
d Krötenkopfagamen; Krötenköpfe
f Agames à tête de crapaud; Phrynocéphales

6320 ***Phrynocephalus acutirostris***
Squamata - Agamidae
e Sharpnose Toadhead Agama

6321 ***Phrynocephalus alpherakii***
Squamata - Agamidae
e Toadhead Agama

6322 ***Phrynocephalus alticola***
Squamata - Agamidae
e Jammu Toadhead Agama

6323 ***Phrynocephalus arabicus***
Squamata - Agamidae
e Arabian Toadhead Agama

6324 ***Phrynocephalus arcellazzii***
Squamata - Agamidae
e Central Asian Toadhead Agama

6325 ***Phrynocephalus axillaris***
Squamata - Agamidae
e Greater Toadhead Agama

6326 ***Phrynocephalus birulai***
Squamata - Agamidae
e Tsarewsky's Toadhead Agama

6327 ***Phrynocephalus blanfordi***
Squamata - Agamidae
e Blanford's Toadhead Agama

6328 ***Phrynocephalus carinilabris***
Squamata - Agamidae
e Lesser Toadhead Agama

6329 ***Phrynocephalus ciliaris***
Squamata - Agamidae
e Cryptic Toadhead Agama

6330 ***Phrynocephalus clarcorum***
Squamata - Agamidae
e Afghan Toadhead Agama

6331 ***Phrynocephalus elegans***
Squamata - Agamidae
e Elegant Toadhead Agama

6332 ***Phrynocephalus erythrurus***
Squamata - Agamidae
e Zugmayer's Toadhead Agama

6333 ***Phrynocephalus euptilopus***
Squamata - Agamidae
e Pakistan Toadhead Agama; Indian Desert Toadhead Agama

6334 ***Phrynocephalus forsythii***
Squamata - Agamidae
e Forsyth's Toadhead Agama

6335 ***Phrynocephalus frontalis***
Squamata - Agamidae
e Shansi Toadhead Agama; Shansi Sand Lizard

6336 ***Phrynocephalus geckoides***
Squamata - Agamidae
e Gecko Toadhead Agama

6337 ***Phrynocephalus grumgrzimailoi***
Squamata - Agamidae
e Grumgrzimail's Toadhead Agama

6338 ***Phrynocephalus guetheri***
Squamata - Agamidae
e Günther's Toadhead Agama

6339 ***Phrynocephalus guttatus***
Squamata - Agamidae
e Spotted Toadhead Agama; Spotted Toad Agama
d Gefleckter Krötenkopf; Schwanzroller
f Phrynocéphale caudivolvue; Phrynocéphale tacheté
i Frinocefallo guttato

6340 ***Phrynocephalus haeckeli***
Squamata - Agamidae
e Haeckel's Toadhead Agama

6341 ***Phrynocephalus helioscopus***
Squamata - Agamidae
e Sunwatcher Toadhead Agama; Sunwatcher
d Sonnengucker
f Phrynocéphale hélioscope
i Frinocefallo elioscopo

6342 ***Phrynocephalus interscapularis***
Squamata - Agamidae
e Lichtenstein's Toadhead Agama
d Stachelkopf

6343 ***Phrynocephalus isseli***
Squamata - Agamidae
e Issel's Toadhead Agama

6344 ***Phrynocephalus koslowi***
Squamata - Agamidae
e Koslow's Toadhead Agama

6345 ***Phrynocephalus kuschakewitschi***
Squamata - Agamidae
e Bédriaga's Toadhead Agama

6346 ***Phrynocephalus laungwalensis***
Squamata - Agamidae
e Rajasthan Toadhead Agama

6347 ***Phrynocephalus lidskii***
Squamata - Agamidae
e Tibetan Toadhead Agama

6348 ***Phrynocephalus ludovici***
Squamata - Agamidae
e Ludovic's Toadhead Agama

6349 ***Phrynocephalus luteoguttatus***
Squamata - Agamidae
e Yellow-speckled Toadhead Agama; Yellow-speckled Toad Agama

6350 ***Phrynocephalus maculatus***
Squamata - Agamidae
e Blacktail Toadhead Agama; Black-tailed Toad Agama
d Gefleckte Krötenkopfagame; Gefleckter Krötenkopf
f Phrynocéphale tacheté

6351 ***Phrynocephalus mystaceus***
Squamata - Agamidae
e Secret Toadhead Agama; Toad-headed Agama
d Bärtiger Krötenkopf; Ohrenkrötenkopf
f Phrynocéphale barbé; Phrynocéphale à oreilles; Phrynocéphale à moustache; Agame des sables
i Frinocefallo mustacchino

6352 ***Phrynocephalus nejdensis***
Squamata - Agamidae
e Haas's Toadhead Agama; Toad-headed Agama
d Arabische Krötenagame

6353 ***Phrynocephalus ornatus***
Squamata - Agamidae
e Striped Toadhead Agama; Striped Toad Agama
d Schmuckkrötenkopf
f Phrynocéphale orné

6354 ***Phrynocephalus parvulus***
Squamata - Agamidae
e Slight Toadhead Agama

6355 ***Phrynocephalus przewalskii***
Squamata - Agamidae
e Przewalski's Toadhead Agama

6356 ***Phrynocephalus pylzowi***
Squamata - Agamidae
e Pylzow's Toadhead Agama

6357 ***Phrynocephalus reldae***
Squamata - Agamidae
e Hidden Toadhead Agama

6358 ***Phrynocephalus reticulatus***
Squamata - Agamidae
e Reticulate Toadhead Agama
d Netzkrötenkopfagame
f Phrynocéphale réticulé

6359 ***Phrynocephalus roborowskii***
Squamata - Agamidae
e Roborowski's Toadhead Agama

6360 ***Phrynocephalus rossikowi***
Squamata - Agamidae
e Uzbekistan Toadhead Agama
d Kisyl-Kum-Krötenkopf

6361 ***Phrynocephalus rostralis***
Squamata - Agamidae
e Tzarewsky's Toadhead Agama

6362 ***Phrynocephalus salenskyi***
Squamata - Agamidae
e Salensky's Toadhead Agama

6363 ***Phrynocephalus scutellatus***
Squamata - Agamidae
e Grey Toadhead Agama

6364 ***Phrynocephalus sogdianus***
Squamata - Agamidae
e Tadjikistan Toadhead Agama

6365 ***Phrynocephalus steindachneri***
Squamata - Agamidae
e Steindachner's Toadhead Agama

6366 ***Phrynocephalus theobaldi***
Squamata - Agamidae

e Himalaya Toadhead Agama
d Himalaja-Krötenkopf
f Phrynocéphale de Théobald

6367 ***Phrynocephalus versicolor***
Squamata - Agamidae
e Tuva Toadhead Agama

6368 ***Phrynocephalus vlangalii***
Squamata - Agamidae
e Ching-Hai Toadhead Agama

6369 ***Phrynops***
Testudines - Chelidae
e Toadhead Turtles; Toad-headed Turtles
d Krötenkopfschildkröten; Froschkopfschildkröten
f Batrachémydes; Platémydes

6370 ***Phrynops dahli***
Testudines - Chelidae
e Dahl's Toadhead Turtle; Dahl's Toad-headed Turtle; Colombian Toad-headed Turtle
d Kolumbianische Froschkopfschildkròte; Kolumbianische Krötenkopfkopfschildkröte
f Batrachémyde de Dahl; Platémyde de Dahl

6371 ***Phrynops geoffroanus***
Testudines - Chelidae
e Geoffroy's Sideneck Turtle; Geoffrey's Side-necked Turtle; Dark Frog-headed Turtle; Geoffroy's Side-necked Turtle; Spotted Toad-headed Turtle
d Dunkle Krötenkopfschildkröte
f Platémyde de Geoffroy; Batrachémyde de Geoffroy
i Tartaruga di Geoffroy

6372 ***Phrynops gibbus***
Testudines - Chelidae
e Gibba Turtle; South American Gibba Turtle; Keel-backed Side-necked Turtle; Lesser Toad-headed Turtle
d Buckelkrötenkopfschildkröte; Buckelschildkröte
f Platémyde bossue; Batrachémyde bossue
i Tartaruga gobbosa dell'America del sud

6373 ***Phrynops hilarii***
Testudines - Chelidae
e Hilaire's Sideneck Turtle; Hilaire's Side-necked Turtle; Spot-bellied Side-necked Turtle
d Hellrandige Krötenkopfschildkröte
f Platémyde de Saint-Hilaire; Batrachémyde de Saint-Hilaire
i Tartaruga di Hilary

6374 ***Phrynops hogei***
Testudines - Chelidae
e Hoge's Sideneck Turtle; Hoge's Side-necked Turtle
f Platémyde de Hoge; Batrachémyde de Hoge

6375 ***Phrynops nasutus***
Testudines - Chelidae
e Common Toadhead Turtle; Common Toad-headed Turtle; American Toad-headed Turtle; Common Toad-headed Side-necked Turtle
d Gewöhnliche Krötenkopfschildkröte; Gewöhnliche Froschkopfschildkröte
f Platémyde de Schweigger; Batrachémyde de Schweigger; Tortue à tête de crapaud
i Tartaruga testa di rospo

6376 ***Phrynops raniceps***
Testudines - Chelidae
e Amazon Toadhead Turtle; Amazon Toad-headed Turtle; Greater Toad-headed Turtle
f Platémyde de l'Amazone; Batrachémyde de l'Amazone

6377 ***Phrynops rufipes***
Testudines - Chelidae
e Red Sideneck Turtle; Red-footed Amazon Side-necked Turtle; Red Turtle; Red Amazon Side-necked Turtle; Red-headed Side-necked Turtle; Red Toad-headed Turtle
d Rote Krötenkopfschildkröte
f Platémyde à pieds rouges; Batrachémyde à pieds rouges

6378 ***Phrynops tuberculatus***
Testudines - Chelidae
e Tuberculate Toadhead Turtle; Tuberculate Toad-headed Turtle
f Platémyde à tubercules; Batrachémyde à tubercules

6379 ***Phrynops tuberosus***
Testudines - Chelidae
e Peters's Side-necked Turtle
d Peters Krötenkopfschildkröte
f Platémyde à dos rugueux; Batrachémyde à dos rugueux

6380 ***Phrynops vanderhaegei***
Testudines - Chelidae
e Vanderhaege's Toadhead Turtle; Vanderhaege's Toad-headed Turtle
f Platémyde de Vanderhaege; Batrachémyde de Vanderhaege

6381 ***Phrynops williams***
Testudines - Chelidae
e Williams's Sideneck Turtle; Williams's Side-necked Turtle; Williams's South American Toad-headed Turtle
f Platémyde de Williams; Batrachémyde de Williams

6382 ***Phrynops zuliae***
Testudines - Chelidae
e Zulia Toadhead Turtle; Zulia Toad-headed Turtle; Zulia Toad-headed Side-necked Turtle
d Zulia-Krötenkopfschildkröte
f Platémyde de Zulia; Batrachémyde de Zulia

6383 ***Phrynosaura***
Squamata - Iguanaidae
e Tree Iguanas

6384 ***Phrynosaura audituvelatus***
Squamata - Iguanaidae
e Tree Iguana

6385 ***Phrynosaura reichei***
Squamata - Iguanaidae
e Reich's Tree Iguana

6386 ***Phrynosaura stolzmanni***
Squamata - Iguanaidae
e Stolzmann'sTree Iguana

6387 ***Phrynosoma***
Squamata - Iguanaidae
e Horned Lizards; American Horned Lizards; Horny Toads (NA)
d Krötenechsen
f Phrynosomes; Lézards cornues; Lézards-crapaud

6388 ***Phrynosoma asio***
Squamata - Iguanaidae
e Giant Horned Lizard; Long-spined Horned Lizard
d Riesenkrötenechse

6389 ***Phrynosoma braconnieri***
Squamata - Iguanaidae
e Short-tail Horned Lizard; Braconnier's Horned Lizard

6390 ***Phrynosoma cerroense***
Squamata - Iguanaidae
e Cedros Island Horned Lizard

6391 ***Phrynosoma cornutum***
Squamata - Iguanaidae
e Texas Horned Lizard

d Texas-Krötenechse
f Phrynosome de Texas

6392 ***Phrynosoma cornutum blainvillei***
Squamata - Iguanaidae
e San Diego Horned Lizard; Pacific Horned Toad (NA)
d Texas-Kröte
f Lézard cornu de San Diego

6393 ***Phrynosoma cornutum frontale***
Squamata - Iguanaidae
e California Horned Lizard

6394 ***Phrynosoma coronatum***
Squamata - Iguanaidae
e Coast Horned Lizard; Pacific Horned Lizard
d Kronenkrötenechse; Nordamerikanische Krötenechse
f Phrynosome de la côte; Phrynosome couronné; Tapaya

6395 ***Phrynosoma douglassii***
Squamata - Iguanaidae
e Short-horned Lizard; Douglass's Horned Lizard (NA); Pygmy Horned Lizard; Pigmy Short-horned Lizard (NA)
d Nordamerikanische Kurzhornkrötenechse; Kurzhornkrötenechse
f Phrynosome de Douglas; Iguane à petities cornues

6396 ***Phrynosoma douglassii brevirostre***
Squamata - Iguanaidae
e Eastern Short-horned Lizard

6397 ***Phrynosoma douglassii douglassii***
Squamata - Iguanaidae
e Pygmy Shorthorned Lizard

6398 ***Phrynosoma douglassii ornatissimum***
Squamata - Iguanaidae
e Desert Short-horned Lizard; Painted Horned Lizard

6399 ***Phrynosoma douglassii ornatum***
Squamata - Iguanaidae
e Salt Lake Short-horned Lizard

6400 ***Phrynosoma hernandesi***
Squamata - Iguanaidae
e Mountain Short-horned Lizard; Greater Short-horned Lizard

6401 ***Phrynosoma hernandesi hernandesi***
Squamata - Iguanaidae
e Hernandez Horned Lizard

6402 ***Phrynosoma mcallii***
Squamata - Iguanaidae
e Flat-tail Horned Lizard; Flat-tailed Horned Lizard; McCall's Horned Lizard
d Flachschwanzkrötenechse
f Phrynosome à queue plate

6403 ***Phrynosoma modestum***
Squamata - Iguanaidae
e Roundtail Horned Lizard; Little Bleached Horned Lizard; Round-tailed Horned Lizard
d Rundschwanzkrötenechse
f Phrynosome à queue ronde

6404 ***Phrynosoma orbiculare***
Squamata - Iguanaidae
e Mountain Horned Lizard; Mexican Horned Lizard; Red Horned Lizard
f Phrynosome orbiculaire

6405 ***Phrynosoma platyrhinhos***
Squamata - Iguanaidae
e Desert Horned Lizard; Smooth Horned Lizard (NA)
d Wüstenkrötenechse
f Phrynosome du désert

6406 ***Phrynosoma platyrhinhos calidiarum***
Squamata - Iguanaidae
e Southern Desert Horned Lizard

6407 ***Phrynosoma platyrhinhos goodei***
Squamata - Iguanaidae
e Goode's Desert Horned Lizard

6408 ***Phrynosoma platyrhinos platyrhinos***
Squamata - Iguanaidae
e Northern Desert Horned Lizard

6409 ***Phrynosoma solare***
Squamata - Iguanaidae
e Regal Horned Lizard; Southern Desert Horned Lizard
d Vierhornkrötenechse; Königskrötenechse
f Phrynosome royal

6410 ***Phrynosoma taurus***
Squamata - Iguanaidae
e Mexican Horned Lizard

6411 ***Phyllodactylus***
Squamata - Gekkonidae
e Leaf-toed Geckos
d Blattfingergeckos
f Phyllodactyles

6412 ***Phyllodactylus angustidigitus***
Squamata - Gekkonidae
e Narrow Leaf-toed Gecko

6413 ***Phyllodactylus ansorgii***
Squamata - Gekkonidae
e Angola Leaf-toed Gecko

6414 ***Phyllodactylus apricus***
Squamata - Gekkonidae
e Las Animas Island Gecko

6415 ***Phyllodactylus barringtonensis***
Squamata - Gekkonidae
e Barrington Leaf-toed Gecko

6416 ***Phyllodactylus baurii***
Squamata - Gekkonidae
e Baur's Leaf-toed Gecko

6417 ***Phyllodactylus bordai***
Squamata - Gekkonidae
e Guerreran Leaf-toed Gecko

6418 ***Phyllodactylus bugastrolepis***
Squamata - Gekkonidae
e Catalina Island Leaf-toed Gecko; Santa Catalina Island Leaf-toed Gecko

6419 ***Phyllodactylus clinatus***
Squamata - Gekkonidae
e Cerro Illescas Gecko

6420 ***Phyllodactylus darwini***
Squamata - Gekkonidae
e Darwin's Leaf-toed Gecko

6421 ***Phyllodactylus davisi***
Squamata - Gekkonidae
e Davis's Leaf-toed Gecko

6422 ***Phyllodactylus delcampoi***
Squamata - Gekkonidae
e Del Campo's Leaf-toed Gecko

6423 ***Phyllodactylus dixoni***
Squamata - Gekkonidae
e Dixons Leaf-toed Gecko

6424 ***Phyllodactylus duellmani***
Squamata - Gekkonidae
e Duellman's Leaf-toed Gecko

6425 ***Phyllodactylus europaeus***
Squamata - Gekkonidae
e European Leaf-toed Gecko; Leaf-

fingered Gecko; European Leaf-fingered Gecko
d Europäischer Blattfingergecko; Europäischer Blattfinger
f Phyllodactyle européen; Phyllodactyle d'Europe; Gecko à doigt en forme de feuille; Phyllodactyle
i Tarantolino; Fillodattilo europeo

6426 ***Phyllodactylus galapagensis***
Squamata - Gekkonidae
e Galapagos Leaf-toed Gecko

6427 ***Phyllodactylus gerrhopygus***
Squamata - Gekkonidae
e South American Leaf-toed Gecko

6428 ***Phyllodactylus gilberti***
Squamata - Gekkonidae
e Gilbert's Leaf-toed Gecko

6429 ***Phyllodactylus heterurus***
Squamata - Gekkonidae
e Tarapaca Leaf-toed Gecko

6430 ***Phyllodactylus homolepidurus***
Squamata - Gekkonidae
e Sonoran Leaf-toed Gecko

6431 ***Phyllodactylus inaequalis***
Squamata - Gekkonidae
e Peru Leaf-toed Gecko

6432 ***Phyllodactylus insularis***
Squamata - Gekkonidae
e Belize Leaf-toed Gecko

6433 ***Phyllodactylus interandinus***
Squamata - Gekkonidae
e Antilles Leaf-toed Gecko

6434 ***Phyllodactylus johnwrighti***
Squamata - Gekkonidae
e Rio Huancabamba Leaf-toed Gecko

6435 ***Phyllodactylus julieni***
Squamata - Gekkonidae
e Aruba Leaf-toed Gecko

6436 ***Phyllodactylus kofordi***
Squamata - Gekkonidae
e Coastal Leaf-toed Gecko

6437 ***Phyllodactylus lanei***
Squamata - Gekkonidae
e Lane's Leaf-toed Gecko

6438 ***Phyllodactylus leei***
Squamata - Gekkonidae
e Chatham Leaf-toed Gecko

6439 ***Phyllodactylus lepidopygus***
Squamata - Gekkonidae
e Western Leaf-toed Gecko

6440 ***Phyllodactylus lineata***
Squamata - Gekkonidae
e Lined Leaf-toed Gecko; Lined Day Gecko

6441 ***Phyllodactylus martini***
Squamata - Gekkonidae
e Dutch Leaf-toed Gecko

6442 ***Phyllodactylus melanostictus***
Squamata - Gekkonidae
e Thai Leaf-toed Gecko

6443 ***Phyllodactylus microlepidotus***
Squamata - Gekkonidae
e Fitzsimons's Leaf-toed Gecko; Small-scaled Leaf-toed Gecko (CSA)

6444 ***Phyllodactylus microphyllus***
Squamata - Gekkonidae
e Central Leaf-toed Gecko

6445 ***Phyllodactylus muralis***
Squamata - Gekkonidae

e Oaxacan Leaf-toed Gecko
i Lucertola degli muragli; Lucertola muraiola

6446 ***Phyllodactylus nocticolus***
Squamata - Gekkonidae
e Insular Leaf-toed Gecko; Peninsular Leaf-toed Gecko

6447 ***Phyllodactylus palmeus***
Squamata - Gekkonidae
e Honduras Leaf-toed Gecko; Bay Islands Leaf-toed Gecko

6448 ***Phyllodactylus partidus***
Squamata - Gekkonidae
e Partida Norte Leaf-toed Gecko; Partida Norte Island Leaf-toed Gecko

6449 ***Phyllodactylus paucituberculatus***
Squamata - Gekkonidae
e Rio Marquez Leaf-toed Gecko

6450 ***Phyllodactylus peringueyi***
Squamata - Gekkonidae
e Péringuey's Leaf-toed Gecko

6451 ***Phyllodactylus porphyreus***
Squamata - Gekkonidae
e Daudin's Leaf-toed Gecko; Marbled Leaf-toed Gecko (CSA); Marbled Gecko
f Phyllodactyle porphyré

6452 ***Phyllodactylus pulcher***
Squamata - Gekkonidae
e Barbados Leaf-toed Gecko
f Phyllodactyle gentille

6453 ***Phyllodactylus pumilus***
Squamata - Gekkonidae
e Ecuador Leaf-toed Gecko

6454 ***Phyllodactylus reissi***
Squamata - Gekkonidae
e Peters's Leaf-toed Gecko

6455 ***Phyllodactylus riebecki***
Squamata - Gekkonidae
e Riebeck's Leaf-toed Gecko

6456 ***Phyllodactylus rutteni***
Squamata - Gekkonidae
e Venezuela Leaf-toed Gecko

6457 ***Phyllodactylus santacruzensis***
Squamata - Gekkonidae
e Santa Cruz Leaf-toed Gecko

6458 ***Phyllodactylus sentosus***
Squamata - Gekkonidae
e Lima Leaf-toed Gecko

6459 ***Phyllodactylus siamensis***
Squamata - Gekkonidae
e Siamese Leaf-toed Gecko
d Siamesischer Blattfingergecko

6460 ***Phyllodactylus tincklei***
Squamata - Gekkonidae
e Raza Island Leaf-toed Gecko

6461 ***Phyllodactylus trachyrhinus***
Squamata - Gekkonidae
e Boulenger's Leaf-toed Gecko

6462 ***Phyllodactylus transversalis***
Squamata - Gekkonidae
e Colombian Leaf-toed Gecko

6463 ***Phyllodactylus tuberculosus***
Squamata - Gekkonidae
e Yellow-belly Gecko; Common Leaf-toed Gecko; Yellowbelly Leaf-toed Gecko
f Phyllodactyle tuberculeux

6464 ***Phyllodactylus unctus***
Squamata - Gekkonidae

e San Lucan Gecko; San Lucan Leaf-toed Gecko

6465 ***Phyllodactylus ventralis***
Squamata - Gekkonidae
e Margarita Leaf-toed Gecko

6466 ***Phyllodactylus wirshingi***
Squamata - Gekkonidae
e Puerto Rican Leaf-toed Gecko; Antillean Banded Gecko

6467 ***Phyllodactylus xanti***
Squamata - Gekkonidae
e Leaf-toed Gecko; Cape Leaf-toed Gecko; Peninsular Leaf-toed Gecko
f Gecko phyllodactyle américain

6468 ***Phyllodactylus xanti nocticolus***
Squamata - Gekkonidae
e Peninsular Leaf-toed Gecko

6469 ***Phyllodactylus xanti xanti***
Squamata - Gekkonidae
e Cape Leaf-toed Gecko; Xantus Leaf-toed Gecko

6470 ***Phyllodactylus xanti zweifeli***
Squamata - Gekkonidae
e Santa Margarita Island Leaf-toed Gecko

6471 ***Phyllopezus***
Squamata - Gekkonidae
e Brazilian Geckos

6472 ***Phyllopezus periosus***
Squamata - Gekkonidae
e Peraiba Gecko

6473 ***Phyllopezus pollicaris***
Squamata - Gekkonidae
e Brazilian Gecko

6474 ***Phyllorhynchus***
Squamata - Colubridae
e Leafnose Snakes; Leaf-nosed Snakes
d Blattnasennattern
f Couleuvres à nez foliacé

6475 ***Phyllorhynchus browni***
Squamata - Colubridae
e Saddled Leafnose Snake; Saddled Leaf-nosed Snake; Saddle Leafnose Snake; Brown Leaf-nosed Snake; Pima Leaf-nosed Snake
d Geringelte Blattnasennatter
f Couleuvre à nez foliacé de Brown

6476 ***Phyllorhynchus browni browni***
Squamata - Colubridae
e Pima Leafnose Snake; Pima Leaf-nosed Snake

6477 ***Phyllorhynchus browni lucidus***
Squamata - Colubridae
e Maricopa Leafnose Snake; Maricopa Leaf-nosed Snake

6478 ***Phyllorhynchus decurtatus***
Squamata - Colubridae
e Spotted Leafnose Snake; Spotted Leaf-nosed Snake; Leaf-nosed Snake
d Gefleckte Blattnasennatter
f Couleuvre à nez foliacé tachetée

6479 ***Phyllorhynchus decurtatus decurtatus***
Squamata - Colubridae
e Baja California Leaf-nosed Snake

6480 ***Phyllorhynchus decurtatus nubilis***
Squamata - Colubridae
e Clouded Leafnose Snake; Clouded Leaf-nosed Snake; Cloudy Leaf-nosed Snake

6481 ***Phyllorhynchus decurtatus perkinsi***
Squamata - Colubridae
e Western Leafnose Snake; Western Leaf-nosed Snake

6482 ***Phyllurus***
Squamata - Gekkonidae
e Leaftail Geckos; Leaf-tailed Geckos
d Blattschwanzgeckos

6483 ***Phyllurus caudiannulatus***
Squamata - Gekkonidae
e Barred Leaftail Gecko

6484 ***Phyllurus cornutus***
Squamata - Gekkonidae
e Northern Leaftail Gecko; Northern Leaf-tailed Gecko (ANZ)
d Hornblattschwanzgecko; Blattschwanzgecko

6485 ***Phyllurus platurus***
Squamata - Gekkonidae
e Southern Leaftail Gecko; Southern Leaf-tailed Gecko (ANZ)
d Breitschwanzgecko
f Gecko du Sud à queue-en-feuille

6486 ***Phyllurus salebrosus***
Squamata - Gekkonidae
e Tuberculate Leaftail Gecko

6487 ***Phymaturus***
Squamata - Iguanaidae
e Mountain Lizards
d Hochgebirgsleguane

6488 ***Phymaturus mallimaccii***
Squamata - Iguanaidae
e Thorntail Mountain Lizard

6489 ***Phymaturus palluma***
Squamata - Iguanaidae
e High Mountain Lizard; Dwarf Chilean Chuckwalla
d Hochgebirgsleguan

6490 ***Phymaturus patagonicus***
Squamata - Iguanaidae
e Patagonia Mountain Lizard

6491 ***Phymaturus punae***
Squamata - Iguanaidae
e Cei's Mountain Lizard

6492 ***Physignathus***
Squamata - Agamidae
e Water Dragons; Water Lizards
d Wasseragamen; Wasserdrachen
f Physignathes
i Fisignati; Draghi d'acqua

6493 ***Physignathus cocincinus***
Squamata - Agamidae
e Water Dragon; Green Water Dragon; Chinese Water Dragon; Oriental Water Dragon
d Grüne Wasseragame; Cochin-China-Wasseragame; Cochin-China-Wasserdrachen
f Physignathe de Cochinchine; Dragon d'eau; Agame aquatique

6494 ***Physignathus lesueuri***
Squamata - Agamidae
e Eastern Water Dragon; Brown Water Dragon; Australian Water Dragon; Lesueur's Water Dragon; Leisure Water Dragon (ANZ)
d Braune Wasseragame; Australische Wasseragame; Wasseragame; Gewöhnlicher Wasserdrachen
f Agame aquatique; Physignathe de Lesueuer

6495 ***Physignathus lesueuri howitti***
Squamata - Agamidae
e Gippsland Water Dragon

6496 ***Physignathus lesueuri lesueri***
Squamata - Agamidae
e Eastern Water Dragon

6497 ***Pituophis***
Squamata - Colubridae
e Gopher Snakes; Bull Snakes; Pine Snakes

d Bullennattern; Kiefernnattern

6498 ***Pituophis catenifer***
Squamata - Colubridae
e Bull Snake; Gopher Snake; Pacific Gopher Snake; Coast Gopher Snake
d Kettenbullennattern

6499 ***Pituophis catenifer affinis***
Squamata - Colubridae
e Sonoran Gopher Snake; Bull Snake
d Bullennatter

6500 ***Pituophis catenifer annectens***
Squamata - Colubridae
e San Diego Gopher Snake; Gopher Snake
d Bullennatter

6501 ***Pituophis catenifer catenifer***
Squamata - Colubridae
e Pacific Gopher Snake; Pacific Bull Snake
d Bullennatter

6502 ***Pituophis catenifer coronalis***
Squamata - Colubridae
e Los Coronados Island Gopher Snake
d Bullennatter

6503 ***Pituophis catenifer deserticola***
Squamata - Colubridae
e Great Basin Gopher Snake
d Bullennatter

6504 ***Pituophis catenifer fulginatus***
Squamata - Colubridae
e San Martin Island Gopher Snake
d Bullennatter

6505 ***Pituophis catenifer pumilis***
Squamata - Colubridae
e Santa Cruz Gopher Snake; Santa Cruz Island Gopher Snake
d Bullennatter

6506 ***Pituophis catenifer sayi***
Squamata - Colubridae
e Bull Snake; Common Bull Snake (NA)
d Bullennatter
i Serpente toro

6507 ***Pituophis deppei***
Squamata - Colubridae
e Mexican Bull Snake; Mexican Pine Snake; Mexican Gopher Snake
d Mexikanische Bullennatter

6508 ***Pituophis deppei jani***
Squamata - Colubridae
e Northern Mexican Pine Snake

6509 ***Pituophis insulanus***
Squamata - Colubridae
e Cedros Island Gopher Snake

6510 ***Pituophis lineaticollis***
Squamata - Colubridae
e Middle American Gopher Snake

6511 ***Pituophis melanoleucus***
Squamata - Colubridae
e Gopher Snake; Pine Snake; Bull Snake; Common Pine Snake
d Kiefernnatter; Kettennatter; Gophernatter; Schwarzweiße Bullennatter; Bullennatter
f Serpent-taureau; Couleuvre à nez mince
i Colubro delle pinete; Serpente dei roditori

6512 ***Pituophis melanoleucus bimaris***
Squamata - Colubridae
e Central Baja California Gopher Snake; Baja Gopher Snake

6513 ***Pituophis melanoleucus lodingi***
Squamata - Colubridae
e Black Pine Snake
d Schwarze Kiefernnatter

6514 ***Pituophis melanoleucus melanoleucus***
Squamata - Colubridae
e Northern Pine Snake; Pine Snake
d Nördliche Kiefernnatter
i Colubro delle pinete settentrionale

6515 ***Pituophis melanoleucus mugitus***
Squamata - Colubridae
e Florida Pine Snake
d Florida-Kiefernnatter

6516 ***Pituophis ruthveni***
Squamata - Colubridae
e Louisiana Pine Snake

6517 ***Pituophis vertebralis***
Squamata - Colubridae
e Baja California Gopher Snake; Cape Gopher Snake

6518 ***Placosoma***
Squamata - Teiidae
e Brazilian Lizards

6519 ***Placosoma cipoense***
Squamata - Teiidae
e Cuha's Brazilian Lizard

6520 ***Placosoma cordylinum***
Squamata - Teiidae
e Tschudi's Brazilian Lizard

6521 ***Placosoma glabellum***
Squamata - Teiidae
e Peters's Brazilian Lizard

6522 ***Plagiopholis***
Squamata - Colubridae
e Chinese Mountain Snakes

6523 ***Plagiopholis blakewayi***
Squamata - Colubridae
e Blakeway's Mountain Snake

6524 ***Plagiopholis nuchalis***
Squamata - Colubridae
e Assam Mountain Snake

6525 ***Plagiopholis styani***
Squamata - Colubridae
e Chinese Mountain Snake

6526 ***Plagiopholis unipostocularis***
Squamata - Colubridae
e Yunnan Mountain Snake

6527 ***Platemys***
Testudines - Chelidae
e Twistneck Turtles; South American Side-necked Swamp Turtle; Twist-necked Turtles
d Plattschildkröten
f Platémydes

6528 ***Platemys platycephala***
Testudines - Chelidae
e Twistneck Turtle; South American Snake-necked Turtle; Flat-headed Flat-shelled Turtle; Twist-necked Turtle; Grooved Side-necked Turtle; South American Flathead Turtle
d Plattschildkröte; Rotkopfplattschildkröte
f Platémyde platycéphale; Platémyde à tête plate; Platémyde à tête orange (Ants)
i Tartaruga girocollo

6529 ***Platemys platycephala melanonota***
Testudines - Chelidae
e Western Twist-necked Turtle
d Platémyde à tête plate occidentale

6530 ***Platemys platycephala platycephala***
Testudines - Chelidae
d Platémyde à tête plate orientale

6531 ***Platyinion***
Squamata - Colubridae

e Blue Racers

6532 ***Platyinion lividum***
Squamata - Colubridae
e Guiacurus Blue Racer

6533 ***Platyplacopus***
Squamata - Lacertidae
e Rock Racers

6534 ***Platyplacopus intermedius***
Squamata - Lacertidae
e Stejneger's Rock Racer

6535 ***Platyplacopus kuehnei***
Squamata - Lacertidae
e Van Denburgh's Rock Racer

6536 ***Platyplecturus***
Squamata - Uropeltidae
e Thorntail Snakes

6537 ***Platyplecturus madurensis***
Squamata - Uropeltidae
e Travancore Hills Thorntail Snake

6538 ***Platyplecturus trilineatus***
Squamata - Uropeltidae
e Lined Thorntail Snake

6539 ***Platysaurus***
Squamata - Cordylidae
e Flat Lizards
d Plattgürtelechechsen; Plattgürtelschweife; Plattechsen
f Platysaures

6540 ***Platysaurus capensis***
Squamata - Cordylidae
e Cape Flat Lizard

6541 ***Platysaurus guttatus***
Squamata - Cordylidae
e Lesser Flat Lizard; Dwarf Flat Lizard (CSA); Transvaal Red-tailed Rock Lizard; Flat Rock Lizard
d Kleine Plattgürtelechse; Kleiner Plattgürtelschweif
f Platysaure à gouttelettes

6542 ***Platysaurus guttatus guttatus***
Squamata - Cordylidae
e Lesser Flat Rock Lizard

6543 ***Platysaurus guttatus minor***
Squamata - Cordylidae
e Lesser Flat Rock Lizard

6544 ***Platysaurus guttatus orientais***
Squamata - Cordylidae
e Lesser Flat Rock Lizard

6545 ***Platysaurus imperator***
Squamata - Cordylidae
e Imperial Flat Lizard; Emperor Flat Lizard (CSA)
d Kaiser-Plattgürtelschweif; Kaiser-Plattgürtelechse
f Platysaure impérial

6546 ***Platysaurus intermedius***
Squamata - Cordylidae
e Common Flat Lizard; Yumba Flat Lizard; Transvaal Flat Rock Lizard (CSA); Common Flat Lizard (CSA)
d Gewöhnliche Plattechse

6547 ***Platysaurus maculatus***
Squamata - Cordylidae
e Spotted Flat Lizard
d Fleckige Plattechse

6548 ***Platysaurus mitchelli***
Squamata - Cordylidae
e Mitchell's Flat Lizard
d Mitchells Plattechse

6549 ***Platysaurus ocellatus***
Squamata - Cordylidae

e Chimanimani Flat Lizard; Ocellated Flat Lizard (CSA)

6550 ***Platysaurus pungweensis***
Squamata - Cordylidae
e Pungwe Flat Lizard

6551 ***Platysaurus relictus***
Squamata - Cordylidae
e Transvaal Flat Lizard; Relict Flat Lizard (CSA)

6552 ***Platysaurus torquatus***
Squamata - Cordylidae
e Transvaal Flat Lizard; Striped Flat Lizard (CSA); Tete Flat Lizard (CSA)

6553 **Platysternidae**
Testudines
e Big-headed Turtles; Bighead Turtles
d Großkopfschildkröten
f Platysternes
i Tartarughe capogrosso

6554 ***Platysternon***
Testudines - Pelomedusidae
e Bighead Turtles; Big-headed Turtles
d Großkopfschildkröten
f Platysternes

6555 ***Platysternon megacephalum***
Testudines - Pelomedusidae
e Bighead Turtle; Big-headed Turtle
d Großkopfschildkröte
f Platysterne
i Platisterno capogrosso

6556 ***Platysternon megacephalum megacephalum***
Testudines - Pelomedusidae
e Chinese Bighead Turtle
f Platysterne de Chine

6557 ***Platysternon megacephalum peguense***
Testudines - Pelomedusidae
e Burmese Bighead Turtle
f Platysterne de Birmanie

6558 ***Platysternon megacephalum shiui***
Testudines - Pelomedusidae
e Vietnam Bighead Turtle
f Platysterne du Vietnam

6559 ***Platysternon megacephalum tristernalis***
Testudines - Pelomedusidae
e Yunnan Bighead Turtle
f Platysterne du Yunnan

6560 ***Platysternon megacephalum vogeli***
Testudines - Pelomedusidae
e Thailand Bighead Turtle
f Platysterne de la Thaïlande

6561 ***Plecturus***
Squamata - Uropeltidae
e African Burrowing Snakes

6562 ***Plecturus aureus***
Squamata - Uropeltidae
e Kerala Burrowing Snake

6563 ***Plecturus canaricus***
Squamata - Uropeltidae
e Karnataka Burrowing Snake

6564 ***Plecturus guentheri***
Squamata - Uropeltidae
e Günther's Burrowing Snake

6565 ***Plecturus perroteti***
Squamata - Uropeltidae
e Nilgiri Burrowing Snake; Nilgiri Uropelt (ISC)

6566 ***Plesiomicrolophus***
Squamata - Iguanaidae

e Frost's Iguanas

6567 ***Plesiomicrolophus kuepehforon***
Squamata - Iguanaidae
e Frost's Iguana

6568 ***Pletholax***
Squamata - Gekkonidae
e Black-striped Legless Geckos

6569 ***Pletholax gracilis***
Squamata - Gekkonidae
e Black-striped Legless Gecko

6570 ***Plica***
Squamata - Iguanaidae
e Racerunners; Treerunners
d Stelzenläufer; Stelzenläuferleguane

6571 ***Plica plica***
Squamata - Iguanaidae
e Treerunner; Tree Racerunner; Collared Tree Lizard

6572 ***Plica umbra***
Squamata - Iguanaidae
e Harlequin Racerunner; Racerunner
d Stelzenläufer

6573 ***Plica umbra ochrocollaris***
Squamata - Iguanaidae
e Blue-lipped Tree Lizard

6574 ***Pliocercus***
Squamata - Colubridae
e Common False Coral Snakes
d Halloween-Nattern; Falsche Korallenschlange

6575 ***Pliocercus annellatus***
Squamata - Colubridae
e Annulated False Coral Snake

6576 ***Pliocercus arubricus***
Squamata - Colubridae
e Aruba False Coral Snake

6577 ***Pliocercus dimidiatus***
Squamata - Colubridae
e Tiny False Coral Snake

6578 ***Pliocercus elapoides***
Squamata - Colubridae
e Common False Coral Snake; Northern Halloween Snake

6579 ***Pliocercus euryzonus***
Squamata - Colubridae
e Cope's False Coral Snake; Southern Halloween Snake

6580 ***Podarcis***
Squamata - Lacertidae
e Wall Lizards
d Mauereidechsen
f Lézards des murailles
i Lucertole

6581 ***Podarcis dugesii***
Squamata - Lacertidae
e Madeira Wall Lizard; Madeira Lizard
d Madeira-Mauereidechse; Madeira-Eidechse
f Lézard de Madère
i Lucertola del Duges

6582 ***Podarcis erhardi***
Squamata - Lacertidae
e Erhard's Wall Lizard; Aegean Wall Lizard
d Ägäische Mauereidechse; Zykladen-Eidechse; Ägäische Eidechse
f Lézard de l'Égée; Lézard d'Erhard; Lézard égéen
i Lucertola di Ehrhardt

6583 ***Podarcis filfolensis***
Squamata - Lacertidae
e Filfola Wall Lizard; Maltese Wall Lizard

d Malta-Eidechse; Filfola-Mauereidechse; Maltesische Mauereidechse
f Lézard maltais; Lézard de Filfola; Lézard de Malte
i Lucertola maltese; Lucertola di Malta

6584 ***Podarcis hispanica***
Squamata - Lacertidae
e Iberian Wall Lizard
d Iberische Mauereidechse; Spanische Mauereidechse; Spitzkopfeidechse
f Lézard hispanique; Lézard espagnol
i Lucertola spagnola

6585 ***Podarcis hispanica vaucheri***
Squamata - Lacertidae
f Lézard hispanique d'Afrique du Nord

6586 ***Podarcis lilfordi***
Squamata - Lacertidae
e Lilford's Wall Lizard; Balearic Wall Lizard
d Balearen-Eidechse
f Lézard des Baléares
i Lucertola di Lilford; Lucertola balearica

6587 ***Podarcis melisellensis***
Squamata - Lacertidae
e Dalmatian Wall Lizard; Karst Lizard; Adriatic Lizard
d Adriatische Mauereidechse; Karstläufer
f Lézard de l'Adriatique
i Lucertola adriatica; Lucertola di Melisello

6588 ***Podarcis milensis***
Squamata - Lacertidae
e Milos Wall Lizard; Wall Lizard
d Milos-Mauereidechse; Milos-Eidechse
f Lézard de Milo
i Lucertola di Milo

6589 ***Podarcis muralis***
Squamata - Lacertidae
e Common Wall Lizard; Wall Lizard
d Mauereidechse; Ungefleckte Mauereidechse; Gestreifte Mauereidechse; Gewöhnliche Mauereidechse
f Lézard des murailles; Lézard gris
i Lucertola dei muri; Lucertola muraiola; Lusèrta; Lusèrtla; Lagromuso

6590 ***Podarcis muralis colossi***
Squamata - Lacertidae
i Lucertola dei muri a dorso verde

6591 ***Podarcis peloponnesiaca***
Squamata - Lacertidae
e Peloponnese Wall Lizard
d Peloponnesos-Eidechse; Peloponnesische Eidechse; Peleponnes-Eidechse
f Lézard des Péloponnèses; Lézard du Péloponnèse
i Lucertola peloponnesiaca

6592 ***Podarcis perspicillata***
Squamata - Lacertidae
e Menorca Wall Lizard; Moroccan Rock Lizard
d Brilleneidechse
f Lézard à lunettes
i Lucertola dagli occhiali

6593 ***Podarcis pityuensis***
Squamata - Lacertidae
e Ibiza Wall Lizard
d Pityusen-Eidechse
f Lézard des Pityuses
i Lucertola delle Pitiuse

6594 ***Podarcis sicula***
Squamata - Lacertidae
e Italian Wall Lizard; Ruin Lizard
d Ruineneidechse
f Lézard des ruines; Lézard sicilien;

Lézard des murailles; Lézard des murailles italien; Lézard de Sicilie
i Lucertola campestre; Luce; Lucertola sicula; Lusèrta; Lusèrtla

6595 ***Podarcis sicula cettii***
Squamata - Lacertidae
i Lucertola campestre sarda

6596 ***Podarcis taurica***
Squamata - Lacertidae
e Crimean Wall Lizard; Meadow Lizard; Taurican Lizard; Bull Lizard; Crimean Lizard
d Taurische Eidechse; Krim-Eidechse
f Lézard du Taurus; Lézard de Crimée; Lézard taurique
i Lucertola taurica

6597 ***Podarcis tiliguerta***
Squamata - Lacertidae
e Tyrrhenian Wall Lizard
d Tyrrhenische Mauereidechse
f Lézard tyrrhénien
i Lucertola tirrenica; Lucertola tiliguerta; Lucertola sarda

6598 ***Podarcis wagleriana***
Squamata - Lacertidae
e Sicilian Wall Lizard
d Sizilianische Mauereidechse
f Lézard des murailles de Sicile
i Lucertola siciliana; Lucertola di Wagner

6599 ***Podocnemis***
Testudines - Pelomedusidae
e South American Sideneck River Turtles; River Turtles; South American River Turtles
d Schienenschildkröten
f Podocnémides; Tortues de l'Amazone; Podocnemies (Ants); Tortues fluviales; Tortues fluviales de l'Amérique du Sud; Tortues fluviatiles ; Tortues fluviatiles d'Amérique du Sud
i Tartarughe dei fiume

6600 ***Podocnemis erythrocephala***
Testudines - Pelomedusidae
e Red-headed River Turtle; Red-head Amazon Side-necked Turtle ; Red-headed Amazon River Turtle; Red-headed Side-necked Turtle
d Rotkopfschienenschildkröte
f Podocnémide à tête rouge

6601 ***Podocnemis expansa***
Testudines - Pelomedusidae
e South American River Turtle; Arrau Sideneck Turtle; Arrau River Turtle; Arrau Sideneck; Giant South American River Turtle; True Amazon Turtle; Arrau Side-necked Turtle
d Arrau-Schienenschildkröte; Arrau-Schildkröte
f Podocnémide élargie; Tortue géante d'Amérique du Sud
i Tartaruga Charapa

6602 ***Podocnemis lewyana***
Testudines - Pelomedusidae
e Magdalena River Turtle; Rio Magdalena River Turtle
d Glattmaulschienenschildkröte
f Podocnémide de Léwy
i Tartaruga Arrau

6603 ***Podocnemis sextuberculata***
Testudines - Pelomedusidae
e Six-tuberculed River Turtle; Six-tuberculed Amazon River Turtle; Amazon River Turtle
d Höckerschienenschildkröte
f Podocnémide tuberculée; Tortue de l'Amazone à six tubercules

6604 ***Podocnemis unifilis***
Testudines - Pelomedusidae
e Yellow-spotted River Turtle; Yellow-spotted Amazon River Turtle; Yellow-spotted Sideneck Turtle;

Yellow-headed Side-necked Turtle
d Terekay-Schienenschildkröte
f Podocnémide de Cayenne; Tortue de l'Amazone à taches jaune; Podocnemie de la Cayenne (Ants)
i Tartaruga aica

6605 ***Podocnemis vogli***
Testudines - Pelomedusidae
e Savanna Sideneck Turtle; Savanna Side-necked Turtle; Savannah Side-necked Turtle; Llanos Side-necked Turtle
d Orinoko-Schienenschildkröte
f Podocnémide de Vogl
i Tartaruga Terecay; Tartaruga di Llanos

6606 ***Poecilopholis***
Squamata - Colubridae
e Cameroon Racers

6607 ***Poecilopholis cameronensis***
Squamata - Colubridae
e Cameroon Racer

6608 ***Pogona***
Squamata - Agamidae
e Bearded Dragons
d Bartagamen
f Lézards barbus

6609 ***Pogona barbata***
Squamata - Agamidae
e Bearded Dragon; Eastern Bearded Dragon (ANZ); Bearded Lizard; Barbed Agama
d Gemeine Bartagame; Östliche Bartagame; Bartagame
f Lézard barbu d'Australie; Agame barbé; Amphibolure barbé
i Anfiboluro dalla barba

6610 ***Pogona barbata henrylawsoni***
Squamata - Agamidae
e Black-soil Bearded Dragon

6611 ***Pogona microlepidota***
Squamata - Agamidae
e Drysdale River Bearded Dragon; Small-scaled Drgon
d Kimberley-Bartagame

6612 ***Pogona minima***
Squamata - Agamidae
e Western Bearded Dragon; Lesser Bearded Dragon
d Kleine Bartagame; Westliche Bartagame
f Petit Amphibolure

6613 ***Pogona minor***
Squamata - Agamidae
e Dwarf Bearded Dragon
d Zwergbartagame

6614 ***Pogona mitchelli***
Squamata - Agamidae
e Mitchell's Bearded Dragon; Mitchell's Dragon
d Mitchells Bartagame
f Amphibolure de Mitchell

6615 ***Pogona nullarbor***
Squamata - Agamidae
e Nullarbor Bearded Dragon
d Nullabor Bartagame

6616 ***Pogona vitticeps***
Squamata - Agamidae
e Inland Bearded Dragon; Central Bearded Dragon; Australian Bearded Dragon; Bearded Dragon; Common Bearded Dragon
d Australische Bartagame; Bartagame
f Agame barbu d'Australie; Agame barbue des terres; Dragon barbu; Agame barbu; Dragon barbu australien

6617 ***Polemon***
Squamata - Colubridae
e Snake-eaters

6618 ***Polemon acanthias***
Squamata - Colubridae
e Reinhardt's Snake-eater

6619 ***Polemon barthii***
Squamata - Colubridae
e Guinea Snake-eater

6620 ***Polemon bocourti***
Squamata - Colubridae
e Bocourt's Snake-eater

6621 ***Polemon christyi***
Squamata - Colubridae
e Eastern Snake-eater

6622 ***Polemon collaris***
Squamata - Colubridae
e Collared Snake-eater ; Pale-collared Snake-eater

6623 ***Polemon fulvicollis***
Squamata - Colubridae
e African Snake-eater

6624 ***Polemon gabonensis***
Squamata - Colubridae
e Gaboon Snake-eater

6625 ***Polemon graciliceps***
Squamata - Colubridae
e Cameroon Snake-eater

6626 ***Polemon gracilis***
Squamata - Colubridae
e Graceful Snake-eater

6627 ***Polemon neuwiedi***
Squamata - Colubridae
e Ivory Coast Snake-eater

6628 ***Polemon notatus***
Squamata - Colubridae
e Peters's Snake-eater

6629 ***Polemon robustus***
Squamata - Colubridae
e Zaire Snake-eater

6630 ***Polychrus***
Squamata - Iguanaidae
e Bush Anoles; Long-legged Lizards; Monkey Lizards
d Buntleguane; Labgbeinleguane
f Polychres

6631 ***Polychrus acutirostris***
Squamata - Iguanaidae
e Brazilian Bush Anole
d Spitzkopfleguan
f Polychre à museau pointu

6632 ***Polychrus femoralis***
Squamata - Iguanaidae
e Werner's Bush Anole; Monkey Lizard

6633 ***Polychrus gutturosus***
Squamata - Iguanaidae
e Berthold's Bush Anole; Canopy Lizard

6634 ***Polychrus liogaster***
Squamata - Iguanaidae
e Boulenger's Bush Anole; Southern Monkey Lizard

6635 ***Polychrus marmoratus***
Squamata - Iguanaidae
e Many-coloured Bush Anole; Bush Lizard; Common Monkey Lizard
d Marmorierter Buntleguan
f Polychre marbré

6636 ***Poromera***
Squamata - Lacertidae
e Striped Lizards

6637 ***Poromera fordii***
Squamata - Lacertidae

e West African Striped Lizard

6638 ***Porthidium***
Squamata - Viperidae
e Hognose Vipers
d Stülpnasenottern

6639 ***Porthidium dunni***
Squamata - Viperidae
e Dunn's Hognose Viper; Dunn's Hog-nosed Viper

6640 ***Porthidium hespere***
Squamata - Viperidae
e Western Hognose Viper

6641 ***Porthidium hyoprora***
Squamata - Viperidae
e Amazon Hognose Viper; Hognose Lancehead

6642 ***Porthidium lansbergi***
Squamata - Viperidae
e Lansberg's Hognose Viper; Lansberg's Hog-nosed Viper

6643 ***Porthidium melanurum***
Squamata - Viperidae
e Blacktail Hognose Viper; Black-tailed Montane Pit Viper; Blacktail Montane Pit Viper

6644 ***Porthidium nasutum***
Squamata - Viperidae
e Rain Forest Hognose Viper; Hog-nosed Pit Viper; Rain Forest Hog-nosed Viper
d Nasenlanzenotter; Stülpnasenlanzenotter

6645 ***Porthidium ophryomegas***
Squamata - Viperidae
e Slender Hognose Viper; Slender Hog-nosed Pit Viper
d Schlanke Stülpnasenlanzenotter

6646 ***Porthidium yucatanicum***
Squamata - Viperidae
e Yucatan Hognose Viper; Yucatan Hog-nosed Pit Viper

6647 ***Prasinohaema***
Squamata - Scincidae
e Green Tree Skinks

6648 ***Prasinohaema flavipes***
Squamata - Scincidae
e Common Green Tree Skink; Green-blooded Skink

6649 ***Prasinohaema parkeri***
Squamata - Scincidae
e Parker's Green Tree Skink

6650 ***Prasinohaema prehensicauda***
Squamata - Scincidae
e Prehensile Green Tree Skink; Prehensile-tailed Skink

6651 ***Prasinohaema semoni***
Squamata - Scincidae
e Semon's Green Tree Skink; Green-blooded Skink

6652 ***Prasinohaema virens***
Squamata - Scincidae
e Green Tree Skink
d Grünblütiger Schlankskink

6653 ***Prionodactylus***
Squamata - Teiidae
e Slender Priondactyluses; Eyed Lizards

6654 ***Prionodactylus argulus***
Squamata - Teiidae
e White-lipped Priondactylus; Elegant Eyed Lizard

6655 ***Prionodactylus dicrus***
Squamata - Teiidae

e Uzzell's Priondactylus

6656 ***Prionodactylus eigenmanni***
Squamata - Teiidae
e Eigenmann's Priondactylus

6657 ***Prionodactylus manicatus***
Squamata - Teiidae
e Slender Priondactylus

6658 ***Prionodactylus vertebralis***
Squamata - Teiidae
e Brown Priondactylus; Striped Prionodactylus

6659 ***Pristidactylus***
Squamata - Iguanaidae
e Primitive Anoles

6660 ***Pristidactylus achalensis***
Squamata - Iguanaidae
e Argentine Anole

6661 ***Pristidactylus alvaroi***
Squamata - Iguanaidae
e Alvaro's Anole

6662 ***Pristidactylus casuhatiensis***
Squamata - Iguanaidae
e Casuhatien Anole

6663 ***Pristidactylus fasciatus***
Squamata - Iguanaidae
e D'Orbigny's Banded Anole

6664 ***Pristidactylus scapulatus***
Squamata - Iguanaidae
e Burmeister's Anole

6665 ***Pristurus***
Squamata - Gekkonidae
e Rock Geckos; Semaphore Geckos
f Pristures

6666 ***Pristurus abdelkuri***
Squamata - Gekkonidae
e Abd el Kuri Rock Gecko

6667 ***Pristurus carteri***
Squamata - Gekkonidae
e Carter's Rock Gecko; Ornate Rock Gecko
f Gecko scorpion

6668 ***Pristurus celerrimus***
Squamata - Gekkonidae
e Oman Rock Gecko

6669 ***Pristurus collaris***
Squamata - Gekkonidae
e Collared Rock Gecko

6670 ***Pristurus crucifer***
Squamata - Gekkonidae
e Valenciennes Rock Gecko

6671 ***Pristurus flavipunctatus***
Squamata - Gekkonidae
e Middle Eastern Rock Gecko
f Pristure à points jaunes

6672 ***Pristurus gasperetti***
Squamata - Gekkonidae
e Wadi Kharrar Rock Gecko

6673 ***Pristurus guichardi***
Squamata - Gekkonidae
e Guichard's Rock Gecko

6674 ***Pristurus insignis***
Squamata - Gekkonidae
e Blanford's Rock Gecko

6675 ***Pristurus insignoides***
Squamata - Gekkonidae
e Haggier Massif Rock Gecko

6676 ***Pristurus minimus***
Squamata - Gekkonidae
e Arnold's Rock Gecko

6677 ***Pristurus ornithocephalus***
Squamata - Gekkonidae
e Birdhead Rock Gecko

6678 ***Pristurus phillipsi***
Squamata - Gekkonidae
e Somali Rock Gecko

6679 ***Pristurus popovi***
Squamata - Gekkonidae
e Saudi Rock Gecko

6680 ***Pristurus rupestris***
Squamata - Gekkonidae
e Persia Rock Gecko

6681 ***Pristurus saada***
Squamata - Gekkonidae
e Yemen Rock Gecko

6682 ***Pristurus simonettai***
Squamata - Gekkonidae
e Coastal Rock Gecko

6683 ***Pristurus sokotranus***
Squamata - Gekkonidae
e Socotra Rock Gecko

6684 ***Proablepharus***
Squamata - Scincidae
e Small Skinks

6685 ***Proablepharus kinghorni***
Squamata - Scincidae
e Kinghorn's Small Skink

6686 ***Proablepharus reginae***
Squamata - Scincidae
e Royal Small Skink

6687 ***Proablepharus tenuis***
Squamata - Scincidae
e Broom's Small Skink

6688 ***Procellosaurinus***
Squamata - Teiidae
e Rodrigues's Teiids

6689 ***Procellosaurinus erythrocercus***
Squamata - Teiidae
e Rodrigues's Red Teiid

6690 ***Procellosaurinus tetradactylus***
Squamata - Teiidae
e Rodrigues's Four-fingered Teiid

6691 ***Proctoporus***
Squamata - Teiidae
e Lightbulb Lizards

6692 ***Proctoporus bogotensis***
Squamata - Teiidae
e Bogota Lightbulb Lizard

6693 ***Proctoporus bolivianus***
Squamata - Teiidae
e Bolivian Lightbulb Lizard

6694 ***Proctoporus cashcaensis***
Squamata - Teiidae
e Kizorian Lightbulb Lizard

6695 ***Proctoporus columbianus***
Squamata - Teiidae
e Colombian Lightbulb Lizard

6696 ***Proctoporus guentheri***
Squamata - Teiidae
e Günther's Lightbulb Lizard

6697 ***Proctoporus hypostictus***
Squamata - Teiidae
e Boulenger's Lightbulb Lizard

6698 ***Proctoporus laevis***
Squamata - Teiidae
e Shiny Lightbulb Lizard

6699 ***Proctoporus luctuosus***
Squamata - Teiidae
e Lightbulb Lizard

6700 ***Proctoporus meleagris***
Squamata - Teiidae
e Brown Lightbulb Lizard

6701 ***Proctoporus oculatus***
Squamata - Teiidae
e Tropical Lightbulb Lizard

6702 ***Proctoporus pachyurus***
Squamata - Teiidae
e Tschudi's Lightbulb Lizard

6703 ***Proctoporus shrevei***
Squamata - Teiidae
e Shreve's Lightbulb Lizard; Luminous Lizard
d Leuchtteju

6704 ***Proctoporus simoterus***
Squamata - Teiidae
e O'Shaughnessy's Lightbulb Lizard

6705 ***Proctoporus striatus***
Squamata - Teiidae
e Striped Lightbulb Lizard

6706 ***Proctoporus unicolor***
Squamata - Teiidae
e Drab Lightbulb Lizard

6707 ***Proctoporus ventrimaculatus***
Squamata - Teiidae
e Spotted Lightbulb Lizard

6708 ***Proctotretus***
Squamata - Iguanaidae
e Shingled Iguanas

6709 ***Proctotretus azureus***
Squamata - Iguanaidae
e Muller's Iguana

6710 ***Proctotretus doellojuradoi***
Squamata - Iguanaidae
e Freiberg's Iguana

6711 ***Proctotretus pectinatus***
Squamata - Iguanaidae
e Brazilian Striped Iguana

6712 ***Proscelotes***
Squamata - Scincidae
e Montane Skinks; Slender Skinks (CSA)

6713 ***Proscelotes aenea***
Squamata - Scincidae
e Montane Skink

6714 ***Proscelotes arnoldi***
Squamata - Scincidae
e Arnold's Montane Skink; Arnold's Skink (CSA)

6715 ***Proscelotes eggeli***
Squamata - Scincidae
e Usambara Five-toed Skink

6716 ***Prosymna***
Squamata - Colubridae
e Shovelsnout Snakes; Shovel-nosed Snakes; Coppery Snakes
d Schaufelnasennattern

6717 ***Prosymna ambigua***
Squamata - Colubridae
e East African Shovelsnout Snake; Spotted Shovel-nosed Snake; Grey Shovelsnout (CSA); Bocage's Shovelsnout (CSA)

6718 ***Prosymna ambigua ambigua***
Squamata - Colubridae
e East African Shovelsnout

6719 ***Prosymna angolensis***
Squamata - Colubridae
e Angola Shovelsnout Snake

6720 ***Prosymna bivittata***
Squamata - Colubridae
e Twin-striped Shovelsnout Snake; Two-striped Shovel-nosed Snake; Two-striped Shovel-snouted Snake; Two-striped Shovelsnout (CSA); Twin-striped Shovelsnout (CSA)

6721 ***Prosymna frontalis***
Squamata - Colubridae
e Southwestern Shovelsnout Snake; Southwestern Shovelsnout

6722 ***Prosymna greigerti***
Squamata - Colubridae
e Greigert's Shovelsnout Snake

6723 ***Prosymna janii***
Squamata - Colubridae
e Mozambique Shovelsnout Snake; Black-headed Coppery Snake

6724 ***Prosymna meleagris***
Squamata - Colubridae
e Reinhardt's Shovelsnout Snake

6725 ***Prosymna ornatissima***
Squamata - Colubridae
e Ornate Shovelsnout Snake

6726 ***Prosymna pitmani***
Squamata - Colubridae
e Pitman's Shovelsnout Snake; Multi-scaled Shovelsnout Snake

6727 ***Prosymna ruspolii***
Squamata - Colubridae
e Ruspoli's Shovelsnout Snake

6728 ***Prosymna somalica***
Squamata - Colubridae
e Somali Shovelsnout Snake

6729 ***Prosymna sundevalli***
Squamata - Colubridae
e Sundevall's Shovelsnout Snake; Sundevall's Shovel-nosed Snake; Coppery Shovelsnout Snake; Ornate Shovelsnout Snake; Sundevall's Shovelsnout (CSA)
d Sundevall-Schaufelnasennatter

6730 ***Prosymna sundevalli lineata***
Squamata - Colubridae
e Lined Shovelsnout

6731 ***Prosymna sundevalli sundevallii***
Squamata - Colubridae
e Lined Shovelsnout

6732 ***Prosymna varius***
Squamata - Colubridae
e Jan's Shovelsnout Snake

6733 ***Psammobates***
Testudines - Testudinidae
e South African Star Tortoises
d Landschildkröten

6734 ***Psammobates geometricus***
Testudines - Testudinidae
e Geometeric Tortoise
d Geometrische Landschildkröte
f Tortue géométrique
i Psammobate geometrico; Testuggine geometrica

6735 ***Psammobates oculifer***
Testudines - Testudinidae
e Serrated Tortoise; Serrated Tent Tortoise (CSA); Kalahari Tent Tortoise (CSA); African Serrated

Tortoise; Kalahari Geometric Tortoise (CSA); Serrated Star Tortoise
d Stachelrandlandschildkröte
f Tortue ocellée
i Psammobate oculifero

6736 ***Psammobates tentorius***
Testudines - Testudinidae
e Tent Tortoise; South African Tent Tortoise; African Tent Toirtoise
d Falsche Spitzkopfschildkröte; Höckerlandschildkröte
f Tortue bocellée
i Psammobate tentorio; Testuggine tentoria

6737 ***Psammodromus***
Squamata - Lacertidae
e Sand Lizards; Psammodromuses; Plated Lacertids; Plated Lizards
d Sandläufer
f Psammodromes
i Psammodromi

6738 ***Psammodromus algirus***
Squamata - Lacertidae
e Algerian Psammodromus; Algerian Sand Lizard; Algerian Sand Racer; Large Psammodromus
d Algerischer Sandläufer
f Psammodrome algire; Psammodrome d'Algérie; Psammodrome
i Lucertola striata comune; Psammodromo algerino; Lucertola della sabbia

6739 ***Psammodromus blanci***
Squamata - Lacertidae
e Blanc's Psammodromus; Blank's Sand Lizard; Blanc's Plated Lizard
f Psammodrome de Blanc

6740 ***Psammodromus hispanicus***
Squamata - Lacertidae
e Spanish Psammodromus; Spanish Sand Lizard; Spanish Sand Racer
d Spanischer Sandläufer
f Lézard des truffiers; Psammodrome cendré; Psammodrome d'Espagne
i Psammodromo spagnolo

6741 ***Psammodromus hispanicus edwardsi***
Squamata - Lacertidae
f Psammodrome d'Edwards; Psammodrome d'Espagne

6742 ***Psammodromus microdactylus***
Squamata - Lacertidae
e Small-fingered Psammodromus; Small-fingered Sand Lizard; Green Psammodromus
f Psammodrome vert

6743 ***Psammodynastes***
Squamata - Colubridae
e Mock Vipers
d Trugnattern; Trugvipern

6744 ***Psammodynastes pictus***
Squamata - Colubridae
e Painted Mock Viper; Günther's Mock Viper

6745 ***Psammodynastes pulverulentus***
Squamata - Colubridae
e Common Mock Viper; Mock Viper (ISC)
d Asiatische Trugviper

6746 ***Psammophilus***
Squamata - Agamidae
e Sand Agamas
d Sandagamen

6747 ***Psammophilus blanfordianus***
Squamata - Agamidae
e Blanford's Sand Agama
d Blanfords Sandagame

6748 ***Psammophilus dorsalis***
Squamata - Agamidae

e Indian Sand Agama; Peninsular Rock Agama (ISC); Peninsular Rock Lizard (ISC)
f Agame dos-a-bandes

6749 ***Psammophis***
Squamata - Colubridae
e Sand Racers; Sand and Grass Snakes (CSA); Grass Snakes; Sand Snakes
d Sandrennn-Nattern
f Couleuvres des sables

6750 ***Psammophis angolensis***
Squamata - Colubridae
e Dwarf Sand Racer; Dwarf Sand Snake (CSA); Pygmy Sand Snake (CSA); Angola Sand Snake

6751 ***Psammophis ansorgii***
Squamata - Colubridae
e Link-marked Sand Racer

6752 ***Psammophis biseriatus***
Squamata - Colubridae
e Two-striped Sand Racer; Link-marked Sand Snake

6753 ***Psammophis brevirostris***
Squamata - Colubridae
e Short-snouted Sand Racer; Short-snouted Sand Snake; Short-snouted Grass Snake (CSA)

6754 ***Psammophis condanarus***
Squamata - Colubridae
e Indian Sand Racer; Indian Sand Snake; Indo-Chinese Sand Snake; Condanarous Snake (ISC)

6755 ***Psammophis crucifer***
Squamata - Colubridae
e Crossed Sand Racer; Cross-marked Grass Snake; Montane Grass Snake (CSA); Cross-marked Sand Snake (CSA); Crossed Grass Snake (CSA)
d Quergebänderte Sandrenn-Natter

6756 ***Psammophis elegans***
Squamata - Colubridae
e Elegant Sand Racer

6757 ***Psammophis jallae***
Squamata - Colubridae
e Jalla's Sand Racer; Jalla's Sand Snake (CSA); Rhodesian Sand Snake

6758 ***Psammophis leightoni***
Squamata - Colubridae
e Cape Sand Snake

6759 ***Psammophis leightoni leightoni***
Squamata - Colubridae
e Cape Sand Snake

6760 ***Psammophis leightoni namibensis***
Squamata - Colubridae
e Namib Sand Snake

6761 ***Psammophis leightoni trinasalis***
Squamata - Colubridae
e Fork-marked Sand Snake

6762 ***Psammophis leithii***
Squamata - Colubridae
e Pakistan Sand Racer; Pakistan Ribbon Snake

6763 ***Psammophis leucoaster***
Squamata - Colubridae
e Sprawl's Sand Racer

6764 ***Psammophis lineolatus***
Squamata - Colubridae
e Steppe Ribbon Racer

6765 ***Psammophis longifrons***
Squamata - Colubridae
e Long Sand Racer

6766 ***Psammophis notostictus***
Squamata - Colubridae

e Karoo Sand Racer; Karoo Sand Snake (CSA); Karoo Whip Snake (CSA); Whip Snake (CSA); Dapple-backed Sand Snake

6767 ***Psammophis phillipsi***
Squamata - Colubridae
e Olive Sand Racer; Olive Grass Snake (CSA); Sun Snake; Olive Sand Snake (CSA)

6768 ***Psammophis pulcher***
Squamata - Colubridae
e Boulenger's Sand Racer

6769 ***Psammophis punctatus***
Squamata - Colubridae
e Speckled Sand Racer; Speckled Sand Snake

6770 ***Psammophis rukwae***
Squamata - Colubridae
e Rukwa Sand Racer

6771 ***Psammophis schokari***
Squamata - Colubridae
e Schokar Sand Racer; Afro-Asian Sand Snake; Desert Sand Snake (ISC); Schokar Sand Snake
d Sandrenn-Natter; Schlanke Sandrenn-Natter; Forskals Sandrenn-Natter
f Couleuvre des sables; Psammophis de Forskal; Couleuvre de Schokar

6772 ***Psammophis sibilans***
Squamata - Colubridae
e African Beauty Racer; African Beauty Snake; Hissing Sand Snake; Olive Grass Snake (CSA); Short-snouted Grass Snake (CSA)
d Zischnatter; Gestreifte Sandrenn-Natter
f Couleuvre sifflante; Psammophis rayé

6773 ***Psammophis sibilans brevirostris***
Squamata - Colubridae
e Short-snouted Grass Snake; Short-snouted Sand Snake (CSA)

6774 ***Psammophis sibilans leopardinus***
Squamata - Colubridae
e Leopard Grass Snake

6775 ***Psammophis subtaeniatus***
Squamata - Colubridae
e Stripebelly Sand Racer; Stripe-bellied Sand Snake (CSA); Stripe-bellied Grass Snake
d Rotgestreifte Sandrenn-Natter

6776 ***Psammophis subtaeniatus subtaeniatus***
Squamata - Colubridae
e Yellow-bellied Sand Snake
d Rotgestreifte Sandrenn-Natter

6777 ***Psammophis tanganicus***
Squamata - Colubridae
e Western Link-marked Sand Racer

6778 ***Psammophis trigrammus***
Squamata - Colubridae
e Western Sand Snake; Namaqualand Sand Snake

6779 ***Psammophylax***
Squamata - Colubridae
e Skaapstekers; Rear-fanged Snake; Grass Snakes
d Shaapsteker (pl)

6780 ***Psammophylax rhombeatus***
Squamata - Colubridae
e Spotted Skaapsteker (CSA); Spotted Rear-fanged Snake; Rhombic Skaapsteker (CSA); Skaapsteker (CSA)
d Gefleckter Shaapsteker

6781 ***Psammophylax tritaeniatus***
Squamata - Colubridae

e Striped Skaapsteker (CSA); Striped Rear-fanged Snake; White-bellied Grass Snake; Striped Grass Snake (CSA); Three-lined Grass Snake (CSA)
d Gestreifter Shaapsteker

6782 ***Psammophylax variabilis***
Squamata - Colubridae
e Greybelly Skaapsteker; Grey-bellied Rear-fanged Snake; Grey-bellied Skaapsteker (CSA); Grey-bellied Grass Snake

6783 ***Pseudablabes***
Squamata - Colubridae
e Burrowing Night Snakes

6784 ***Pseudablabes agassizzii***
Squamata - Colubridae
e Burrowing Night Snake

6785 ***Pseudaspis***
Squamata - Colubridae
e Mole Snakes; Sand Snakes
d Maulwurfsnattern; Maulwurfnattern

6786 ***Pseudaspis cana***
Squamata - Colubridae
e Mole Snake (CSA); Sand Snake; Molslang (CSA); African Mole Snake
d Maulwurfsnatter; Maulwurfnatter
i Serpente talpa

6787 ***Pseudechis***
Squamata - Elapidae
e King Brown Snakes; Australian Black Snakes; Mulga Snakes; Mulga Brown Snakes
d Schwarzottern
f Serpents noirs

6788 ***Pseudechis australis***
Squamata - Elapidae
e King Brown Snake; Mulga Brown Snake; Mulga
d Mulga-Schlange; Schwarzotter

6789 ***Pseudechis butleri***
Squamata - Elapidae
e Butler's Black Snake

6790 ***Pseudechis colletti***
Squamata - Elapidae
e Collett's Black Snake; Collett's Snake (ANZ)
d Collett-Giftnatter; Collet-Schwarzotter
f Serpent de Collet
i Serpente di Collett

6791 ***Pseudechis guttatus***
Squamata - Elapidae
e Spotted Black Snake (ANZ); Blue-bellied Black Snake
d Gefleckte Schwarzotter

6792 ***Pseudechis papuanus***
Squamata - Elapidae
e Papua Brown Snake; Papuan Black Snake

6793 ***Pseudechis porphyriacus***
Squamata - Elapidae
e Redbelly Black Snake; Red-bellied Black Snake (ANZ); Papuan Black Snake
d Rotbäuchige Schwarzotter; Rotbauchschwarzotter

6794 ***Pseudemoia***
Squamata - Scincidae
e Window-eyed Skinks

6795 ***Pseudemoia baudini***
Squamata - Scincidae
e Baudin's Window-eyed Skink

6796 ***Pseudemoia conventryi***
Squamata - Scincidae

e Coventry's Window-eyed Skink

6797 ***Pseudemoia duperreyi***
Squamata - Scincidae
e Duperrey's Window-eyed Skink

6798 ***Pseudemoia entrecasteauxii***
Squamata - Scincidae
e Entrecasteaux's Window-eyed Skink; Entrecasteaux's Leiolopisma; Southern Grass Skink (ANZ)

6799 ***Pseudemoia greeni***
Squamata - Scincidae
e Green Window-eyed Skink

6800 ***Pseudemoia jigurru***
Squamata - Scincidae
e Shiny Window-eyed Skink

6801 ***Pseudemoia lichenigerum***
Squamata - Scincidae
e Lord Howe Island Skink
d Lord-Howe-Insel-Schlankskink

6802 ***Pseudemoia metallica***
Squamata - Scincidae
e Metallic Skink

6803 ***Pseudemoia microlepidota***
Squamata - Scincidae
e Tiny-scaled Window-eyed Skink

6804 ***Pseudemoia ocellata***
Squamata - Scincidae
e Ocellated Window-eyed Skink; Ocellated Skink (ANZ)

6805 ***Pseudemoia orocrypta***
Squamata - Scincidae
e Highland Window-eyed Skink

6806 ***Pseudemoia pagenstecheri***
Squamata - Scincidae
e Tusssock Skink

6807 ***Pseudemoia palfreymani***
Squamata - Scincidae
e Palfreyman's Window-eyed Skink
d Australischer Skink

6808 ***Pseudemoia platynotum***
Squamata - Scincidae
e Red-throated Skink

6809 ***Pseudemoia pretiosa***
Squamata - Scincidae
e Bass Strait Skink

6810 ***Pseudemoia rawlinsoni***
Squamata - Scincidae
e Rawlinson's Window-eyed Skink; Glossy Grass Skink (ANZ)

6811 ***Pseudemoia spenceri***
Squamata - Scincidae
e Spencer's Window-eyed Skink; Spencer's Skink (ANZ)

6812 ***Pseudemoia trilineatum***
Squamata - Scincidae
e Three-lined Skink

6813 ***Pseudemoia zia***
Squamata - Scincidae
e Flecked Window-eyed Skink

6814 ***Pseudemydura***
Testudines - Chelidae
e Swamp Turtles; Short-necked Turtles
d Falsche Spitzkopfschildkröten

6815 ***Pseudemydura umbrina***
Testudines - Chelidae
e Swamp Turtle; Western Swamp Turtle (ANZ); Short-necked Turtle; Swamp Tortoise; Western Turtle; Western Short-necked Swamp Tortoise
d Falsche Spitzkopfschildkröte

f Tortue à col de serpent de l'Ouest
i Tartaruga di palude occidentale

6816 ***Pseudemys***
Testudines - Emydidae
e River Cooters; Pseudemid Turtles; Cooters
d Schmuckschildkröten
f Tortues friandes de soleil; Tortues d'ornement; Pseudémydes

6817 ***Pseudemys alabamensis***
Testudines - Emydidae
e Alabama Redbelly Turtle; Alabama Red-bellied Turtle; Redbelly Turtle; Alamaba Red-bellied Turtle
f Tortue à ventre rouge d'Alabama; Pseudémyde de l'Alabama

6818 ***Pseudemys concinna***
Testudines - Emydidae
e River Cooter; Cooter Turtle; Cooter; Common Cooter
d Hieroglyphenschmuckschildkröte
f Tortue conchoïde des rivières; Pseuémyde concinne
i Cooter di fiume

6819 ***Pseudemys concinna concinna***
Testudines - Emydidae
e Eastern River Cooter

6820 ***Pseudemys concinna floridana***
Testudines - Emydidae
e Aquatic Cooter; Mobile Turtle; Florida Cooter; Cooter (NA); Coastal Plain Cooter
d Florida-Schmuckschildkröte
f Tortue conchoïde; Pseudémyde de Floride; Pseudémyde de Floride septentrionale
i Cooter comune

6821 ***Pseudemys concinna hieroglyphica***
Testudines - Emydidae
e Hieroglyphic River Cooter

6822 ***Pseudemys concinna metteri***
Testudines - Emydidae
e Missouri River Cooter

6823 ***Pseudemys gorzugi***
Testudines - Emydidae
e Rio Grande River Cooter; Rio Grande Cooter
f Pseudémyde du Rio Grande

6824 ***Pseudemys nelsoni***
Testudines - Emydidae
e Florida Red-bellied Turtle; Florida Redbelly Turtle; Florida Red-bellied Cooter
d Florida-Rotbauchschmuckschildkröte
f Tortue à ventre rouge de Floride; Pseudémyde de Nelson
i Cooter di Nelson; Testuggine della Florida

6825 ***Pseudemys peninsularis***
Testudines - Emydidae
e Peninsula Cooter; Western River Cooter
d Peninsula-Schmuckschildkröte
f Pseudémyde peninsulaire

6826 ***Pseudemys rubriventris***
Testudines - Emydidae
e Eastern Red-bellied Turtle; Red-bellied Turtle; American Redbelly Turtle; American Red-bellied Turtle; Redbelly Turtle (NA); Northern Red-bellied Cooter
d Rotbauchschmuckschildkröte
f Tortue à ventre rouge
i Testuggine americana dal ventre rosso

6827 ***Pseudemys suwanniensis***
Testudines - Emydidae
e River Cooter; Suwanee River Turtle; Suwanee River Cooter; Suwanee Cooter

6828 ***Pseudemys texana***
Testudines - Emydidae
e Texas River Cooter; Texas Cooter
f Pseudémyde du Texas

6829 ***Pseuderemias***
Squamata - Lacertidae
e Greater Racerunners; Sand Lizards

6830 ***Pseuderemias brenneri***
Squamata - Lacertidae
e Brenner's Racerunner

6831 ***Pseuderemias erythrosticta***
Squamata - Lacertidae
e Boulenger's Racerunner

6832 ***Pseuderemias mucronata***
Squamata - Lacertidae
e Sinai Racerunner

6833 ***Pseuderemias smithii***
Squamata - Lacertidae
e Smith's Racerunner

6834 ***Pseuderemias striata***
Squamata - Lacertidae
e Peters's Racerunner; Peters's Sand Lizard

6835 ***Pseudoacontias***
Squamata - Scincidae
e Giant Madagascar Skinks

6836 ***Pseudoacontias madagascariensis***
Squamata - Scincidae
e Giant Madagascar Skink

6837 ***Pseudoboa***
Squamata - Colubridae
e False Boas; Scarlet Snakes

6838 ***Pseudoboa coronata***
Squamata - Colubridae
e Crowned False Boa; Amazon Scarlet Snake

6839 ***Pseudoboa haasi***
Squamata - Colubridae
e Parana False Boa

6840 ***Pseudoboa neuwiedi***
Squamata - Colubridae
e Neuwied's False Boa; Windward Pseudoboa (WI)

6841 ***Pseudoboa nigra***
Squamata - Colubridae
e Black False Boa

6842 ***Pseudoboodon***
Squamata - Colubridae
e Banded Snakes

6843 ***Pseudoboodon lemniscatus***
Squamata - Colubridae
e Banded Snake

6844 ***Pseudocalotes brevipes***
Squamata - Agamidae
e Vietnam False Bloodsucker

6845 ***Pseudocalotes flavigula***
Squamata - Agamidae
e Malaya False Bloodsucker

6846 ***Pseudocalotes floweri***
Squamata - Agamidae
e Thai False Bloodsucker
d Flowers Schönechse

6847 ***Pseudocalotes microlepis***
Squamata - Agamidae
e Burmese False Bloodsucker

6848 ***Pseudocalotes poilani***
Squamata - Agamidae
e Laotioan False Bloodsucker

6849 ***Pseudocalotes tympanystriga***
Squamata - Agamidae
e Indonesian False Bloodsucker
d Kuhls Spitzkopfagame
f Bronchocèle à tympan strié

6850 ***Pseudocerastes***
Squamata - Viperidae
e False Horned Vipers
d Trughornvipern; Persische Trughornvipern; Asiatische Hornvipern
f Pseudocérastes

6851 ***Pseudocerastes fieldi***
Squamata - Viperidae
e Field's Horned Viper
d Field's Trughornviper

6852 ***Pseudocerastes latirostris***
Squamata - Viperidae
e Iranian Horned Viper

6853 ***Pseudocerastes persicus***
Squamata - Viperidae
e Persian Horned Viper; False Cerastes
d Persische Trughornviper; Westasiatische Hornviper
f Pseudocéraste de Perse

6854 ***Pseudoclotes***
Squamata - Agamidae
e False Bloodsuckers; Slender Beauty Lizards

6855 ***Pseudocordylus***
Squamata - Cordylidae
e False Girdled Lizards; Crag Lizards (CSA)
d Scheingürtelschweife; Falsche Gürtelschweife; Unechte Gürtelschweife
f Pseudocordyle

6856 ***Pseudocordylus capensis***
Squamata - Cordylidae
e False Girdled Lizard; Graceful Crag Lizard (CSA)
d Unechter Gürtelschweif

6857 ***Pseudocordylus langi***
Squamata - Cordylidae
e Lang's Girdled Lizard; Lang's Crag Lizard (CSA)
d Langs Gürtelschweif
f Pseudocordyle de Lang

6858 ***Pseudocordylus melanotus***
Squamata - Cordylidae
e Drakensberg Crag Lizard
d Drakensberg-Gürtelschweif

6859 ***Pseudocordylus microlepidotus***
Squamata - Cordylidae
e Crag Lizard; Cape Crag Lizard (CSA)
d Kleinschuppiger Scheingürtelschweif; Falscher Kap-Gürtelschweif
f Pseudocordyle microlépidote; Zonure microlépidote

6860 ***Pseudocordylus robertsi***
Squamata - Cordylidae
e Roberts's Girdled Lizard

6861 ***Pseudocordylus spinosus***
Squamata - Cordylidae
e Prickly Girdled Lizard; Spiny Crag Lizard (CSA)
d Dorniger Gürtelschweif

6862 ***Pseudocordylus spinosus subviridis***
Squamata - Cordylidae
e Drakensberg Crag Lizard

6863 ***Pseudocordylus spinosus transvaalensis***
Squamata - Cordylidae
e Drakensberg Crag Lizard

6864 ***Pseudocyclophis***
Squamata - Colubridae
e Dwarf Racers

6865 ***Pseudocyclophis persicus***
Squamata - Colubridae
e Dark-headed Dwarf Racer; Persian Dwarf Snake
f Couleuvre naine de Perse
i Irene persiana

6866 ***Pseudoeryx***
Squamata - Colubridae
e Pond Snakes; Eel Snakes

6867 ***Pseudoeryx plicatilis***
Squamata - Colubridae
e South American Pond Snake; Eel Snake

6868 ***Pseudoficimia***
Squamata - Colubridae
e False Ficimias

6869 ***Pseudoficimia frontalis***
Squamata - Colubridae
e False Ficimia

6870 ***Pseudogecko***
Squamata - Gekkonidae
e False Gecko

6871 ***Pseudogecko brevipes***
Squamata - Gekkonidae
e Luzon False Gecko

6872 ***Pseudogecko compressicorpus***
Squamata - Gekkonidae
e Philippine False Gecko

6873 ***Pseudogecko labialis***
Squamata - Gekkonidae
e Mindanao False Gecko

6874 ***Pseudogecko smaragdinus***
Squamata - Gekkonidae
e Polillo False Gecko

6875 ***Pseudogonatodes***
Squamata - Gekkonidae
e South American Clawed Geckos; Pygmy Geckos

6876 ***Pseudogonatodes barbouri***
Squamata - Gekkonidae
e Barbour's Clawed Gecko

6877 ***Pseudogonatodes furvus***
Squamata - Gekkonidae
e Colombian Clawed Gecko

6878 ***Pseudogonatodes guanensis***
Squamata - Gekkonidae
e Guyana Clawed Gecko; Amazon Pygmy Gecko

6879 ***Pseudogonatodes lunulatus***
Squamata - Gekkonidae
e Venezuela Clawed Gecko

6880 ***Pseudogonatodes peruvianus***
Squamata - Gekkonidae
e Peru Clawed Gecko

6881 ***Pseudohaje***
Squamata - Elapidae
e Forest Cobras; Tree Cobras
d Waldkobras

6882 ***Pseudohaje goldii***
Squamata - Elapidae
e Black Forest Cobra; Gold's Forest Cobra; Gold's Tree Cobra; Gold's Cobra
d Waldkobra; Golds Waldkobra
f Cobra de Gold

6883 ***Pseudohaje nigra***
Squamata - Elapidae

e Hoodless Cobra
d Schwarze Waldkobra; Australische Scheinkobra

6884 ***Pseudoleptodeira***
Squamata - Colubridae
e False Cat-eyed Snakes

6885 ***Pseudoleptodeira latifasciata***
Squamata - Colubridae
e False Cat-eyed Snake

6886 ***Pseudoleptodeira uribei***
Squamata - Colubridae
e False Cat-eyed Snake

6887 ***Pseudonaja***
Squamata - Elapidae
e Venomous Brown Snakes; Brown Snakes
d Braunschlangen; Scheinkobras

6888 ***Pseudonaja affinis***
Squamata - Elapidae
e Dugite
d Dugite

6889 ***Pseudonaja guttata***
Squamata - Elapidae
e Speckled Brown Snake; Spotted Brown Snake; Banded Brown Snake
d Gefleckte Scheinkobra

6890 ***Pseudonaja inframacula***
Squamata - Elapidae
e Pensinsula Brown Snake

6891 ***Pseudonaja ingrami***
Squamata - Elapidae
e Ingram's Brown Snake
d Ingrams Scheinkobra

6892 ***Pseudonaja modesta***
Squamata - Elapidae
e Ringed Brown Snake

6893 ***Pseudonaja nuchalis***
Squamata - Elapidae
e Western Brown Snake; Gwardar (ANZ)
d Westliche Braunschlange

6894 ***Pseudonaja textilis***
Squamata - Elapidae
e Eastern Brown Snake
d Gewöhnliche Braunschlange; Östliche Braunschlange; Braunschlange; Östliche Scheinkobra

6895 ***Pseudorabdion***
Squamata - Colubridae
e Dwarf Reed Snakes

6896 ***Pseudorabdion collaris***
Squamata - Colubridae
e Common Collared Snake

6897 ***Pseudorabdion longiceps***
Squamata - Colubridae
e Cantor's Dwarf Reed Snake

6898 ***Pseudorabdion oxycephalum***
Squamata - Colubridae
e Günther's Dwarf Reed Snake

6899 ***Pseudotomodon***
Squamata - Colubridae
e False Tomodon Snakes

6900 ***Pseudotomodon trigonatus***
Squamata - Colubridae
e False Tomodon Snake

6901 ***Pseudotrapelus***
Squamata - Agamidae
e Sinai Agamas
d Sinai-Agamen
f Agames du Sinaï

6902 ***Pseudotrapelus sinaitus***
Squamata - Agamidae
e Sinai Agama
d Sinai-Agame
f Agame du Sinaï

6903 ***Pseudotyphlops***
Squamata - Uropeltidae
e Large Shieldtail Snakes

6904 ***Pseudotyphlops philippinus***
Squamata - Uropeltidae
e Large Shieldtail Snake; Sri Lanka Shield-tailed Snake

6905 ***Pseudoxenodon***
Squamata - Colubridae
e Bamboo Snakes; Hooded Tree Snakes; Keelbacks

6906 ***Pseudoxenodon bambusicola***
Squamata - Colubridae
e Bamboo Snake

6907 ***Pseudoxenodon baramensis***
Squamata - Colubridae
e Baramen Bamboo Snake

6908 ***Pseudoxenodon dorsalis***
Squamata - Colubridae
e Night Bamboo Snake

6909 ***Pseudoxenodon fukiensis***
Squamata - Colubridae
e Fukien Bamboo Snake

6910 ***Pseudoxenodon inornatus***
Squamata - Colubridae
e Dull Bamboo Snake

6911 ***Pseudoxenodon jacobsoni***
Squamata - Colubridae
e Jacobson's Bamboo Snake

6912 ***Pseudoxenodon karlschmidti***
Squamata - Colubridae
e Chinese Bamboo Snake

6913 ***Pseudoxenodon macrops***
Squamata - Colubridae
e Big-eyed Bamboo Snake; Mock Cobra; Hooded Tree Snake; Big-eyed Keelback; Big-eyed Mountain Keelback

6914 ***Pseudoxenodon stejnegeri***
Squamata - Colubridae
e Stejneger's Bamboo Snake

6915 ***Pseudoxyrhopus***
Squamata - Colubridae
e Brook Snakes

6916 ***Pseudoxyrhopus ambreensis***
Squamata - Colubridae
e Ambre Brook Snake

6917 ***Pseudoxyrhopus dubius***
Squamata - Colubridae
e Mocquard's Brook Snake

6918 ***Pseudoxyrhopus heterurus***
Squamata - Colubridae
e Night Brook Snake

6919 ***Pseudoxyrhopus imerinae***
Squamata - Colubridae
e Plateau Brook Snake

6920 ***Pseudoxyrhopus microps***
Squamata - Colubridae
e Brown Brook Snake

6921 ***Pseudoxyrhopus occipitalis***
Squamata - Colubridae
e Madagascar Brook Snake

6922 ***Pseudoxyrhopus quinquelineatus***
Squamata - Colubridae
e Striped Brook Snake

6923 ***Pseustes***
Squamata - Colubridae
e Puffing Snakes; Bird Snakes
d Vogelnattern

6924 ***Pseustes poecilonotus***
Squamata - Colubridae
e Puffing Snake; Neotropical Bird Snake

6925 ***Pseustes poecilonotus polylepis***
Squamata - Colubridae
e Common Bird Snake

6926 ***Pseustes sexcarinatus***
Squamata - Colubridae
e Wagler's Puffing Snake

6927 ***Pseustes shropshirei***
Squamata - Colubridae
e Shropshire's Puffing Snake

6928 ***Pseustes sulphureus***
Squamata - Colubridae
e Amazon Puffing Snake

6929 ***Pseustes sulphureus***
Squamata - Colubridae
e Giant Bird Snake

6930 ***Psilothalmus***
Squamata - Teiidae
e Psilothalmuses

6931 ***Psilothalmus paeminosus***
Squamata - Teiidae
e Psilothalmus

6932 ***Psomophis***
Squamata - Colubridae
e Spirit Ground Snakes

6933 ***Psomophis genimaculatus***
Squamata - Colubridae
e Spirit Ground Snake

6934 ***Psomophis jobertsi***
Squamata - Colubridae
e Joberts's Ground Snake

6935 ***Psomophis obtusus***
Squamata - Colubridae
e Wide Ground Snake

6936 ***Ptenopus***
Squamata - Gekkonidae
e Chirping Geckos; Barking Geckos (CSA)

6937 ***Ptenopus carpi***
Squamata - Gekkonidae
e Namib Chirping Gecko; Carp's Barking Gecko (CSA)

6938 ***Ptenopus garrulus***
Squamata - Gekkonidae
e Whistling Gecko; Barking Gecko; Common Barking Gecko
d Sandgecko; Grabgecko; Pfeifgecko

6939 ***Ptenopus garrulus garrulus***
Squamata - Gekkonidae
e Common Barking Gecko (CSA); Barking Gecko (CSA)

6940 ***Ptenopus garrulus maculatus***
Squamata - Gekkonidae
e Barking Gecko (CSA)

6941 ***Ptenopus kochi***
Squamata - Gekkonidae
e Koch's Chirping Gecko; Koch's Barking Gecko (CSA)

6942 ***Ptyas***
Squamata - Colubridae
e Indian Rat Snakes; Asian Rat Snakes; Dhaman Snakes
d Asiatische Rattenschlangen
f Serpents-ratiers

6943 ***Ptyas korros***
Squamata - Colubridae
e Indian Rat Snake; Chinese Rat Snake; Schlegel's Rat Snake; Indo-Chinese Rat Snake; Slender Dhaman
d Gelbbäuchige Rattenschlange; Chinesische Rattenschlange
f Petit serpent-ratier d'Inde

6944 ***Ptyas mucosus***
Squamata - Colubridae
e Oriental Rat Snake; Dhaman Rat Snake; Greater Indian Rat Snake; Dhaman (ISC); Common Rat Snake; Asian Rat Snake
d Dhaman; Orientalische Rattennatter; Asiatische Rattenschlange; Rattennatter; Rattenschlange
f Grand serpent-ratier d'Inde; Serpent-ratier; Serpent-ratier asiatique

6945 ***Ptychoglossus***
Squamata - Teiidae
e Largescale Lizard; Swamp Lizards

6946 ***Ptychoglossus bicolor***
Squamata - Teiidae
e Werner's Largescale Lizard

6947 ***Ptychoglossus brevifrontalis***
Squamata - Teiidae
e Boulenger's Largescale Lizard; Slender Swamp Lizard

6948 ***Ptychoglossus danieli***
Squamata - Teiidae
e Daniel's Largescale Lizard

6949 ***Ptychoglossus eurylepis***
Squamata - Teiidae
e Largescale Lizard

6950 ***Ptychoglossus festae***
Squamata - Teiidae
e Peracca's Largescale Lizard

6951 ***Ptychoglossus gorgonas***
Squamata - Teiidae
e Rueda's Largescale Lizard

6952 ***Ptychoglossus grandisquamatus***
Squamata - Teiidae
e Common Largescale Lizard

6953 ***Ptychoglossus kugleri***
Squamata - Teiidae
e Kugler's Largescale Lizard

6954 ***Ptychoglossus nicefori***
Squamata - Teiidae
e Loveridge's Largescale Lizard

6955 ***Ptychoglossus picticeps***
Squamata - Teiidae
e Cope's Largescale Lizard

6956 ***Ptychoglossus plicatus***
Squamata - Teiidae
e Taylor's Largescale Lizard

6957 ***Ptychoglossus romulus***
Squamata - Teiidae
e King Largescale Lizard

6958 ***Ptychophis***
Squamata - Colubridae
e Fanged Water Snakes; Big Oriental Rat Snakes; Oriental Rat Snakes

6959 ***Ptychophis flavovirgatus***
Squamata - Colubridae
e Fanged Water Snake; Big Oriental Rat Snake; Oriental Rat Snake

6960 ***Ptychozoon***
Squamata - Gekkonidae
e Flying Geckos; Parachuting Geckos; Fringed Geckos
d Faltengeckos; Fächerfinger; Fächerfußsgeckos
f Geckos volants

6961 ***Ptychozoon horsfieldi***
Squamata - Gekkonidae
e Horsfield's Flying Gecko; Horsfield's Parachuting Gecko; Horsfield's Fringed Gecko

6962 ***Ptychozoon intermedium***
Squamata - Gekkonidae
e Intermediate Flying Gecko; Intermediate Parachuting Gecko; Philippines Fringed Gecko

6963 ***Ptychozoon kuhli***
Squamata - Gekkonidae
e Kuhl's Flying Gecko; Kuhl's Parachuting Gecko; Flying Gecko; Parachuting Gecko; Kuhl's Fringed Gecko; Gliding Gecko (ISC)
d Kuhls Faltengecko; Faltengecko
f Gecko volant; Gecko parachute

6964 ***Ptychozoon lionotum***
Squamata - Gekkonidae
e Common Flying Gecko; Common Parachuting Gecko; Burmese Flying Gecko; Flying Gecko; Gliding Gecko
d Gewöhnlicher Faltengecko; Faltengecko

6965 ***Ptychozoon rhacophorus***
Squamata - Gekkonidae
e Sabah Flying Gecko; Sabah Parachuting Gecko

6966 ***Ptyodactylus***
Squamata - Gekkonidae
e Fan-fingered Geckos; Fan-footed Geckos
d Fächerfingergeckos; Fächerfinger; Fächerzäher
f Ptyodactyles

6967 ***Ptyodactylus guttatus***
Squamata - Gekkonidae
e Sinai Fan-fingered Gecko

6968 ***Ptyodactylus hasselquistii***
Squamata - Gekkonidae
e Yellow Fan-fingered Gecko; Fan-toed Rock Gecko; Fan-toed Gecko; Fan-footed Gecko
d Gelber Fächerfingergecko; Östlicher Fächerfußgecko; Gemeiner Fächerfuß; Gemeiner Fächerzeher; Fächerfingergecko; Fächerfußgecko
f Gecko des maisons; Ptyodactyle de Hasselquist; Ptyodactyle; Gecko à doigts en forme d'éventail

6969 ***Ptyodactylus hasselquistii hasselquistii***
Squamata - Gekkonidae
d Östlicher Fächerfußgecko

6970 ***Ptyodactylus hasselquistii oudrii***
Squamata - Gekkonidae
d Wüstenfächerfußgecko

6971 ***Ptyodactylus hasselquistii togoensis***
Squamata - Gekkonidae
d Westlicher Fächerfußgecko

6972 ***Ptyodactylus homolepis***
Squamata - Gekkonidae
e Pakistan Fan-fingered Gecko

6973 ***Ptyodactylus oudrii***
Squamata - Gekkonidae
e Algerian Fan-fingered Gecko; Oudri's Fan-footed Gecko
d Oudris Flachfingergecko
f Ptyodactyle d'Oudri

6974 ***Ptyodactylus puiseuxi***
Squamata - Gekkonidae

e Israeli Fan-fingered Gecko; Fan-toed Gecko

6975 ***Ptyodactylus ragazzii***
Squamata - Gekkonidae
e Ragazzi's Fan-footed Gecko
d Ragazzis Fächerfingergecko
f Gecko de Ragazzi

6976 ***Pyctolaemus***
Squamata - Agamidae
e Throated Agamas

6977 ***Pyctolaemus gularis***
Squamata - Agamidae
e Throated Agama; Blue-throated Lizard (ISC)

6978 ***Pygomeles***
Squamata - Scincidae
e Short Skinks

6979 ***Pygomeles braconnieri***
Squamata - Scincidae
e Braconnier's Short Skink

6980 ***Pygomeles petteri***
Squamata - Scincidae
e Petter's Short Skink

6981 ***Pygopus***
Squamata - Gekkonidae
e Australian Scalyfoots; Scalyfoots
d Flossenfüße
f Pygopodes

6982 ***Pygopus lepidopodus***
Squamata - Gekkonidae
e Common Scalyfoot
d Gewöhnlicher Flossenfuß; Schlangenflossenfuss
f Pygopode lépidopode

6983 ***Pygopus nigriceps***
Squamata - Gekkonidae
e Hooded Scalyfoot; Western Scalyfoot
d Westlicher Flossenfuß

6984 ***Python***
Squamata - Boidae
e Common Pythons; Pythons; Typical Pythons
d Echte Pythone; Pythone
f Pythons

6985 ***Python anchietae***
Squamata - Boidae
e Angola Python; Anchieta's Dwarf Python (CSA); Anchieta's Python; Angolan Python
d Angola-Python
f Python d'Angola

6986 ***Python curtus***
Squamata - Boidae
e Blood Python; Short-tailed Python; Short Python; Malayan Python; Lesser Python; Red Python; Borneo Blood Python
d Buntpython; Blutpython; Kurzpython; Kurzschwanzpython
f Python malais; Python à queue courte
i Pitone corto

6987 ***Python curtus breitensteini***
Squamata - Boidae
e Bornean Short-tailed Python; Borneo Short-tailed Python

6988 ***Python curtus brongersmai***
Squamata - Boidae
e Blood Python; Red Blood Python
d Buntpython; Roter Blutpython
f Python tacheté

6989 ***Python curtus curtus***
Squamata - Boidae
e Sumatran Short-tailed Python; Black Blood Python
d Kurzschwanzpython; Blutpython

6990 ***Python molurus***
Squamata - Boidae
e Indian Python ; Burmese Python; Indian Rock Python; Black-tailed Python
d Tigerpython; Burma-Python; Burma-Tigerpython; Tigerschlange
f Python molure; Python indien; Python de l'Inde
i Pitone moluro birmani; Pitone moluro; Pitone delle rocce indiano

6991 ***Python molurus bivittatus***
Squamata - Boidae
e Burmese Python
d Dunkler Tigerpython; Burma-Python
f Python birman
i Pitone delle rocce

6992 ***Python molurus molurus***
Squamata - Boidae
e Indian Python ; Indian Rock Python
d Heller Tigerpython

6993 ***Python regius***
Squamata - Boidae
e Ball Python; Royal Python; King Python
d Königspython; Ballschlange; Ballpython
f Python royal
i Pitone reale; Pitone palla

6994 ***Python reticulatus***
Squamata - Boidae
e Reticulate Python; Reticulated Python; Malay Python
d Netzpython; Gitterschlange; Netzschlange; Borneo-Riesenschlange
f Python réticulé
i Pitone reticolato

6995 ***Python sebae***
Squamata - Boidae
e African Rock Python; African Python (CSA); Common African Python; Rock Python
d Felsenpython; Hieroglyphschlange; Assala
f Python de Séba
i Pitone di Seba; Pitone delle rocce; Pitone delle rocce africano

6996 ***Python sebae natalensis***
Squamata - Boidae
e Lesser Rock Python; Natal Rock Python; Southern African Python; Southern African Rock Python
d Südlicher Felsenpython

6997 ***Python timoriensis***
Squamata - Boidae
e Timor Python
d Timor-Python

6998 ***Pythonodipsas***
Squamata - Colubridae
e Western Keeled Snakes; Keeled Snakes

6999 ***Pythonodipsas carinata***
Squamata - Colubridae
e Western Keeled Snake; Keeled Snake

7000 ***Pyxidea***
Testudines - Emydidae
e Keeled Box Turtles
d Indische Dornschildkröten
f Tortues-boîte à trois carènes

7001 ***Pyxidea mouhotii***
Testudines - Emydidae
e Keeled Box Turtle (ISC); Jaggedshell Turtle; Jaggedshell Box Turtle; Jagged-shelled Turtle
d Indische Dornschildkröte
f Tortue-boîte à trois carènes
i Tartaruga scatola carinata

7002 ***Pyxis***
Testudines - Testudinidae
e Spider Tortoises
d Spinnenschildkröten
f Pyxides

7003 ***Pyxis arachnoides***
Testudines - Testudinidae
e Spider Tortoise; Common Spider Tortoise; Madagascan Spider Tortoise
d Spinnenschildkröte; Madagaskar-Spinnenschildkröte
f Pyxide arachnoide
i Pisside; Testuggine aracnoide comune

7004 ***Pyxis planicauda***
Testudines - Testudinidae
e Flat-shelled Spider Tortoise; Madagascan Flat-tailed Tortoise; Madagascar Flat-shelled Tortoise; Madagascar Flat-tailed Spider Tortoise
d Madagasische Flachschwanzschildkröte; Madagasische Flachrückenschildkröte
f Tortue à queue plate
i Acinisside; Testuggine aracnoide a guscio platto

Q

7005 ***Quedenfeldtia***
Squamata - Gekkonidae
e Atlas Day Geckos
d Atlas-Taggeckos

7006 ***Quedenfeldtia trachyblepharus***
Squamata - Gekkonidae
e Atlas Day Gecko; Thorny-eyelidded Gecko; High Atlas Moroccan Day Gecko
d Atlas-Taggecko
f Gecko à paupière épineuse; Gecko à paupières hérissées; Gecko à paupières épineuses du Haut Atlas

R

7007 ***Rafetus***
Testudines - Trionychidae
e Billacosites; Billacosite Softshells

7008 ***Rafetus euphraticus***
Testudines - Trionychidae
e Euphrates Softshell; Euphrates Soft-shelled Turtle; Billacosite Softshell Turtle; Euphrates Softshell Turtle
d Euphrat-Weichschildkröte
f Trionyx à deux callositées; Trionyx de l'Euphrate
i Trionice eufratico; Tartaruga a guscio molle dell'Eufrate

7009 ***Rafetus swinhoei***
Testudines - Trionychidae
e Shanghai Softshell; Swinhoe's Soft-shelled Turtle; Shanghai Softshell Turtle; Swinhoe's Softshell Turtle; Swinhoe's Softshell
f Trionyx de Swinhoe

7010 ***Ramphotyphlops***
Squamata - Typhlopidae
e Common Blind Snakes; Australasian Blind Snakes; Australian Thread Snakes; Brahminy Blind Snakes
d Blindschlangen
i Rantotiflopi

7011 ***Ramphotyphlops acuticauda***
Squamata - Typhlopidae
e Palau Island Blind Snake

7012 ***Ramphotyphlops affinis***
Squamata - Typhlopidae
e Brown Blind Snake

7013 ***Ramphotyphlops albiceps***
Squamata - Typhlopidae
e White-headed Blind Snake

7014 ***Ramphotyphlops australis***
Squamata - Typhlopidae
e Australian Blind Snake

7015 ***Ramphotyphlops batillus***
Squamata - Typhlopidae
e Batillus Blind Snake

7016 ***Ramphotyphlops bituberculatus***
Squamata - Typhlopidae
e Anglenose Blind Snake

7017 ***Ramphotyphlops braminus***
Squamata - Typhlopidae
e Brahminy Blind Snake; Bootlace Snake; Flowerpot Snake; Conmmon Worm Snake (ISC); Common Blind Snake (ISC); Island Blind Snake (NA); Flowerpot Blind Snake; Russel's Blind Snake
d Gewöhnliche Blindschlange; Blindschlange; Schnürsenkelschlange; Blumentopfschlange
f Typhlops commun; Serpent aveugle du Brahminy
i Rantotiflope partenogenico

7018 ***Ramphotyphlops broomi***
Squamata - Typhlopidae
e Broom's Blind Snake; Striated Blind Snake; Striate Blind Snake
d Gestreifte Blindschlange

7019 ***Ramphotyphlops centralis***
Squamata - Typhlopidae
e Central Blind Snake

7020 ***Ramphotyphlops cunningii***
Squamata - Typhlopidae
e Philippine Blind Snake

7021 ***Ramphotyphlops diversus***
Squamata - Typhlopidae
e Diverse Blind Snake

7022 ***Ramphotyphlops endotera***
Squamata - Typhlopidae
e Reddish Blind Snake

7023 ***Ramphotyphlops exocoeti***
Squamata - Typhlopidae
e Pale Blind Snake

7024 ***Ramphotyphlops glavipenter***
Squamata - Typhlopidae
e Indonesian Blind Snake

7025 ***Ramphotyphlops gryphus***
Squamata - Typhlopidae
e Blacktail Blind Snake

7026 ***Ramphotyphlops guentheri***
Squamata - Typhlopidae
e Günther's Blind Snake

7027 ***Ramphotyphlops hamatus***
Squamata - Typhlopidae
e Hooked Blind Snake

7028 ***Ramphotyphlops howi***
Squamata - Typhlopidae
e Wall Point Blind Snake

7029 ***Ramphotyphlops kimberleyensis***
Squamata - Typhlopidae
e Kimberley Blind Snake

7030 ***Ramphotyphlops leptosoma***
Squamata - Typhlopidae
e Purple Blind Snake

7031 ***Ramphotyphlops leucoproctus***
Squamata - Typhlopidae
e Island Blind Snake

7032 ***Ramphotyphlops ligatus***
Squamata - Typhlopidae
e Eastern Blind Snake

7033 ***Ramphotyphlops lineatus***
Squamata - Typhlopidae
e Striped Blind Snake

7034 ***Ramphotyphlops margaretae***
Squamata - Typhlopidae
e Lake Throssell Blind Snake

7035 ***Ramphotyphlops micromma***
Squamata - Typhlopidae
e Leopold Downs Blind Snake

7036 ***Ramphotyphlops minimus***
Squamata - Typhlopidae
e Groote Eylandt Blind Snake

7037 ***Ramphotyphlops multilineatus***
Squamata - Typhlopidae
e Many-lined Blind Snake

7038 ***Ramphotyphlops nigrescens***
Squamata - Typhlopidae
e Night Blind Snake

7039 ***Ramphotyphlops olivaceus***
Squamata - Typhlopidae
e Olive Blind Snake

7040 ***Ramphotyphlops pinguis***
Squamata - Typhlopidae
e Fat Blind Snake

7041 ***Ramphotyphlops polygrammicus***
Squamata - Typhlopidae
e Yellow-tipped Blind Snake

7042 ***Ramphotyphlops proximus***
Squamata - Typhlopidae
e Chocolate Blind Snake

7043 ***Ramphotyphlops tovelli***
Squamata - Typhlopidae
e Darwin Blind Snake

7044 ***Ramphotyphlops troglodytes***
Squamata - Typhlopidae
e Tunnel Cave Blind Snake

7045 ***Ramphotyphlops unguirostris***
Squamata - Typhlopidae
e Peters's Blind Snake
d Peters Blindschlange

7046 ***Ramphotyphlops waitii***
Squamata - Typhlopidae
e Waite's Blind Snake

7047 ***Ramphotyphlops wiedii***
Squamata - Typhlopidae
e Wied's Blind Snake

7048 ***Ramphotyphlops yampiensis***
Squamata - Typhlopidae
e Kooland Island Blind Snake

7049 ***Ramphotyphlops yirrikala***
Squamata - Typhlopidae
e Darwin Land Blind Snake

7050 ***Regina***
Squamata - Colubridae
e Crayfish Snakes; Queen Snakes
d Königinnatter; Krebsnattern
f Couleuvres-écrevisse

7051 ***Regina alleni***
Squamata - Colubridae
e Striped Crayfish Snake; Allen's Mud Snake; Allen's Snake; Striped Swamp Snake
d Gestreifte Sumpfnatter; Gestreifte Krebsnatter
f Couleuvre-écrevisse des marais

7052 ***Regina grahamii***
Squamata - Colubridae
e Graham's Crayfish Snake; Graham's Water Snake; Graham's Queen Snake
d Grahams Wassernatter; Grahams Krebsnatter
f Couleuvre-écrevisse de Graham; Couleuvre de Graham

7053 ***Regina rigida***
Squamata - Colubridae
e Glossy Crayfish Snake; Glossy Crayfish Water Snake; Glossy Water Snake; Stiff Snake; Striped Water Snake
d Gestreifte Wassernatter; Glanzkrebsnatter
f Couleuvre-écrevisse luisante; Couleuvre rayée

7054 ***Regina rigida deltae***
Squamata - Colubridae
e Delta Crayfish Snake

7055 ***Regina rigida rigida***
Squamata - Colubridae
e Glossy Crayfish Snake

7056 ***Regina rigida sinicola***
Squamata - Colubridae
e Gulf Crayfish Snake

7057 ***Regina septemvittata***
Squamata - Colubridae
e Queen Snake; Leather King Water Snake
d Königinennatter; Königinnatter; Königinschlange; Königsnatter; StreifenwWassernatter
f Couleuvre-écrevisse royale

7058 ***Rhabderemias***
Squamata - Lacertidae
e Lesser Racerunners

7059 ***Rhabderemias fasciata***
Squamata - Lacertidae
e Sistan Racerunner

7060 ***Rhabderemias lineolata***
Squamata - Lacertidae
e Striped Racerunner

7061 ***Rhabderemias pleskei***
Squamata - Lacertidae
e Pleske's Racerunner
d Transkaukasischer Wüstenrenner; Pleskes Wüstenrenner
f Érémias de Pleske
i Eremia di Pleske

7062 ***Rhabderemias scripta***
Squamata - Lacertidae
e Sand Racerunner

7063 ***Rhabdophidium***
Squamata - Colubridae
e Pointed Snake

7064 ***Rhabdophidium forsteni***
Squamata - Colubridae
e Forsten's Pointed Snake

7065 ***Rhabdophis***
Squamata - Colubridae
e Keelbacks; Common Keelbacks

7066 ***Rhabdophis auriculatus***
Squamata - Colubridae
e Eared Keelback

7067 ***Rhabdophis ceylonensis***
Squamata - Colubridae
e Ceylonese Keelback; Ceylon Keelback

7068 ***Rhabdophis chrysargoides***
Squamata - Colubridae
e Günther's Keelback

7069 ***Rhabdophis chrysargus***
Squamata - Colubridae
e Specklebelly Keelback; Schlegel's Keelback; Speckle-bellied Keelback

7070 ***Rhabdophis himalayanus***
Squamata - Colubridae
e Orange-collared Keelback; Gamboge-collared Keelback

7071 ***Rhabdophis leonardi***
Squamata - Colubridae
e Leonards Keelback

7072 ***Rhabdophis lineata***
Squamata - Colubridae
e Lined Keelback

7073 ***Rhabdophis murudensis***
Squamata - Colubridae
e Muruden Keelback

7074 ***Rhabdophis nigrocinctus***
Squamata - Colubridae
e Green Keelback

7075 ***Rhabdophis nuchalis***
Squamata - Colubridae
e Hubei Keelback

7076 ***Rhabdophis spilogaster***
Squamata - Colubridae
e Boie's Keelback

7077 ***Rhabdophis subminiatus***
Squamata - Colubridae
e Redneck Keelback; Red-necked Keelback; Malayan Bush Snake; Vermilion-necked Keelback
d Rothalskielrückennatter; Rotnackenwassernatter

7078 ***Rhabdophis swinhonis***
Squamata - Colubridae
e Taiwan Keelback

7079 ***Rhabdophis tigrinus***
Squamata - Colubridae

e Asian Tiger Keelback; Tiger Keelback; Asian Tiger Snake; Yamakagashi
d Tigernatter; Tigerwassernatter

7080 ***Rhabdops***
Squamata - Colubridae
e Trapezoid Snakes

7081 ***Rhabdops bicolor***
Squamata - Colubridae
e Brown Trapezoid Snake

7082 ***Rhabdops olivaceus***
Squamata - Colubridae
e Olive Trapezoid Snake

7083 ***Rhachidelus***
Squamata - Colubridae
e Brazilian Bird Snakes

7084 ***Rhachidelus brazili***
Squamata - Colubridae
e Brazilian Bird Snake

7085 ***Rhacodactylus***
Squamata - Gekkonidae
e Giant Geckos
d Riesengeckos

7086 ***Rhacodactylus auriculatus***
Squamata - Gekkonidae
e New Caledonia Bumpy Gecko
d Höckerkopfgecko
f Gecko gargouille

7087 ***Rhacodactylus australis***
Squamata - Gekkonidae
e Australian Giant Gecko

7088 ***Rhacodactylus cavaticus***
Squamata - Gekkonidae
e Kimberley Giant Cave Gecko

7089 ***Rhacodactylus chahoua***
Squamata - Gekkonidae
e Bavay's Giant Gecko; Bavay's New Caledonian Gecko
d Neukaledonischer Riesengecko; Flechtengecko

7090 ***Rhacodactylus ciliatus***
Squamata - Gekkonidae
e Guichenot's Giant Gecko
d Kronengecko; Kronenkopfgecko
f Gecko à crête

7091 ***Rhacodactylus leachianus***
Squamata - Gekkonidae
e New Caledonia Giant Gecko; New Caledonian Giant Forest Gecko; New Caledonian Gecko; New Caledonian Giant Gecko
d Neukaledonischer Riesengecko; Kaledonischer Riesengecko

7092 ***Rhacodactylus lindneri***
Squamata - Gekkonidae
e Giant Cave Gecko (ANZ)

7093 ***Rhacodactylus sarasinorum***
Squamata - Gekkonidae
e Roux's Giant Gecko

7094 ***Rhacodactylus trachyrhynchus***
Squamata - Gekkonidae
e Tough-snouted Giant Gecko

7095 ***Rhadinaea***
Squamata - Colubridae
e Graceful Brown Snakes; Pine Woods Snakes; Litter Snakes

7096 ***Rhadinaea altamontana***
Squamata - Colubridae
e Mountain Graceful Brown Snake

7097 ***Rhadinaea anachoreta***
Squamata - Colubridae
e Litter Snake

7098 ***Rhadinaea bogertorum***
Squamata - Colubridae
e Oaxacan Graceful Brown Snake

7099 ***Rhadinaea calligaster***
Squamata - Colubridae
e Thick Graceful Brown Snake; Spotbelly Litter Snake

7100 ***Rhadinaea cuneata***
Squamata - Colubridae
e Veracruz Graceful Brown Snake

7101 ***Rhadinaea decipiens***
Squamata - Colubridae
e American Graceful Brown Snake

7102 ***Rhadinaea decorata***
Squamata - Colubridae
e Adorned Graceful Brown Snake; Pink-bellied Litter Snake

7103 ***Rhadinaea dumerilii***
Squamata - Colubridae
e Duméril's Graceful Brown Snake

7104 ***Rhadinaea flavilata***
Squamata - Colubridae
e Pine Wood Snake; Pine Woods Snake; Yellow-lipped Snake; Brown-headed Snake; Pine Woods Litter Snake
d Kiefernwaldschlange
f Couleuvre des pinèdes

7105 ***Rhadinaea forbesi***
Squamata - Colubridae
e Forbes's Graceful Brown Snake

7106 ***Rhadinaea fulviceps***
Squamata - Colubridae
e Graceful Brown Snake

7107 ***Rhadinaea fulvivittis***
Squamata - Colubridae
e Ribbon Graceful Brown Snake

7108 ***Rhadinaea gaigeae***
Squamata - Colubridae
e Gaige's Pine Forest Snake

7109 ***Rhadinaea godmani***
Squamata - Colubridae
e Godman's Graceful Brown Snake; Godman's Litter Snake

7110 ***Rhadinaea guentheri***
Squamata - Colubridae
e Günther's Graceful Brown Snake

7111 ***Rhadinaea hannsteini***
Squamata - Colubridae
e Hannstein's Spot-lipped Snake; Hannstein's Litter Snake

7112 ***Rhadinaea hempsteadae***
Squamata - Colubridae
e Hempstead's Pine Woods Snake; Hempstead's Litter Snake

7113 ***Rhadinaea hesperia***
Squamata - Colubridae
e Western Graceful Brown Snake

7114 ***Rhadinaea kanalchutchan***
Squamata - Colubridae
e Litter Snake

7115 ***Rhadinaea kinkelini***
Squamata - Colubridae
e Kinkelin's Graceful Brown Snake; Kinkelin's Litter Snake

7116 ***Rhadinaea lachrymams***
Squamata - Colubridae
e Tearful Pine-Oak Snake

7117 ***Rhadinaea lateristriga***
Squamata - Colubridae
e Berthold's Graceful Brown Snake

7118 ***Rhadinaea laureata***
Squamata - Colubridae
e Crowned Graceful Brown Snake

7119 ***Rhadinaea macdougalli***
Squamata - Colubridae
e MacDougall's Graceful Brown Snake; MacDougall's Litter Snake

7120 ***Rhadinaea marcellae***
Squamata - Colubridae
e Marcella's Graceful Brown Snake

7121 ***Rhadinaea montana***
Squamata - Colubridae
e Nuevo Leon Graceful Brown Snake

7122 ***Rhadinaea montecristi***
Squamata - Colubridae
e Monte Cristi Graceful Brown Snake; Monte Cristi Litter Snake

7123 ***Rhadinaea multilineata***
Squamata - Colubridae
e Many-lined Graceful Brown Snake

7124 ***Rhadinaea myersi***
Squamata - Colubridae
e Myers's Graceful Brown Snake

7125 ***Rhadinaea omiltemana***
Squamata - Colubridae
e Guerreran Pine Woods Snake

7126 ***Rhadinaea pachyura***
Squamata - Colubridae
e Cope's Graceful Brown Snake

7127 ***Rhadinaea pilonaorum***
Squamata - Colubridae
e Stuart's Graceful Brown Snake; Slender Litter Snake

7128 ***Rhadinaea posadasi***
Squamata - Colubridae
e Posada's Graceful Brown Snake; Slevin's Litter Snake

7129 ***Rhadinaea pulveriventris***
Squamata - Colubridae
e Common Graceful Brown Snake; Common Litter Snake

7130 ***Rhadinaea quinquelineata***
Squamata - Colubridae
e Pueblan Graceful Brown Snake

7131 ***Rhadinaea rogerromani***
Squamata - Colubridae
e Orange-bellied Litter Snake

7132 ***Rhadinaea sargenti***
Squamata - Colubridae
e Sargent's Graceful Brown Snake

7133 ***Rhadinaea schistosa***
Squamata - Colubridae
e Broken-collar Graceful Brown Snake

7134 ***Rhadinaea serperaster***
Squamata - Colubridae
e Southern Graceful Brown Snake; Striped Litter Snake

7135 ***Rhadinaea stadelmani***
Squamata - Colubridae
e Stadelman's Graceful Brown Snake; Stadelman's Litter Snake

7136 ***Rhadinaea taeniata***
Squamata - Colubridae
e Pine-Oak Snake

7137 ***Rhadinaea vermiculaticeps***
Squamata - Colubridae
e Vermiculate Graceful Brown Snake; Vermiculate Litter Snake

7138 ***Rhadinophanes***
Squamata - Colubridae

e Graceful Mountain Snakes; Pine Wood Snakes

7139 ***Rhadinophanes monticola***
Squamata - Colubridae
e Graceful Mountain Snake; Pine Wood Snake

7140 ***Rhamphiophis***
Squamata - Colubridae
e Beaked Snakes; Beak-nosed Snakes
d Schnabelnasennattern

7141 ***Rhamphiophis acutus***
Squamata - Colubridae
e Striped Beaked Snake

7142 ***Rhamphiophis oxyrhychus***
Squamata - Colubridae
e Hook-nosed Beaked Snake
d Rötliche Schnabelnasennatter

7143 ***Rhamphiophis oxyrhychus rostratus***
Squamata - Colubridae
e Rufous Beaked Snake (CSA)

7144 ***Rhamphiophis rubropunctatus***
Squamata - Colubridae
e Red-spotted Beaked Snake; Red Beaked Snake
d Rotrückenschnabelnasennatter

7145 ***Rhampholeon***
Squamata - Chamaeleonidae
e Stumptail Chameleons; Leaf Chameleons (CSA); Pygmy Chameleons
d Stummelschwanzchamäleone; Zwergchamäleone
f Caméléons nains

7146 ***Rhampholeon brachura***
Squamata - Chamaeleonidae
e African Stumptail Chameleon; Beardless Pygmy Chameleon

7147 ***Rhampholeon brevicaudata***
Squamata - Chamaeleonidae
e Tanzania Stumptail Chameleon; Bearded Pygmy Chameleon

7148 ***Rhampholeon chapmani***
Squamata - Chamaeleonidae
e Chapman's Stumptail Chameleon

7149 ***Rhampholeon kersteni***
Squamata - Chamaeleonidae
e Kenya Stumptail Chameleon; Kenya Pygmy Chameleon; Somali Pygmy Chameleon; Dwarf Chameleon (CSA)
d Somali-Zwergchamäleon
f Caméléon nain de Somali

7150 ***Rhampholeon marshalli***
Squamata - Chamaeleonidae
e Marshall's Stumptail Chameleon; Marshall's Leaf Chameleon (CSA); Marshall's Chameleon; Marshall's Dwarf Forest Chameleon (CSA); Marshall's Dwarf Chameleon (CSA)
d Marshalls Zwergchamäleon
f Caméléon nain de Marshall; Caméléon de Marshall

7151 ***Rhampholeon mclachlani***
Squamata - Chamaeleonidae
e McLachlan's Stumptail Chameleon

7152 ***Rhampholeon nchisiensis***
Squamata - Chamaeleonidae
e South African Stumptail Chameleon; Pitless Pygmy Chameleon

7153 ***Rhampholeon platyceps***
Squamata - Chamaeleonidae
e Malawi Stumptail Chameleon; Flat-headed Chameleon
d Flachkopfblattchamäleon

7154 ***Rhampholeon spectrum***
Squamata - Chamaeleonidae
e Cameroon Stumptail Chameleon; Cameroon Dwarf Chameleon; Spectral Pygmy Chameleon
d Stummelschwanzchamäleon; Blattchamäleon

7155 ***Rhampholeon temporalis***
Squamata - Chamaeleonidae
e Usambara Stumptail Chameleon; Pitted Pygmy Chameleon

7156 ***Rheodytes***
Testudines - Chelidae
e Fitzroy River Turtles; Fitzroy Turtles; Fitzroy Tortoises
d Fitzroy-Schildkröten

7157 ***Rheodytes leukops***
Testudines - Chelidae
e Fitzroy River Turtle; Fitzroy Turtle; Fitzroy Tortoise; White-eyed River Diver
d Fitzroy-Schildkröte
i Tartaruga del fiume Fitzroy

7158 ***Rhineura***
Squamata - Amphisbaenidae
e Florida Worm Lizards; Wide-snouted Wall Lizards (NA)
d Florida-Doppelschleichen
f Amphisbènes de Floride; Rhineures ; Rhineures de Floride

7159 ***Rhineura floridana***
Squamata - Amphisbaenidae
e Florida Worm Lizard
d Florida-Doppelschleiche
f Amphisbène de Floride; Rhineure de Floride

7160 ***Rhinobothryum***
Squamata - Colubridae
e Banded Snakes; Ringed Snakes

7161 ***Rhinobothryum bovallii***
Squamata - Colubridae
e Honduras Banded Snake; Big-headed False Coral Snake

7162 ***Rhinobothryum lentiginosum***
Squamata - Colubridae
e Amazon Banded Snake; Amazon Ringed Snake

7163 ***Rhinocalamus***
Squamata - Colubridae
e Tanganyika Snakes

7164 ***Rhinocalamus dimidiatus***
Squamata - Colubridae
e Tanganyika Red-banded Snake

7165 ***Rhinocheilus***
Squamata - Colubridae
e Longnose Snakes; Long-nosed Snakes
d Langnasennattern; Nasennattern

7166 ***Rhinocheilus lecontei***
Squamata - Colubridae
e Long-nosed Snake; Le Conte's Snake; Longnose Snake
d Nasennatter
f Couleuvre à long nez

7167 ***Rhinocheilus lecontei lecontei***
Squamata - Colubridae
e Western Longnose Snake; Western Long-nosed Snake

7168 ***Rhinocheilus lecontei tesselatus***
Squamata - Colubridae
e Texas Longnose Snake; Texas Long-nosed Snake

7169 ***Rhinoclemmys***
Testudines - Emydidae
e Neotropical Wood Turtles
d Amerikanische Erdschildkröten
f Rhinoclemmydes

7170 ***Rhinoclemmys annulata***
Testudines - Emydidae
e Brown Wood Turtle
d Braune Erdschildkröte
f Rhinoclemmyde annelée

7171 ***Rhinoclemmys areolata***
Testudines - Emydidae
e Furrowed Wood Turtle
d Gefurchte Erdschildkröte
f Rhinoclemmyde aréolée
i Tartaruga rugosa di foresta

7172 ***Rhinoclemmys diademata***
Testudines - Emydidae
e Maracaibo Wood Turtle
d Venezuela-Erdschildkröte
f Rhinoclemmyde à diadème

7173 ***Rhinoclemmys funerea***
Testudines - Emydidae
e Black Wood Turtle; Black River Turtle
d Bauchstreifenerdschildkröte
f Rhinoclemmyde funèbre

7174 ***Rhinoclemmys melanosterna***
Testudines - Emydidae
e Colombian Wood Turtle
f Rhinoclemmyde à plastron noir

7175 ***Rhinoclemmys nasuta***
Testudines - Emydidae
e Largenose Wood Turtle; Large-nosed Wood Turtle
f Rhinoclemmyde à museau pointu

7176 ***Rhinoclemmys pulcherrima***
Testudines - Emydidae
e Painted Wood Turtle; American Geomyda; Mexican Wood Turtle; Mexican Painted Wood Turtle; Mexican Red Turtle; Ornate Wood Turtle
d Prachterdschildkröte
f Géoémyde américaine; Rhinoclemmyde peinte
i Tartaruga dipinta di foresta

7177 ***Rhinoclemmys pulcherrima incisa***
Testudines - Emydidae
e Salvador Painted Wood Turtle
f Rhinoclemmyde peinte du Salvador

7178 ***Rhinoclemmys pulcherrima manni***
Testudines - Emydidae
e Costa Rica Painted Wood Turtle
f Rhinoclemmyde peinte du Costa Rica

7179 ***Rhinoclemmys pulcherrima rogerbarbouri***
Testudines - Emydidae
e Sonora Painted Wood Turtle
f Rhinoclemmyde peinte de Sonora

7180 ***Rhinoclemmys punctularia***
Testudines - Emydidae
e Spot-legged Wood Turtle; Spotted-legged Wood Turtle; Spotted-legged Turtle; Spot-legged Turtle; Guiana Wood Turtle
d Südamerikanische Erdschildkröte; Rückenfleckenerdschildkröte
f Rhinoclemmyde ponctulaire
i Tartaruga dalle zampe penteggiate

7181 ***Rhinoclemmys rubida***
Testudines - Emydidae
e Mexican Spotted Wood Turtle; Oaxaca Wood Turtle
d Rückenfleckenerdschildkröte
f Rhinoclemmyde brune d'Oaxaca

7182 ***Rhinoclemmys rubida perixantha***
Testudines - Emydidae
e Colima Spotted Wood Turtle
f Rhinoclemmyde brune de Colima

7183 ***Rhinoleptus***
Squamata - Leptotyphlopidae

e Villier's Blind Snakes

7184 ***Rhinoleptus koniagui***
Squamata - Leptotyphlopidae
e Villier's Blind Snake

7185 ***Rhinophis***
Squamata - Uropeltidae
e Rhinoceros Earth Snakes; Earth Snakes

7186 ***Rhinophis blythii***
Squamata - Uropeltidae
e Blyth's Earth Snake
f Serpent de queue armée de Blyth

7187 ***Rhinophis dorsimaculatus***
Squamata - Uropeltidae
e Polkadot Earth Snake

7188 ***Rhinophis drummondhayi***
Squamata - Uropeltidae
e Drummond-Hay's Earth Snake; Drummond-Hay's Shieldtail

7189 ***Rhinophis fergusonius***
Squamata - Uropeltidae
e Cardamon Hills Earth Snake

7190 ***Rhinophis oxyrhynchys***
Squamata - Uropeltidae
e Schneider's Earth Snake
d Ceylonesischer Schildschwanz

7191 ***Rhinophis philippinus***
Squamata - Uropeltidae
e Peters's Philippine Earth Snake; Cuvier's Earth Snake

7192 ***Rhinophis punctatus***
Squamata - Uropeltidae
e Müller's Earth Snake

7193 ***Rhinophis sanguineus***
Squamata - Uropeltidae
e Salty Earth Snake

7194 ***Rhinophis travancoricus***
Squamata - Uropeltidae
e Tamil Nadu Earth Snake

7195 ***Rhinophis trevelyana***
Squamata - Uropeltidae
e Trevelyan's Earth Snake

7196 ***Rhinoplocephalus***
Squamata - Elapidae
e Deadly Whip Snakes

7197 ***Rhinoplocephalus bicolor***
Squamata - Elapidae
e Southwestern Whip Snake

7198 ***Rhinoplocephalus boschmai***
Squamata - Elapidae
e Carpentaria Whip Snake

7199 ***Rhinoplocephalus nigrescens***
Squamata - Elapidae
e Eastern Small-eyed Snake; Australian Small-eyed Snake
d Östliche Kleinaugenotter

7200 ***Rhinoplocephalus nigrostriatus***
Squamata - Elapidae
e Australian Black-striped Snake; Black-striped Snake (ANZ)

7201 ***Rhinoplocephalus pallidiceps***
Squamata - Elapidae
e Northern Small-eyed Snake
d Nördliche Kleinaugenotter

7202 ***Rhinotyphlops***
Squamata - Typhlopidae
e Beaked Snakes

7203 ***Rhinotyphlops acutus***
Squamata - Typhlopidae

e Beaked Blind Snake

7204 ***Rhinotyphlops anomalus***
Squamata - Typhlopidae
e Anomalous Beaked Snake

7205 ***Rhinotyphlops caecus***
Squamata - Typhlopidae
e Gaboon Beaked Snake; Benin Beaked Snake

7206 ***Rhinotyphlops crossii***
Squamata - Typhlopidae
e Cross's Beaked Snake

7207 ***Rhinotyphlops feae***
Squamata - Typhlopidae
e St. Thomas Beaked Snake

7208 ***Rhinotyphlops gracilis***
Squamata - Typhlopidae
e Urungu Beaked Snake

7209 ***Rhinotyphlops graueri***
Squamata - Typhlopidae
e Sternfeld's Beaked Snake

7210 ***Rhinotyphlops kibarae***
Squamata - Typhlopidae
e Katanga Beaked Snake

7211 ***Rhinotyphlops lalandei***
Squamata - Typhlopidae
e Delalande's Beaked Snake; Delalande's Blind Snake (CSA); Lalande's Blind Snake (CSA)

7212 ***Rhinotyphlops leucocephalus***
Squamata - Typhlopidae
e Whiteheaded Beaked Snake

7213 ***Rhinotyphlops lumbriciformis***
Squamata - Typhlopidae
e Wormlike Beaked Snake; Wormlike Blind Snake

7214 ***Rhinotyphlops newtonii***
Squamata - Typhlopidae
e Newton's Beaked Snake

7215 ***Rhinotyphlops pallidus***
Squamata - Typhlopidae
e Zanzibar Beaked Snake; Zanzibar Blind Snake

7216 ***Rhinotyphlops praeocularis***
Squamata - Typhlopidae
e Leopoldville Beaked Snake

7217 ***Rhinotyphlops rufescens***
Squamata - Typhlopidae
e Haut-Oubangui Beaked Snake

7218 ***Rhinotyphlops schinzi***
Squamata - Typhlopidae
e Beaked Snake; Beaked Blind Snake (CSA)

7219 ***Rhinotyphlops schlegelii***
Squamata - Typhlopidae
e Schlegel's Beaked Snake; Schlegel's Beaked Blind Snake; Schlegel's Blind Snake (CSA); Southeastern Blind Snake
d Afrikanische Blindschlange
f Typhlops de Schlegel

7220 ***Rhinotyphlops schlegelii schlegelii***
Squamata - Typhlopidae
e Giant Blind Snake

7221 ***Rhinotyphlops scorteccii***
Squamata - Typhlopidae
e Scortecci's Beaked Snake

7222 ***Rhinotyphlops simoni***
Squamata - Typhlopidae
e Simon's Beaked Snake
d Israelische Wurmschlange
f Typhlops de Simon; Typhlops d'Israël

7223 ***Rhinotyphlops somalicus***
Squamata - Typhlopidae
e Highland Beaked Snake

7224 ***Rhinotyphlops stejnegeri***
Squamata - Typhlopidae
e Stejneger's Beaked Snake

7225 ***Rhinotyphlops sudeanensis***
Squamata - Typhlopidae
e Sudan Beaked Snake

7226 ***Rhinotyphlops unitaeniatus***
Squamata - Typhlopidae
e Kenya Beaked Snake

7227 ***Rhinotyphlops wittei***
Squamata - Typhlopidae
e De Witte's Beaked Snake

7228 ***Rhoptropus***
Squamata - Gekkonidae
e Slender Geckos; Namib Day Geckos (CSA)

7229 ***Rhoptropus afer***
Squamata - Gekkonidae
e Olive Slender Gecko; Namib Day Gecko (CSA); Common Namib Day Gecko

7230 ***Rhoptropus barnardi***
Squamata - Gekkonidae
e Barnard's Slender Gecko; Barnard's Namib Day Gecko (CSA)
d Namib-Taggecko

7231 ***Rhoptropus biporosus***
Squamata - Gekkonidae
e Orupembe Slender Gecko; Kaokoveld Namib Day Gecko (CSA)

7232 ***Rhoptropus boultoni***
Squamata - Gekkonidae
e Boulton's Slender Gecko; Boulton's Namib Day Gecko (CSA)

7233 ***Rhoptropus bradfieldi***
Squamata - Gekkonidae
e Namib Slender Gecko; Bradfield's Namib Day Gecko (CSA)

7234 ***Rhoptropus taeniostictus***
Squamata - Gekkonidae
e Angola Slender Gecko

7235 ***Rhynchocalamus***
Squamata - Colubridae
e Kukri Snakes

7236 ***Rhynchocalamus arabicus***
Squamata - Colubridae
e Aden Kukri Snake

7237 ***Rhynchocalamus dorsolateralis***
Squamata - Colubridae
e Wall's Kukri Snake

7238 ***Rhynchocalamus eberhardti***
Squamata - Colubridae
e Eberhardt's Kukri Snake

7239 ***Rhynchocalamus melanocephalus***
Squamata - Colubridae
e Palestine Kukri Snake
d Schwarzkopfzwergnatter
f Couleuvre à tête noire

7240 ***Rhynchocalamus phaenochalinus***
Squamata - Colubridae
e Cope's Kukri Snake

7241 ***Rhynchocalamus propinquis***
Squamata - Colubridae
e Jan's Kukri Snake

7242 ***Rhynchocalamus satunini***
Squamata - Colubridae
f Rhynchocalame de Satunin

i Rincocalamo di Satunin

7243 ***Rhynchocalamus violaceus***
Squamata - Colubridae
e Violet Kukri Snake

7244 **Rhynchocephalia**
e Tuataras
d Brückenechsen
f Rhynchcéphales
i Rincocefalli

7245 ***Rhynchoedura***
Squamata - Gekkonidae
e Beaked Geckos

7246 ***Rhynchoedura ornata***
Squamata - Gekkonidae
e Beaked Gecko

7247 ***Rhynchophis***
Squamata - Colubridae
e Rhinoceros Snakes

7248 ***Rhynchophis boulengeri***
Squamata - Colubridae
e Rhinoceros Snake

7249 ***Riolama***
Squamata - Teiidae
e White-spotted Riolamas

7250 ***Riolama leucostictus***
Squamata - Teiidae
e White-spotted Riolama

7251 ***Ristella***
Squamata - Scincidae
e Ristellas
d Katzenskinke

7252 ***Ristella beddomei***
Squamata - Scincidae
e Beddome's Ristella

7253 ***Ristella guentheri***
Squamata - Scincidae
e Günther's Ristella

7254 ***Ristella rurkii***
Squamata - Scincidae
e Rurk's Ristella

7255 ***Ristella travancorica***
Squamata - Scincidae
e Travancore Ristella
d Travancore-Schlankskink

S

7256 ***Sacalia***
Testudines - Emydidae
e Eyed Turtles
d Pfauenaugenwasserschildkröten; Pfauenaugeschildkröten

7257 ***Sacalia bealei***
Testudines - Emydidae
e Beale's Eyed Turtle; Eyed Turtle
d Pfauenaugeschildkröte; Pfauenaugenwasserschildkröte
f Émyde chinoise à ocelles

7258 ***Sacalia bealei bealei***
Testudines - Emydidae
f Émyde chinoise à deux ocelles

7259 ***Sacalia pseudocellata***
Testudines - Emydidae
e Chinese False Eyed Turtle; False Eyed Turtle

7260 ***Sacalia quadriocellata***
Testudines - Emydidae
e Four-eyed Turtle
d Vieraugenschildkröte
i Tartaruga quattr'occhi

7261 ***Saiphos***
Squamata - Scincidae
e Three-toed Skinks

7262 ***Saiphos equalis***
Squamata - Scincidae
e Three-toed Skink

7263 ***Salea***
Squamata - Agamidae
e Asia Minor Agamas; Salea Lizards (ISC)

7264 ***Salea anamallayana***
Squamata - Agamidae
e Anamally Agama; Anamalai Salea (ISC); Anamalai Spiny Lizard (ISC)

7265 ***Salea brachydactyla***
Squamata - Agamidae
e Rendahl's Agama

7266 ***Salea horsfieldi***
Squamata - Agamidae
e Nilgiri Agama; Nilgiri Slea (ISC); Nilgiri Spiny Lizard (ISC)

7267 ***Salea kakhienensis***
Squamata - Agamidae
e Kakhien Agama

7268 ***Salomonelaps***
Squamata - Elapidae
e Solomon Islands Brown Snakes

7269 ***Salomonelaps par***
Squamata - Elapidae
e Solomon Islands Brown Snake

7270 ***Salvadora***
Squamata - Colubridae
e Patchnose Snakes; Patch-nosed Snakes
d Pflasternasennattern; Pflasterzahnnattern
f Couleuvres à nez plaqué

7271 ***Salvadora bairdi***
Squamata - Colubridae
e Baird's Patchnose Snake; Baird's Patch-nosed Snake
d Bairds Pflasternasennatter

7272 ***Salvadora deserticola***
Squamata - Colubridae

e Big Bend Patchnose Snake; Big Bend Patch-nosed Snake
d Wüstenpflasternasennatter
f Couleuvre à nez plaqué de Big Bend

7273 ***Salvadora grahamiae***
Squamata - Colubridae
e Mountain Patchnose Snake; Mountain Patch-nosed Snake; Flat-nosed Snake (NA); Eastern Patchnose Snake; Eastern Patch-nosed Snake
d Bergpflasternasennatter; Gebirgs-Pflasternasennatter
f Couleuvre à nez plaqué des montagnes

7274 ***Salvadora grahamiae grahamiae***
Squamata - Colubridae
e Platean Patch-nosed Snake; Mountain Patchnose Snake; Mountain Patch-nosed Snake

7275 ***Salvadora grahamiae lineata***
Squamata - Colubridae
e Texas Patchnose Snake; Texas Patch-nosed Snake

7276 ***Salvadora hexalepis***
Squamata - Colubridae
e Western Patchnose Snake; Western Patch-nosed Snake
d Westliche Pflasternasennatter
f Couleuvre à nez plaqué de l'Ouest

7277 ***Salvadora hexalepis hexalepis***
Squamata - Colubridae
e Desert Patchnose Snake; Desert Patch-nosed Snake
d Wüstenpflasternasennatter

7278 ***Salvadora hexalepis klauberi***
Squamata - Colubridae
e Baja California Patch-nosed Snake

7279 ***Salvadora hexalepis mohavensis***
Squamata - Colubridae
e Mojave Patch-nosed Snake; Mojave Patchnose Snake; Western Patch-nosed Snake

7280 ***Salvadora hexalepis virgultea***
Squamata - Colubridae
e Coast Patchnose Snake; Coast Patch-nosed Snake

7281 ***Salvadora intermedia***
Squamata - Colubridae
e Oaxacan Patchnose Snake; Oaxacan Patch-nosed Snake

7282 ***Salvadora lemniscata***
Squamata - Colubridae
e Pacific Patchnose Snake; Pacific Patch-nosed Snake

7283 ***Salvadora lineata***
Squamata - Colubridae
e Texas Patchnose Snake; Texas Patch-nosed Snake; Patch-nosed Snake (NA); Rio Grande Patch-nosed Snake (NA)

7284 ***Salvadora mexicana***
Squamata - Colubridae
e Mexican Patchnose Snake; Mexican Patch-nosed Snake

7285 ***Saphenophis***
Squamata - Colubridae
e Saphenophis Snakes

7286 ***Saphenophis antioquiensis***
Squamata - Colubridae
e Dunn's Saphenophis Snake

7287 ***Saphenophis atahuallpae***
Squamata - Colubridae
e Atahuallpa Saphenophis Snake

7288 ***Saphenophis boursieri***
Squamata - Colubridae

e Boursier's Saphenophis Snake

7289 ***Saphenophis sneiderni***
Squamata - Colubridae
e Saphenophis Snake

7290 ***Saphenophis tristriatus***
Squamata - Colubridae
e Striped Saphenophis Snake

7291 ***Saproscincus***
Squamata - Scincidae
e Basilisk Skinks; Shade Skinks

7292 ***Saproscincus basiliscus***
Squamata - Scincidae
e Basilisk Skink; Basal Shade Skink

7293 ***Saproscincus challengeri***
Squamata - Scincidae
e Challenger's Skink; Challenger's Shade Skink; Orange-tailed Shade Skink

7294 ***Saproscincus czechurai***
Squamata - Scincidae
e Czechura's Skink; Czechura's Shade Skink

7295 ***Saproscincus mustelinus***
Squamata - Scincidae
e Weasel Skink

7296 ***Saproscincus tetradactylus***
Squamata - Scincidae
e Marked Skink

7297 ***Sator***
Squamata - Iguanaidae
e Sators

7298 ***Sator angustus***
Squamata - Iguanaidae
e Santa Cruz Island Sator

7299 ***Sator grandaevus***
Squamata - Iguanaidae
e Cerralvo Island Sator

7300 ***Sauresia***
Squamata - Anguidae
e Sauresias; Sauresians

7301 ***Sauresia agasepsoides***
Squamata - Anguidae
e Thomas's Sauresia; Serpentine Four-toed Galliwasp

7302 ***Sauresia sepsoides***
Squamata - Anguidae
e Common Sauresia; Hispaniolan Four-toed Galliwasp
d Ägyptische Erzschleiche

7303 ***Saurodactylus***
Squamata - Gekkonidae
e Lizard-fingered Geckos; Lizard-toed Geckos
d Echsenfingergeckos
f Saurodactyles

7304 ***Saurodactylus brosseti***
Squamata - Gekkonidae
e Brosset's Lizard-toed Gecko
d Brossets Echsenfingergecko
f Saurodactyle de Brosset

7305 ***Saurodactylus fasciatus***
Squamata - Gekkonidae
e Banded Lizard-fingered Gecko; Banded Lizard-toed Gecko; Banded Toed Gecko
d Echsenfingergecko; Gebänderter Echsenfingergecko
f Saurodactyle à bandes; Saurodactyle fascié

7306 ***Saurodactylus mauritanicus***
Squamata - Gekkonidae
e Moroccan Lizard-fingered Gecko; Mauritanian Lizard-toed Gecko;

Toed Gecko
d Marokkanischer Zwerggecko; Echsenfingergecko; Mauritanischer Echsenfingergecko
f Saurodactyle maurétanique; Saurodactyle de Maurétanie

7307 ***Sauromalus***
Squamata - Iguanaidae
e Chuckwallas
d Chuckwallas

7308 ***Sauromalus ater***
Squamata - Iguanaidae
e Northern Chuckwalla; Chuckwalla; Common Chuckwalla
d Chuckwalla

7309 ***Sauromalus australis***
Squamata - Iguanaidae
e Peninsula Chuckwalla

7310 ***Sauromalus hispidus***
Squamata - Iguanaidae
e Spiny Chuckwalla

7311 ***Sauromalus klauberi***
Squamata - Iguanaidae
e Spotted Chuckwalla; Klauber's Chuckwalla (NA)

7312 ***Sauromalus obesus***
Squamata - Iguanaidae
e Chuckwalla; Common Chuckwalla
d Chuckwalla
f Chuckwalla

7313 ***Sauromalus obesus multiforminatus***
Squamata - Iguanaidae
e Glen Canyon Chuckwalla

7314 ***Sauromalus obesus obesus***
Squamata - Iguanaidae
e Western Chuckwalla; Great Basin Chuckwalla

7315 ***Sauromalus obesus tumidus***
Squamata - Iguanaidae
e Chuckwalla; Arizona Chuckwalla

7316 ***Sauromalus slevini***
Squamata - Iguanaidae
e Monserrat Chuckwalla; Slevin's Chuckwalla

7317 ***Sauromalus varius***
Squamata - Iguanaidae
e Piebald Chuckwalla; San Esteban Island Chuckwalla
d St.Esteban-Chuckwalla

7318 ***Scaphiodontophis***
Squamata - Colubridae
e Neckband Snakes; Neck-banded Snakes
d Amerikanische Vielzahnnatter

7319 ***Scaphiodontophis annulatus***
Squamata - Colubridae
e Guatemala Neckband Snake; Neck-banded Snake; Shovel-toothed Snake

7320 ***Scaphiodontophis carpicinctus***
Squamata - Colubridae
e Mexican Neckband Snake

7321 ***Scaphiodontophis venustissimus***
Squamata - Colubridae
e Common Neckband Snake

7322 ***Scaphiodontophis zeteki***
Squamata - Colubridae
e Zetek's Neckband Snake

7323 ***Scaphiophis***
Squamata - Colubridae
e Grey Beaked Agkistrodon Snakes; Grey Beaked Snakes; African Shovel-nosed Snakes
d Afrikanische Schaufelnasennattern; Schaufelnattern

7324 ***Scaphiophis albopunctatus***
Squamata - Colubridae
e Grey Beaked Snake; African Shovel-nosed Snake
d Afrikanische Schaufelnasennatter; Schaufelnattern

7325 ***Sceloporus***
Squamata - Iguanaidae
e Spiny Lizards; Rough-scaled Lizards; Spiny Swifts; Bluebellies (NA)
d Stachelleguane; Zaunleguane; Stachelschuppenleguane
f Lézards épineux; Scélopores

7326 ***Sceloporus acanthinus***
Squamata - Iguanaidae
e Bocourt's Spiny Lizard

7327 ***Sceloporus adleri***
Squamata - Iguanaidae
e Adler's Spiny Lizard

7328 ***Sceloporus aeneus***
Squamata - Iguanaidae
e Southern Bunchgrass Lizard

7329 ***Sceloporus anahuacus***
Squamata - Iguanaidae
e Anahuacan Bunchgrass Lizard

7330 ***Sceloporus arenicolus***
Squamata - Iguanaidae
e Sand Dune Lizard; Dunes Sagebrush Lizard

7331 ***Sceloporus bicanthalis***
Squamata - Iguanaidae
e Trans-Volcanic Spiny Lizard

7332 ***Sceloporus bulleri***
Squamata - Iguanaidae
e Buller's Spiny Lizard

7333 ***Sceloporus carinatus***
Squamata - Iguanaidae
e Keeled Spiny Lizard

7334 ***Sceloporus cautus***
Squamata - Iguanaidae
e Shy Spiny Lizard

7335 ***Sceloporus chaneyi***
Squamata - Iguanaidae
e Chaney's Spiny Lizard

7336 ***Sceloporus chrysostictus***
Squamata - Iguanaidae
e Yellow-spotted Spiny Lizard

7337 ***Sceloporus clarkii***
Squamata - Iguanaidae
e Clark's Spiny Lizard (NA); Clark's Swift (NA); Clark Spiny Lizard
d Clarks Stacheschuppenleguan; Clark's Stachelleguan
f Lézard de Clark

7338 ***Sceloporus clarkii clarkii***
Squamata - Iguanaidae
e Sonoran Spiny Lizard

7339 ***Sceloporus clarkii vallaris***
Squamata - Iguanaidae
e Plateau Spiny Lizard

7340 ***Sceloporus couchii***
Squamata - Iguanaidae
e Couch's Spiny Lizard

7341 ***Sceloporus cozumelae***
Squamata - Iguanaidae
e Cozumel Spiny Lizard

7342 ***Sceloporus cryptus***
Squamata - Iguanaidae
e Sierra Juarez Spiny Lizard

7343 ***Sceloporus dugesii***
Squamata - Iguanaidae
e Duges's Spiny Lizard

7344 ***Sceloporus edwardtaylori***
Squamata - Iguanaidae
e Taylor's Spiny Lizard

7345 ***Sceloporus exsul***
Squamata - Iguanaidae
e Queretaran Spiny Lizard

7346 ***Sceloporus formosus***
Squamata - Iguanaidae
e Emerald Spiny Lizard

7347 ***Sceloporus gadovae***
Squamata - Iguanaidae
e Gadow's Spiny Lizard

7348 ***Sceloporus goldmani***
Squamata - Iguanaidae
e Goldman's Bunchgrass Lizard

7349 ***Sceloporus graciosus***
Squamata - Iguanaidae
e Sagebrush Lizard; Spiny Lizard; Fence Swift; Sand-dune Lizard; Common Sagebrush Lizard
f Lézard des armoises

7350 ***Sceloporus graciosus gracilis***
Squamata - Iguanaidae
e Western Sagebrush Lizard

7351 ***Sceloporus graciosus graciosus***
Squamata - Iguanaidae
e Northern Sagebrush Lizard

7352 ***Sceloporus graciosus vandenburgianus***
Squamata - Iguanaidae
e Southern Sagebrush Lizard; Western Sagebrush Lizard

7353 ***Sceloporus grammicus***
Squamata - Iguanaidae
e Mesquite Lizard; Common Sagebrush Lizard; Northern Sagebrush Lizard; Graphic Spiny Lizard
d Mesquite-Stachelschuppenleguan
f Lézard des mesquites

7354 ***Sceloporus grammicus disparilis***
Squamata - Iguanaidae
e Common Mesquite Lizard

7355 ***Sceloporus grammicus microlepidotus***
Squamata - Iguanaidae
e Mesquite Lizard

7356 ***Sceloporus heterolepis***
Squamata - Iguanaidae
e Dorsal-keel Spiny Lizard

7357 ***Sceloporus horridus***
Squamata - Iguanaidae
e Horrible Spiny Lizard; Northern Mesquite Lizard

7358 ***Sceloporus hunsakeri***
Squamata - Iguanaidae
e Hunsaker's Spiny Lizard; Hunsaker Spiny Lizard

7359 ***Sceloporus insignis***
Squamata - Iguanaidae
e Michoacan Spiny Lizard

7360 ***Sceloporus jalapae***
Squamata - Iguanaidae
e Jalapa Spiny Lizard

7361 ***Sceloporus jarrovii***
Squamata - Iguanaidae
e Yarrow's Spiny Lizard; Yarrow's Scaly Lizard; Mountain Spiny Lizard (NA)

d Stachelleguan
f Lézard de Yarrow; Lézard des pallisades

7362 ***Sceloporus jarrovii jarovii***
Squamata - Iguanaidae
e Yarrow's Spiny Lizard
f Lézard de Yarrow; Lézard des pallisades

7363 ***Sceloporus licki***
Squamata - Iguanaidae
e Cape Arboreal Spiny Lizard; Mountain Spiny Lizard; Cape Spiny Lizard (NA)

7364 ***Sceloporus lineatulus***
Squamata - Iguanaidae
e Santa Catalina Spiny Lizard; Santa Catalina Island Spiny Lizard

7365 ***Sceloporus lundelli***
Squamata - Iguanaidae
e Lundell's Spiny Lizard

7366 ***Sceloporus macdougalli***
Squamata - Iguanaidae
e MacDougall's Spiny Lizard

7367 ***Sceloporus maculosus***
Squamata - Iguanaidae
e Spotted Spiny Lizard

7368 ***Sceloporus magister***
Squamata - Iguanaidae
e Desert Spiny Lizard; Twin-spotted Spiny Lizard
d Wüstenstachelleguan; Wüstenstachelschuppenleguan
f Lézard épineux du désert; Scélopore du désert

7369 ***Sceloporus magister bimaculosus***
Squamata - Iguanaidae
e Twin-spotted Spiny Lizard

7370 ***Sceloporus magister cephaloflavus***
Squamata - Iguanaidae
e Orange-headed Spiny Lizard

7371 ***Sceloporus magister magister***
Squamata - Iguanaidae
e Purple-backed Spiny Lizard

7372 ***Sceloporus magister monserratensis***
Squamata - Iguanaidae
e Southern Spiny Lizard

7373 ***Sceloporus magister rufidorsum***
Squamata - Iguanaidae
e Red-striped Spiny Lizard

7374 ***Sceloporus magister transversus***
Squamata - Iguanaidae
e Barred Spiny Lizard

7375 ***Sceloporus magister uniformis***
Squamata - Iguanaidae
e Yellowback Spiny Lizard; Yellow-backed Spiny Lizard (NA)

7376 ***Sceloporus malachiticus***
Squamata - Iguanaidae
e Green Spiny Lizard
d Malachitgrüner Stachelleguan; Stachelleguan

7377 ***Sceloporus megalepidurus***
Squamata - Iguanaidae
e Largescale Spiny Lizard

7378 ***Sceloporus melanorhinus***
Squamata - Iguanaidae
e Pastel Spiny Lizard

7379 ***Sceloporus merriami***
Squamata - Iguanaidae
e Canyon Lizard; Merriam's Canyon Lizard
f Lézard des canyons

7380 ***Sceloporus merriami annulatus***
Squamata - Iguanaidae
e Big Bend Canyon Lizard

7381 ***Sceloporus merriami longipunctatus***
Squamata - Iguanaidae
e Presidio Canyon Lizard

7382 ***Sceloporus merriami merriami***
Squamata - Iguanaidae
e Merriam's Canyon Lizard

7383 ***Sceloporus monserratensis***
Squamata - Iguanaidae
e Monserrat Island Spiny Lizard

7384 ***Sceloporus mucronatus***
Squamata - Iguanaidae
e Southern Crevice Spiny Lizard

7385 ***Sceloporus nelsoni***
Squamata - Iguanaidae
e Nelson's Spiny Lizard

7386 ***Sceloporus occidentalis***
Squamata - Iguanaidae
e Western Fence Lizard; Pacific Swift (NA); Pacific Fence Lizard; Blue-belly Lizard
d Westlicher Zaunleguan; Westlicher Stachelschuppenleguan
f Lézard des palissade de l'Ouest; Scélopore du désert

7387 ***Sceloporus occidentalis becki***
Squamata - Iguanaidae
e Island Fence Lizard

7388 ***Sceloporus occidentalis biseriatus***
Squamata - Iguanaidae
e San Joaquin Fence Lizard

7389 ***Sceloporus occidentalis bocourtii***
Squamata - Iguanaidae
e Coast Range Fence Lizard

7390 ***Sceloporus occidentalis longipes***
Squamata - Iguanaidae
e Great Basin Fence Lizard

7391 ***Sceloporus occidentalis occidentalis***
Squamata - Iguanaidae
e Northwestern Fence Lizard

7392 ***Sceloporus occidentalis taylori***
Squamata - Iguanaidae
e Sierra Fence Lizard

7393 ***Sceloporus ochoterenae***
Squamata - Iguanaidae
e Queretaran Spiny Lizard

7394 ***Sceloporus olivaceus***
Squamata - Iguanaidae
e Texas Spiny Lizard
d Texas-Stachelschuppenleguan
f Lézard épineux de Texas

7395 ***Sceloporus orcutti***
Squamata - Iguanaidae
e Granite Spiny Lizard
f Lézard granite

7396 ***Sceloporus ornatus***
Squamata - Iguanaidae
e Ornate Spiny Lizard

7397 ***Sceloporus palaciosi***
Squamata - Iguanaidae
e Palacios's Bunchgrass Lizard

7398 ***Sceloporus parvus***
Squamata - Iguanaidae
e Bluebelly Lizard

7399 ***Sceloporus poinsettii***
Squamata - Iguanaidae
e Crevice Spiny Lizard (NA); Crevice Scaly Lizard; Red Scaly Lizard

d Stachelleguan
f Lézard des crevasses; Lézard épineux

7400 ***Sceloporus poinsettii poinsettii***
Squamata - Iguanaidae
e Northern Crevice Spiny Lizard; Crevice Spiny Lizard (NA)

7401 ***Sceloporus pyrocephalus***
Squamata - Iguanaidae
e Boulder Spiny Lizard

7402 ***Sceloporus rufidorsum***
Squamata - Iguanaidae
e Mexican Spiny Lizard

7403 ***Sceloporus salvini***
Squamata - Iguanaidae
e Salvin's Spiny Lizard

7404 ***Sceloporus scalaris***
Squamata - Iguanaidae
e Bunchgrass Lizard
d Zaunleguan
f Lézard scalaire

7405 ***Sceloporus scalaris slevini***
Squamata - Iguanaidae
e Bunchgrass Lizard

7406 ***Sceloporus serrifer***
Squamata - Iguanaidae
e Blue Spiny Lizard

7407 ***Sceloporus serrifer cyanogenys***
Squamata - Iguanaidae
e Blue Spiny Lizard
f Lézard bleu

7408 ***Sceloporus siniferus***
Squamata - Iguanaidae
e Longtail Spiny Lizard

7409 ***Sceloporus slevinii***
Squamata - Iguanaidae
e Slevin's Bunchgrass Lizard

7410 ***Sceloporus smithi***
Squamata - Iguanaidae
e Smith's Rosebelly Lizard

7411 ***Sceloporus spinosus***
Squamata - Iguanaidae
e Spiny Lizard; Spiny Swift (NA)

7412 ***Sceloporus squamosus***
Squamata - Iguanaidae
e Mexican Spiny Lizard

7413 ***Sceloporus stejnegeri***
Squamata - Iguanaidae
e Stejneger's Blackcollar Spiny Lizard

7414 ***Sceloporus subniger***
Squamata - Iguanaidae
e Tolucan Bunchgrass Lizard Lizard

7415 ***Sceloporus subpictus***
Squamata - Iguanaidae
e Paintbelly Spiny Lizard

7416 ***Sceloporus taeniocnemis***
Squamata - Iguanaidae
e Guatemalan Emerald Spiny Lizard

7417 ***Sceloporus teapensis***
Squamata - Iguanaidae
e Teapan Rosebelly Lizard; Teapa Spiny Lizard

7418 ***Sceloporus torquatus***
Squamata - Iguanaidae
e Crevice Swift; Collared Swift (NA)
d Blauer Stachelleguan

7419 ***Sceloporus undulatus***
Squamata - Iguanaidae
e Eastern Fence Lizard; Undulated

Swift (NA); Eastern Fence Swift; Prairie Lizard; Fence Lizard
d Östlicher Zaunleguan; Gewöhnlicher Zaunleguan; Zaunleguan
f Lézard des palissades de l'Est; Scélopore ondulé

7420 ***Sceloporus undulatus consobrinus***
Squamata - Iguanaidae
e Southern Prairie Lizard; Banded Swift (NA)

7421 ***Sceloporus undulatus cowlesi***
Squamata - Iguanaidae
e White Sands Prairie Lizard; Cowles Prairie Lizard

7422 ***Sceloporus undulatus elongatus***
Squamata - Iguanaidae
e Northern Plateau Lizard

7423 ***Sceloporus undulatus erythrocheilus***
Squamata - Iguanaidae
e Red-lipped Prairie Lizard; Red-lipped Plateau Lizard

7424 ***Sceloporus undulatus garmani***
Squamata - Iguanaidae
e Northern Prairie Lizard

7425 ***Sceloporus undulatus hyacinthinus***
Squamata - Iguanaidae
e Northern Fence Lizard

7426 ***Sceloporus undulatus speari***
Squamata - Iguanaidae
e Cabeza de Vaca Prairie Lizard

7427 ***Sceloporus undulatus tristichus***
Squamata - Iguanaidae
e Southern Plateau Lizard

7428 ***Sceloporus undulatus undulatus***
Squamata - Iguanaidae
e Southern Fence Lizard

7429 ***Sceloporus utiformis***
Squamata - Iguanaidae
e Cope's Largescale Spiny Lizard; Largescale Spiny Lizard

7430 ***Sceloporus variabilis***
Squamata - Iguanaidae
e Rosebelly Lizard; Rose-bellied Lizard
f Lézard à ventre rose

7431 ***Sceloporus variabilis marmoratus***
Squamata - Iguanaidae
e Texas Rose-bellied Lizard; Rosebelly Lizard

7432 ***Sceloporus virgatus***
Squamata - Iguanaidae
e Striped Plateau Lizard
f Scélopore ligné des plateaux

7433 ***Sceloporus woodi***
Squamata - Iguanaidae
e Florida Scrub Lizard; Scrub Pine Lizard
f Lézard de Wood

7434 ***Sceloporus zosteromus***
Squamata - Iguanaidae
e San Lucan Spiny Lizard; Baja California Spiny Lizard (NA)

7435 ***Scelotes***
Squamata - Scincidae
e Burrowing Skinks; Dwarf Burrowing Skinks (CSA)
d Scelotesskinke
f Scélotes

7436 ***Scelotes anguina***
Squamata - Scincidae
e Boulenger's Burrowing Skink; Algoa Dwarf Burrowing Skink (CSA)

7437 ***Scelotes arenicolor***
Squamata - Scincidae

e Sandy Burrowing Skink; Zululand Dwarf Burrowing Skink (CSA)

7438 ***Scelotes bicolor***
Squamata - Scincidae
e Two-coloured Burrowing Skink

7439 ***Scelotes bidigittata***
Squamata - Scincidae
e Two-toed Burrowing Skink; Lowveld Dwarf Burrowing Skink (CSA)

7440 ***Scelotes bipes***
Squamata - Scincidae
e Common Burrowing Skink; Silvery Dwarf Burrowing Skink (CSA); Two-legged Burrowing Skink
d Zweizehenscelotesskink; Zweibeinskink; Scelotesskink

7441 ***Scelotes brevipes***
Squamata - Scincidae
e Shortfoot Burrowing Skink; Hewit's Dwarf Burrowing Skink (CSA); Short-footed Burrowing Skink (CSA)

7442 ***Scelotes caffer***
Squamata - Scincidae
e Peters's Burrowing Skink; Cape Dwarf Burrowing Skink (CSA)

7443 ***Scelotes capensis***
Squamata - Scincidae
e Cape Burrowing Skink; Western Dwarf Burrowing Skink (CSA)
d Kap-Scelotesskink

7444 ***Scelotes gronovii***
Squamata - Scincidae
e Gronovi's Burrowing Skink; Gronovi's Dwarf Burrowing Skink (CSA)

7445 ***Scelotes guentheri***
Squamata - Scincidae
e Günther's Burrowing Skink; Günther's Dwarf Burrowing Skink (CSA)

7446 ***Scelotes inornatus***
Squamata - Scincidae
e Legless Burrowing Skink; Smith's Dwarf Burrowing Skink (CSA); Olain Burrowing Skink

7447 ***Scelotes kasneri***
Squamata - Scincidae
e Kasner's Burrowing Skink; Kasner's Dwarf Burrowing Skink (CSA)

7448 ***Scelotes limpopoensis***
Squamata - Scincidae
e Limpopo Burrowing Skink; Limpopo Dwarf Burrowing Skink (CSA); Limpopo Skink (CSA); Limpopo Sand Skink (CSA)

7449 ***Scelotes mira***
Squamata - Scincidae
e Montane Burrowing Skink; Montane Dwarf Burrowing Skink (CSA)

7450 ***Scelotes uluguriensis***
Squamata - Scincidae
e Uluguru Burrowing Skink; Uluguru Five-toed Scelotes

7451 ***Scincella***
Squamata - Scincidae
e Ground Skinks; Dwarf Skinks
d Schlankskinke

7452 ***Scincella barbouri***
Squamata - Scincidae
e Barbour's Ground Skink; Barbour's Dwarf Skink

7453 ***Scincella beddomi***
Squamata - Scincidae
e Beddome's Ground Skink; Beddome's Dwarf Skink

7454 ***Scincella bilineata***
Squamata - Scincidae
e Two-lined Ground Skink
d Zeilinienschlankskink

7455 ***Scincella boettgeri***
Squamata - Scincidae
e Boettger's Ground Skink; Boettger's Dwarf Skink

7456 ***Scincella caudaequinae***
Squamata - Scincidae
e Horsehead Ground Skink

7457 ***Scincella doriae***
Squamata - Scincidae
e Doria's Ground Skink; Doria's Dwarf Skink

7458 ***Scincella exigua***
Squamata - Scincidae
e Anderson's Ground Skink

7459 ***Scincella formosa***
Squamata - Scincidae
e Formosa Ground Skink

7460 ***Scincella formosensis***
Squamata - Scincidae
e Van Denburgh's Ground Skink

7461 ***Scincella gemmingeri***
Squamata - Scincidae
e Forest Ground Skink

7462 ***Scincella himalayana***
Squamata - Scincidae
e Himalaya Ground Skink; Himalayan Dwarf Skink
d Himalaja-Schlankskink

7463 ***Scincella inconspicua***
Squamata - Scincidae
e Hidden Ground Skink

7464 ***Scincella kohtaoensis***
Squamata - Scincidae
e Kohtaoen Ground Skink

7465 ***Scincella ladacensis***
Squamata - Scincidae
e Ladak Ground Skink; Ladak Dwarf Skink
d Ladak-Schlankskink

7466 ***Scincella lateralis***
Squamata - Scincidae
e Ground Skink; Little Brown Skink
f Scinque terrestre; Scinque à bandes latérales

7467 ***Scincella laterimaculata***
Squamata - Scincidae
e Spotted Ground Skink
d Seitenfleckschlankskink

7468 ***Scincella melanosticta***
Squamata - Scincidae
e Black Ground Skink

7469 ***Scincella modesta***
Squamata - Scincidae
e Modest Ground Skink

7470 ***Scincella ochracea***
Squamata - Scincidae
e Bourret's Ground Skink

7471 ***Scincella reevesi***
Squamata - Scincidae
e Reeves's Ground Skink; Reeves's Skink

7472 ***Scincella schmidti***
Squamata - Scincidae
e Schmidt's Ground Skink; Schmidt's Dwarf Skink

7473 ***Scincella siamensis***
Squamata - Scincidae

e Siamese Ground Skink
d Siam-Schlankskink

7474 ***Scincella sikimensis***
Squamata - Scincidae
e Sikkim Ground Skink
d Sikkim-Schlankskink

7475 ***Scincella sylvicola***
Squamata - Scincidae
e Taylor's Ground Skink

7476 **Scincidae**
Squamata
e Skinks
d Skinke
f Scincidés; Scinques

7477 ***Scincopus***
Squamata - Scincidae
e Banded Skinks
d Nachtskinke
f Scinques à bandes; Scinques fasciés

7478 ***Scincopus fasciatus***
Squamata - Scincidae
e Peters's Banded Skink; Banded Skink
d Nachtskink
f Scinque à bandes; Scinque fascié

7479 ***Scincus***
Squamata - Scincidae
e Sand Skinks; Sandfishes
d Sandskinke
f Scinques; Poissons du sable

7480 ***Scincus conirostris***
Squamata - Scincidae
e Iranian Sand Skink; Iranian Desert-dwelling Skink; Iranian Skink
d Iranischer Sandskink

7481 ***Scincus mitranus***
Squamata - Scincidae
e Eastern Sand Skink; Mitre Sandfish; Arabian Skink
d Östlicher Sandskink
f Scinque de l'Est

7482 ***Scincus scincus***
Squamata - Scincidae
e Sand Skink; Burrowing Sand Skink; Desert Skink; Egyptian Skink; Sandfish; Common Skink; North African Sandfish
d Apothekerskink
f Scinque officinal; Scinque des sables; Scinque des boutiques

7483 ***Scolecophis***
Squamata - Colubridae
e Black-banded Snakes

7484 ***Scolecophis atrocinctus***
Squamata - Colubridae
e Black-banded Snake; Harlequin Snake

7485 ***Scolecoseps***
Squamata - Scincidae
e Limbless Skinks

7486 ***Scolecoseps acontias***
Squamata - Scincidae
e Sandy Limbless Skink; Sandy-soil Limbless Skink

7487 ***Scolecoseps boulengeri***
Squamata - Scincidae
e Boulenger's Limbless Skink

7488 ***Seminatrix***
Squamata - Colubridae
e Swamp Snakes; Swamp Mud Snakes; Red-bellied Mud Snakes
d Schwarze Sumpfnattern; Sumpfnatter
f Couleuvres noires des marais

7489 ***Seminatrix pygaea***
Squamata - Colubridae

e Black Swamp Snake; Swamp Snake; Red-bellied Mud Snake
d Schwarze Sumpfnatter
f Couleuvre noire des marais

7490 ***Seminatrix pygaea cyclas***
Squamata - Colubridae
e South Florida Swamp Snake

7491 ***Seminatrix pygaea paludis***
Squamata - Colubridae
e Carolina Swamp Snake; Black Swamp Snake

7492 ***Seminatrix pygaea pygaea***
Squamata - Colubridae
e North Florida Swamp Snake

7493 ***Senticolis***
Squamata - Colubridae
e Green Rat Snakes

7494 ***Senticolis trisaspis***
Squamata - Colubridae
e Green Rat Snake; Olivaceous Tropical Rat Snake

7495 ***Senticolis trisaspis trisaspis***
Squamata - Colubridae
e Green Rat Snake

7496 ***Sepsina***
Squamata - Scincidae
e Bocage's Skinks; Savanna Burrowings Skink (CSA)

7497 ***Sepsina alberti***
Squamata - Scincidae
e Albert's Skink; Albert's Burrowing Skink (CSA)

7498 ***Sepsina angolensis***
Squamata - Scincidae
e Angola Skink; Angola Burrowing Skink (CSA)

7499 ***Sepsina bayoni***
Squamata - Scincidae
e Bayon's Skink

7500 ***Sepsina copei***
Squamata - Scincidae
e Sepsina Skink

7501 ***Sepsina tetradactyla***
Squamata - Scincidae
e Four-fingered Skink; Four-toed Sepsina

7502 ***Shinisaurus***
Squamata - Xantusiidae
e Crocodile Lizards
d Krokodilschwanzhöckerechsen
f Shinisaures

7503 ***Shinisaurus crocodilurus***
Squamata - Xantusiidae
e Chinese Crocodile Lizard; Crocodile Lizard
d Krokodilschwanzhöckerechse
f Shinisaure crocodilure

7504 ***Sibon***
Squamata - Colubridae
e Snail-suckers; Slug-eaters; Snail-eating Snakes

7505 ***Sibon annulata***
Squamata - Colubridae
e Ringed Snail-sucker; Ringed Slug-eater
d Schneckennatter

7506 ***Sibon annuliferus***
Squamata - Colubridae
e Western Snail-sucker

7507 ***Sibon anthracops***
Squamata - Colubridae
e Cope's Snail-sucker; Coral Mimic Slug-eater

7508 ***Sibon argus***
Squamata - Colubridae
e Argus Snail-sucker; Argus Slug-eater

7509 ***Sibon carri***
Squamata - Colubridae
e Carr's Snail-sucker; Pacific Slug-eater

7510 ***Sibon dimidiatus***
Squamata - Colubridae
e Slender Snail-sucker; Banded Slug-eater

7511 ***Sibon dunni***
Squamata - Colubridae
e Dunn's Snail-sucker

7512 ***Sibon longifrenis***
Squamata - Colubridae
e Stejneger's Snail-sucker; Lichen-coloured Slug-eater

7513 ***Sibon nebulatus***
Squamata - Colubridae
e Cloudy Snail-sucker; Common Slug-eater; Cloudy Snail-eating Snake

7514 ***Sibon neilli***
Squamata - Colubridae
e Neill's Snail-sucker

7515 ***Sibon philippi***
Squamata - Colubridae
e Philippi's Snail-sucker

7516 ***Sibon sannicola***
Squamata - Colubridae
e Pygmy Snail-sucker; Yucatan Slug-eater

7517 ***Sibon zweifeli***
Squamata - Colubridae
e Zweifel's Snail-sucker

7518 ***Sibynomorphus***
Squamata - Colubridae
e South American Tree Snakes

7519 ***Sibynomorphus inaequifasciata***
Squamata - Colubridae
e Equator Tree Snake

7520 ***Sibynomorphus mikani***
Squamata - Colubridae
e Brazilian Tree Snake

7521 ***Sibynomorphus neuwiedi***
Squamata - Colubridae
e Neuwied's Tree Snake

7522 ***Sibynomorphus oneilli***
Squamata - Colubridae
e O'Neill's Tree Snake

7523 ***Sibynomorphus turgidus***
Squamata - Colubridae
e Bolivian Tree Snake

7524 ***Sibynomorphus vagrans***
Squamata - Colubridae
e Dunn's Tree Snake

7525 ***Sibynomorphus vagus***
Squamata - Colubridae
e Jan's Tree Snake

7526 ***Sibynomorphus ventrimaculatus***
Squamata - Colubridae
e Boulenger's Tree Snake

7527 ***Sibynomorphus williamsi***
Squamata - Colubridae
e Williams's Tree Snake

7528 ***Sibynophis***
Squamata - Colubridae
e Many-toothed Snakes; Manytooth Snakes; Polyondont Snakes

d Asiatische Vielzahnnattern

7529 ***Sibynophis bistrigatus***
Squamata - Colubridae
e Günther's Many-toothed Snake; Günther's Manytooth Snake

7530 ***Sibynophis chinensis***
Squamata - Colubridae
e Chinese Many-toothed Snake; Chinese Manytooth Snake; Chinese Collared Snake

7531 ***Sibynophis collaris***
Squamata - Colubridae
e Common Many-toothed Snake; Common Manytooth Snake; Collared Polyodont; Gray's Polyodont; Collared Snake; Collared Dwarf Snake

7532 ***Sibynophis geminatus***
Squamata - Colubridae
e Boie's Many-toothed Snake; Boie's Manytooth Snake

7533 ***Sibynophis melanocephalus***
Squamata - Colubridae
e Malayan Many-toothed Snake; Malayan Manytooth Snake; Malayan Polyodont

7534 ***Sibynophis subpunctatus***
Squamata - Colubridae
e Jerdon's Many-toothed Snake; Jerdon's Manytooth Snake; Jerdon's Polyodont

7535 ***Sibynophis triangularis***
Squamata - Colubridae
e Triangle Many-toothed Snake; Triangle Manytooth Snake

7536 ***Siebenrockiella***
Testudines - Emydidae
e Black Marsh Turtles
d Schwarze Dickkopfschildkröten
f Émydes dentelées à trois carènes

7537 ***Siebenrockiella crassicollis***
Testudines - Emydidae
e Black Marsh Turtle; Siamese Temple Turtle; Thick-necked Turtle
d Schwarze Dickkopfschildkröte
f Émyde dentelée à trois carènes
i Tartaruga nera di palude

7538 ***Simophis***
Squamata - Colubridae
e Snouted Snakes

7539 ***Simophis rhinostoma***
Squamata - Colubridae
e Snouted Snake; Sao Paulo False Coral Snake; Sao Paolo False Coral Snake
d Sao- Paulo-Falsche Korallenschlange

7540 ***Simophis rhodei***
Squamata - Colubridae
e Rhode's Snouted Snake

7541 ***Simoselaps***
Squamata - Elapidae
e Australian Coral Snakes
d Australische Korallenottern

7542 ***Simoselaps anomalus***
Squamata - Elapidae
e Morthwestern Black-striped Snake

7543 ***Simoselaps approximans***
Squamata - Elapidae
e West Australian Black-striped Snake

7544 ***Simoselaps australis***
Squamata - Elapidae
e Australian Coral Snake; Coral Snake (ANZ)
d Australische Korallenotter

7545 ***Simoselaps bertholdi***
Squamata - Elapidae
e Desert Banded Snake

7546 ***Simoselaps bimaculatus***
Squamata - Elapidae
e Western Black-naped Snake

7547 ***Simoselaps calonotus***
Squamata - Elapidae
e Western Black-striped Snake

7548 ***Simoselaps fasciolatus***
Squamata - Elapidae
e Narrow-banded Snake; Banded Racer
d Gebänderte Zornnatter; Australische Wüstenkorallenotter

7549 ***Simoselaps incinctus***
Squamata - Elapidae
e Norther Desert Banded Snake

7550 ***Simoselaps littoralis***
Squamata - Elapidae
e Yellow-banded Snake

7551 ***Simoselaps minimus***
Squamata - Elapidae
e White-banded Snake

7552 ***Simoselaps roperi***
Squamata - Elapidae
e Roper's Banded Snake

7553 ***Simoselaps semifasciatus***
Squamata - Elapidae
e Half-girdled Snake; Half-banded Snake
d Halbgebänderte Australische Korallenotter

7554 ***Simoselaps warro***
Squamata - Elapidae
e Orange Coral Snake; Warro's Australian Coral Snake

7555 ***Sinonatrix***
Squamata - Colubridae
e Asiatic Water Snakes
d Dreieckschwimmnattern; Asiatische Dreieckschwimmnatter; Asiatische Schwimmnatter

7556 ***Sinonatrix aequifasciata***
Squamata - Colubridae
e Asiatic Water Snake

7557 ***Sinonatrix annularis***
Squamata - Colubridae
e Ringed Water Snake; Red-bellied Water Snake

7558 ***Sinonatrix percarinata***
Squamata - Colubridae
e Eastern Water Snake; Boulenger's Eastern Water Snake; Boulenger's Asian Water Snake

7559 ***Siphlophis***
Squamata - Colubridae
e Spotted Night Snakes; Liana Snakes

7560 ***Siphlophis cervinus***
Squamata - Colubridae
e Panama Spotted Night Snake; Spotted Night Snake; Common Liana Snake

7561 ***Siphlophis leucocephalus***
Squamata - Colubridae
e Common Spotted Night Snake

7562 ***Siphlophis longicaudtus***
Squamata - Colubridae
e Brazilian Spotted Night Snake

7563 ***Siphlophis pulcher***
Squamata - Colubridae
e Guanabara Spotted Night Snake

7564 ***Siphlophis worontzowi***
Squamata - Colubridae
e Worontzow's Spotted Night Snake

7565 ***Sistrurus***
Squamata - Viperidae
e Pygmy Rattlesnakes; Ground Rattlesnakes (NA); Massasaugas
d Zwergklapperschlangen
f Crotales à plaques

7566 ***Sistrurus catenatus***
Squamata - Viperidae
e Massasauga; Black Snapper (NA); Swamp Rattler (NA); Swamp Rattlesnake; Western Massasauga
d Westliche Massasauga; Kettenklapperschlange; Massasauga
f Massasauga

7567 ***Sistrurus catenatus catenatus***
Squamata - Viperidae
e Eastern Massasauga
d Östliche Massasauga

7568 ***Sistrurus catenatus edwardsii***
Squamata - Viperidae
e Desert Massasauga
d Wüstenmassasauga

7569 ***Sistrurus catenatus tergeminus***
Squamata - Viperidae
e Western Massasauga ; Prairie Pygmy Rattlesnake
d Westliche Massasauga

7570 ***Sistrurus miliarius***
Squamata - Viperidae
e Pygmy Rattlesnake; Ground Rattlesnake (NA)
d Zwergklapperschlange; Eigentliche Zwergklapperschlange
f Crotale nain; Crotale pygmée; Serpent à sonnettes pygmée

7571 ***Sistrurus miliarius barbouri***
Squamata - Viperidae
e Dusky Pygmy Rattlesnake
d Wolkige Zwergklapperschlange

7572 ***Sistrurus miliarius miliarius***
Squamata - Viperidae
e Carolina Pygmy Rattlesnake
d Carolina-Zwergklapperschlange

7573 ***Sistrurus miliarius streckeri***
Squamata - Viperidae
e Western Pygmy Rattlesnake
d Westliche Klapperschlange

7574 ***Sistrurus ravus***
Squamata - Viperidae
e Mexican Pygmy Rattlesnake; Mexican Massasauga
d Mexikanische Zwergklapperschlange
f Crotale pygmée de Mexique

7575 ***Sitana***
Squamata - Agamidae
e Sita Lizards; Four-toed Agamas
d Sita-Agamen
f Sitanes

7576 ***Sitana ponticeriana***
Squamata - Agamidae
e Sita Lizard; Fan-throated Lizard (ISC)
d Sita-Agame
f Sitane de Pondichéry

7577 ***Sonora***
Squamata - Colubridae
e Desert Ground Snakes; Ground Snakes; North american Ground Snakes
d Erdnattern
f Couleuvres de terre

7578 ***Sonora aemula***
Squamata - Colubridae
e Filetail Ground Snake

7579 ***Sonora michoacaensis***
Squamata - Colubridae
e Michoacan Ground Snake

7580 ***Sonora semiannulata***
Squamata - Colubridae
e Ground Snake; Arizona Earth Snake; Western Ground Snake
f Couleuvre de terre de l'Ouest; Couleuvre de terre

7581 ***Sordellina***
Squamata - Colubridae
e Dotted Brown Snakes; Ground Snakes

7582 ***Sordellina punctata***
Squamata - Colubridae
e Dotted Brown Snake; Ground Snake

7583 ***Spalerosophis***
Squamata - Colubridae
e Diadem Snakes
d Diademnattern
f Couleuvres diadèmes

7584 ***Spalerosophis arenarius***
Squamata - Colubridae
e Red-spotted Diadem Snake

7585 ***Spalerosophis atriceps***
Squamata - Colubridae
e Royal Diadem Snake; Royal Snake (ISC)

7586 ***Spalerosophis diadema***
Squamata - Colubridae
e Diadem Snake
d Pakistanische Diademnatter; Diademnatter
f Couleuvre diadème; Couleuvre à diadème; Couleuvre diadème de Clifford

7587 ***Spalerosophis diadema cliffordi***
Squamata - Colubridae
e Clifford's Diadem Snake; Clifford's Snake; Clifford's Rat Snake; Diadem Snake; Diadem Rat Snake
f Couleuvre diadème de Clifford

7588 ***Spalerosophis dolichospilos***
Squamata - Colubridae
e Werner's Diadem Snake; Mograbin Diadem Snake
f Couleuvre diadème du Maghreb

7589 ***Spalerosophis josephcorteccii***
Squamata - Colubridae
e Scortecci's Diadem Snake

7590 ***Spalerosophis microlepis***
Squamata - Colubridae
e Zebra Snake

7591 ***Spehenodon***
Rhynchocephalia - Sphenodontidae
e Tuataras; Tuateras
d Tuataras; Brückenechsen
f Rhynchocéphales

7592 ***Sphaerodactylus***
Squamata - Gekkonidae
e Least Geckos; Dwarf Geckos (NA); Sphaeros
d Kugelfingergeckos; Kugelfinger
f Sphérodactyles

7593 ***Sphaerodactylus alayoi***
Squamata - Gekkonidae
e Cuban Least Gecko; Cuban Dwarf Gecko

7594 ***Sphaerodactylus altavelensis***
Squamata - Gekkonidae
e Alto Velo Least Gecko; Alto Velo Dwarf Gecko; Alto Velo Sphaero

7595 ***Sphaerodactylus anasae***
Squamata - Gekkonidae
e Jaragua Sphaero

7596 ***Sphaerodactylus argivus***
Squamata - Gekkonidae
e Cayman Least Gecko; Cayman Dwarf Gecko; Cayman Islands Sphaero (WI)

7597 ***Sphaerodactylus argus***
Squamata - Gekkonidae
e Ocellated Gecko; Ocellated Dwarf Gecko; Jamaican Stippled Sphaero (WI)
d Antillen-Kugelfingergecko
f Gecko ocellé

7598 ***Sphaerodactylus argus argus***
Squamata - Gekkonidae
e Common Ocellated Gecko

7599 ***Sphaerodactylus armasi***
Squamata - Gekkonidae
e Guantanamo Least Gecko; Guantanamo Dwarf Gecko; Spotted-head Sphaero

7600 ***Sphaerodactylus armstrongi***
Squamata - Gekkonidae
e Armstrong's Least Gecko; Armstrong's Dwarf Gecko; Southern Forest Sphaero

7601 ***Sphaerodactylus asterulus***
Squamata - Gekkonidae
e Haitian Least Gecko; Haitian Dwarf Gecko; Haitian Stippled Sphaero

7602 ***Sphaerodactylus beattyi***
Squamata - Gekkonidae
e Beatty's Least Gecko; Beatty's Dwarf Gecko; Cotton Ginner ; St. Croix Sphaero (NA)

7603 ***Sphaerodactylus becki***
Squamata - Gekkonidae
e Beck's Least Gecko; Beck's Dwarf Gecko; Navassa Sphaero

7604 ***Sphaerodactylus bromelarium***
Squamata - Gekkonidae
e El Yunque Least Gecko; El Yunque Dwarf Gecko; Cuban Bromeliad Sphaero
d Kubanischer Kugelfingergecko

7605 ***Sphaerodactylus caicosensis***
Squamata - Gekkonidae
e Caicos Least Gecko; Caicos Dwarf Gecko; Turks and Caicos Banded Sphaero (WI)

7606 ***Sphaerodactylus callocricus***
Squamata - Gekkonidae
e Callous Least Gecko; Callous Dwarf Gecko; Rough-banded Sphaero

7607 ***Sphaerodactylus celicara***
Squamata - Gekkonidae
e Baracoan Eyespot Sphaero

7608 ***Sphaerodactylus cinereus***
Squamata - Gekkonidae
e Grey Least Gecko; Grey Dwarf Gecko; Ashy Gecko; Haitian Broad-banded Sphaero
d Aschgrauer Kugelfingergecko; Grauer Kugelfinger
f Gecko cendré; Sphérodactyle cendré

7609 ***Sphaerodactylus clenchi***
Squamata - Gekkonidae
e Peninsula Least Gecko; Peninsula Dwarf Gecko; Orange-spotted Sphaero

7610 ***Sphaerodactylus cochranae***
Squamata - Gekkonidae
e Cochran's Least Gecko; Cochran's Dwarf Gecko; Haiti Striped Sphaero

7611 ***Sphaerodactylus copei***
Squamata - Gekkonidae
e Cope's Least Gecko; Cope's Dwarf Gecko; Haitian Big-scaled Sphaero

(WI)

7612 ***Sphaerodactylus corticola***
Squamata - Gekkonidae
e Rum Cay Least Gecko; Rum Cay Dwarf Gecko; Central Bahamas Sphaero (WI)

7613 ***Sphaerodactylus cricoderus***
Squamata - Gekkonidae
e Turquino Collared Sphaero

7614 ***Sphaerodactylus cryphius***
Squamata - Gekkonidae
e Baoruco Least Gecko; Baoruco Dwarf Gecko; Desert Pygmy Sphaero

7615 ***Sphaerodactylus darlingtoni***
Squamata - Gekkonidae
e Darlington's Least Gecko; Darlington's Dwarf Gecko; Northern Forest Sphaero

7616 ***Sphaerodactylus difficilis***
Squamata - Gekkonidae
e Difficult Least Gecko; Difficult Dwarf Gecko; Hispaniolan Eyespot Sphaero

7617 ***Sphaerodactylus docimus***
Squamata - Gekkonidae
e Cabo Cruz Banded Sphaeroi

7618 ***Sphaerodactylus dunni***
Squamata - Gekkonidae
e Dunns Least Gecko; Dunns Dwarf Gecko

7619 ***Sphaerodactylus elasmorhynchus***
Squamata - Gekkonidae
e Marche Leon Least Gecko; Marche Leon Dwarf Gecko; Snout-shield Sphaero

7620 ***Sphaerodactylus elegans***
Squamata - Gekkonidae
e Macleay's Ashy Gecko; Ashy Gecko; Ashy Sphaero

7621 ***Sphaerodactylus elegans elegans***
Squamata - Gekkonidae
e Cuban Ashy Gecko

7622 ***Sphaerodactylus elegantulus***
Squamata - Gekkonidae
e Antigua Least Gecko; Antigua Dwarf Gecko; Antigua Bank Sphaero (WI)

7623 ***Sphaerodactylus epiurus***
Squamata - Gekkonidae
e Hispaniolan Tailspot Sphaero

7624 ***Sphaerodactylus fantasticus***
Squamata - Gekkonidae
e Fantastic Least Gecko; Fantastic Dwarf Gecko; Southern Leeward Sphaero (WI)
f Sphérodactyle bizarre

7625 ***Sphaerodactylus gaigeae***
Squamata - Gekkonidae
e Gaige's Least Gecko; Gaige's Dwarf Gecko; Pandura Gecko ; Chevronate Gecko (NA)

7626 ***Sphaerodactylus gilvitorques***
Squamata - Gekkonidae
e Jamaica Least Gecko; Jamaica Dwarf Gecko; Jamaican Collared Sphaero (WI)

7627 ***Sphaerodactylus glaucus***
Squamata - Gekkonidae
e Least Gecko; Dwarf Gecko; House Gecko

7628 ***Sphaerodactylus goniorhynchus***
Squamata - Gekkonidae
e Cabarita Least Gecko; Cabarita Dwarf Gecko; Jamaican Forest Sphaero (WI)

7629 ***Sphaerodactylus graptolaemus***
Squamata - Gekkonidae
e Costa Rica Least Gecko; Costa Rica Dwarf Gecko

7630 ***Sphaerodactylus helioconia***
Squamata - Gekkonidae
e Santa Marta Least Gecko; Santa Marta Dwarf Gecko

7631 ***Sphaerodactylus homolepis***
Squamata - Gekkonidae
e Caribbean Least Gecko; Caribbean Dwarf Gecko

7632 ***Sphaerodactylus inaguae***
Squamata - Gekkonidae
e Inagua Least Gecko; Inagua Dwarf Gecko; Inagua Sphaero (WI)

7633 ***Sphaerodactylus intermedius***
Squamata - Gekkonidae
e Mantanzas Least Gecko; Mantanzas Dwarf Gecko; North Coast Banded Sphaero

7634 ***Sphaerodactylus kirbyi***
Squamata - Gekkonidae
e Bequia Sphaero

7635 ***Sphaerodactylus klauberi***
Squamata - Gekkonidae
e Klauber's Least Gecko; Klauber's Dwarf Gecko; Upland Gecko ; Puerto Rican Upland Sphaero

7636 ***Sphaerodactylus ladae***
Squamata - Gekkonidae
e Martin Garcia Least Gecko; Martin Garcia Dwarf Gecko; Spotted Agave Sphaero

7637 ***Sphaerodactylus lazelli***
Squamata - Gekkonidae
e Cap Haitien Least Gecko; Cap Haitien Dwarf Gecko; Cape Haitien Sphaero

7638 ***Sphaerodactylus leucaster***
Squamata - Gekkonidae
e Eastern Least Gecko; Eastern Dwarf Gecko; Broken-stripe Sphaero

7639 ***Sphaerodactylus levinsi***
Squamata - Gekkonidae
e Isla Desecheo Least Gecko; Isla Desecheo Dwarf Gecko; Desecheo Gecko ; Desecheo Sphaero

7640 ***Sphaerodactylus lineolatus***
Squamata - Gekkonidae
e Panama Least Gecko; Panama Dwarf Gecko

7641 ***Sphaerodactylus macrolepis***
Squamata - Gekkonidae
e Big-scaled Least Gecko; Big-scaled Dwarf Gecko; Common Dwarf Gecko ; Puerto Rican Eyespot Sphaero (WI)

7642 ***Sphaerodactylus mariguanae***
Squamata - Gekkonidae
e Mayaguana Least Gecko; Mayaguana Dwarf Gecko; Southern Bahamas Sphaero (WI)

7643 ***Sphaerodactylus microlepis***
Squamata - Gekkonidae
e Little-scaled Least Gecko; Little-scaled Dwarf Gecko; St. Lucia Sphaero (WI)

7644 ***Sphaerodactylus micropithecus***
Squamata - Gekkonidae
e Isla Monita Least Gecko; Isla Monita Dwarf Gecko; Monita Sphaero

7645 ***Sphaerodactylus millepunctatus***
Squamata - Gekkonidae
e Spotted Least Gecko; Spotted Dwarf Gecko

7646 ***Sphaerodactylus molei***
Squamata - Gekkonidae
e Tobago Least Gecko; Tobago Dwarf Gecko

7647 ***Sphaerodactylus monensis***
Squamata - Gekkonidae
e Mona Least Gecko; Mona Dwarf Gecko; Mona Gecko ; Mona Sphaero

7648 ***Sphaerodactylus nicholsi***
Squamata - Gekkonidae
e Nichols Least Gecko; Nichols Dwarf Gecko; Puerto Rican Crescent Sphaero

7649 ***Sphaerodactylus nigropunctatus***
Squamata - Gekkonidae
e Black-spotted Least Gecko; Black-spotted Dwarf Gecko; Cuban Three-spotted Sphaero; Cuban Three-banded Sphaero (WI)

7650 ***Sphaerodactylus notatus***
Squamata - Gekkonidae
e Reef Gecko; Brown-speckled Sphaero (WI)
d Riffgecko
f Gecko des récifs

7651 ***Sphaerodactylus notatus notatus***
Squamata - Gekkonidae
e Florida Reef Gecko

7652 ***Sphaerodactylus nycteropus***
Squamata - Gekkonidae
e Morne Dubois Least Gecko; Morne Dubois Dwarf Gecko; Haitian Flathead Sphaero

7653 ***Sphaerodactylus ocoae***
Squamata - Gekkonidae
e Peravia Least Gecko; Peravia Dwarf Gecko; Ocoa Sphaero

7654 ***Sphaerodactylus oliveri***
Squamata - Gekkonidae
e Huventud Least Gecko; Huventud Dwarf Gecko; Escambray Spotted Sphaero

7655 ***Sphaerodactylus omoglaux***
Squamata - Gekkonidae
e Fond Parisien Least Gecko; Fond Parisien Dwarf Gecko; Haitian Border Sphaero

7656 ***Sphaerodactylus oxyrhinus***
Squamata - Gekkonidae
e Sharp-snouted Least Gecko; Sharp-snouted Dwarf Gecko; Jamaican Sharp-nosed Sphaero (WI)

7657 ***Sphaerodactylus pacificus***
Squamata - Gekkonidae
e Pacific Least Gecko; Pacific Dwarf Gecko

7658 ***Sphaerodactylus parkeri***
Squamata - Gekkonidae
e Parker's Least Gecko; Parker's Dwarf Gecko; Parker's Banded Sphaero (WI)

7659 ***Sphaerodactylus parthenopion***
Squamata - Gekkonidae
e Virgin Gorda Least Gecko; Virgin Gorda Dwarf Gecko; Virgin Gorda Gecko ; Virgin Islands Dwarf Sphaero

7660 ***Sphaerodactylus perissodactylius***
Squamata - Gekkonidae
e Dominican Least Gecko; Dominican Dwarf Gecko; Martin Garcia Sphaero

7661 ***Sphaerodactylus pimienta***
Squamata - Gekkonidae
e Cuban Pepper Sphaero

7662 ***Sphaerodactylus plummeri***
Squamata - Gekkonidae
e Barahona Big-scaled Sphaero

7663 ***Sphaerodactylus ramsdeni***
Squamata - Gekkonidae
e Ramsden's Least Gecko; Ramsden's Dwarf Gecko; Cuban Forest Sphaero

7664 ***Sphaerodactylus randi***
Squamata - Gekkonidae
e Pedernales Least Gecko; Pedernales Dwarf Gecko; Barahoona Speckled Sphaero

7665 ***Sphaerodactylus rhabdotus***
Squamata - Gekkonidae
e Valle de Neiba Least Gecko; Valle de Neiba Dwarf Gecko; Hispaniolan Two-stripe Sphaero

7666 ***Sphaerodactylus richardi***
Squamata - Gekkonidae
e Richard's Banded Sphaero

7667 ***Sphaerodactylus richardsoni***
Squamata - Gekkonidae
e Richardson's Least Gecko; Richardson's Dwarf Gecko; Montego Banded Sphaero (WI)

7668 ***Sphaerodactylus roosevelti***
Squamata - Gekkonidae
e Roosevelt's Least Gecko; Roosevelt's Dwarf Gecko; Littoral Gecko (WI}; Guanica Khaki Sphaero

7669 ***Sphaerodactylus rosaurae***
Squamata - Gekkonidae
e Bay Islands Least Gecko; Bay Island Dwarf Gecko

7670 ***Sphaerodactylus ruibali***
Squamata - Gekkonidae
e Ruibal's Least Gecko; Ruibal's Dwarf Gecko; Guantanamo Bay Sphaero

7671 ***Sphaerodactylus sabanus***
Squamata - Gekkonidae
e Saba Least Gecko; Saba Dwarf Gecko; Northern Leeward Sphaero (WI)

7672 ***Sphaerodactylus samanensis***
Squamata - Gekkonidae
e Samana Least Gecko; Samana Dwarf Gecko; Haiti Banded Sphaero

7673 ***Sphaerodactylus savagei***
Squamata - Gekkonidae
e Savage's Least Gecko; Savage's Dwarf Gecko; Altagracia Speckled Sphaero

7674 ***Sphaerodactylus scaber***
Squamata - Gekkonidae
e Camaguey's Least Gecko; Camaguey's Dwarf Gecko; Double-collared Sphaero

7675 ***Sphaerodactylus scapularis***
Squamata - Gekkonidae
e Boulenger's Least Gecko; Boulenger's Dwarf Gecko

7676 ***Sphaerodactylus schuberti***
Squamata - Gekkonidae
e Neiba Agave Sphaero

7677 ***Sphaerodactylus schwartzi***
Squamata - Gekkonidae
e Guantanamo Bay Sphaero

7678 ***Sphaerodactylus semasiops***
Squamata - Gekkonidae
e Cockpit Least Gecko; Cockpit Dwarf Gecko; Cockpit Eyespot Sphaero (WI)

7679 ***Sphaerodactylus shrevei***
Squamata - Gekkonidae
e Shreve's Least Gecko; Shreve's Dwarf Gecko; Northwest Haiti Blotched Sphaero

7680 ***Sphaerodactylus sommeri***
Squamata - Gekkonidae
e Terre Nueve Least Gecko; Terre Nueve Dwarf Gecko; Northwest Haiti Banded Sphaero

7681 ***Sphaerodactylus sputator***
Squamata - Gekkonidae
e Least Gecko; Dwarf Geckol; Leeward Banded Sphaero (WI)
f Sphérodactyle sputateur

7682 ***Sphaerodactylus stejnegeri***
Squamata - Gekkonidae
e Stejneger's Least Gecko; Stejneger's Dwarf Gecko

7683 ***Sphaerodactylus storeyae***
Squamata - Gekkonidae
e Isles of Pines Sphaero

7684 ***Sphaerodactylus streptophorus***
Squamata - Gekkonidae
e Hispaniola Least Gecko; Hispaniola Dwarf Gecko; Hispaniolan Small-eared Sphaero

7685 ***Sphaerodactylus thompsoni***
Squamata - Gekkonidae
e Thompson's Least Gecko; Thompson's Dwarf Gecko; Barahona Limestone Sphaero

7686 ***Sphaerodactylus torrei***
Squamata - Gekkonidae
e Barbour's Least Gecko; Barbour's Dwarf Gecko; Cuban Broad-banded Sphaero

7687 ***Sphaerodactylus townsendi***
Squamata - Gekkonidae
e Townsend's Least Gecko; Townsend's Dwarf Gecko; Puerto Rican Sandy Sphaero

7688 ***Sphaerodactylus underwoodi***
Squamata - Gekkonidae
e Underwood's Least Gecko; Underwood's Dwarf Gecko; Turks Islands Sphaero (WI)

7689 ***Sphaerodactylus vincenti***
Squamata - Gekkonidae
e Vincent's Least Gecko; Vincent's Dwarf Gecko; Windward Sphaero (WI)

7690 ***Sphaerodactylus williamsi***
Squamata - Gekkonidae
e Williams's Least Gecko; Williams's Dwarf Gecko; Haitian Striped Sphaero

7691 ***Sphaerodactylus zygaena***
Squamata - Gekkonidae
e Dame Marie Least Gecko; Dame Marie Dwarf Gecko; Tiburon Coastal Sphaero

7692 ***Sphenodon guentheri***
Rhynchocephalia - Sphenodontidae
e Brother Island Tuatara

7693 ***Sphenodon punctatus***
Rhynchocephalia - Sphenodontidae
e Tuatara
d Brückenechse
i Tuatara

7694 **Sphenodontidae**
Rhynchocephalia
e Tuataras
d Brückenechsen
f Sphénodontidés

7695 ***Sphenomorphus***
Squamata - Scincidae
e Forest Skinks; Forest-dweller Skinks; Tree Skinks
d Waldskinke
f Sphénomorphes

7696 ***Sphenomorphus assatus***
Squamata - Scincidae
e Red Forest Skink

7697 ***Sphenomorphus cherriei***
Squamata - Scincidae
e Brown Forest Skink; Ground Skink; Striped Litter Skink

7698 ***Sphenomorphus derooyae***
Squamata - Scincidae
e De Rooij's Forest Skink; De Rooij's Tree Skink

7699 ***Sphenomorphus dussumieri***
Squamata - Scincidae
e Dussumier's Forest Skink; Dussumier's Forest-dweller
f Sphénomorphe de Dussumier

7700 ***Sphenomorphus incerta***
Squamata - Scincidae
e Stuart's Forest Skink

7701 ***Sphenomorphus tenuis***
Squamata - Scincidae
e Fawn Forest Skink

7702 ***Sphenops***
Squamata - Scincidae
e Wedge-snouted Skinks
d Keilskinke; Keilkopfschleichen; Keilkopfskinke
f Sphénops

7703 ***Sphenops boulengeri***
Squamata - Scincidae
e Boulenger's Wedge-snouted Skink; Boulenger's Sand Skink
f Sphénops de Boulenger

7704 ***Sphenops delislei***
Squamata - Scincidae
e De l'Isle's Wedge-snouted Skink; De l'Isle's Sand Skink
f Sphénops de De l'Isle

7705 ***Sphenops sepsoides***
Squamata - Scincidae
e Wedge-snouted Skink
d Keilskink; Keilkopfschleiche; Keilkopfskink; Keilschleiche

7706 ***Sphenops sphenopsiformis***
Squamata - Scincidae
e Demuril's Wedge-snouted Skink; Senegal Sand Skink
f Sphénops occidental

7707 ***Spilotes***
Squamata - Colubridae
e Tropical Rat Snakes; Chicken Snakes; Black-and-yellow Rat Snakes; Tiger Snakes; Tiger Rat Snakes
d Hühnerfresser; Hühneresser
f Serpents des poulaillers

7708 ***Spilotes pullatus***
Squamata - Colubridae
e Tropical Rat Snake; Chicken Snake; Tiger Rat Snake; Black-and-yellow Rat Snake; Tiger Tree Snake; Tiger Snake
d Hühnerfresser; Hühneresser
f Serpent des poulaillers

7709 ***Spilotes pullatus pullatus***
Squamata - Colubridae
e Common Tiger Rat Snake

7710 **Squamata**
e Lizards, Worm Lizards and Snakes
d Squamaten; Echsen, Brückenechsen und Schlangen
f Reptiles écailleux
i Squamati

7711 ***Staurotypus***
Testudines - Kinosternidae
e Giant Musk Turtles

d Kreuzbrustschildkröten
f Staurotypes

7712 ***Staurotypus salvinii***
Testudines - Kinosternidae
e Chiapas Giant Musk Turtle; Crucilla Turtle; Salvin's Musk Turtle; Mexican Giant Musk Turtle; Pacific Coast Giant Musk Turtle; Central American Musk Turtle
d Kreuzbrustschildkröte; Salvins Kreuzbrustschildkröte; Salvin-Kreuzbrustschildkröten
f Staurotype de Salvin

7713 ***Staurotypus triporcatus***
Testudines - Kinosternidae
e Mexican Giant Musk Turtle; Guau Musk Turtle; Giant Musk Turtle
d Mexikanische Kreuzbrustschildkröte; Große Kreuzbrustschildkröte
f Staurotype tricariné; Staurotype de l'Atlantique
i Tartaruga del muschio gigante

7714 ***Stegonotus***
Squamata - Colubridae
e Frog-eating Snakes

7715 ***Stegonotus batjanensis***
Squamata - Colubridae
e Batjan Frog-eating Snake

7716 ***Stegonotus borneensis***
Squamata - Colubridae
e Borneo Frog-eating Snake

7717 ***Stegonotus cucullatus***
Squamata - Colubridae
e Murray Frog-eating Snake; Slaty-grey Snake (ANZ); Slate-grey Snake
d Graue Schieferenatter

7718 ***Stegonotus dumerili***
Squamata - Colubridae
e Duméril's Frog-eating Snake

7719 ***Stegonotus heterurus***
Squamata - Colubridae
e Günther's Frog-eating Snake

7720 ***Stegonotus modestus***
Squamata - Colubridae
e New Guinea Frog-eating Snake

7721 ***Stegonotus muelleri***
Squamata - Colubridae
e Müller's Frog-eating Snake

7722 ***Stegonotus parvus***
Squamata - Colubridae
e Meyer's Frog-eating Snake

7723 ***Stellio***
Squamata - Agamidae
e Rock Agamas
d Wirtelschwanzagamen
f Stellions

7724 ***Stellio adramitanus***
Squamata - Agamidae
e Anderson's Rock Agama

7725 ***Stellio agrorensis***
Squamata - Agamidae
e Stoliczka's Rock Agama

7726 ***Stellio annectens***
Squamata - Agamidae
e Eritrea Rock Agama; Eritrean Rock Agama

7727 ***Stellio atricollis***
Squamata - Agamidae
e Blue-throated Rock Agama; Black-necked Agama; Tree Agama; Blue-throated Agama; Blue-headed Tree Agama
d Blaukehlagame
f Agame à cou noir; Agame a gorge bleue

7728 ***Stellio badakhshana***
Squamata - Agamidae
e Badakhshana Rock Agama

7729 ***Stellio blanfordi***
Squamata - Agamidae
e Blanford's Rock Agama

7730 ***Stellio caucasius***
Squamata - Agamidae
e Northern Rock Agama; Caucasian Agama
d Kaukasus-Agame
f Agame du Caucase
i Agama caucasica

7731 ***Stellio chernovi***
Squamata - Agamidae
e Chernov Rock Agama

7732 ***Stellio cyanogaster***
Squamata - Agamidae
e Bluebelly Rock Agama; Tree Agama; Black-necked Arboreal Agama

7733 ***Stellio erythrogaster***
Squamata - Agamidae
e Redbelly Rock Agama

7734 ***Stellio himalayanus***
Squamata - Agamidae
e Himalaya Rock Agama; Himalayan Rock Agama

7735 ***Stellio kirmanensis***
Squamata - Agamidae
e Nikolsky's Rock Agama

7736 ***Stellio lehmanni***
Squamata - Agamidae
e Turkestan Rock Agama; Turkestan Agama; Turkmenian Agama
d Wirtelschwanzagame; Turkestanische Agame; Turkmenische Agame

7737 ***Stellio melanurus***
Squamata - Agamidae
e Black Rock Agama
d Schwarze Pakistan-Agame

7738 ***Stellio microlepis***
Squamata - Agamidae
e Small-scaled Rock Agama

7739 ***Stellio nupta***
Squamata - Agamidae
e Yellow-headed Rock Agama; Yellow-headed Agama; Large-scaled Rock Agama
d Riesenagame

7740 ***Stellio nuristanicus***
Squamata - Agamidae
e Leviton's Rock Agama

7741 ***Stellio phillipsii***
Squamata - Agamidae
e Phillips's Rock Agama

7742 ***Stellio sacra***
Squamata - Agamidae
e Annan's Rock Agama

7743 ***Stellio stellio***
Squamata - Agamidae
e Roughtail Rock Agama; Hardun; Rough-tailed Agama; Rhode Dragon; Golden Agama; Short-toed Agama; Clown Agama; Painted Agama; Painted Dragon
d Hardun; Schleuderschwanz
f Hardoun stellion; Stellion
i Stellione; Agama stellione

7744 ***Stellio stoliczkanus***
Squamata - Agamidae
e Mongolia Rock Agama; Mongolean Agama

d Gobi-Agame

7745 ***Stellio tarimensis***
Squamata - Agamidae
e Zugmayer's Rock Agama

7746 ***Stellio trachypleurus***
Squamata - Agamidae
e Peters's Rock Agama

7747 ***Stellio tuberculatus***
Squamata - Agamidae
e Kashmir Rock Agama
d Kaschmir-Agame
f Agame tuberculeux

7748 ***Stellio yemmenensis***
Squamata - Agamidae
e Yemen Rock Agama

7749 ***Stellio zonurus***
Squamata - Agamidae
e Boulenger's Rock Agama

7750 ***Stenocercus***
Squamata - Iguanaidae
e Whorltail Iguanas; Narrow-tailed Lizards; Leaf Lizards
d Kurzschwanzleguane
f Sténocerques

7751 ***Stenocercus apuricamus***
Squamata - Iguanaidae
e Fritt's Whorltail Iguana

7752 ***Stenocercus boettgeri***
Squamata - Iguanaidae
e Boettger's Whorltail Iguana

7753 ***Stenocercus bolivarensis***
Squamata - Iguanaidae
e Bolivian Whorltail Iguana

7754 ***Stenocercus carrioni***
Squamata - Iguanaidae
e Parker's Whorltail Iguana

7755 ***Stenocercus chrysopagus***
Squamata - Iguanaidae
e Golden Whorltail Iguana

7756 ***Stenocercus crassicaudatus***
Squamata - Iguanaidae
e Spiny Whorltail Iguana

7757 ***Stenocercus cupreus***
Squamata - Iguanaidae
e Copper Whorltail Iguana

7758 ***Stenocercus empetrus***
Squamata - Iguanaidae
e Rock Whorltail Iguana

7759 ***Stenocercus festae***
Squamata - Iguanaidae
e Peracca's Whorltail Iguana

7760 ***Stenocercus formosus***
Squamata - Iguanaidae
e Formosa Whorltail Iguana

7761 ***Stenocercus guentheri***
Squamata - Iguanaidae
e Günther's Whorltail Iguana

7762 ***Stenocercus haenschi***
Squamata - Iguanaidae
e Haensch's Whorltail Iguana

7763 ***Stenocercus humeralis***
Squamata - Iguanaidae
e Patterned Whorltail Iguana

7764 ***Stenocercus ivitus***
Squamata - Iguanaidae
e Ivy Whorltail Iguana

7765 ***Stenocercus marmoratus***
Squamata - Iguanaidae
e Spotted Whorltail Iguana

7766 ***Stenocercus melanopygus***
Squamata - Iguanaidae
e Dark Whorltail Iguana

7767 ***Stenocercus modestus***
Squamata - Iguanaidae
e Modest Whorltail Iguana

7768 ***Stenocercus nigromaculatus***
Squamata - Iguanaidae
e Black-spotted Whorltail Iguana

7769 ***Stenocercus nubicola***
Squamata - Iguanaidae
e Cloud Whorltail Iguana

7770 ***Stenocercus ochoai***
Squamata - Iguanaidae
e Ochoa Whorltail Iguana

7771 ***Stenocercus orientalis***
Squamata - Iguanaidae
e Eastern Whorltail Iguana

7772 ***Stenocercus ornatissimus***
Squamata - Iguanaidae
e Lesser Ornate Whorltail Iguana

7773 ***Stenocercus ornatus***
Squamata - Iguanaidae
e Girard's Whorltail Iguana

7774 ***Stenocercus praeornatus***
Squamata - Iguanaidae
e Greater Ornate Whorltail Iguana

7775 ***Stenocercus rhodomelas***
Squamata - Iguanaidae
e Red-black Whorltail Iguana

7776 ***Stenocercus roseiventris***
Squamata - Iguanaidae
e Rose Whorltail Iguana; Pink-bellied Leaf Lizard
f Sténocerque à ventre rose

7777 ***Stenocercus simonsii***
Squamata - Iguanaidae
e Simons's Whorltail Iguana

7778 ***Stenocercus trachycephalus***
Squamata - Iguanaidae
e Duméril's Whorltail Iguana

7779 ***Stenocercus variabilis***
Squamata - Iguanaidae
e Variable Whorltail Iguana

7780 ***Stenocercus varius***
Squamata - Iguanaidae
e Keeled Whorltail Iguana

7781 ***Stenodactylus***
Squamata - Gekkonidae
e Short-fingered Geckos
d Dünnfingergeckos
f Sténodactyles

7782 ***Stenodactylus affinis***
Squamata - Gekkonidae
e Iranian Short-fingered Gecko

7783 ***Stenodactylus arabicus***
Squamata - Gekkonidae
e Arabian Short-fingered Gecko

7784 ***Stenodactylus doriae***
Squamata - Gekkonidae
e Middle Eastern Short-fingered Gecko
d Engfingergecko

7785 ***Stenodactylus grandiceps***
Squamata - Gekkonidae
e Jordan Short-fingered Gecko; Large-

headed Ground Gecko

7786 ***Stenodactylus khobarensis***
Squamata - Gekkonidae
e Gulf Short-fingered Gecko

7787 ***Stenodactylus leptocosymbotes***
Squamata - Gekkonidae
e Southern Short-fingered Gecko

7788 ***Stenodactylus petrii***
Squamata - Gekkonidae
e Anderson's Short-fingered Gecko; Petrie's Gecko; Sand Gecko
d Petris Dünnfingergecko; Dünnfingergecko
f Sténodactyle de Petrie; Gecko de Petrie

7789 ***Stenodactylus pulcher***
Squamata - Gekkonidae
e Beautiful Short-fingered Gecko

7790 ***Stenodactylus slevini***
Squamata - Gekkonidae
e Slevin's Short-fingered Gecko

7791 ***Stenodactylus sthenodactylus***
Squamata - Gekkonidae
e Lichtenstein's Short-fingered Gecko; Sand Gecko; Sand-eye; Whiptail Gecko; Israeli Sand Gecko
d Wüstendünnfingergecko
f Sténodactyle tacheté; Sténodactyle commun
i Stenodattilo

7792 ***Stenodactylus yemenensis***
Squamata - Gekkonidae
e Yemen Short-fingered Gecko

7793 ***Stenolepis***
Squamata - Teiidae
e Pernambuco Teiids

7794 ***Stenolepis ridleyi***
Squamata - Teiidae
e Pernambuco Teiid

7795 ***Stenorrhina***
Squamata - Colubridae
e Blood Snakes; South American Tree Snakes

7796 ***Stenorrhina degenhardtii***
Squamata - Colubridae
e Degenhardt's Scorpion-eating Snake; Degenhardt Scorpion-eater

7797 ***Stenorrhina freminvillii***
Squamata - Colubridae
e Blood Snake; Freminville's Scorpion-eater

7798 ***Stilosoma***
Squamata - Colubridae
e Short-tail Snakes; Short-tailed Snakes
d Kurzschwanznattern
f Couleuvres à queue courte

7799 ***Stilosoma extenuatum***
Squamata - Colubridae
e Short-tail Snake; Short-tailed Snake
d Kurzschwanznatter
f Couleuvre à queue courte

7800 ***Stoliczkaia***
Squamata - Colubridae
e Red Snakes

7801 ***Stoliczkaia borneensis***
Squamata - Colubridae
e Borneo Red Snake

7802 ***Stoliczkaia khasiensis***
Squamata - Colubridae
e Khasi Red Snake

7803 ***Storeria***
Squamata - Colubridae
e Brown Snakes; Red-bellied Snakes; North American Brown Snakes
d Braunschlangen; Braunnattern
f Couleuvres brunes

7804 ***Storeria dekayi***
Squamata - Colubridae
e DeKay's Brown Snake; Common Brown Snake; DeKay's Snake (NA); Brown Snake; Tropical Brown Snake
d Gewöhnliche Braunschlange; Braunnatter; Nördliche Braunnatter; Gewöhnliche Braunnatter
f Couleuvre brune

7805 ***Storeria dekayi dekayi***
Squamata - Colubridae
e Northern Brown Snake
d Nördliche Braunnatter

7806 ***Storeria dekayi limnetes***
Squamata - Colubridae
e Marsh Brown Snake; Brown Snake

7807 ***Storeria dekayi texana***
Squamata - Colubridae
e Texas Brown Snake; Northern Brown Snake

7808 ***Storeria dekayi victa***
Squamata - Colubridae
e Florida Brown Snake

7809 ***Storeria dekayi wrightorum***
Squamata - Colubridae
e Midland Brown Snake

7810 ***Storeria hidalgoensis***
Squamata - Colubridae
e Mexican Yellowbelly Brown Snake

7811 ***Storeria occipitomaculata***
Squamata - Colubridae
e Red-bellied Snake; Redbelly Snake; Storer's Snake
d Rotbäuchige Braunschlange; Rotbauchnatter; Rotbauchbraunnatter
f Couleuvre à ventre rouge

7812 ***Storeria occipitomaculata obscura***
Squamata - Colubridae
e Florida Redbelly Snake; Florida Red-bellied Snake

7813 ***Storeria occipitomaculata occipitomaculata***
Squamata - Colubridae
e Northern Redbelly Snake; Northern Red-bellied Snake

7814 ***Storeria occipitomaculata pahasapae***
Squamata - Colubridae
e Black Hills Redbelly Snake; Black Hills Red-bellied Snake

7815 ***Storeria storerioides***
Squamata - Colubridae
e Mexican Brown Snake

7816 ***Suta***
Squamata - Elapidae
e Little Snakes; Myall Snakes; Curl Snakes

7817 ***Suta fasciata***
Squamata - Elapidae
e Rosen's Snake
d Rosens Schlange; Rosen-Bänderotter

7818 ***Suta flagellum***
Squamata - Elapidae
e Little Whip Snake

7819 ***Suta gouldii***
Squamata - Elapidae
e Gould's Black-headed Snake; Black-headed Snake

7820 ***Suta monachus***
Squamata - Elapidae
e Hooded Snake

7821 ***Suta nigriceps***
Squamata - Elapidae
e Günther's Black-headed Snake

7822 ***Suta ordensis***
Squamata - Elapidae
e Storr's Cross-banded Snake

7823 ***Suta punctatus***
Squamata - Elapidae
e Little Spotted Snake
d Gefleckte Bänderotter

7824 ***Suta spectabilis***
Squamata - Elapidae
e Little Night Snake

7825 ***Suta suta***
Squamata - Elapidae
e Curl Snake; Myall Snake (ANZ); Myall (ANZ)

7826 ***Symphimus***
Squamata - Colubridae
e White-lipped Snakes

7827 ***Symphimus leucostomus***
Squamata - Colubridae
e Isthmian White-lipped Snake

7828 ***Symphimus mayae***
Squamata - Colubridae
e Yucatan White-lipped Snake

7829 ***Sympholis***
Squamata - Colubridae
e Mexican Short-tail Snakes; Mexican Short-tailed Snake

7830 ***Sympholis lippiens***
Squamata - Colubridae
e Mexican Short-tail Snake; Mexican Short-tailed Snake

7831 ***Synophis***
Squamata - Colubridae
e Fishing Snakes

7832 ***Synophis bicolor***
Squamata - Colubridae
e Two-coloured Fishing Snake

7833 ***Synophis lasallei***
Squamata - Colubridae
e Lasalle's Fishing Snake

7834 ***Synophis miops***
Squamata - Colubridae
e Fishing Snake
f Couleuvre pêcheuse

T

7835 ***Tachymenis***
Squamata - Colubridae
e Slender Snakes

7836 ***Tachymenis affinis***
Squamata - Colubridae
e Boulenger's Slender Snake

7837 ***Tachymenis attenuata***
Squamata - Colubridae
e Walkers Slender Snake

7838 ***Tachymenis chilensis***
Squamata - Colubridae
e Chilean Slender Snake

7839 ***Tachymenis elongata***
Squamata - Colubridae
e Depax's Slender Snake

7840 ***Tachymenis peruviana***
Squamata - Colubridae
e Peru Slender Snake

7841 ***Tachymenis tarmensis***
Squamata - Colubridae
e Slender Snake

7842 ***Taeniophallus***
Squamata - Colubridae
e Forest Snakes; Ground Snakes

7843 ***Taeniophallus affinis***
Squamata - Colubridae
e Günther's Forest Snake

7844 ***Taeniophallus bilineatus***
Squamata - Colubridae
e Two-lined Forest Snake

7845 ***Taeniophallus brevirostris***
Squamata - Colubridae
e Peters's Forest Snake; Short-nosed Ground Snake

7846 ***Taeniophallus occipitalis***
Squamata - Colubridae
e Jan's Forest Snake; Spotted Ground Snake

7847 ***Taeniophallus persimilis***
Squamata - Colubridae
e Dunn's Forest Snake

7848 ***Taeniophallus poecilopogon***
Squamata - Colubridae
e Cope's Forest Snake

7849 ***Takydromus***
Squamata - Lacertidae
e Grass Lizards; Oriental Grass Lizards; Tachydromes
d Schnellläufer; Langschwanzeidechsen
f Tachydromes

7850 ***Takydromus amurensis***
Squamata - Lacertidae
e Amur Grass Lizard; Northern Grass Lizard
d Amur-Schnellläufer

7851 ***Takydromus formosanus***
Squamata - Lacertidae
e Formosa Grass Lizard

7852 ***Takydromus haughtonianus***
Squamata - Lacertidae
e Goalpora Grass Lizard

7853 ***Takydromus khasiensis***
Squamata - Lacertidae
e Java Grass Lizard

7854 ***Takydromus sauteri***
Squamata - Lacertidae
e Koshun Grass Lizard

7855 ***Takydromus septentrionalis***
Squamata - Lacertidae
e China Grass Lizard

7856 ***Takydromus sexlineatus***
Squamata - Lacertidae
e Asian Grass Lizard; Six-lined Lizard
d Sechsstreifiger Schnellläufer; Sechsstreifige Langschwanzeidechse
f Tachydrome à six raies

7857 ***Takydromus smaragdinus***
Squamata - Lacertidae
e Green Grass Lizard
d Smaragdschnellläufer

7858 ***Takydromus tachydromoides***
Squamata - Lacertidae
e Japanese Grass Lizard; Japan Grass Lizard

7859 ***Takydromus wolteri***
Squamata - Lacertidae
e Mountain Grass Lizard
d Koreanischer Schnellläufer

7860 ***Tantalophis***
Squamata - Colubridae
e Oaxacan Cat-eyed Snakes

7861 ***Tantalophis discolor***
Squamata - Colubridae
e Oaxacan Cat-eyed Snake

7862 ***Tantilla***
Squamata - Colubridae
e Centipede Snakes; Black-headed Snakes (NA); Centipede-eaters; American Black-headed Snakes; Crowned Snakes
d Schwarzkopfschlangen; Schwarzkopfnattern
f Serpents à tête plate

7863 ***Tantilla albiceps***
Squamata - Colubridae
e Barbour's Centipede Snake; Panamanian Centipede-eater

7864 ***Tantilla alticola***
Squamata - Colubridae
e Boulenger's Centipede Snake; Highland Centipede-eater

7865 ***Tantilla andanista***
Squamata - Colubridae
e Andes Centipede Snake

7866 ***Tantilla atriceps***
Squamata - Colubridae
e Mexican Black-headed Snake; Mexican Blackhead Snake
f Serpent à tête noire mexicain

7867 ***Tantilla bairdi***
Squamata - Colubridae
e Baird's Black-headed Snake; Baird's Centipede-eater

7868 ***Tantilla bocourti***
Squamata - Colubridae
e Bocourt's Black-headed Snake

7869 ***Tantilla brevicauda***
Squamata - Colubridae
e Mertens's Centipede Snake; Short-tailed Centipede-eater

7870 ***Tantilla brevis***
Squamata - Colubridae
e Short Centipede Snake

7871 ***Tantilla briggsi***
Squamata - Colubridae
e Briggs's Centipede Snake

7872 ***Tantilla calamarina***
Squamata - Colubridae
e Pacific Coast Centipede Snake

7873 ***Tantilla canula***
Squamata - Colubridae
e Cope's Centipede Snake

7874 ***Tantilla capistrata***
Squamata - Colubridae
e Capistrata Centipede Snake

7875 ***Tantilla cascadae***
Squamata - Colubridae
e Michoacan Centipede Snake

7876 ***Tantilla coronadoi***
Squamata - Colubridae
e Guerreran Centipede Snake

7877 ***Tantilla coronata***
Squamata - Colubridae
e Southeastern Crowned Snake; Blackhead Snake; Crowned Snake (NA)
d Gekrönte Schwarzkopfschlange
f Serpent diadème du Sud-est

7878 ***Tantilla cucullata***
Squamata - Colubridae
e Blackhood Snake; Big Bend Blackhead Snake; Trans-Pecos Black-headed Snake

7879 ***Tantilla cuesta***
Squamata - Colubridae
e Wilson's Centipede Snake

7880 ***Tantilla cuniculato***
Squamata - Colubridae
e Peten Centipede Snake; Collared Centipede-eater

7881 ***Tantilla deppei***
Squamata - Colubridae
e Deppe's Centipede Snake

7882 ***Tantilla equatoriana***
Squamata - Colubridae
e Equator Centipede Snake

7883 ***Tantilla excubitor***
Squamata - Colubridae
e South American Centipede Snake

7884 ***Tantilla flavilineata***
Squamata - Colubridae
e Yellow-lined Centipede Snake

7885 ***Tantilla gracilis***
Squamata - Colubridae
e Flathead Snake; Flat-headed Snake; Miter Snake (NA); Southern Slender Black-headed Snake; Graceful Tantilla
d Schlanke Schwarzkopfschlange
f Serpent à tête plate

7886 ***Tantilla hobartsmithi***
Squamata - Colubridae
e Southwestern Black-headed Snake; Southwestern Blackhead Snake; Smith's Black-headed Snake; Smith's Black-head Snake

7887 ***Tantilla impensa***
Squamata - Colubridae
e Centipede-eater

7888 ***Tantilla insulamontani***
Squamata - Colubridae
e Mountain Centipede Snake

7889 ***Tantilla jani***
Squamata - Colubridae

e Jan's Centipede Snake

7890 ***Tantilla lempira***
Squamata - Colubridae
e Mena's Centipede Snake

7891 ***Tantilla melanocephala***
Squamata - Colubridae
e Black-headed Snake; Black-headed Centipede-eater
d Dunkelköpfige Schwarzkopfschlange

7892 ***Tantilla melanocephala melanocephala***
Squamata - Colubridae
e Common Black-headed Snake

7893 ***Tantilla miniata***
Squamata - Colubridae
e Veracruz Centipede Snake

7894 ***Tantilla moesta***
Squamata - Colubridae
e Blackbelly Centipede Snake; Black-bellied Centipede-eater

7895 ***Tantilla morgani***
Squamata - Colubridae
e Morgan's Black-headed Snake

7896 ***Tantilla nigra***
Squamata - Colubridae
e Black Centipede Snake

7897 ***Tantilla nigriceps***
Squamata - Colubridae
e Plains Black-headed Snake; Plains Blackhead Snake; Great Plains Black-headed Snake
d Flachlandschwarzkopfschlange
f Serpent à tête noire des plaines

7898 ***Tantilla nigriceps nigriceps***
Squamata - Colubridae
e Western Plains Black-headed Snake

7899 ***Tantilla oaxacae***
Squamata - Colubridae
e Oaxacan Centipede Snake

7900 ***Tantilla oolitica***
Squamata - Colubridae
e Rim Rock Crowned Snake
f Serpent diadème oolitique

7901 ***Tantilla petersi***
Squamata - Colubridae
e Peters's Black-headed Snake

7902 ***Tantilla planiceps***
Squamata - Colubridae
e Western Black-headed Snake; Western Blackhead Snake; California Black-headed Snake
d Westliche Schwarzkopfschlange

7903 ***Tantilla relicta***
Squamata - Colubridae
e Florida Crowned Snake; Southwestern Blackhead Snake; Central Florida Crowned Snake
f Serpent didadème de Floride

7904 ***Tantilla relicta neilli***
Squamata - Colubridae
e Central Florida Crowned Snake

7905 ***Tantilla relicta pamlica***
Squamata - Colubridae
e Central Dunes Crowned Snake; Coastal Dunes Crowned Snake

7906 ***Tantilla relicta relicta***
Squamata - Colubridae
e Peninsula Crowned Snake

7907 ***Tantilla reticulata***
Squamata - Colubridae
e Reticulate Centipede Snake; Reticulated Centipede-eater

7908 ***Tantilla rubra***
Squamata - Colubridae
e Red Centipede-eater; Big Bend Blackhead Snake; Big Bend Black-headed Snake; Red Blackhead Snake
d Große Schwarzkopfschlange
f Serpent à tête noire de Big Bend

7909 ***Tantilla rubra diabola***
Squamata - Colubridae
e Devil's River Black-headed Snake

7910 ***Tantilla schistosa***
Squamata - Colubridae
e Red Earth Centipede Snake; Red-tailed Centipede-eater

7911 ***Tantilla semicincta***
Squamata - Colubridae
e Ringed Centipede Snake

7912 ***Tantilla shawi***
Squamata - Colubridae
e Potosi Centipede Snake

7913 ***Tantilla slaveni***
Squamata - Colubridae
e Slaven's Centipede Snake

7914 ***Tantilla striata***
Squamata - Colubridae
e Striped Centipede Snake

7915 ***Tantilla supracincta***
Squamata - Colubridae
e Banded Centipede Snake; Tri-coloured Centipede-eater

7916 ***Tantilla taeniata***
Squamata - Colubridae
e Central American Centipede Snake; Striped Centipede-eater

7917 ***Tantilla tayrae***
Squamata - Colubridae
e Volcan Tacana Centipede Snake; Tayra's Centipede-eater

7918 ***Tantilla tecta***
Squamata - Colubridae
e White-striped Centipede-eater

7919 ***Tantilla trilineta***
Squamata - Colubridae
e Three-lined Centipede Snake

7920 ***Tantilla triseriata***
Squamata - Colubridae
e Isthmian Centipede-eater

7921 ***Tantilla tritaeniata***
Squamata - Colubridae
e Three-banded Centipede Snake; Guanaja Centipede-eater

7922 ***Tantilla vermiformis***
Squamata - Colubridae
e Hallowell's Centipede Snake; Wormlike Centipede-eater

7923 ***Tantilla virgata***
Squamata - Colubridae
e Common Centipede Snake

7924 ***Tantilla vulcani***
Squamata - Colubridae
e Vulcan Centipede-eater

7925 ***Tantilla wilcoxi***
Squamata - Colubridae
e Chihuahuan Black-headed Snake; Chihuahuan Blackhead Snake; Arizona Tantilla
f Serpent à tête noire Chihuahua

7926 ***Tantilla wilcoxi wilcoxi***
Squamata - Colubridae
e Huachaca Black-headed Snake

7927 ***Tantilla yaquia***
Squamata - Colubridae
e Yaqui Black-headed Snake; Yaqui Blackhead Snake
f Serpent à tête noire Yaqui

7928 ***Tantillita***
Squamata - Colubridae
e Dwarf Short-tail Snakes

7929 ***Tantillita brevissima***
Squamata - Colubridae
e Speckled Dwarf Short-tail Snake; Speckled Dwarf Centipede-eater

7930 ***Tantillita canula***
Squamata - Colubridae
e Yucatan Dwarf Short-tail Snake; Yucatan Dwarf Centipede-eater

7931 ***Tantillita lintoni***
Squamata - Colubridae
e Linton's Dwarf Short-tail Snake; Linton's Dwarf Centipede-eater

7932 ***Tarentola***
Squamata - Gekkonidae
e Wall Geckos
d Mauergeckos
f Tarentes
i Tarantole

7933 ***Tarentola albertschwartzi***
Squamata - Gekkonidae
e Jamaican Giant Gecko

7934 ***Tarentola americana***
Squamata - Gekkonidae
e American Wall Gecko; Cuban Giant Gecko (WI)

7935 ***Tarentola angustimentalis***
Squamata - Gekkonidae
e Canary Wall Gecko
d Kanarischer Mauergecko

7936 ***Tarentola annularis***
Squamata - Gekkonidae
e Ringed Wall Gecko; Egyptian Gecko; White-spotted Gecko
d Ringgecko
f Tarante annelée

7937 ***Tarentola annularis annularis***
Squamata - Gekkonidae
e House Gecko
f Tarante annelée

7938 ***Tarentola boehmei***
Squamata - Gekkonidae
e Morocco Wall Gecko; Böhme's Gecko
d Böhmes Gecko
f Tarante de Boehne; Tarente du Maroc

7939 ***Tarentola boettgeri***
Squamata - Gekkonidae
e Boettger's Wall Gecko
d Gestreifter Mauergecko

7940 ***Tarentola boreensis***
Squamata - Gekkonidae
e Branco Wall Gecko

7941 ***Tarentola caboverdianus***
Squamata - Gekkonidae
e Cape Verde Wall Gecko

7942 ***Tarentola clypyeata***
Squamata - Gekkonidae
e Gray's Wall Gecko

7943 ***Tarentola darwini***
Squamata - Gekkonidae
e Darwin's Wall Gecko

7944 ***Tarentola delalandii***
Squamata - Gekkonidae
e Tenerife Wall Gecko
d Tenerife-Mauergecko

7945 ***Tarentola deserti***
Squamata - Gekkonidae
e Desert Wall Gecko; Moorish Desert Gecko; Desert Gecko
d Wüstenmauergecko
f Tarante du désert

7946 ***Tarentola ephippiata***
Squamata - Gekkonidae
e African Wall Gecko
f Tarente du Hoggar

7947 ***Tarentola ephippiata hoggarensis***
Squamata - Gekkonidae
e Hoggar Gecko
f Tarente de Hoggar

7948 ***Tarentola gomerensis***
Squamata - Gekkonidae
e Gomera Wall Gecko
d Gomera-Mauergecko; Gomera-Gecko

7949 ***Tarentola mauritanica***
Squamata - Gekkonidae
e Common Wall Gecko; Common Gecko; Wall Gecko; Moorish Gecko; Crocodile Gecko; Moorish Wall Gecko (NA)
d Mauergecko
f Tarente de Mauritanie; Gecko des murs; Gecko des murailles; Tarente des murailles; Gecko de Mauritanie; Tarente; Tarente du Midi; Tarente commune
i Tarantola mauritanica; Geco comune; Geco muraiolo; Tarentola muraiola; Geco

7950 ***Tarentola mauritanica juliae***
Squamata - Gekkonidae
e Moorish Gecko
f Tarente commune

7951 ***Tarentola mauritanica mauritanica***
Squamata - Gekkonidae
e Moorish Gecko
f Tarente commune

7952 ***Tarentola neglecta***
Squamata - Gekkonidae
e Algerian Wall Gecko; Neglected Gecko
f Tarente négligée; Tarente dédaignée

7953 ***Tarentola parvicarinata***
Squamata - Gekkonidae
e Sierra Leone Wall Gecko

7954 **Teiidae**
Squamata
e Tegus; Whiptails (NA); Racerunners
d Schienenechsen; Tejus; Macroteteiide Echsen
f Téjus; Téiidés

7955 ***Teius***
Squamata - Teiidae
e Four-toed Tegus

7956 ***Teius teyou***
Squamata - Teiidae
e Four-toed Tegu; Lagarija Verde (NA)
d Vierzehenameive

7957 ***Telescopus***
Squamata - Colubridae
e Tiger Snakes; Cat Snakes
d Katzennattern
f Couleuvres-chat

7958 ***Telescopus beetzi***
Squamata - Colubridae
e Namib Tiger Snake

7959 ***Telescopus dhara***
Squamata - Colubridae
e Arabian Tiger Snake; Large-eyed Snake; Israeli Cat Snake
d Israelische Katzennatter

f Couleuvre-chat d'Israël

7960 ***Telescopus dhara obtusus***
Squamata - Colubridae
e North African Cat Snake
f Couleuvre-chat d'Afrique du Nord

7961 ***Telescopus fallax***
Squamata - Colubridae
e Southern Tiger Snake; Mediterranean Cat Snake; European Cat Snake; Cat Snake
d Europäische Katzennatter
f Couleuvre-chat d'Europe; Couleuvre-chat; Telescopus
i Serpente gatto; Colubro gatto; Telescopo vivace; Telescopo europeo; Colubro carsico

7962 ***Telescopus guentheri***
Squamata - Colubridae
e Iberian Tiger Snake; Southern Large-eyed Snake

7963 ***Telescopus iberus***
Squamata - Colubridae
e Indian Desert Tiger Snake

7964 ***Telescopus rhynopoma***
Squamata - Colubridae
e Eastern Tiger Snake; Indian Desert Cat Snake

7965 ***Telescopus semiannulatus***
Squamata - Colubridae
e Soosan Tiger Snake; Tiger Cat Snake; Eastern Tiger Snake; Tiger Snake (CSA)
d Getigerte Katzennatter; Tigerschlange
f Serpent tigre

7966 ***Telescopus variegatus***
Squamata - Colubridae
e Variegated Tiger Snake

7967 ***Tenuidactylus***
Squamata - Gekkonidae
e Grasping Gecko
d Nacktfingergeckos; Bogenfingergeckos

7968 ***Tenuidactylus caspicus***
Squamata - Gekkonidae
e Caspian Grasping Gecko; Caspian Gecko; Caspian Naked-fingered Gecko
d Kaspischer Nacktfingergecko; Kaspischer Bogenfingergecko; Kaspischer Nacktfinger
f Cyrtopodion de la Caspienne
i Geco caspico

7969 ***Tenuidactylus feldschenkoi***
Squamata - Gekkonidae
e Feldtschenko's Grasping Gecko; Turkistan Rock Gecko; Central Asian Naked-fingered Gecko

7970 ***Tenuidactylus longipes***
Squamata - Gekkonidae
e Iranian Grasping Gecko

7971 ***Tenuidactylus turcmenicus***
Squamata - Gekkonidae
e Turkish Grasping Gecko
d Turkmenischer Gecko

7972 ***Tenuidactylus voraginosus***
Squamata - Gekkonidae
e Afghan Grasping Gecko

7973 ***Teratolepis***
Squamata - Gekkonidae
e Viper Geckos; Sind Broad-tailed Geckos; Scaled Geckos (ISC)
d Viperngeckos; Rübenschwanzviperngeckos
f Geckos vipère

7974 ***Teratolepis albofasciata***
Squamata - Gekkonidae

e White-striped Viper Gecko
d Rübenschwanzviperngecko

7975 ***Teratolepis fasciata***
Squamata - Gekkonidae
e Carrot-tail Viper Gecko; Sind Broad-tailed Gecko; Banded Scaled Gecko (ISC); Viper Gecko
d Rübenschwanzviperngecko
f Gecko vipère

7976 ***Teratoscincus***
Squamata - Gekkonidae
e Wonder Geckos; Plate-tailed Geckos
d Wundergecko

7977 ***Teratoscincus bedriagai***
Squamata - Gekkonidae
e Bédriaga's Wonder Gecko

7978 ***Teratoscincus microlepis***
Squamata - Gekkonidae
e Small-scaled Wonder Gecko; Baluch Plate-tailed Gecko
d Zwergwundergecko

7979 ***Teratoscincus przewalskii***
Squamata - Gekkonidae
e Przewalski's Wonder Gecko
d Przewalskis Wundergecko

7980 ***Teratoscincus scincus***
Squamata - Gekkonidae
e Wonder Gecko; Turkestan Plate-tailed Gecko; Sand Gecko; Frog-eyed Gecko
d Wundergecko; Sandgecko
f Scinque monstrueux

7981 ***Teratoscincus scincus keyserlingii***
Squamata - Gekkonidae
d Großer Wundergecko

7982 ***Teratoscincus tripolitanus***
Squamata - Gekkonidae
e Common Wonder Gecko

7983 ***Tereturus***
Squamata - Uropeltidae
e Purple-red Earth Snakes

7984 ***Tereturus sanguineus***
Squamata - Uropeltidae
e Purple-red Earth Snake

7985 ***Terrapene***
Testudines - Emydidae
e Box Turtles; American Box Turtles
d Dosenschildkröten
f Tortues-boîte; Tortue-tabatières
i Tartarughe scatola

7986 ***Terrapene carolina***
Testudines - Emydidae
e Box Turtle; Common Box Turtle (NA); Eastern Box Turtle; American Box Turtle
d Carolina-Dosenschildkröte
f Tortue-boîte de Caroline; Tortue-boîte orientale; Tortue-boîte commune; Tortue tabatière; Tortue boîte de l'Est
i Tartaruga scatola americana; Terrapene a scatola comune

7987 ***Terrapene carolina bauri***
Testudines - Emydidae
e Florida Box Turtle
f Tortue-boîte de Floride

7988 ***Terrapene carolina carolina***
Testudines - Emydidae
e Eastern Box Turtle
d Carolina-Dosenschildkröte
f Tortue-boîte commune

7989 ***Terrapene carolina major***
Testudines - Emydidae
e Gulf Coast Box Turtle
f Tortue-boîte du Golfe

7990 ***Terrapene carolina mexicana***
Testudines - Emydidae
e Mexican Box Turtle
f Tortue-boîte du Mexique

7991 ***Terrapene carolina triunguis***
Testudines - Emydidae
e Three-toed Box Turtle
f Tortue-boîte à trois griffes
i Tartaruga scatola triungulata

7992 ***Terrapene carolina yucatana***
Testudines - Emydidae
e Yucatan Box Turtle
f Tortue-boîte du Yucatan

7993 ***Terrapene coahuila***
Testudines - Emydidae
e Coahuilan Box Turtle; Coahuila Box Turtle; Aquatic Box Turtle
d Wasserdosenschildkröte
f Tortue-boîte de Coahuila; Tortue de Mexique
i Terrapene di Coahula

7994 ***Terrapene nelsoni***
Testudines - Emydidae
e Spotted Box Turtle; Nelson's Box Turtle; Southern Spotted Box Turtle; Mexican Black Box Turtle
d Nelson-Dosenschildkröte
f Tortue-boîte ponctuée; Tortue-boîte de Nelson

7995 ***Terrapene ornata***
Testudines - Emydidae
e Ornate Box Turtle; Western Box Turtle
d Schmuckdosenschildkröte
f Tortue-boîte ornée
i Tartaruga scatola ornata

7996 ***Terrapene ornata luteola***
Testudines - Emydidae
e Desert Box Turtle

7997 ***Terrapene ornata ornata***
Testudines - Emydidae
e Ornate Box Turtle

7998 **Testudines**
e Turtles and Tortoises
d Wasser-und Landschildkröten
f Tortues
i Tartarughe e Testuggine

7999 **Testudinidae**
Testudines
e Tortoises; Terrestrial Turtles (NA)
d Landschildkröten
f Tortues terrestres; Tortues vraies; Testudinidés
i Testudini

8000 ***Testudo***
Testudines - Testudinidae
e Palearctic Tortoises
d Landschildkröten; Schildkröten; Eigentliche Landschildkröten
f Tortue terrestres

8001 ***Testudo graeca***
Testudines - Testudinidae
e Mediterranean Spur-thighed Tortoise; Spur-thighed Tortoise; Spurred Tortoise; Greek Tortoise
d Maurische Landschildkröte
f Tortue grecque; Tortue mauresque; Tortue turque
i Testuggine moresca; Testuggine greca

8002 ***Testudo hermanni***
Testudines - Testudinidae
e Hermann's Tortoise; Greek Tortoise
d Griechische Landschildkröte; Landschildkròte
f Tortue d'Hermann
i Testuggine comune; Testuggine di Hermann; Tartaruga

8003 ***Testudo hermanni hermanni***
Testudines - Testudinidae
e Western Hermann's Tortoise
f Tortue d'Hermann occidentale

8004 ***Testudo hermanni noettgeri***
Testudines - Testudinidae
e Eastern Hermann's Tortoise
f Tortue d'Hermann orientale

8005 ***Testudo hermanni robertmertensi***
Testudines - Testudinidae
e Robert Mertens's Tortoise
d Robert-Mertens-Landschildkröte

8006 ***Testudo horsfieldii***
Testudines - Testudinidae
e Central Asia Tortoise; Horsfield's Tortoise; Russian Tortoise; Afghan Tortoise
d Vierzehenlandschildkröte; Steppenschildkröte
f Tortue d'Horsfield; Tortue des steppes
i Testuggine di Horsfield; Testuggine delle steppe

8007 ***Testudo ibera***
Testudines - Testudinidae
e Asia Minor Spur-thighed Tortoise; Algerian Tortoise
d Maurische Landschildkröte

8008 ***Testudo kleinmanni***
Testudines - Testudinidae
e Egyptian Tortoise; Kleinmann's Tortoise; Leith's Tortoise
d Ägyptische Landschildkröte
f Tortue de Kleinmann; Tortue d'Égypte
i Testuggine egiziana

8009 ***Testudo marginata***
Testudines - Testudinidae
e Marginated Tortoise
d Griechische Breitrandschildkröte; Breitrandschildkröte
f Tortue bordée
i Testuggine marginata

8010 ***Testudo terrestris***
Testudines - Testudinidae
e Middle Eastern Spur-thighed Tortoise

8011 ***Testudo weissingeri***
Testudines - Testudinidae
e Peloponnese Dwarf Tortoise
f Tortue naine du Péloponnèse

8012 ***Tetradactylus***
Squamata - Cordylidae
e Whip Lizards; Plated Snake Lizards (CSA); Seps (CSA)
d Geißelschildechsen
f Tétradactyles

8013 ***Tetradactylus africanus***
Squamata - Cordylidae
e African Whip Lizard; African Long-tailed Seps (CSA)
d Große Geißelschildechse
f Tétradactyle africain

8014 ***Tetradactylus breyeri***
Squamata - Cordylidae
e Breyer's Whip Lizard; Breyer's Long-tailed Seps (CSA); Breyer's Plated Lizard (CSA)
d Breyers Geißelschildechse
f Tétradactyle de Breyer

8015 ***Tetradactylus eastwoodae***
Squamata - Cordylidae
e Eastwood's Whip Lizard; Eastwood's Long-tailed Seps (CSA); Eastwood's Plated Lizard (CSA)
d Eastwoods Geißelschildechse
f Tétradactyle de Eastwood

8016 ***Tetradactylus ellenbergeri***
Squamata - Cordylidae
e Ellenberger's Whip Lizard; Ellenberger's Long-tailed Seps (CSA)
d Ellenbergers Geißelschildechse
f Tétradactyle d'Ellenberger

8017 ***Tetradactylus fitzsimonsi***
Squamata - Cordylidae
e Fitzsimons's Whip Lizard; Congo Long-tailed Seps
f Tétradactyle de Fitzsimons

8018 ***Tetradactylus seps***
Squamata - Cordylidae
e Five-toed Whip Lizard; Short-legged Seps (CSA); Short-legged Plated Lizard
d Fünfzehige Geißelschildechse
f Tétradactyle à cinque doigts

8019 ***Tetradactylus tetradactylus***
Squamata - Cordylidae
e Longtail Whip Lizard; Common Long-tailed Seps (CSA); Long-tailed Seps
d Vierzehige Geißelschildechse
f Tétradactyle à quatre doigts

8020 ***Tetralepis***
Squamata - Colubridae
e Java Snakes

8021 ***Tetralepis frustorferi***
Squamata - Colubridae
e Bluebelly Java Snake

8022 ***Teuchocercus***
Squamata - Teiidae
e Key Tegus

8023 ***Teuchocercus keyi***
Squamata - Teiidae
e Key Tegu

8024 ***Thalassophis***
Squamata - Elapidae
e Viperine Sea Snakes

8025 ***Thalassophis anomalus***
Squamata - Elapidae
e Anomolous Sea Snake

8026 ***Thalassophis viperina***
Squamata - Elapidae
e Viperine Sea Snake

8027 ***Thamnodynastes***
Squamata - Colubridae
e Coastal House Snakes; Mock Vipers

8028 ***Thamnodynastes chimanta***
Squamata - Colubridae
e Roze's House Snake

8029 ***Thamnodynastes pallidus***
Squamata - Colubridae
e Amazon Coastal House Snake

8030 ***Thamnodynastes rutilus***
Squamata - Colubridae
e Prado's Coastal House Snake

8031 ***Thamnodynastes strigilis***
Squamata - Colubridae
e Northern Coastal House Snake
d Graue Laubnatter

8032 ***Thamnodynastes strigtus***
Squamata - Colubridae
e Coastal House Snake

8033 ***Thamnophis***
Squamata - Colubridae
e Garter Snakes; Ribbon Snakes; North American Garter Snakes
d Strumpfbandnattern
f Serpents-jarretière et serpents-ruban; Serpents-ruban

i Serpenti giarrettiera

8034 ***Thamnophis atratus***
Squamata - Colubridae
e California Garter Snake; Middle California Garter Snake; Western Coast Garter Snake; Pacific Coast Garter Snake; Black Garter Snake; Western Garter Snake; Western Aquatic Garter Snake; Central Coast Garter Snake; Aquatic Garter Snake; Santa Cruz Garter Snake
d Westliche Strumpfbandnatter

8035 ***Thamnophis atratus aquaticus***
Squamata - Colubridae
e Aquatic Garter Snake; Northern California Coast Garter Snake

8036 ***Thamnophis atratus atratus***
Squamata - Colubridae
e Santa Cruz Garter Snake; Single-striped Garter Snake
d Einstreifige Strumpfbandnatter

8037 ***Thamnophis atratus hydrophilus***
Squamata - Colubridae
e Oregon Garter Snake; Oregon Grey Garter Snake
d Oregon-Strumpfbandnatter

8038 ***Thamnophis atratus zaxanthus***
Squamata - Colubridae
e Diablo Garter Snake; Diablo Range Garter Snake

8039 ***Thamnophis brachystoma***
Squamata - Colubridae
e Shorthead Garter Snake; Short-headed Garter Snake; Small-headed Garter Snake (NA); New York Dwarf Garter Snake
d Kurzkopfstrumpfbandnatter
f Serpent-jarretière à petite tête

8040 ***Thamnophis butleri***
Squamata - Colubridae
e Butler's Garter Snake
d Butlers Strumpfbandnatter
f Serpent-jarretière de Butler; Couleuvre à petite tête

8041 ***Thamnophis chrysocephalus***
Squamata - Colubridae
e Goldenhead Garter Snake

8042 ***Thamnophis couchii***
Squamata - Colubridae
e Western Aquatic Garter Snake; Sierra Garter Snake; Sierra Nevada Garter Snake; Couch's Garter Snake; West Coast Garter Snake; Black Water Snake (NA); Couch's Water Snake; Fish Snake; Mocassin Garter Snake
d Westliche Strumpfbandnatter; Couchs Strumpfbandnatter
f Serpent-jarretière aquatique de l'Ouest

8043 ***Thamnophis couchii aquaticus***
Squamata - Colubridae
e Aquatic Garter Snake

8044 ***Thamnophis couchii couchii***
Squamata - Colubridae
e Western Aquatic Garter Snake

8045 ***Thamnophis cyrtopsis***
Squamata - Colubridae
e Blackneck Garter Snake; Black-necked Garter Snake; White-bellied Garter Snake; Spotted Garter Snake; Brown Garter Snake; Checkerboard Garter Snake; Black-necked Garter Snake; Green-bellied Garter Snake; Reuss's Garter Snake
d Gefleckte Strumpfbandnatter; Schwarznackenstrumpfbandnatter; Mexikanische Strumpfbandnatter
f Serpent-jarretière à cou noir

8046 ***Thamnophis cyrtopsis collaris***
Squamata - Colubridae
e Collared Garter Snake
d Halsbandstrumpfbandnatter

8047 ***Thamnophis cyrtopsis cyrtopsis***
Squamata - Colubridae
e Western Blackneck Garter Snake; Western Black-necked Garter Snake; Western Blackhead Garter Snake
d Westliche Schwarznackenstrumpfbandnatter; Westliche Gefleckte Strumpfbandnatter

8048 ***Thamnophis cyrtopsis ocellatus***
Squamata - Colubridae
e Eastern Blackneck Garter Snake; Eastern Black-necked Garter Snake; Ocellated Garter Snake
d Ocellenstrumpfbandnatter; Östliche Schwarznackenstrumpfbandnatter; Östliche Gefleckte Strumpfbandnatter

8049 ***Thamnophis cyrtopsis postremus***
Squamata - Colubridae
e Tepalcatepec Valley Garter Snake

8050 ***Thamnophis cyrtopsis pulchrilatus***
Squamata - Colubridae
e Mexican Highland Garter Snake

8051 ***Thamnophis elegans***
Squamata - Colubridae
e Western Terrestrial Garter Snake; Terrestial Garter Snake; Elegant Garter Snake; Boyd's Garter Snake; Three-striped Garter Snake
d Bergstrumpfbandnatter; Westliche Strumpfbandnatter
f Serpent-jarretière élégant; Couleuvre de l'Ouest

8052 ***Thamnophis elegans arizonae***
Squamata - Colubridae
e Arizona Garter Snake; Arizona Terrestial Garter Snake; Arizona Wandering Garter Snake

8053 ***Thamnophis elegans biscutatus***
Squamata - Colubridae
e Klamath Garter Snake; Northwestern Garter Snake; Washington Garter Snake

8054 ***Thamnophis elegans elegans***
Squamata - Colubridae
e Mountain Garter Snake
d Westliche Strumpfbandnatter

8055 ***Thamnophis elegans hueyi***
Squamata - Colubridae
e San Pedro Martir Garter Snake

8056 ***Thamnophis elegans nigrecens***
Squamata - Colubridae
e Dusky Garter Snake

8057 ***Thamnophis elegans terrestris***
Squamata - Colubridae
e Coast Garter Snake; Terrestial Garter Snake
d Westliche Strumpfbandnatter; Bergstrumpfbandnatter

8058 ***Thamnophis elegans vagrans***
Squamata - Colubridae
e Wandering Garter Snake; Intermountain Wandering Garter Snake
d Wandernde Strumpfbandnatter; Bergstrumpfbandnatter

8059 ***Thamnophis elegans vascotanneri***
Squamata - Colubridae
e Upper Basin Garter Snake

8060 ***Thamnophis eques***
Squamata - Colubridae
e Mexican Garter Snake; Arizona Garter Snake; Emory's Garter Snake
d Mexikanische Strumpfbandnatter;

Braune Strumpfbandnatter
f Serpent-jarretière mexicain

8061 ***Thamnophis eques eques***
Squamata - Colubridae
e Mexican Garter Snake

8062 ***Thamnophis eques megalops***
Squamata - Colubridae
e Northern Mexican Garter Snake

8063 ***Thamnophis exsul***
Squamata - Colubridae
e Montane Garter Snake

8064 ***Thamnophis fulvus***
Squamata - Colubridae
e Highland Garter Snake; Guatemalan Garter Snake
d Guatemala-Strumpfbandnatter

8065 ***Thamnophis gigas***
Squamata - Colubridae
e Giant Garter Snake
d Riesenstrumpfbandnatter

8066 ***Thamnophis godmani***
Squamata - Colubridae
e Godman's Garter Snake
d Godmans Strumpfbandnatter; Treppenstrumpfbandnatter

8067 ***Thamnophis hammondii***
Squamata - Colubridae
e Two-striped Garter Snake; Southern California Garter Snake; Hammond's Garter Snake; Pacific Garter Snake; Water Snake ; Garter Snake; Ribbon Snake
d Zweistreifenstrumpfbandnatter
i Serpente giarrettiera

8068 ***Thamnophis hammondii digueti***
Squamata - Colubridae
e Diguet Two-striped Garter Snake

8069 ***Thamnophis hammondii hammondii***
Squamata - Colubridae
e Hammond Two-striped Garter Snake

8070 ***Thamnophis marcianus***
Squamata - Colubridae
e Chequered Garter Snake; Checkered Garter Snake (NA); Northern Mexican Garter Snake; Marcy's Garter Snake; Spotted Garter Snake; Desert Garter Snake; Mouse Snake; Marcy Garter Snake
d Karierte Strumpfbandnatter; Texas-Strumpfbandnatter
f Serpent-jarretière réticulé; Serpent-jarretière à échiquier
i Serpente giarrettiera a scacchi

8071 ***Thamnophis marcianus marcianus***
Squamata - Colubridae
e Sonoran Garter Snake; Marcy's Checkered Garter Snake; Checkered Garter Snake (NA)
d Gefleckte Strumpfbandnatter

8072 ***Thamnophis melanogaster***
Squamata - Colubridae
e Blackbelly Garter Snake; Mexican Blackbelly Water Snake
d Schwarzbauchstrumpfbandnatter; Mexikanische Strumpfbandnatter

8073 ***Thamnophis mendax***
Squamata - Colubridae
e Tamaulipan Montane Garter Snake

8074 ***Thamnophis ordinoides***
Squamata - Colubridae
e Northwestern Garter Snake; Gray Garter Snake (NA); Grey Garter Snake; Northwest Garter Snake; Puget Garter Snake; Red-striped Garter Snake; Black Garter Snake; Pacific Coast Garter Snake; Western Garter Snake
d Nordwestliche Strumpfbandnatter

f Serpent-jarretière du Nord-ouest

8075 ***Thamnophis proximus***
Squamata - Colubridae
e Western Ribbon Snake; Long's Garter Snake
d Westliche Streifenstrumpfbandnatter; Westliche Bändernatter
f Serpent-ruban de l'Ouest

8076 ***Thamnophis proximus diabolicus***
Squamata - Colubridae
e Arid Land Ribbon Snake

8077 ***Thamnophis proximus orarius***
Squamata - Colubridae
e Gulf Coast Ribbon Snake

8078 ***Thamnophis proximus proximus***
Squamata - Colubridae
e Western Ribbon Snake; Orange-striped Ribbon Snake

8079 ***Thamnophis proximus rubrilineatus***
Squamata - Colubridae
e Red-striped Ribbon Snake; Redstripe Ribbon Snake
d Rotstreifenbändernatter

8080 ***Thamnophis radix***
Squamata - Colubridae
e Plains Garter Snake; Prairie Garter Snake; Transition Garter Snake; Red-striped Garter Snake
d Tiefland Strumpfbandnatter; Prairie-Strumpfbandnatter; Südwestliche Strumpfbandnatter
f Serpent-jarretière des plaines; Couleuvre des plaines; Serpent-jarretière
i Serpente giarrettiera di pianuara

8081 ***Thamnophis radix haydeni***
Squamata - Colubridae
e Western Plains Garter Snake

8082 ***Thamnophis radix radix***
Squamata - Colubridae
e Eastern Plains Garter Snake

8083 ***Thamnophis rufipunctatus***
Squamata - Colubridae
e Narrowhead Garter Snake; Narrow-headed Garter Snake; Long-nosed Garter Snake; Brown-spotted Garter Snake; Red-spotted Garter Snake; Mexican Highland Garter Snake; Red-spotted Chilopoma
d Schmalköpfige Strumpfbandnatter; Braunfleckenstrumpfbandnatter
f Serpent-jarretière à tête étroite

8084 ***Thamnophis sauritus***
Squamata - Colubridae
e Eastern Ribbon Snake; Riband Snake; North American Ribbon Snake; Slender Garter Snake; Saurite Snake; Slim Garter Snake; Spotted Ribbon Snake; Striped Water Snake; Swift Garter Snake; Swift Streaked Garter Snake; Water Garter Snake; Yellow-headed Garter Snake
d Bändernatter; Östliche Strumpfbandnatter; Östliche Streifenstrumpfbandnatter; Östliche Bändernatter
f Serpent-ruban de l'Est; Serpent-jarretière

8085 ***Thamnophis sauritus nitae***
Squamata - Colubridae
e Bluestripe Ribbon Snake; Blue-striped Ribbon Snake
d Blaugestreifte Bändernatter
f Couleuvre mince

8086 ***Thamnophis sauritus sackeni***
Squamata - Colubridae
e Peninsula Ribbon Snake; Southern Ribbon Snake (NA)
d Florida-Bändernatter

8087 ***Thamnophis sauritus sauritus***
Squamata - Colubridae
e Eastern Ribbon Snake; Common Ribbon Snake
d Östliche Bändernatter

8088 ***Thamnophis sauritus septentrionalis***
Squamata - Colubridae
e Northern Ribbon Snake
d Nördliche Bändernatter

8089 ***Thamnophis scalaris***
Squamata - Colubridae
e Longtail Alpine Garter Snake; Mexican Garter Snake; Highland Garter Snake
d Treppenstrumpfbandnatter

8090 ***Thamnophis scaliger***
Squamata - Colubridae
e Short-tail Alpine Garter Snake; Mexican Highland Garter Snake
d Hochlandstrumpfbandnatter

8091 ***Thamnophis sirtalis***
Squamata - Colubridae
e Striped Garter Snake; Spotted Garter Snake (NA); North American Garter Snake; Sirtal Snake; Spotted Snake; Streaked Snake; Striped Grass Snake; Striped Adder; Eastern Garter Snake; Common Garter Snake
d Gewöhnliche Strumpfbandnatter; Strumpfbandnatter; Rotstreifenstrumpfbandnatter; Gemeine Strumpfbandnatter
f Serpent-jarretière rayé; Couleuvre rayé (Qué)
i Serpente giarrettiera comune; Serpente giarrettiera

8092 ***Thamnophis sirtalis annectens***
Squamata - Colubridae
e Striped Garter Snake; Texas Garter Snake
d Texanische Strumpfbandnatter

8093 ***Thamnophis sirtalis concinnus***
Squamata - Colubridae
e Red-spotted Garter Snake; Northwestern Garter Snake; Red-barred Garter Snake; Pacific Garter Snake; One-striped Garter Snake

8094 ***Thamnophis sirtalis dorsalis***
Squamata - Colubridae
e New Mexico Garter Snake; New Mexican Garter Snake

8095 ***Thamnophis sirtalis fitchi***
Squamata - Colubridae
e Valley Garter Snake; Red-barred Garter Snake; California Garter Snake; Cascade Garter Snake; Borthwestern Garter Snake; Pacific Garter Snake

8096 ***Thamnophis sirtalis infernalis***
Squamata - Colubridae
e California Red-sided Garter Snake; San Francisco Garter Snake; One-striped Garter Snake

8097 ***Thamnophis sirtalis lowei***
Squamata - Colubridae
e Chihuaha Garter Snake

8098 ***Thamnophis sirtalis pallidulus***
Squamata - Colubridae
e Maritime Garter Snake; Quebec Garter Snake; Spotted Garter Snake

8099 ***Thamnophis sirtalis parietalis***
Squamata - Colubridae
e Red-sided Garter Snake; Green-striped Garter Snake; Parietal Garter Snake; Red Garter Snake; Say's Garter Snake; Western Garter Snake; Prairie Garter Snake; Rocky Mountain Garter Snake
d Rotgefleckte Strumpfbandnatter; Rotseitige Strumpfbandnatter; Rotseitenstrumpfbandnatter
f Serpent-jarretière à flancs rouges

8100 ***Thamnophis sirtalis pickeringii***
Squamata - Colubridae
e Puget Sound Garter Snake; Pickering's Garter Snake; Nisqually Garter Snake; Northwestern Garter Snake

8101 ***Thamnophis sirtalis semifasciatus***
Squamata - Colubridae
e Chicago Garter Snake
d Chicago-Strumpfbandnatter

8102 ***Thamnophis sirtalis similis***
Squamata - Colubridae
e Bluestripe Garter Snake; Florida Blue-striped Garter Snake; Blue-striped Garter Snake; Florida Garter Snake; Eastern Garter Snake
d Blaue Strumpfbandnatter; Florida-Strumpfbandnatter; Blaustreifen Florida-Strumpfbandnatter

8103 ***Thamnophis sirtalis sirtalis***
Squamata - Colubridae
e Eastern Garter Snake; Slow Garter Snake; Yellow Snake; Spotted Garter Snake; Ordinary Garter Snake; Common Garter Snake
d Gewöhnliche Strumpfbandnatter; Östliche Strumpfbandnatter
f Serpent-jarretière
i Serpente giarettiera comune

8104 ***Thamnophis sirtalis tetrataenia***
Squamata - Colubridae
e San Francisco Garter Snake

8105 ***Thamnophis sumichrasti***
Squamata - Colubridae
e Sumichrast's Garter Snake

8106 ***Thamnophis vicinus***
Squamata - Colubridae
e Smith's Garter Snake

8107 ***Thecadactylus***
Squamata - Gekkonidae
e Turniptail Geckos; Smooth Geckos; Turnip-tailed Geckos
d Rübenschwanzgeckos
f Thécadactyles

8108 ***Thecadactylus rapicauda***
Squamata - Gekkonidae
e Turniptail Gecko; Turnip-tailed Gecko; Smooth Gecko; Giant Woodslave ; Thick-tailed Gecko (NA)
d Rübenschwanzgecko
f Thécadactyle

8109 ***Thelotornis***
Squamata - Colubridae
e Bird Snakes; Vine Snakes; Twig Snakes; Kirtland's Snakes
d Vogelnattern; Baumnattern
i Serpenti degli uccelli

8110 ***Thelotornis capensis***
Squamata - Colubridae
e Bird Snake; Twig Snake; Savanna Twig Snake; Southern Vine Snake (CSA); Savanna Vine Snake (CSA); Vine Snake (CSA)
d Kap Baumnatter; Südafrikanische Baumnatter

8111 ***Thelotornis kirtlandi***
Squamata - Colubridae
e Northern Vine Snake; Twig Snake; Forest Twig Snake; Northern Vine Snake; Kirtland's Tree Snake; Bird Snake; Voelslang (CSA)
d Vogelnatter; Graue Baumnatter; Lianennatter
i Serpente degli uccelli

8112 ***Thermophis***
Squamata - Colubridae
e Bailey's Snakes

8113 ***Thermophis baileyi***
Squamata - Colubridae

e Bailey's Snake

8114 ***Thrasops***
Squamata - Colubridae
e Black Tree Snakes
d Schwarze Baumschlangen

8115 ***Thrasops aethiopissa***
Squamata - Colubridae
e Black Tree Snake

8116 ***Thrasops jacksoni***
Squamata - Colubridae
e Jackson's Black Tree Snake; Black Tree Snake
d Schwarze Baumschlange

8117 ***Thrasops occidentalis***
Squamata - Colubridae
e Western Black Tree Snake; Parker's Black Tree Snake

8118 ***Tiliqua***
Squamata - Scincidae
e Bluetongue Skinks; Blue-tonged Skinks; Tiliquine Skinks; Giant Skinks
d Blauzungenskinke; Riesenskinkverwandte; Blauzungen
f Scinques géantes; Scinques à langue bleue
i Tilique

8119 ***Tiliqua adelaidensis***
Squamata - Scincidae
e Adelaide Scrub Skink

8120 ***Tiliqua branchialis***
Squamata - Scincidae
e Günther's Bluetongue Skink

8121 ***Tiliqua casuarinae***
Squamata - Scincidae
e Scrub Skink; Oak Skink (ANZ); She-oak Skink
f Scinque de la Casuarine

8122 ***Tiliqua gerrardii***
Squamata - Scincidae
e Pinktongue Skink; Pink-tongued Lizard (ANZ); Blue-tongued Skink; Pink-tongued Skink
d Rosazungenskink; Schneckenskink; Schneckenblauzungenskink; Gerrards Skink; Gerrards Blauzunge
f Scinque à langue rose

8123 ***Tiliqua gigas***
Squamata - Scincidae
e Giant Bluetongue Skink; Giant Blue-tongued Skink
d Riesenblauzungenskink; Riesenskink; Oaouanische Blauzunge

8124 ***Tiliqua maxima***
Squamata - Scincidae
e Storr's Scrub Skink

8125 ***Tiliqua multifaciata***
Squamata - Scincidae
e Central Bluetongue Skink; Centralian Blue-tongued Lizard (ANZ)
d Nördlicher Blauzungenskink

8126 ***Tiliqua nigrolutea***
Squamata - Scincidae
e Blotched Bluetongue Skink; Blotched Blue-tongued Lizard (ANZ)
d Gefleckter Blauzungenskink; Schwarzgelbe Glattechse; Blauzunge
f Scinque noir-jaune

8127 ***Tiliqua occipitalis***
Squamata - Scincidae
e Western Bluetongue Skink; Western Blue-tongued Lizard (ANZ); Western Blue-tongued Skink (ANZ)
d Westliche Blauzunge
f Scinque à la langue bleue

8128 ***Tiliqua rugosa***
Squamata - Scincidae
e Shingleback Skink; Shingleback (ANZ); Sleepy Lizard; Stump-tailed Skink; Rugose Stump-tail
d Tannenzapfenechse; Stutzechse
f Lézard pomme-de-pin; Trachydosaure rugeux; Scinque à queue tronquée

8129 ***Tiliqua scincoides***
Squamata - Scincidae
e Bluetongue Skink; Blue-tongued Skink; Eastern Blue-tongued Skink; Eastern Blue-tongued Lizard (ANZ); Common Bluetongue; Common Blue-tongued Lizard (ANZ)
d Blauzungenskink; Gewöhnliche Blauzunge; Gemeiner Blauzungenskink
f Scinque d'Australie à langue bleue; Scinque à langue bleue
i Tiliqua della lingua azzurra; Scinco

8130 ***Tiliqua scincoides gigas***
Squamata - Scincidae
d Papuanische Riesenblauzunge

8131 ***Tiliqua scincoides intermedia***
Squamata - Scincidae
d Nördliche Blauzunge

8132 ***Toluca***
Squamata - Colubridae
e Tolucan Ground Snakes

8133 ***Toluca amphistricha***
Squamata - Colubridae
e Twin-spotted Tolucan Ground Snake

8134 ***Toluca conica***
Squamata - Colubridae
e Large-blotched Tolucan Ground Snake

8135 ***Toluca lineata***
Squamata - Colubridae
e Lined Tolucan Ground Snake

8136 ***Toluca megalodon***
Squamata - Colubridae
e San Felipe Ground Snake

8137 ***Tomistoma***
Crocodylia - Crocodylidae
e False Gharials; False Gavials; Tomistomas
d Sunda-Gaviale
f Faux-gavials; Faux-gavials malais

8138 ***Tomistoma schlegeli***
Crocodylia - Crocodylidae
e False Gharial; False Gavial; Malayan Gavial; Malayan Gharial; Malayan False Gharial; Sunda Ghavial; Malay Gavial; Tomistoma
d Sunda-Gavial; Sunda-Krokodil
f Faux-gavial; Faux-gavial malais

8139 ***Tomodon***
Squamata - Colubridae
e Pampas Snakes

8140 ***Tomodon***
Squamata - Colubridae
e Pampas Snake

8141 ***Tomodon dorsatus***
Squamata - Colubridae
e Pampas Snake

8142 ***Tomodon ocellatus***
Squamata - Colubridae
e Ocellated Pampas Snake

8143 ***Tomuropeltis***
Squamata - Amphisbaenidae
e Horntail Worm Lizards

8144 ***Tomuropeltis ellenbergeri***
Squamata - Amphisbaenidae

e Lealui Worm Lizard

8145 ***Tomuropeltis gigantea***
Squamata - Amphisbaenidae
e Kwango Worm Lizard

8146 ***Tomuropeltis granti***
Squamata - Amphisbaenidae
e Beira Worm Lizard

8147 ***Tomuropeltis jallae***
Squamata - Amphisbaenidae
e Northern Rhodesia Worm Lizard

8148 ***Tomuropeltis luluae***
Squamata - Amphisbaenidae
e Sandoa Worm Lizard

8149 ***Tomuropeltis pistillum***
Squamata - Amphisbaenidae
e Horntail Worm Lizard; Blunt-tailed Worm Lizard (CSA); Square-tailed Wedge-snouted Amphisbaenian (CSA); Horny-tailed Worm Lizard (CSA)

8150 ***Toxicocalamus***
Squamata - Elapidae
e Venomous Island Snakes

8151 ***Toxicocalamus buergersi***
Squamata - Elapidae
e Toricelli Tropical Snake

8152 ***Toxicocalamus grandis***
Squamata - Elapidae
e Iran Jaya Tropical Snake

8153 ***Toxicocalamus holopelturus***
Squamata - Elapidae
e Rossel Island Snake

8154 ***Toxicocalamus longissimus***
Squamata - Elapidae
e Woodlark Island Snake

8155 ***Toxicocalamus loriae***
Squamata - Elapidae
e New Guinea Tropical Snake

8156 ***Toxicocalamus misimae***
Squamata - Elapidae
e Misima Island Snake

8157 ***Toxicocalamus preussi***
Squamata - Elapidae
e Northern Tropical Snake

8158 ***Toxicocalamus spilolepidotus***
Squamata - Elapidae
e Highland Tropical Snake

8159 ***Toxicocalamus stanleyanus***
Squamata - Elapidae
e Southern Tropical Snake

8160 ***Tracheloptychus***
Squamata - Cordylidae
e Keeled Cordylids
d Kielschildechsen

8161 ***Tracheloptychus madagascariensis***
Squamata - Cordylidae
e Madagascar Keeled Cordylid
d Madagassische Kielschildechse
f Scinque crocodile

8162 ***Tracheloptychus petersi***
Squamata - Cordylidae
e Peters's Keeled Cordylid
d Peters Kielschildechse

8163 ***Trachemys***
Testudines - Emydidae
e Sliders
d Schmuckschildkröten
f Trachémydes

8164 ***Trachemys decorata***
Testudines - Emydidae
e Hispaniola Slider; Hispaniolan Slider; Haitian Slider; Hispaniolan Elegant Slider
d Kinnfleckpfauenaugenschildkröte; Kinnfleckpfauenaugeschildkröte
f Trachémyde d'Hispaniola

8165 ***Trachemys decussata***
Testudines - Emydidae
e North Antilles Slider; Slider Turtle; North Antillean Slider; Antillean Painted Turtle ; Cuban Slider (WI)
d Kubanische Schmuckschildkröte
f Trachémyde de Cuba
i Tartaruga delle Antille

8166 ***Trachemys dorbigni***
Testudines - Emydidae
e D'Orbigny's Slider; South American Slider; Black-bellied Slider; Brazil Pseudemyd Slider
d Brasilianische Schmuckschildkröte
f Pseudémyde du Brésil; Trachémyde de d'Orbigny

8167 ***Trachemys gaigeae***
Testudines - Emydidae
e Big Bend Slider

8168 ***Trachemys scripta***
Testudines - Emydidae
e Slider; Pond Slider; American Red-eared Terrapin ; Red-eared Turtle; South American Red-lined Turtle; Slider Turtle; American Slider Turtle; Red-eared Slider (WI)
d Buchstabenschmuckschildkröte; Rotwangenschildkröte; Gelbwangenschildkröte
f Tortue à dessins; Tortue d'eau douce américaine; Trachémyde écrite
i Testuggine palustre comune

8169 ***Trachemys scripta callirostris***
Testudines - Emydidae
d Kinnfleckschmuckschildkröte

8170 ***Trachemys scripta elegans***
Testudines - Emydidae
e Red-eared Slider; Texas Red-eared Terrapin (CSA)
d Rotwangenschmuckschildkröte
f Tortue aux oreilles rouges; Tortue de Floride
i Testuggine scritta elegante; Testuggine d'acqua dipinta dalle orecchie rosse; Testuggine dalle orecchie rosse; Tartaruga dalle orecchie rosse; Terrapene dalle orecchie rosse

8171 ***Trachemys scripta hiltoni***
Testudines - Emydidae
e Fuerte Slider

8172 ***Trachemys scripta nebulosa***
Testudines - Emydidae
e Baja California Slider

8173 ***Trachemys scripta ornata***
Testudines - Emydidae
e Mesoamerican Slider
d Pfauenaugenschmuckschildkröte

8174 ***Trachemys scripta scripta***
Testudines - Emydidae
e Yellowbelly Slider; Yellow-bellied Terrapin; Yellow-bellied Slider
d Gelbbauchschildkröte
i Testuggine d'acqua dipinta dalle orecchie gialle

8175 ***Trachemys scripta troostii***
Testudines - Emydidae
e Cumberland Slider
d Cumberland-Zierschildkröte
i Testuggine d'acqua dipinta di Troosti

8176 ***Trachemys stejnegeri***
Testudines - Emydidae
e Central Antillian Slider; Puerto Rican

Slider; Caiman Island Terrapin; Antillean Slider (WI); Dominican Slider (WI)
f Trachémyde de Stejneger

8177 ***Trachemys terrapen***
Testudines - Emydidae
e Slider; Jamaican Slider (WI); Jamaican Terrapin
d Jamaika-Schmuckschildkröte; Antillen-Schmuckschildkröte
f Trachémyde de la Jamaïque

8178 ***Trachischium***
Squamata - Colubridae
e Worm-eating Snakes; Roughsides
d Orientalische Wurmnattern

8179 ***Trachischium fuscum***
Squamata - Colubridae
e Blackbelly Worm-eating Snake; Black-bellied Roughside; Black-bellied Worm-eating Snake; Black-bellied Vermiphage

8180 ***Trachischium guentheri***
Squamata - Colubridae
e Rosebelly Worm-eating Snake; Rose-bellied Roughside

8181 ***Trachischium monticola***
Squamata - Colubridae
e Mountain Worm-eating Snake

8182 ***Trachischium tenuiceps***
Squamata - Colubridae
e Yellowbelly Worm-eating Snake; Canary-bellied Worm-eating Snake; Yellow-bellied Roughside

8183 ***Trachyboa***
Squamata - Boidae
e Eyelash Boas; Rough Boas
d Rauhboas; Rauhschuppenboas

8184 ***Trachyboa boulengeri***
Squamata - Boidae
e Northern Eyelash Boa; Eyelash Boa; Rough Boa
d Rauhboa; Rauhschuppenboa

8185 ***Trachyboa gularis***
Squamata - Boidae
e Southern Eyelash Boa

8186 ***Trapelus***
Squamata - Agamidae
e Middle-Eastern Agamas; Steppes Agamas
d Agamen
f Trapelus

8187 ***Trapelus agilis***
Squamata - Agamidae
e Brilliant Agama
d Schlankagame
f Agame agile

8188 ***Trapelus blanfordi***
Squamata - Agamidae
e Anderson's Agama; Blanford's Agama; Persian Agama
d Blanfords Agame

8189 ***Trapelus flavimaculatus***
Squamata - Agamidae
e Yellow-spotted Agama

8190 ***Trapelus isolepis***
Squamata - Agamidae
e Similar-scaled Agama

8191 ***Trapelus jaykari***
Squamata - Agamidae
e Oman Agama

8192 ***Trapelus megalonyx***
Squamata - Agamidae
e Afghan Ground Agama

8193 ***Trapelus mutabilis***
Squamata - Agamidae
e Desert Agama; Changeable Agama; Pale Agama; Variable Agamid
d Wüstenagame
f Agame changeant; Agame variable; Agame inerme; Agame du désert

8194 ***Trapelus persicus***
Squamata - Agamidae
e Persia Agama

8195 ***Trapelus rubrigularis***
Squamata - Agamidae
e Red-throated Agama

8196 ***Trapelus ruderatus***
Squamata - Agamidae
e Horn-scaled Agama; Horny-scaled Agama; Oliver's Agama; Baluch Ground Agama
d Syrische Agame; Ruinenagame; Schuttagame; Dornschuppenhautagame
f Agame à peau rugueuse; Agame des ruines
i Agama ruderale; Agama dei ruderi

8197 ***Trapelus sanguinolentus***
Squamata - Agamidae
e Steppe Agama
d Steppenagame; Asiatische Steppenagame
f Agame des steppes
i Agama sanguinolenta; Agama delle steppe

8198 ***Trapelus tournevillei***
Squamata - Agamidae
e Sahara Agama; Erg Agama; Tourneville's Agama
d Dünenagame; Gescheckte Agame
f Agame de Tourneville; Agame des sables

8199 ***Tretanorhinus***
Squamata - Colubridae
e Striped Swamp Snakes
d Sumpfnattern; Karibische Wasserschlangen

8200 ***Tretanorhinus mocquardi***
Squamata - Colubridae
e Mocquard's Swamp Snake; Panamanian Swamp Snake

8201 ***Tretanorhinus nigroluteus***
Squamata - Colubridae
e Orangebelly Swamp Snake; Middle American Swamp Snake

8202 ***Tretanorhinus taeniatus***
Squamata - Colubridae
e Striped Swamp Snake

8203 ***Tretanorhinus variabilis***
Squamata - Colubridae
e Variable Swamp Snake; Caribbean Water Snake; Cuban Water Snake (WI)
d Karibische Wasserschlange

8204 ***Tretanorhinus variabilis lewisi***
Squamata - Colubridae
e Cuba Swamp Snake
d Kuba-Wasserenatter

8205 ***Tretioscincus***
Squamata - Gymnophthalmidae
e Smooth Tegus
d Skinktejus

8206 ***Tretioscincus agilis***
Squamata - Gymnophthalmidae
e Smooth Tegu
d Skinkteju

8207 ***Tretioscincus bifasciatus***
Squamata - Gymnophthalmidae
e Rio Magdalena Tegu

8208 ***Tribolonotus***
Squamata - Scincidae
e Helmet Skinks; Spiny-tailed Skinks; Casque-headed Skinks; Tribolonotes
d Helmskinke; Buschkrokodile
f Tribolonotes; Scinques à tête casquée

8209 ***Tribolonotus annectens***
Squamata - Scincidae
e Zweifel's Helmet Skink

8210 ***Tribolonotus blanchardi***
Squamata - Scincidae
e Blanchard's Helmet Skink

8211 ***Tribolonotus brongersmai***
Squamata - Scincidae
e Brongersma's Helmet Skink; Brongersma's Casque-headed Skink

8212 ***Tribolonotus gracilis***
Squamata - Scincidae
d Krokodilskink
f Scinque à tête casquée; Scinque crocodile aux yeux oranges

8213 ***Tribolonotus novaeguinea***
Squamata - Scincidae
e New Guinea Helmet Skink; Casque-headed Skink
d Neuguinea-Helmskink; Buschkrokodil
f Tribolonote de la Nouvelle Guinée

8214 ***Tribolonotus ponceletei***
Squamata - Scincidae
e Poncelet's Helmet Skink

8215 ***Tribolonotus pseudoponceleti***
Squamata - Scincidae
e False Poncelet's Helmet Skink

8216 ***Tribolonotus schmidti***
Squamata - Scincidae
e Schmidt's Helmet Skink
d Schmidts Helmskink
f Tribolonote de Schmidt

8217 ***Trimeresurus***
Squamata - Viperidae
e Asian Pit Vipers; Asiatic Lancehead Snakes; Asian Lancehead Snakes
d Lanzenottern; Asiatische Lanzenottern; Bambusottern
f Fer-de-lances asiatiques
i Trimeresuri

8218 ***Trimeresurus albolabris***
Squamata - Viperidae
e White-lipped Pit Viper; White-ranged Green Pit Viper; Bamboo Viper; White-lipped Tree Viper; Seven-step Snake
d Weißlippenbambusotter; Weißlippenlanzenotter
i Trimeresuro labra bianche; Vipera del bamboo

8219 ***Trimeresurus borneensis***
Squamata - Viperidae
e Borneo Pit Viper; Bornean Pit Viper; Bornean Flat-nosed Viper
d Borneo-Bambusotter

8220 ***Trimeresurus brongersmai***
Squamata - Viperidae
e Brongersma's Pit Viper

8221 ***Trimeresurus cantori***
Squamata - Viperidae
e Cantor's Pit Viper

8222 ***Trimeresurus cornutus***
Squamata - Viperidae
e Horned Pit Viper; Fan-si-pan Horned Pit Viper

8223 ***Trimeresurus elegans***
Squamata - Viperidae
e Elegant Pit Viper; Sakishima Habu

8224 ***Trimeresurus erythrurus***
Squamata - Viperidae
e Redtail Pit Viper; Red-tailed Pit Viper; Bengal Pit Viper; Bengal Green Lancehead Snake
d Smaragdlanzenotter; Smaragdbambusotter

8225 ***Trimeresurus fasciatus***
Squamata - Viperidae
e Banded Pit Viper

8226 ***Trimeresurus flavomaculatus***
Squamata - Viperidae
e Philippine Pit Viper; Habu
d Philippinische Bambusotter

8227 ***Trimeresurus flavoviridis***
Squamata - Viperidae
e Yellow-spotted Pit Viper; Yellow-spotted Lancehead Snake; Okinawa Habu; Habu
d Habu-Schlange
f Habu

8228 ***Trimeresurus gracilis***
Squamata - Viperidae
e Kikuchi Habu

8229 ***Trimeresurus gramineus***
Squamata - Viperidae
e Indian Green Tree Viper; Common Bamboo Viper (ISC); Bamboo Pit Viper (ISC); Green Pit Viper (ISC)
d Gewöhnliche Bambusotter; Grüne Bambusotter; Bambusotter
f Vipère du bambou; Crotale des bambous

8230 ***Trimeresurus hageni***
Squamata - Viperidae
e Indonesian Pit Viper

8231 ***Trimeresurus huttoni***
Squamata - Viperidae
e Madura Pit Viper

8232 ***Trimeresurus jerdoni***
Squamata - Viperidae
e Oriental Pit Viper

8233 ***Trimeresurus kanburiensis***
Squamata - Viperidae
e Kanburi Pit Viper; Kanchanaburi Pit Viper
d Kanchanaburi-Bambusotter; Kanchanaburi-Lanzenotter

8234 ***Trimeresurus kaulbacki***
Squamata - Viperidae
e Burmese Pit Viper

8235 ***Trimeresurus labialis***
Squamata - Viperidae
e Island Pit Viper

8236 ***Trimeresurus macrolepis***
Squamata - Viperidae
e Large-scaled Pit Viper

8237 ***Trimeresurus macrops***
Squamata - Viperidae
e Kramer's Pit Viper; Large-eyed Pit Viper
d Großäugige Bambusotter

8238 ***Trimeresurus malabaricus***
Squamata - Viperidae
e Malabarian Pit Viper; Malabar Pit Viper (ISC)

8239 ***Trimeresurus mangshanensis***
Squamata - Viperidae
e Mangshanen Pit Viper

8240 ***Trimeresurus medoensis***
Squamata - Viperidae
e Green Bamboo-leaf Pit Viper

8241 ***Trimeresurus mucrosquamatus***
Squamata - Viperidae
e Chinese Habu; Pointed-scaled Pit

Viper (ISC); Turtle-designed Snake; Taiwan Habu
d Taiwanesische Habuschlange; Glattschuppige Lanzenotter

8242 ***Trimeresurus occidentalis***
Squamata - Viperidae
e Green Pit Viper

8243 ***Trimeresurus popeorum***
Squamata - Viperidae
e Pope's Tree Viper; Pope's Pit Viper; Pope's Lancehead Viper; Pope's Bamboo Snake
d Popes Lanzenotter; Pope-Lanzenotter; Pope-Bambusotter

8244 ***Trimeresurus puniceus***
Squamata - Viperidae
e Ashy Pit Viper; Flat-nosed Pit Viper
d Java-Palmenotter; Braune Bambusotter

8245 ***Trimeresurus purpureomaculatus***
Squamata - Viperidae
e Mangrove Viper; Mangrove Pit Viper; Shore Pit Viper (ISC); Gray's Shore Pit Viper; Purple-spotted Pit Viper
d Mangrovenotter; Mangrovenviper; Mangrovengrubenotter

8246 ***Trimeresurus schultzei***
Squamata - Viperidae
e Schultz's Pit Viper; Schultze's Pit Viper

8247 ***Trimeresurus stejnegeri***
Squamata - Viperidae
e Chinese Green Tree Viper; Stejneger's Pit Viper; Taiwan Bamboo Pit Viper; Ao Habu
d Chinesische Bambusotter; Chinesische Baumviper

8248 ***Trimeresurus strigatus***
Squamata - Viperidae
e Gray's Pit Viper; Horseshoe Pit Viper

8249 ***Trimeresurus sumatranaus***
Squamata - Viperidae
e Sumatra Pit Viper; Sumatran Pit Viper
d Sumatra-Lanzenotter; Sumatra-Grubenviper
i Trimeresuro di Sumatra

8250 ***Trimeresurus tibetanus***
Squamata - Viperidae
e Tibetan Pit Viper

8251 ***Trimeresurus tokarensis***
Squamata - Viperidae
e Tokaren Pit Viper; Tokara Habu Pit Viper; Tokara Habu
d Tokara-Habugrubenotter

8252 ***Trimeresurus trigonocephalus***
Squamata - Viperidae
e Ceylon Pit Viper
d Ceylon-Bambusotter; Ceylon-Lanzenotter

8253 ***Trimeresurus venustus***
Squamata - Viperidae
e Vogal's Pit Viper
d Schöne Lanzenotter; Bunte Lanzenotter

8254 ***Trimeresurus xiangchengensis***
Squamata - Viperidae
e Szechwan Pit Viper

8255 ***Trimetopon***
Squamata - Colubridae
e Tropical Ground Snakes

8256 ***Trimetopon barbouri***
Squamata - Colubridae
e Barbour's Tropical Ground Snake ; Barbour's Dwarf Snake

8257 ***Trimetopon gracile***
Squamata - Colubridae
e Günther's Tropical Ground Snake; Slender Dwarf Snake

8258 ***Trimetopon pliolepis***
Squamata - Colubridae
e Cope's Tropical Ground Snake; Faded Dwarf Snake

8259 ***Trimetopon simile***
Squamata - Colubridae
e Dunn's Tropical Ground Snake; White-bellied Dwarf Snake

8260 ***Trimetopon slevini***
Squamata - Colubridae
e Slevin's Tropical Ground Snake; Slevin's Dwarf Snake

8261 ***Trimetopon viquezi***
Squamata - Colubridae
e Viquez's Tropical Ground Snake; Viquez's Dwarf Snake

8262 ***Trimorphodon***
Squamata - Colubridae
e Lyre Snakes
d Lyraschlangen
f Couleuvres-lyres

8263 ***Trimorphodon biscutatus***
Squamata - Colubridae
e Lyre Snake; Coastal Lyre Snake; Western Lyre Snake
d Küstenlyraschlange
f Couleuvre-lyre

8264 ***Trimorphodon biscutatus lambda***
Squamata - Colubridae
e Sonoran Lyre Snake
d Sonora-Lyraschlange; Sonora-Leierschlange

8265 ***Trimorphodon biscutatus lyrophanes***
Squamata - Colubridae
e Baja California Lyre Snake (NA); Western Lyre Snake; Arizona Lyre Snake (NA); Southern Baja California Lyre Snake; California Lyre Snake

8266 ***Trimorphodon biscutatus vandenburghi***
Squamata - Colubridae
e California Lyre Snake; Lyre Snake (NA); Vandenburgh's Lyre Snake
d Van den Burghs Lyraschlange; Leiernatter; Van den Burghs Leiernatter

8267 ***Trimorphodon biscutatus vilkinsonii***
Squamata - Colubridae
e Texas Lyre Snake; Wilkinson's Lyre Snake
d Wilkinsons Lyrachlange

8268 ***Trimorphodon tau***
Squamata - Colubridae
e Mexican Lyre Snake
d Mexikanische Leiernatter

8269 **Trionychidae**
Testudines
e Softshell Turtles; Softshells
d Echte Weichschildkröten; Holarktische und Paläotropische Weichschildkröten
f Tortues molles; Tortues à carapace molle; Trionychidés
i Tartarughe dal guscio molle

8270 ***Trionyx***
Testudines - Trionychidae
e African Softshells
d Afrikanische Weichschildkröten; Dreiklauenweichschildkröten
f Trionyx; Tortues de l'eau douce

8271 ***Trionyx triunguis***
Testudines - Trionychidae
e African Softshell; African Soft-

shelled Turtle; Nile Soft-shelled Turtle; Nile Soft-shelled Terrapin (CSA); Nile Softshell Turtle; African Softshell Turtle; Three-clawed Turtle

d Afrikanische Dreiklauenweichschildkröte; Afrikanische Weichschildkröte

f Trionyx du Nil; Trionyx d'Afrique

i Trionice triungue; Tartaruga a guscio molle africana

8272 ***Tripanurgos***
Squamata - Colubridae
e Tropical Flat Snakes; Red-eyed Tree Snakes

8273 ***Tripanurgos compressus***
Squamata - Colubridae
e Tropical Flat Snake; Banded Tree Snake; Red-eyed Tree Snake

8274 **Trogonophidae**
Squamata
e Trogonophids
d Spitzschwanzdoppelschleichen; Spitzzahndoppelschleichen
f Trogonophidés

8275 ***Trogonophis***
Squamata - Trogonophidae
e Sharp-tailed Worm Lizards
d Schachbrettdoppelschleichen; Spitzschwanzdoppelschleichen
f Trogonophides; Trogonophis

8276 ***Trogonophis wiegmanni***
Squamata - Trogonophidae
e Chequerboard Worm Lizard; Wiegmann's Worm Lizard; Checkerboard Worm Lizard (NA)
d Schachbrettdoppelschleiche; Spitzschwanzdoppelschleiche
f Trogonophide de Wiegmann; Trogonophis de Wiegmann

8277 ***Trogonophis wiegmanni elegans***
Squamata - Trogonophidae
e Sharp-tailed Worm Lizard
f Trogonophide mauve; Trogonophis mauve; Trogonophis de Wiegmann

8278 ***Trogonophis wiegmanni wiegmanni***
Squamata - Trogonophidae
e Sharp-tailed Worm Lizard
f Trogonophide jaune; Trogonophis jaune; Trogonophis de Wiegmann

8279 ***Tropidechis***
Squamata - Elapidae
e Rough-scaled Snakes

8280 ***Tropidechis carinatus***
Squamata - Elapidae
e Rough-scaled Snake (ANZ); Clarence River Snake

8281 ***Tropidoclonion***
Squamata - Colubridae
e Lined Snakes; Ribbon Snakes
f Serpents lignés

8282 ***Tropidoclonion lineatum***
Squamata - Colubridae
e Lined Snake
d Streifennatter
f Serpent ligné

8283 ***Tropidoclonion lineatum annectens***
Squamata - Colubridae
e Central Lined Snake

8284 ***Tropidoclonion lineatum lineatum***
Squamata - Colubridae
e Northern Lined Snake

8285 ***Tropidoclonion lineatum texanum***
Squamata - Colubridae
e Texas Lined Snake

8286 ***Tropidodipsas***
Squamata - Colubridae

e Snail-eaters

8287 ***Tropidodipsas fasciata***
Squamata - Colubridae
e Banded Snail-sucker; Banded Snail-eater

8288 ***Tropidodipsas fischeri***
Squamata - Colubridae
e Fischer's Snail-eater

8289 ***Tropidodipsas sartorii***
Squamata - Colubridae
e Terrestial Snail-sucker; Ringed Snail-eater

8290 ***Tropidodryas***
Squamata - Colubridae
e Serra Snakes

8291 ***Tropidodryas pseudoserra***
Squamata - Colubridae
e False Serra Snake

8292 ***Tropidodryas serra***
Squamata - Colubridae
e Serra Snake

8293 ***Tropidolaemus***
Squamata - Viperidae
e Wagler's Palm Vipers; Wagler's Pit Vipers
d Tempelottern; Waglers Lanzenottern; Wagler-Lanzenottern
f Vipères de Wagler

8294 ***Tropidolaemus wagleri***
Squamata - Viperidae
e Wagler's Palm Viper; Wagler's Temple Pit Viper; Wagler's Pit Viper
d Waglers Tempelotter; Waglers Lanzentotter; Wagler-Lanzenotter
f Vipère de Wagler

8295 ***Tropidonophis***
Squamata - Colubridae
e Eastern Keelbacks
d Kielrücken

8296 ***Tropidonophis elongata***
Squamata - Colubridae
e Moluccan Keelback

8297 ***Tropidonophis mairii***
Squamata - Colubridae
e Common Keelback; Keelback; Freshwater Snake (ANZ)
d Kielrückenwassernattter

8298 ***Tropidonophis montana***
Squamata - Colubridae
e Lorentz River Keelback

8299 ***Tropidonophis novaeguinea***
Squamata - Colubridae
e New Guinea Keelback

8300 ***Tropidonophis picturata***
Squamata - Colubridae
e Lobo Bay Keelback

8301 ***Tropidonophis punctiventris***
Squamata - Colubridae
e Halmahera Keelback

8302 ***Tropidonophis truncata***
Squamata - Colubridae
e Indonesian Keelback

8303 ***Tropidophis***
Squamata - Boidae
e Dwarf Boas; Wood Snakes; West Indian Boas; Tropes (WI)
d Zwergboas; Erdboas
f Boas terrestres
i Tropidofidi

8304 ***Tropidophis battersbyi***
Squamata - Boidae
e Battersby's Dwarf Boa

8305 ***Tropidophis canus***
Squamata - Boidae
e Great Inagua Island Dwarf Boa; Inagua Trope (WI)
d Bahama-Zwergboa

8306 ***Tropidophis caymanensis***
Squamata - Boidae
e Cayman Islands Dwarf Boa; Grand Cayman Trope (WI)
d Kuba-Zwergboa

8307 ***Tropidophis feicki***
Squamata - Boidae
e Feick's Dwarf Boa

8308 ***Tropidophis greenwayi***
Squamata - Boidae
e Ambergris Cay Dwarf Boa; Caicos Trope (WI)

8309 ***Tropidophis haetianus***
Squamata - Boidae
e Haitian Dwarf Boa; Haitian Wood Snake; Hispaniolan Trope (WI)
d Haiti-Zwergboa
i Tropidofide di Haiti; Pitone di Haiti

8310 ***Tropidophis maculatus***
Squamata - Boidae
e Spotted Dwarf Boa
d Gefleckte Zwergboa

8311 ***Tropidophis melanurus***
Squamata - Boidae
e Navassa Island Blacktail Dwarf Boa; Cuban Dwarf Boa; Cuban Wood Snake; Wood Snake; Navassa Trope
d Kubanische Zwergboa
i Tropidofide di Cuba

8312 ***Tropidophis nigriventris***
Squamata - Boidae
e Blackbelly Dwarf Boa

8313 ***Tropidophis pardalis***
Squamata - Boidae
e Leopard Dwarf Boa

8314 ***Tropidophis paucisquamis***
Squamata - Boidae
e Brazilian Dwarf Boa

8315 ***Tropidophis pilsbryi***
Squamata - Boidae
e Pilsbry's Dwarf Boa

8316 ***Tropidophis semicinctus***
Squamata - Boidae
e Banded Dwarf Boa

8317 ***Tropidophis taczanowskyi***
Squamata - Boidae
e Taczanowsky's Dwarf Boa

8318 ***Tropidophis wrighti***
Squamata - Boidae
e Wright's Dwarf Boa

8319 ***Tropidophorus***
Squamata - Scincidae
e Keeled Skinks; Waterside Skinks
d Kielskinke; Wasserskinke; Krokodilskinke; Kielschuppenskinke
f Tropidophores

8320 ***Tropidophorus baviensis***
Squamata - Scincidae
e Bavay's Keeled Skink

8321 ***Tropidophorus beccarii***
Squamata - Scincidae
e Beccari's Keeled Skink; Beccari's Water Skink

8322 ***Tropidophorus grayi***
Squamata - Scincidae
e Gray's Keeled Skink

8323 ***Tropidophorus laotus***
Squamata - Scincidae
e Laotian Keeled Skink
d Laos-Kielskink

8324 ***Tropidophorus robinsoni***
Squamata - Scincidae
e Robinson's Keeled Skink

8325 ***Tropidosaura***
Squamata - Lacertidae
e Mountain Lizards
d Gebirgseidechsen
f Tropidosaures

8326 ***Tropidosaura cottrelli***
Squamata - Lacertidae
e Cottrell's Mountain Lizard

8327 ***Tropidosaura essexi***
Squamata - Lacertidae
e Essex's Mountain Lizard

8328 ***Tropidosaura gularis***
Squamata - Lacertidae
e Cape Mountain Lizard; Yellow-striped Mountain Lizard

8329 ***Tropidosaura montana***
Squamata - Lacertidae
e Green-striped Mountain Lizard; Common Mountain Lizard (CSA)
d Glattkehlgebirgseidechse

8330 ***Tropidurus***
Squamata - Iguanaidae
e Lava Lizards; Keel-tailed Lizards; Thornytails
d Kielechsen; Kielschwänze; Kielschwanzleguane

8331 ***Tropidurus amathites***
Squamata - Iguanaidae
e Amathites Lava Lizard

8332 ***Tropidurus arenarius***
Squamata - Iguanaidae
e Tschudi's Lava Lizard

8333 ***Tropidurus bogerti***
Squamata - Iguanaidae
e Keeled Lava Lizard

8334 ***Tropidurus catalensis***
Squamata - Iguanaidae
e Catalen Lava Lizard

8335 ***Tropidurus etheridgei***
Squamata - Iguanaidae
e Etheridge's Lava Lizard

8336 ***Tropidurus heterolepis***
Squamata - Iguanaidae
e Many-scaled Lava Lizard

8337 ***Tropidurus hispidus***
Squamata - Iguanaidae
e Peters's Lava Lizard; Guianan Lava Lizard; Lagartixa de pedra (NA)

8338 ***Tropidurus hygomi***
Squamata - Iguanaidae
e Reinhardt's Lava Lizard

8339 ***Tropidurus melanopleurus***
Squamata - Iguanaidae
e Black Lava Lizard

8340 ***Tropidurus nanuzae***
Squamata - Iguanaidae
e Rodrigues's Lava Lizard

8341 ***Tropidurus semitaeniatus***
Squamata - Iguanaidae
e Striped Lava Lizard
d Gelbrückenkielschwanz

8342 ***Tropidurus spinulosus***
Squamata - Iguanaidae

e Spiny Lava Lizard

8343 ***Tropidurus torquatus***
Squamata - Iguanaidae
e Amazon Lava Lizard
d Halsbandkielschwanz; Kielschwanzleguan

8344 ***Tropidurus yanezi***
Squamata - Iguanaidae
e Yanez's Lava Lizard; Tree Lizard

8345 ***Tropiocolotes***
Squamata - Gekkonidae
e Sand Geckos; Dwarf Geckos; Pygmy Geckos
d Mikrogeckos; Zwerggeckos
f Tropiocolotes

8346 ***Tropiocolotes nattereri***
Squamata - Gekkonidae
e Natterer's Gecko
d Natterers Zwerggecko
f Tropiocolote de Natterer

8347 ***Tropiocolotes persicus***
Squamata - Gekkonidae
e Persia Sand Gecko

8348 ***Tropiocolotes scorteccii***
Squamata - Gekkonidae
e Scortecci's Sand Gecko

8349 ***Tropiocolotes steudneri***
Squamata - Gekkonidae
e Algerian Sand Gecko; Steudner's Pygmy Gecko; Steudner's Gecko
d Steudners Zwerggecko
f Tropiocolote de Steudner

8350 ***Tropiocolotes tripolitanus***
Squamata - Gekkonidae
e Northern Sand Gecko
d Zwergwüstengecko
f Tropiocolote d'Algérie; Gecko nain; Gecko nain de Tripoli

8351 ***Tropiocolotes tripolitanus algericus***
Squamata - Gekkonidae
e Tripoli Dwarf Gecko; Tripoli Pygmy Gecko; Tripoli Gecko
f Gecko à écailles carénées

8352 ***Tropiocolotes tripolitanus occidentalis***
Squamata - Gekkonidae
e Tripoli Pygmy Gecko
f Gecko à écailles carénées

8353 ***Tupinambis***
Squamata - Teiidae
e Tegus
d Großtejus; Echte Tejus
f Téjus; Tégus

8354 ***Tupinambis rufescens***
Squamata - Teiidae
e Red Tegu
d Roter Teju; Roter Großteju
f Téju rouge

8355 ***Tupinambis teguixin***
Squamata - Teiidae
e Black Tegu; Northern Black Tegu; Common Tegu; Black-and-yellow Tegu; Golden Tegu
d Bänderteju; Schwarzweißer Teju; Salompenter
f Téju commun; Loup du poulailler; Téju nain

8356 ***Tympanocryptis***
Squamata - Agamidae
e Australian Earless Agamas; Australian Earless Lizards; Deaf Agamas
d Taubagamen; Australische Bodenagamen

8357 ***Tympanocryptis adelaidensis***
Squamata - Agamidae
e Adelaide Earless Lizard; Adelaide Earless Agama

8358 ***Tympanocryptis cephalus***
Squamata - Agamidae
e Red Earless Lizard; Red Earless Agama

8359 ***Tympanocryptis dimiensis***
Squamata - Agamidae
e Mountain Earless Lizard; Mountain Earless Agama; Mountain Dragon (ANZ)

8360 ***Tympanocryptis intima***
Squamata - Agamidae
e Interior Earless Lizard; Interior Earless Agama

8361 ***Tympanocryptis lineata***
Squamata - Agamidae
e Lined Earless Lizard; Lined Earless Agama; Lined Earless Dragon

8362 ***Tympanocryptis lineata lineata***
Squamata - Agamidae
e Lined Earless Dragon (ANZ); Dragon Lizard (ANZ)

8363 ***Tympanocryptis tetraporophora***
Squamata - Agamidae
e Australian Earless Lizard; Australian Earless Agama; Long-tailed Earless Dragon

8364 ***Tympanocryptis uniformis***
Squamata - Agamidae
e Uniform Earless Lizard; Uniform Earless Agama

8365 ***Typhlacontias***
Squamata - Scincidae
e Blind Dart Skinks; Western Burrowing Skinks (CSA)

8366 ***Typhlacontias bogerti***
Squamata - Scincidae
e Bogert's Blind Dart Skink; Bogert's Burrowing Skink (CSA)

8367 ***Typhlacontias brevipes***
Squamata - Scincidae
e Short Blind Dart Skink; Fitzsimons's Burrowing Skink (CSA)

8368 ***Typhlacontias gracilis***
Squamata - Scincidae
e Roux's Blind Dart Skink; Kalahari Burrowing Skink (CSA); Gracile Legless Skink (CSA)

8369 ***Typhlacontias ngamiensis***
Squamata - Scincidae
e Ngamien Blind Dart Skink; Ngami Limbless Skink (CSA)

8370 ***Typhlacontias punctatissimus***
Squamata - Scincidae
e Dotted Blind Dart Skink

8371 ***Typhlacontias rohani***
Squamata - Scincidae
e Rohan's Blind Dart Skink

8372 ***Typhlophis***
Squamata - Anomalepididae
e Trinidad Blind Snakes; Trinidad Worm Snakes

8373 ***Typhlophis squamosus***
Squamata - Anomalepididae
e Trinidad Blind Snake; Trinidad Worm Snake

8374 **Typhlopidae**
Squamata
e Blind Snakes; Blind Worm Snakes; Worm Snakes; Blind Burrowing Snakes
d Gewöhnliche Blindschlangen; Typische Blindschlangen; Eigentliche Blindschlangen

f Typhlopidés; Serpents à deux têtes
i Tiflopidi

8375 ***Typhlops***
Squamata - Typhlopidae
e Common Worm Snakes; Blind Snakes; Glass Snakes; Double-ended Snakes
d Eigentliche Blindschlangen
f Typhlops vermiculaires; Serpents à deux têtes

8376 ***Typhlops acutus***
Squamata - Typhlopidae
e Beaked Worm Snake; Beaked Blind Snake

8377 ***Typhlops andamanensis***
Squamata - Typhlopidae
e Andaman Worm Snake

8378 ***Typhlops angolensis***
Squamata - Typhlopidae
e Angola Worm Snake

8379 ***Typhlops annae***
Squamata - Typhlopidae
e St. Barts Blind Snake

8380 ***Typhlops arenarius***
Squamata - Typhlopidae
e Sand Worm Snake

8381 ***Typhlops ater***
Squamata - Typhlopidae
e Irian Jayan Worm Snake

8382 ***Typhlops beddomi***
Squamata - Typhlopidae
e Beddome's Worm Snake

8383 ***Typhlops bibronii***
Squamata - Typhlopidae
e Bibron's Worm Snake; Bibron's Blind Snake (CSA); South African Blind Snake; Two-headed Snake; Ground Snake

8384 ***Typhlops biminensis***
Squamata - Typhlopidae
e Bimini Worm Snake; Bahamian Pink Blind Snake (WI)

8385 ***Typhlops bipartitus***
Squamata - Typhlopidae
e Tidore Island Worm Snake

8386 ***Typhlops bisubocularis***
Squamata - Typhlopidae
e Western Java Worm Snake

8387 ***Typhlops boylei***
Squamata - Typhlopidae
e Boyle's Worm Snake; Boyle's Blind Snake (CSA)

8388 ***Typhlops brongersmianus***
Squamata - Typhlopidae
e Borngersma's Worm Snake; Black Blind Snake

8389 ***Typhlops canlaoensis***
Squamata - Typhlopidae
e Taylor's Worm Snake

8390 ***Typhlops capitulatus***
Squamata - Typhlopidae
e Richmond's Worm Snake; Haitian Pale-lipped Blind Snake

8391 ***Typhlops cariei***
Squamata - Typhlopidae
e Hoffstetter's Worm Snake

8392 ***Typhlops castanotus***
Squamata - Typhlopidae
e Wynn's Worm Snake

8393 ***Typhlops catapontus***
Squamata - Typhlopidae

e Erica's Worm Snake; British Virgins Blind Snake

8394 ***Typhlops caymanensis***
Squamata - Typhlopidae
e Cayman Worm Snake; Grand Cayman Blind Snake (WI)

8395 ***Typhlops ceylonicus***
Squamata - Typhlopidae
e Ceylon Worm Snake

8396 ***Typhlops coecatus***
Squamata - Typhlopidae
e Ivory Coast Worm Snake

8397 ***Typhlops collaris***
Squamata - Typhlopidae
e Collared Worm Snake

8398 ***Typhlops comorensis***
Squamata - Typhlopidae
e Comoro Worm Snake

8399 ***Typhlops conradi***
Squamata - Typhlopidae
e Conrad's Worm Snake

8400 ***Typhlops costaricensis***
Squamata - Typhlopidae
e Costa Rica Worm Snake; Costa Rican Blind Snake

8401 ***Typhlops cuneirostris***
Squamata - Typhlopidae
e Wedge-nosed Worm Snake

8402 ***Typhlops decorosus***
Squamata - Typhlopidae
e Cameroon Worm Snake

8403 ***Typhlops decorsei***
Squamata - Typhlopidae
e Mocquard's Worm Snake

8404 ***Typhlops depressiceps***
Squamata - Typhlopidae
e Sepik River Worm Snake

8405 ***Typhlops diardi***
Squamata - Typhlopidae
e Indo-Chinese Worm Snake; Diard's Worm Snake (ISC); Diard's Blind Snake (ISC); Burmese Blind Snake
d Diards Blindschlange

8406 ***Typhlops disparilis***
Squamata - Typhlopidae
e Mystery Worm Snake

8407 ***Typhlops dominicanus***
Squamata - Typhlopidae
e Dominican Worm Snake; Dominica Blind Snake (WI)

8408 ***Typhlops elegans***
Squamata - Typhlopidae
e Elegant Worm Snake

8409 ***Typhlops epactia***
Squamata - Typhlopidae
e Cayman Brac Worm Snake; Cayman Brac Blind Snake (WI)

8410 ***Typhlops exiguus***
Squamata - Typhlopidae
e Belgaum Worm Snake

8411 ***Typhlops filiformis***
Squamata - Typhlopidae
e File Worm Snake

8412 ***Typhlops floweri***
Squamata - Typhlopidae
e Flower's Worm Snake

8413 ***Typhlops fornasinii***
Squamata - Typhlopidae
e Fornasini's Worm Snake; Fornasini's Blind Snake (CSA);

West African Blind Snake

8414 ***Typhlops fuscus***
Squamata - Typhlopidae
e Brown Worm Snake

8415 ***Typhlops giadinhensis***
Squamata - Typhlopidae
e Vietnam Worm Snake

8416 ***Typhlops gierrai***
Squamata - Typhlopidae
e Usambara Spotted Worm Snake; Usambara Spotted Blind Snake

8417 ***Typhlops gonavensis***
Squamata - Typhlopidae
e Gonave Island Worm Snake; Gonave Blind Snake

8418 ***Typhlops grandidieri***
Squamata - Typhlopidae
e Grandidier's Worm Snake

8419 ***Typhlops granti***
Squamata - Typhlopidae
e Grant's Worm Snake; Dryland Worm Snake ; Guanica Blind Snake

8420 ***Typhlops grivensis***
Squamata - Typhlopidae
e Grey Worm Snake

8421 ***Typhlops guadeloupensis***
Squamata - Typhlopidae
e Guadeloupe Worm Snake; Guadeloupe Blind Snake (WI)

8422 ***Typhlops hectus***
Squamata - Typhlopidae
e Thomas's Worm Snake; La Hotte Blind Snake

8423 ***Typhlops hedraeus***
Squamata - Typhlopidae
e Negros Island Worm Snake

8424 ***Typhlops hypogius***
Squamata - Typhlopidae
e Cebu Island Worm Snake

8425 ***Typhlops hypomethes***
Squamata - Typhlopidae
e Coastal Dwarf Blind Snake

8426 ***Typhlops hypsphothrius***
Squamata - Typhlopidae
e Sumatra Worm Snake

8427 ***Typhlops inornatus***
Squamata - Typhlopidae
e Dull Worm Snake

8428 ***Typhlops jamaicensis***
Squamata - Typhlopidae
e Jamaica Worm Snake; Jamaican Blind Snake (WI}

8429 ***Typhlops jerdoni***
Squamata - Typhlopidae
e Jerdon's Worm Snake

8430 ***Typhlops khoratensis***
Squamata - Typhlopidae
e Khorat Worm Snake

8431 ***Typhlops klemmeri***
Squamata - Typhlopidae
e Kuala Lumpur Worm Snake

8432 ***Typhlops koekkoeki***
Squamata - Typhlopidae
e Boenjoe Island Worm Snake

8433 ***Typhlops koshunensis***
Squamata - Typhlopidae
e Koshun Worm Snake

8434 ***Typhlops kraali***
Squamata - Typhlopidae
e Kei Island Worm Snake

8435 ***Typhlops lankaensis***
Squamata - Typhlopidae
e Sri Lanka Worm Snake

8436 ***Typhlops lehneri***
Squamata - Typhlopidae
e Estado Falcon Worm Snake

8437 ***Typhlops leucomelas***
Squamata - Typhlopidae
e Pied Worm Snake; Pied Blind Snake

8438 ***Typhlops leucostictus***
Squamata - Typhlopidae
e Liberia Worm Snake

8439 ***Typhlops lioneolatus***
Squamata - Typhlopidae
e Common Lined Worm Snake

8440 ***Typhlops longissimus***
Squamata - Typhlopidae
e Long Worm Snake

8441 ***Typhlops lorenzi***
Squamata - Typhlopidae
e Lorenz's Worm Snake

8442 ***Typhlops loveridgei***
Squamata - Typhlopidae
e Loveridge's Worm Snake

8443 ***Typhlops lumbricalis***
Squamata - Typhlopidae
e Earthworm Worm Snake; Earthworm Blind Snake; Bahamian Brown Blind Snake (WI)
d Regenwurmblindschlange

8444 ***Typhlops luzonensis***
Squamata - Typhlopidae
e Luzon Worm Snake

8445 ***Typhlops macrurus***
Squamata - Typhlopidae
e Secret Worm Snake

8446 ***Typhlops madagascariensis***
Squamata - Typhlopidae
e Madagascar Worm Snake

8447 ***Typhlops malcolmi***
Squamata - Typhlopidae
e Malcolm's Worm Snake

8448 ***Typhlops manilae***
Squamata - Typhlopidae
e Manila Worm Snake

8449 ***Typhlops manni***
Squamata - Typhlopidae
e Mann's Worm Snake

8450 ***Typhlops marxi***
Squamata - Typhlopidae
e Marx's Worm Snake

8451 ***Typhlops melanocephalus***
Squamata - Typhlopidae
e Black-headed Worm Snake

8452 ***Typhlops microstomus***
Squamata - Typhlopidae
e Yucatan Worm Snake; Yucatan Blind Snake

8453 ***Typhlops minuisquamus***
Squamata - Typhlopidae
e Basin Worm Snake; White-nosed Blind Snake

8454 ***Typhlops mirus***
Squamata - Typhlopidae
e Jan's Worm Snake; Jan's Ceylon Blind Snake

8455 ***Typhlops monastus***
Squamata - Typhlopidae
e Montserrat Worm Snake; Montserrat Blind Snake (WI)

8456 ***Typhlops monensis***
Squamata - Typhlopidae
e Mona Worm Snake; Mona Blind Snake
d Mona-Blindschlange

8457 ***Typhlops mucronatus***
Squamata - Typhlopidae
e Boettger's Worm Snake

8458 ***Typhlops mutilatus***
Squamata - Typhlopidae
e Werner's Worm Snake

8459 ***Typhlops oatesii***
Squamata - Typhlopidae
e Andaman Islands Worm Snake

8460 ***Typhlops obtusus***
Squamata - Typhlopidae
e Malawi Worm Snake; Slender Blind Snake (CSA); Blunt-snouted Blind Snake (CSA)

8461 ***Typhlops ocularis***
Squamata - Typhlopidae
e Parker's Worm Snake

8462 ***Typhlops oligolepis***
Squamata - Typhlopidae
e Few-scaled Worm Snake; Few-scaled Blind Snake; Wall's Blind Snake

8463 ***Typhlops pammeces***
Squamata - Typhlopidae
e South India Worm Snake

8464 ***Typhlops paucisquamus***
Squamata - Typhlopidae
e Pernambuco Worm Snake

8465 ***Typhlops platycephalus***
Squamata - Typhlopidae
e Flathead Worm Snake; Puerto Rican Forest Blind Snake (WI)

8466 ***Typhlops platyrhynchus***
Squamata - Typhlopidae
e Tanga Worm Snake; Tanga Blind Snake

8467 ***Typhlops porrectus***
Squamata - Typhlopidae
e Slender Worm Snake; Stoliczka's Blind Snake

8468 ***Typhlops pseudosaurus***
Squamata - Typhlopidae
e Guam Worm Snake

8469 ***Typhlops psittacus***
Squamata - Typhlopidae
e Parrot Worm Snake

8470 ***Typhlops punctatus***
Squamata - Typhlopidae
e Leach's Worm Snake; Spotted Blind Snake

8471 ***Typhlops pusillus***
Squamata - Typhlopidae
e Hispaniola Worm Snake; Hispaniolan Common Blind Snake

8472 ***Typhlops reticulatus***
Squamata - Typhlopidae
e Reticulate Worm Snake; Giant Blind Snake
f Serpent à deux têtes

8473 ***Typhlops reuteri***
Squamata - Typhlopidae
e Reuter's Worm Snake; Reuter's Blind Snake
d Reuters Blindschlange
f Typhlops de Reuter

8474 ***Typhlops richardi***
Squamata - Typhlopidae
e Richard's Worm Snake; Richard's Blind Snake; Virgin Islands Blind Snake

8475 ***Typhlops rondoensis***
Squamata - Typhlopidae
e Rondo Worm Snake; Rondo Blind Snake

8476 ***Typhlops rostellatus***
Squamata - Typhlopidae
e Puerto Rican Worm Snake; Tailspot Worm Snake ; Puerto Rican Brown Blind Snake

8477 ***Typhlops ruber***
Squamata - Typhlopidae
e Red Worm Snake

8478 ***Typhlops ruficaudus***
Squamata - Typhlopidae
e Red-headed Worm Snake

8479 ***Typhlops schmutzi***
Squamata - Typhlopidae
e Schmutz's Worm Snake

8480 ***Typhlops schwartzi***
Squamata - Typhlopidae
e Schwartz's Worm Snake; Eastern Short-tailed Blind Snake

8481 ***Typhlops siamensis***
Squamata - Typhlopidae
e Thailand Worm Snake

8482 ***Typhlops socotranus***
Squamata - Typhlopidae
e Socotra Worm Snake

8483 ***Typhlops stadelmani***
Squamata - Typhlopidae
e Stadelman's Worm Snake

8484 ***Typhlops steinhausi***
Squamata - Typhlopidae
e Steinhaus's Worm Snake

8485 ***Typhlops sulcatus***
Squamata - Typhlopidae
e Island Worm Snake; Bi-coloured Blind Snake

8486 ***Typhlops syntherus***
Squamata - Typhlopidae
e Barahona Worm Snake; Barahona Blind Snake

8487 ***Typhlops tasymicris***
Squamata - Typhlopidae
e Grenada Worm Snake; Grenada Blind Snake (WI)

8488 ***Typhlops tenebrarum***
Squamata - Typhlopidae
e Gloomy Worm Snake

8489 ***Typhlops tenuicollis***
Squamata - Typhlopidae
e Saragutin Worm Snake

8490 ***Typhlops tenuis***
Squamata - Typhlopidae
e Coffee Worm Snake; Coffee Blind Snake

8491 ***Typhlops tetrathyreus***
Squamata - Typhlopidae
e Haitian Worm Snake; Cul-de-sac Blind Snake

8492 ***Typhlops textilis***
Squamata - Typhlopidae
e Rare Worm Snake

8493 ***Typhlops thurstoni***
Squamata - Typhlopidae
e Thurston's Worm Snake

8494 ***Typhlops tindali***
Squamata - Typhlopidae
e Nilgiri Hills Worm Snake

8495 ***Typhlops titanops***
Squamata - Typhlopidae
e Titan Worm Snake; Big-eyed Blind Snake

8496 ***Typhlops trangensis***
Squamata - Typhlopidae
e Trang Worm Snake

8497 ***Typhlops trinitatus***
Squamata - Typhlopidae
e Trinidad Worm Snake

8498 ***Typhlops ulugurensis***
Squamata - Typhlopidae
e Uluguru Worm Snake

8499 ***Typhlops unilineatus***
Squamata - Typhlopidae
e One-lined Worm Snake

8500 ***Typhlops veddae***
Squamata - Typhlopidae
e Vedda Worm Snake

8501 ***Typhlops vermicularis***
Squamata - Typhlopidae
e Eurasian Worm Snake; Eurasian Blind Snake; European Blind Snake; Worm Snake; Vermiform Blind Snake
d Europäische Wurmschlange; Blödauge; Wurmschlange
f Typhlops vermiculaire; Serpent d'Aristotèle; Typhlops vermiculé
i Tiflope vermiculare

8502 ***Typhlops verticalis***
Squamata - Typhlopidae
e Smith's Worm Snake; Yellow-brown Blind Snake

Squamata - Typhlopidae
e Violet Worm Snake

8504 ***Typhlops wilsoni***
Squamata - Typhlopidae
e Iranian Worm Snake

8505 ***Typhlops yonenagae***
Squamata - Typhlopidae
e Yonenga Worm Snake

8506 ***Typhlops zenkeri***
Squamata - Typhlopidae
e Zenker's Worm Snake

8507 ***Typhlosaurus***
Squamata - Scincidae
e Legless Skinks; Blind Legless Skinks (CSA); Blind Skinks
d Afrikanische Blindskinke

8508 ***Typhlosaurus aurantiacus***
Squamata - Scincidae
e Golden Blind Legless Skink; Golden Limbless Skink (CSA); Golden Blind Skink (CSA)

8509 ***Typhlosaurus aurantiacus fitzimonsi***
Squamata - Scincidae
e Fitzsimons's Blind Legless Skink; Fitzsimons's Legless Skink; Golden Legless Skink (CSA)

8510 ***Typhlosaurus braini***
Squamata - Scincidae
e Haacke's Legless Skink; Brain's Blind Legless Skink (CSA)

8511 ***Typhlosaurus caecus***
Squamata - Scincidae
e Cuvier's Legless Skink; Cuvier's Blind Legless Skink (CSA)

8512 ***Typhlosaurus cregoi***
Squamata - Scincidae

e Cregoe's Legless Skink; Cregoe's Blind Legless Skink (CSA); Cregoe's Blind Skink (CSA); Cregoe's Limbless Skink (CSA)

8513 ***Typhlosaurus gariepensis***
Squamata - Scincidae
e Gariep Blind Legless Skink

8514 ***Typhlosaurus lineatus***
Squamata - Scincidae
e Striped Blind Legless Skink (CSA)

8515 ***Typhlosaurus lineatus subtaeniatus***
Squamata - Scincidae
e Striped Legless Skink

8516 ***Typhlosaurus meyeri***
Squamata - Scincidae
e Meyer's Legless Skink; Meyer's Blind Legless Skink (CSA)

8517 ***Typhlosaurus plowesi***
Squamata - Scincidae
e Plowes's Legless Skink; Plowes's Blind Skink

8518 ***Typhlosaurus vermis***
Squamata - Scincidae
e Boulenger's Legless Skink; Boulenger's Blind Legless Skink (CSA)

U

8519 ***Uma***
Squamata - Iguanaidae
e Fringe-toed Lizards; Umas
d Fransenzehenleguane
f Lézards à doigts frangés; Umas

8520 ***Uma exsul***
Squamata - Iguanaidae
e Fringe-toed Sand Lizard

8521 ***Uma inornata***
Squamata - Iguanaidae
e Coachella Valley Fringe-toed Lizard; Coachella Uma
f Lézard à doigts frangés Coachella

8522 ***Uma notata***
Squamata - Iguanaidae
e Sonoran Desert Fringe-toed Lizard; Colorado Desert Fringe-toed Lizard; Colorado Uma; Fringe-toed Lizard
d Colorado-Fransenzehenleguan
f Lézard à doigts frangés ; Uma du Colorado

8523 ***Uma notata notata***
Squamata - Iguanaidae
e Colorado Desert Fringe-toed Lizard

8524 ***Uma notata rufopunctata***
Squamata - Iguanaidae
e Desert Fringe-toed Lizard; Yuman Desert Fringe-toed Lizard

8525 ***Uma paraphygas***
Squamata - Iguanaidae
e Chihuahan Fringe-toed Lizard; Colorado Desert Fringe-toed Lizard
d Colorado-Fransenzehenleguan

8526 ***Uma scoparia***
Squamata - Iguanaidae
e Mojave Fringe-toed Lizard; Crescent Uma
f Lézard à doigts frangés du Mojave

8527 ***Umbrivaga***
Squamata - Colubridae
e Tropical Forest Snakes; Moss Snakes

8528 ***Umbrivaga mertensi***
Squamata - Colubridae
e Mertens's Tropical Forest Snake

8529 ***Umbrivaga pyburni***
Squamata - Colubridae
e Pyburn's Tropical Forest Snake

8530 ***Umbrivaga pygmaea***
Squamata - Colubridae
e Amazon Tropical Forest Snake; Pygmy Moss Snake

8531 ***Ungaliophis***
Squamata - Boidae
e Banana Boas
d Bananenboas; Zwergboas

8532 ***Ungaliophis continentalis***
Squamata - Boidae
e Isthmian Dwarf Boa
d Mittelamereikanische Zwergboa

8533 ***Ungaliophis panamensis***
Squamata - Boidae
e Panamanian Dwarf Boa
d Panamische Boa; Panama-Zwergboa

8534 ***Uracentron***
Squamata - Iguanaidae
e Thornytail Iguanas; Spiny-tailed

Iguanids
d Dornschwanzleguane

8535 ***Uracentron azureum***
Squamata - Iguanaidae
e Green Thornytail Iguana; Thornytail Lizard
d Azurblauer Dornschwanzleguan

8536 ***Uracentron azureun guentheri***
Squamata - Iguanaidae
e Southern Green Thornytail; Southern Green Thornytail Iguana

8537 ***Uracentron flaviceps***
Squamata - Iguanaidae
e Tropical Thornytail Iguana; Amazon Thornytail; Amazon Thornytail Iguana

8538 ***Uranoscodon***
Squamata - Iguanaidae
e Mophead Iguanas
d Mopskopfleguane

8539 ***Uranoscodon superciliosa***
Squamata - Iguanaidae
e Mophead Iguana; Mop-headed Lizard; Diving Lizard
d Mopskopfleguan

8540 ***Urocotyledon***
Squamata - Gekkonidae
e Surprise Geckos

8541 ***Urocotyledon inexpectata***
Squamata - Gekkonidae
e Seychelles Surprise Gecko

8542 ***Urocotyledon palmata***
Squamata - Gekkonidae
e Congo Palm Gecko

8543 ***Urocotyledon weileri***
Squamata - Gekkonidae
e Weiler's Gecko

8544 ***Urocotyledon wolterstorffi***
Squamata - Gekkonidae
e Wolterstorff's Gecko

8545 ***Uromacer***
Squamata - Colubridae
e Pointed Snakes; Haitian Vine Snakes
d Spitzkopfnattern; Haiti-Spitzkopfnattern; Hispanida-Spizkopfnattern

8546 ***Uromacer catesbyi***
Squamata - Colubridae
e Catesby's Pointed Snake; Blunt-headed Green Tree Snake

8547 ***Uromacer frenatus***
Squamata - Colubridae
e Island Pointed Snake; Sharp-nosed Vine Snake

8548 ***Uromacer oxyrhynchus***
Squamata - Colubridae
e Pointed Snake; Sharp-nosed Green Treesnake
d Haiti-Spitzkopfnatter

8549 ***Uromacer wetmorei***
Squamata - Colubridae
e Wetmore's Pointed Snake

8550 ***Uromacerina***
Squamata - Colubridae
e Sao Paolo Sharp Snakes; Sao Paulo Sharp Snakes

8551 ***Uromacerina ricardinii***
Squamata - Colubridae
e Sao Paolo Sharp Snake; Sao Paulo Sharp Snake

8552 ***Uromastyx***
Squamata - Agamidae
e Mastigures; Spiny-tailed Agamids;

Spiny-tailed Lizards; Thorny-tailed Lizards; Dobb Lizards; Dab Lizards; Spiny-tailed Agamas
d Dornschwanzagamen; Dornschwänze
f Lézards à queue épineuse; Fouettes-queue; Uromastyx
i Uromastice

8553 ***Uromastyx acanthinurus***
Squamata - Agamidae
e Dabbs Mastigure; Bell's Mastigure; African Spiny-tailed Lizard; Bell's Dab-Lizard; Bell's Dobb-Lizard; Spiny-tailed Agama; Mali Uromastyx; North African Uromastyx; Moroccan Uromastyx
d Dornschwanzagame; Veränderlicher Dornschwanz; Afrikanischer Dornschwanz; Nordafrikanischer Dornschwanz
f Fouette-queue acanthinure; Fouette-queue d'Afrique du Nord; Lézard des palmiers; Fouette-queue; Fouette queue commun; Uromastyx
i Uromastice maggiore; Uromastice acantinuro; Dab; Uromastice Dab

8554 ***Uromastyx aegyptius***
Squamata - Agamidae
e Egyptian Mastigure; Egyptian Spiny-tailed Lizard; Egyptian Spiny-tailed Agamid; Egyptian Uromastyx
d Ägyptischer Dornschwanz; Ägyptische Dornschwanzagame
f Uromastyx d'Égypte; Fouette-queue d'Égypte; Fouette-queue spinipède; Agame à queue épineuse; Agame fouette-queue d'Égypte; Lézard fouette-queue d'Égypte

8555 ***Uromastyx asmussi***
Squamata - Agamidae
e Iranian Mastigure; Iranian Spiny-tailed Lizard

8556 ***Uromastyx benti***
Squamata - Agamidae
e Bent's Mastigure; Arabian Uromastyx; Red Uromastyx

8557 ***Uromastyx dispar***
Squamata - Agamidae
e Sudan Mastigure
f Fouette-queue du Mali

8558 ***Uromastyx geyri***
Squamata - Agamidae
e Sahara Mastigure; Dab Lizard; Geyr's Dab-Lizard
d Geyrs Dornschwanz; Dornschwanzagame
f Fouette-queue de Geyr

8559 ***Uromastyx hardwickii***
Squamata - Agamidae
e Indian Mastigure; Indian Spiny-tailed Lizard; Spiny-tailed Lizard (ISC); Harwick's Uromastyx
d Indischer Dornschwanz
f Fouette-queue d'Inde; Fouette-queue d'Harwicke

8560 ***Uromastyx loricatus***
Squamata - Agamidae
e Mesopotamian Mastigure; Mesopotamian Spiny-tailed Lizard; Small-scaled Spiny-tailed Lizard

8561 ***Uromastyx macfadyeni***
Squamata - Agamidae
e Somali Mastigure

8562 ***Uromastyx ocellatus***
Squamata - Agamidae
e Ocellated Mastigure; Ocellated Uromastyx
d Dornschwanzagame

8563 ***Uromastyx ornatus***
Squamata - Agamidae
e Ornate Mastigure; Ornate Uromastyx
d Geschmückte Dornschwanzagame
f Uromastyx orné; Uromastix du Proche Orient; Fouette-queue orné

8564 ***Uromastyx princeps***
Squamata - Agamidae
e Princely Mastigure

8565 ***Uromastyx scorteccii***
Squamata - Agamidae
e Scortecci's Mastigure

8566 ***Uromastyx thomasi***
Squamata - Agamidae
e Thomas's Mastigure

8567 **Uropeltidae**
Squamata
e Burrowing Snakes; Short-tail Snakes; Shieldtail Snakes; Shieldtails; Roughtails; Shield-tailed Snakes
d Schildschwanzschlangen; Schildschwänze; Wühlschlangen
f Uropeltidés

8568 ***Uropeltis***
Squamata - Uropeltidae
e Indian Earth Snakes; Uropelts; Earth Snakes
d Rauhschwanzschlangen
f Uropeltis

8569 ***Uropeltis arcticeps***
Squamata - Uropeltidae
e Madura Earth Snake

8570 ***Uropeltis beddomii***
Squamata - Uropeltidae
e Beddome's Earth Snake

8571 ***Uropeltis broughami***
Squamata - Uropeltidae
e Brougham's Earth Snake

8572 ***Uropeltis ceylanicus***
Squamata - Uropeltidae
e Ceylon Earth Snake; Ceylon Shield-tailed Snake
d Ceylon-Schildschwanz
f Uropeltis de Ceylan

8573 ***Uropeltis dindigalensis***
Squamata - Uropeltidae
e Sirumalai Hills Earth Snake

8574 ***Uropeltis ellioti***
Squamata - Uropeltidae
e Elliot's Earth Snake

8575 ***Uropeltis liura***
Squamata - Uropeltidae
e Günther's Earth Snake; Ashambu Roughtail (ISC)

8576 ***Uropeltis macrolepis***
Squamata - Uropeltidae
e Bombay Earth Snake

8577 ***Uropeltis macrorhynchus***
Squamata - Uropeltidae
e Anamally Earth Snake

8578 ***Uropeltis maculatus***
Squamata - Uropeltidae
e Spotted Earth Snake

8579 ***Uropeltis melanogaster***
Squamata - Uropeltidae
e Gray's Earth Snake

8580 ***Uropeltis myhendrae***
Squamata - Uropeltidae
e Boulenger's Earth Snake

8581 ***Uropeltis nitidus***
Squamata - Uropeltidae
e Southern Earth Snake

8582 ***Uropeltis ocellatus***
Squamata - Uropeltidae
e Ocellated Earth Snake; Ocellate Shieldtail (ISC); Ocellate Uropelt
d Gefleckter Schildschwanz

8583 ***Uropeltis petersi***
Squamata - Uropeltidae
e Shieldtail Earth Snake

8584 ***Uropeltis phillipsi***
Squamata - Uropeltidae
e Phillips's Earth Snake

8585 ***Uropeltis pulneyensis***
Squamata - Uropeltidae
e Indian Earth Snake

8586 ***Uropeltis rubrolineatus***
Squamata - Uropeltidae
e Red-lined Earth Snake

8587 ***Uropeltis rubromaculatus***
Squamata - Uropeltidae
e Red-spotted Earth Snake
f Uropeltis à taches rouges

8588 ***Uropeltis ruhane***
Squamata - Uropeltidae
e Deraniyagala's Earth Snake

8589 ***Uropeltis smithi***
Squamata - Uropeltidae
e Smith's Earth Snake

8590 ***Uropeltis woodmasoni***
Squamata - Uropeltidae
e Woodmason's Earth Snake

8591 ***Uroplatus***
Squamata - Gekkonidae
e Flat-tail Geckos; Flat-tailed Geckos
d Blattschwanzgeckos; Plattschwanzgeckos

8592 ***Uroplatus alluaudi***
Squamata - Gekkonidae
e Northern Flat-tail Gecko

8593 ***Uroplatus ebenaui***
Squamata - Gekkonidae
e Nosy Be Flat-tail Gecko; Spear-pointed Leaf-tailed Gecko

8594 ***Uroplatus fimbriatus***
Squamata - Gekkonidae
e Common Flat-tail Gecko; Common Leaf-tailed Gecko; Flat-tailed Gecko; Madagascan Leaf-tiled Gecko
d Blattschwanzgecko; Plattschwanzgecko; Madagassischer Plattschwanzleguan; Madagassischer Plattschwanzgecko

8595 ***Uroplatus guentheri***
Squamata - Gekkonidae
e Günther's Flat-tail Gecko

8596 ***Uroplatus henkeli***
Squamata - Gekkonidae
e Henkel's Leaf-tailed Gecko
f Gecko à queue-en-feuille de Henkel

8597 ***Uroplatus lineatus***
Squamata - Gekkonidae
e Lined Flat-tail Gecko
d Bambusblattschwanzgecko

8598 ***Uroplatus phantasticus***
Squamata - Gekkonidae
e Fantastic Leaf-tailed Gecko; Leaf-tailed Gecko
d Plattschwanzgecko
f Gecko satanique

8599 ***Uroplatus siskorae***
Squamata - Gekkonidae
e Southern Flat-tail Gecko; Mossy Leaftail Gecko

8600 ***Urosaurus***
Squamata - Iguanaidae
e Tree Lizards; Tree and Climbing Lizards; Climbing Utas; Tree and Brush Lizards
f Lézards arboricoles; Urosaures

8601 ***Urosaurus auriculatus***
Squamata - Iguanaidae
e Socorro Island Tree Lizard

8602 ***Urosaurus bicarinatus***
Squamata - Iguanaidae
e Tropical Tree Lizard

8603 ***Urosaurus clarionensis***
Squamata - Iguanaidae
e Clarion Island Tree Lizard

8604 ***Urosaurus gadovi***
Squamata - Iguanaidae
e Gadow's Tree Lizard

8605 ***Urosaurus graciosus***
Squamata - Iguanaidae
e Brush Lizard; Long-tailed Brush Lizard (NA); Long-tailed Uta
f Lézard des brousailles

8606 ***Urosaurus graciosus graciosus***
Squamata - Iguanaidae
e Western Brush Lizard

8607 ***Urosaurus graciosus shanoni***
Squamata - Iguanaidae
e Arizona Brush Lizard; Long-tailed Brush Lizard; Arizona Long-tailed Brush Lizard

8608 ***Urosaurus irregularis***
Squamata - Iguanaidae
e Mexican Brush Lizard

8609 ***Urosaurus lahtelai***
Squamata - Iguanaidae
e Baja California Lizard; Baja California Brush Lizard

8610 ***Urosaurus microscutatus***
Squamata - Iguanaidae
e Small-scaled Lizard; Small-scaled Uta; Small-scaled Tree Lizard

8611 ***Urosaurus nigricaudus***
Squamata - Iguanaidae
e Blacktail Brush Lizard; Black-tailed Brush Lizard
f Lézard à queue noire

8612 ***Urosaurus ornatus***
Squamata - Iguanaidae
e Northern Tree Lizard; Ornate Tree Lizard; Common Tree Lizard; Texas Tree Uta; Tree Lizard (NA)
d Baumleguan
f Lézard arboricole; Urosaure orné

8613 ***Urosaurus ornatus levis***
Squamata - Iguanaidae
e Canyon Tree Lizard; Smooth Tree Lizard

8614 ***Urosaurus ornatus linearis***
Squamata - Iguanaidae
e Lined Tree Lizard; Ornate Tree Lizard

8615 ***Urosaurus ornatus ornatus***
Squamata - Iguanaidae
e Eastern Tree Lizard; Texas Tree Lizard

8616 ***Urosaurus ornatus schmidti***
Squamata - Iguanaidae
e Big Bend Tree Lizard

8617 ***Urosaurus ornatus schottii***
Squamata - Iguanaidae
e Schott's Tree Lizard

8618 ***Urosaurus ornatus symmetricus***
Squamata - Iguanaidae
e Colorado River Tree Lizard

8619 ***Urosaurus ornatus wrighti***
Squamata - Iguanaidae
e Northern Tree Lizard; Cliff Tree Lizard

8620 ***Urostrophus***
Squamata - Iguanaidae
e Steppe Iguanas
f Urostrophes

8621 ***Urostrophus torquatus***
Squamata - Iguanaidae
e Philippi's Steppe Iguana

8622 ***Urostrophus valeriae***
Squamata - Iguanaidae
e Donoso's Steppe Iguana

8623 ***Urostrophus vautieri***
Squamata - Iguanaidae
e Brazilian Steppe Iguana
f Urostrophe de Vautier

8624 ***Urotheca***
Squamata - Colubridae
e Twin Coral Snakes

8625 ***Urotheca elapoides***
Squamata - Colubridae
e Twin Coral Snake

8626 ***Uta***
Squamata - Iguanaidae
e Side-blotched Lizards; Ground Utas; Small-scaled Swifts
d Seitenfleckenleguane
f Utas

8627 ***Uta antiqua***
Squamata - Iguanaidae
e San Lorenzo Island Side-blotched Lizard

8628 ***Uta concinna***
Squamata - Iguanaidae
e Dickerson's Side-blotched Lizard

8629 ***Uta encantadae***
Squamata - Iguanaidae
e Enchanted Side-blotched Lizard

8630 ***Uta lowei***
Squamata - Iguanaidae
e Dead Side-blotched Lizard

8631 ***Uta nolascensis***
Squamata - Iguanaidae
e San Pedro Nolasco Lizard; San Pedro Nolasco Side-blotched Lizard; San Pedro Nolasco Island Spinytail Iguana (NA)

8632 ***Uta palmeri***
Squamata - Iguanaidae
e San Pedro Side-blotched Lizard; San Pedro Martir Side-blotched Iguana

8633 ***Uta squamata***
Squamata - Iguanaidae
e Santa Catalina Island Side-blotched Lizard

8634 ***Uta stansburiana***
Squamata - Iguanaidae
e Side-blotched Lizard; Common Side-blotched Iguana; Stansbury's Swift; Common Side-blotched Lizard
d Seitenfleckenleguan
f Lézard à côtes tachetés

8635 ***Uta stansburiana elegans***
Squamata - Iguanaidae
e California Side-blotched Lizard; California Side-blotched Iguana; Western Side-blotched Lizard

8636 ***Uta stansburiana nevadensis***
Squamata - Iguanaidae
e Nevada Side-blotched Lizard; Nevada Side-blotched Iguana

8637 ***Uta stansburiana stansburiana***
Squamata - Iguanaidae
e Northern Side-blotched Lizard; Stansbury's Swift (NA)

8638 ***Uta stansburiana stejnegeri***
Squamata - Iguanaidae
e Eastern Side-blotched Lizard; Desert Side-blotched Lizard; Desert Side-blotched Iguana

8639 ***Uta stansburiana uniformis***
Squamata - Iguanaidae
e Colorado Side-blotched Lizard; Colorado Side-blotched Iguana

8640 ***Uta stellata***
Squamata - Iguanaidae
e San Benito Island Side-blotched Lizard; San Benito Island Side-blotched Iguana

8641 ***Uta tumidarostra***
Squamata - Iguanaidae
e Swollen-nose Side-blotched Lizard; Swollen-nose Side-blotched Iguana

V

8642 ***Vanzosaura***
Squamata - Gymnophthalmidae
e Redtail Tegus

8643 ***Vanzosaura rubricauda***
Squamata - Gymnophthalmidae
e Redtail Tegu

8644 **Varanidae**
Squamata
e Monitors; Monitor Lizards; Goannas
d Warane
f Varanidés; Varans

8645 ***Varanus***
Squamata - Varanidae
e Monitors
d Warane
f Varans; Psammosaures
i Varani

8646 ***Varanus acanthurus***
Squamata - Varanidae
e Ridgetail Monitor; Ridge-tailed Monitor (ANZ)
d Stachelschwanzwaran

8647 ***Varanus albigularis***
Squamata - Varanidae
e White-throated Monitor; White-throated Monitor Lizard; White-throated Savanna Monitor; Whitethroat Monitor
d Weißkehlwaran; Kap-Waran

8648 ***Varanus baritji***
Squamata - Varanidae
e Stone Monitor; White's Monitor

8649 ***Varanus beccarii***
Squamata - Varanidae
e Black Tree Monitor

8650 ***Varanus bengalensis***
Squamata - Varanidae
e Bengal Monitor; Common Indian Monitor (ISC); Indian Monitor
d Bengal-Waran; Bengalen-Waran; Bengalischer Waran
f Varan du Bengale

8651 ***Varanus bogerti***
Squamata - Varanidae
e Bogert's Monitor

8652 ***Varanus brevicauda***
Squamata - Varanidae
e Short-tail Monitor; Short-tailed Monitor
d Kurzschwanzwaran; Zwergwaran
f Varan à queue courte
i Varano dalla coda breve

8653 ***Varanus caudolineatus***
Squamata - Varanidae
e Stripetail Monitor; Stripe-tailed Monitor
d Schwanzstrichwaran; Streifenschwanzwaran
f Varan à queue rayé

8654 ***Varanus doreanus***
Squamata - Varanidae
e Bluetail Monitor; Blue-tailed Monitor
d Blauschwanzwaran

8655 ***Varanus dumerili***
Squamata - Varanidae
e Duméril's Monitor; Duméril's Monitor Lizard; Duméril's Water Monitor

d Duméril-Waran
f Varan de Duméril

8656 ***Varanus eremius***
Squamata - Varanidae
e Spinifex Monitor; Pygmy Monitor
d Höhlenwaran

8657 ***Varanus exanthematicus***
Squamata - Varanidae
e Savanna Monitor; Rock Monitor (CSA); Bosc Monitor; Bosc's Monitor; Cape Monitor; White-throated Iguana; Rock Iguana; Land Leguan (CSA); Rock Leguan; Savannah Monitor; African Savannah Monitor; Savannah Monitor
d Steppenwaran; Schwarzkehlwaran; Kap-Waran
f Varan du Cap; Varan des savannes; Varan de savanne
i Varano della savanna; Varano; Varano della steppa

8658 ***Varanus flavescens***
Squamata - Varanidae
e Yellow Monitor; Golden Monitor
d Gelbwaran
f Varan jaune

8659 ***Varanus giganteus***
Squamata - Varanidae
e Perentie; Giant Monitor; Perenty
d Australischer Riesenwaran; Großwaran; Perenty
f Varan gigantesque

8660 ***Varanus gilleni***
Squamata - Varanidae
e Pygmy Monitor; Pygmy Mulga Monitor (ANZ); Gillen's Monitor; Gillen's Dwarf Monitor
d Gillen-Waran; Gillens Zwergwaran; Gillens Waran
f Varan de Gillen

8661 ***Varanus glauerti***
Squamata - Varanidae
e Glauert's Monitor; Long-tailed Rock Monitor (ANZ)
d Glauerts Waran
f Varan de Glauert

8662 ***Varanus glebopalma***
Squamata - Varanidae
e Longtail Monitor; Long-tailed Rock Monitor

8663 ***Varanus gouldi horni***
Squamata - Varanidae
d Arguswaran

8664 ***Varanus gouldii***
Squamata - Varanidae
e Sand Monitor; Gould's Goanna (ANZ); Gould's Monitor; Sand Goanna; Monitor (ANZ); Rosenberg's Goanna (ANZ)
d Goulds Waran
f Varan de Gould

8665 ***Varanus griseus***
Squamata - Varanidae
e Desert Monitor; Trans-Caspian Desert Monitor
d Wüstenwaran; Mittelasiatischer Wüstenwaran
f Varan du désert; Varan des sables
i Varano grigio; Varano del deserto

8666 ***Varanus indicus***
Squamata - Varanidae
e Mangrove Monitor; Indian Monitor; Indian Water Monitor; Indian Monitor Lizard; Pacific Monitor
d Pazifik-Waran; Mangrovenwaran
f Varan des Indes

8667 ***Varanus jobiensis***
Squamata - Varanidae
e Peach-throated Monitor; Peachthroat Monitor
d Rothalswaran

8668 ***Varanus kingorum***
Squamata - Varanidae
e Crevice Monitor

8669 ***Varanus komodoensis***
Squamata - Varanidae
e Komodo Dragon; Komodo Monitor; Komodo Island Monitor
d Komodo-Waran
f Varan des Kommodos
i Drago di Komodo

8670 ***Varanus mertensi***
Squamata - Varanidae
e Mertens's Water Monitor; Mertens's Monitor
d Mertens-Wasserwaran
f Varan de Mertens

8671 ***Varanus mitchelli***
Squamata - Varanidae
e Mitchell's Water Monitor; Mitchell's Monitor
d Mitchells Wasserwaran

8672 ***Varanus niloticus***
Squamata - Varanidae
e Nile Monitor; African Monitor; Water Monitor (CSA); Water Leguan (CSA)
d Nil-Waran
f Varan du Nil
i Varano del Nilo

8673 ***Varanus olivaceus***
Squamata - Varanidae
e Gray's Monitor; Nile Water Monitor
d Grays Waran

8674 ***Varanus pilbarensis***
Squamata - Varanidae
e Pilbara Monitor

8675 ***Varanus prasinus***
Squamata - Varanidae
e Emerald Monitor; Emerald Tree Monitor Lizard; Green Tree Monitor; Emerald Tree Monitor
d Smaragdwaran

8676 ***Varanus primordius***
Squamata - Varanidae
e Brown Monitor

8677 ***Varanus rosenbergi***
Squamata - Varanidae
e Black Sand Monitor

8678 ***Varanus rudicollis***
Squamata - Varanidae
e Roughneck Monitor; Rough-necked Monitor
d Rauhnackenwaran
f Varan cou-rude
i Varano rudicollo

8679 ***Varanus salvadorii***
Squamata - Varanidae
e Crocodile Monitor; Papuan Monitor
d Papua-Waran

8680 ***Varanus salvator***
Squamata - Varanidae
e Indo-Malay Water Monitor Lizard; Water Monitor (ISC); Asian Water Monitor (ANZ); Two-banded Monitor
d Bindenwaran
f Varan aquatique; Varan malais; Varan sauvegarde; Varan à deux bandes
i Varano fasciato; Varano della Malesia

8681 ***Varanus scalaris***
Squamata - Varanidae
e Spotted Monitor; Australian Spotted Tree Monitor

8682 ***Varanus semiremex***
Squamata - Varanidae

e Rusty Monitor
d Rostkopfwaran; Rostroter Waran; Rotkopfwaran

8683 ***Varanus spenceri***
Squamata - Varanidae
e Spencer's Monitor
d Spencer-Waran; Spencers Waran
f Varan de Spencer

8684 ***Varanus spinulosus***
Squamata - Varanidae
e Solomon Islands Monitor

8685 ***Varanus storri***
Squamata - Varanidae
e Dwarf Monitor
d Storrs Zwergwaran; Zwergwaran
f Varan nain

8686 ***Varanus telenestses***
Squamata - Varanidae
e Rossel Island Monitor

8687 ***Varanus teriae***
Squamata - Varanidae
e Queensland Tree Monitor; Queensland Blue-nosed Monitor Lizard

8688 ***Varanus timorensis***
Squamata - Varanidae
e Spotted Tree Monitor; Timor Monitor
d Timor-Waran
f Varan de Timor

8689 ***Varanus tristis***
Squamata - Varanidae
e Arid Monitor
d Trauerwaran

8690 ***Varanus varius***
Squamata - Varanidae
e Lace Monitor; Lace Monitor Lizard; Variegated Lizard; Tree Goanna (ANZ); Variegated Monitor
d Buntwaran
f Varan bigarré

8691 ***Varanus yemenensis***
Squamata - Varanidae
e Yemen Monitor

8692 ***Vermicella***
Squamata - Elapidae
e Bandy-Bandy Snakes
d Bandy-Bandys

8693 ***Vermicella annulata***
Squamata - Elapidae
e Bandy-Bandy (ANZ); Ringed Snake (ANZ); Australian Burrowing Snake
d Bandy-Bandy; Australische Giftnatter

8694 ***Vermicella multifasciata***
Squamata - Elapidae
e Northern Bandy-Bandy

8695 ***Vilcunia***
Squamata - Iguanaidae
e Forest Iguanas

8696 ***Vilcunia silvanae***
Squamata - Iguanaidae
e Donoso's Forest Iguana

8697 ***Vipera***
Squamata - Viperidae
e Sand Vipers; Oriental Pit Vipers; True Vipers
d Ottern; Echte Ottern; Echte Vipern
f Vipères
i Vipere

8698 ***Vipera ammodytes***
Squamata - Viperidae
e Sand Viper; Nose-horned Viper; Horned Viper; Long-nosed Viper;

European Nose-horned Viper
d Europäische Sandotter; Europäische Hornotter; Sandotter; Hornotter; Sandviper; Transkaukasische Hornviper; Europäische Hornviper
f Vipère ammodyte; Vipère des sables; Vipère unicorne
i Vipera dal corno; Vipera ammodite

8699 ***Vipera ammodytes montadoni***
Squamata - Viperidae
e Bulgarian Viper
d Bulgarische Hornotter

8700 ***Vipera aspis***
Squamata - Viperidae
e Asp Viper; Aspis Viper; Asp; European Asp
d Aspisviper; Aspisotter; Jura-Viper
f Vipère commune; Vipère aspic; Aspic
i Vipera comune; Aspide; Vipera aspide; Vèpra; Vipra

8701 ***Vipera aspis aspis***
Squamata - Viperidae
e Asp Viper
d Aspisviper

8702 ***Vipera aspis atra***
Squamata - Viperidae
e Alpine Viper
d Alpen-Viper

8703 ***Vipera aspis francisciredi***
Squamata - Viperidae
e Redi's Viper
d Redi-Viper

8704 ***Vipera aspis hugyi***
Squamata - Viperidae
e Sicilian Viper
d Sizilianische Viper

8705 ***Vipera aspis zinnikeri***
Squamata - Viperidae
e Spanish Asp Viper
d Spanische Aspisviper

8706 ***Vipera barani***
Squamata - Viperidae
e Turkish Viper; Black Viper
d Barans Otter
f Vipère de Baran
i Vipera di Baran

8707 ***Vipera berus***
Squamata - Viperidae
e Common Viper; Viper; Adder; Northern Viper; Northern Cross Adder; European Adder; Common Northern Viper; Common European Northern Viper; Common Adder
d Kreuzotter
f Vipère; Péliade; Vipère péliade; Vipère du Nord; Vipère bérus; Vipère commune; Vipère vulgaire; Vipère noire
i Marasso; Marasso palustre

8708 ***Vipera berus bosniensis***
Squamata - Viperidae
f Vipère péliade de Yougoslavie

8709 ***Vipera bornmulleri***
Squamata - Viperidae
e Lebanon Viper
d Libanesische Bergotter
f Vipère du Liban

8710 ***Vipera bulgardaghica***
Squamata - Viperidae
e Taurus Viper; Bulgar Dagh Viper
d Taurische Bergotter
f Vipère du Taurus; Vipère du Mont Bolkar
i Vipera dei Monti Bulgar

8711 ***Vipera darevskii***
Squamata - Viperidae

e Darevsky's Viper
d Darevski-Kreuzotter
f Vipère de Darevski
i Vipera di Darevsky

8712 ***Vipera dinniki***
Squamata - Viperidae
e Dinnik's Viper
d Westliche Kaukasus-Otter; Westkaukasische Kreuzotter
f Vipère du Caucase occidental; Vipère de Dinnik
i Vipera di Dinnick

8713 ***Vipera kaznakovi***
Squamata - Viperidae
e Kaznakow's Viper; Caucasus Viper; Red Viper; Caucasian Viper
d Kaukasus-Otter
f Vipère du Caucase
i Vipera di Kaznakov

8714 ***Vipera latastei***
Squamata - Viperidae
e Lataste's Viper; Snub-nosed Viper
d Stülpnasenotter
f Vipère de Lataste
i Vipera di Lataste

8715 ***Vipera latifii***
Squamata - Viperidae
e Latifi's Viper
d Elburs-Otter; Latifis Otter; Latifi-Otter
f Vipère de Latifi

8716 ***Vipera lebetina***
Squamata - Viperidae
e Bluntnose Viper; Levantine Viper; Blunt-nosed Viper; Snub-nosed Viper; Lebetina Viper
d Levante-Otter; Gjurza; Levante-Viper
f Vipère du Levant; Vipère lébétine
i Vipera levantina; Vipera lebetina

8717 ***Vipera lebetina lebetina***
Squamata - Viperidae
d Zypriotische Levante-Otter

8718 ***Vipera lebetina turanica***
Squamata - Viperidae
e Russian Viper
d Russische Levante-Otter

8719 ***Vipera monticola***
Squamata - Viperidae
e Mountain Viper
d Atlas-Zwergotter
f Vipère naine de l'Atlas; Vipère de l'Atlas

8720 ***Vipera nikolskii***
Squamata - Viperidae
e Nikolsky's Viper
d Waldsteppenotter

8721 ***Vipera pontica***
Squamata - Viperidae
e Black Sea Viper

8722 ***Vipera russellii***
Squamata - Viperidae
e Russell's Viper; Tic Polonga; Tic Polonga Viper; Chain Viper
d Kettenviper; Daboia
f Vipère de Russell

8723 ***Vipera russellii siamensis***
Squamata - Viperidae
d Siamesische Kettenviper

8724 ***Vipera schweizeri***
Squamata - Viperidae
e Milos Blunt-nosed Viper; Cyclades Blunt-nosed Viper
d Zykladen-Otter; Milos-Otter; Zykladen-Levante-Otter
f Vipère des Cyclades
i Vipera di Schweizer; Vipera delle Cicladi; Vipera di Milo

8725 ***Vipera seoanei***
Squamata - Viperidae
e Portuguese Viper; Iberian Adder; Pyrenean Viper; Spanish Viper
d Iberien-Otter; Iberische Kreuzotter; Pyrenäen-Otter; Nordiberische Kreuzotter
f Vipère de Séoane; Vipère des Pyrénées
i Vipera di Seoane; Vipera iberica

8726 ***Vipera ursinii***
Squamata - Viperidae
e Orsini's Viper; Meadow Viper
d Wiesenotter; Spitzkopfotter; Karst-Otter
f Vipère d'Orsini
i Vipera dell'Orisini; Vipera di Orsini

8727 ***Vipera wagneri***
Squamata - Viperidae
e Wagner's Viper
d Wagners Bergotter
f Vipère de Wagner
i Vipera di Wagner

8728 **Viperidae**
Squamata
e Vipers; Adders
d Vipern; Ottern
f Vipéridés

8729 ***Virginia***
Squamata - Colubridae
e Earth Snakes; Ground Snakes; North american Earth Snakes
d Erdschlangen; Rauhnattern
f Couleuvres fouisseuses

8730 ***Virginia striatula***
Squamata - Colubridae
e Rough Earth Snake; Brown Snake (NA); Southern Ground Snake
d Rauhe Erdschlange; Rauhnatter
f Couleuvre fouisseuse rugueuse

8731 ***Virginia valeriae***
Squamata - Colubridae
e Smooth Earth Snake; Eastern Ground Snake
d Glatte Erdschlange
f Couleuvre fouisseuse lisse

8732 ***Virginia valeriae elegans***
Squamata - Colubridae
e Western Earth Snake; Western Smooth Earth Snake

8733 ***Virginia valeriae pulchra***
Squamata - Colubridae
e Mountain Earth Snake

8734 ***Virginia valeriae valeriae***
Squamata - Colubridae
e Eastern Earth Snake; Brown Snake (NA); Worm Snake (NA); Ground Snake (NA); Eastern Smooth Earth Snake

8735 ***Voeltzkovia***
Squamata - Scincidae
e Burrowing Blind Skinks

8736 ***Voeltzkovia fierinensis***
Squamata - Scincidae
e Common Burrowing Blind Skink

8737 ***Voeltzkovia mira***
Squamata - Scincidae
e Burrowing Blind Skink

8738 ***Voeltzkovia petiti***
Squamata - Scincidae
e Tiny Burrowing Blind Skink

8739 ***Voeltzkovia rubrocaudata***
Squamata - Scincidae
e Redtail Burrowing Blind Skink

W

8740 ***Waglerophis***
Squamata - Colubridae
e Wagler's Snakes
d Haubennattern

8741 ***Waglerophis merremii***
Squamata - Colubridae
e Wagler's Snake
d Merrems Haubennattern

8742 ***Walterinnesia***
Squamata - Elapidae
e Black Desert Cobras; Desert Cobras; Desert Black Cobra
d Wüstenkobras

8743 ***Walterinnesia aegyptia***
Squamata - Elapidae
e Black Desert Cobra; Desert Cobra; Desert Black Cobra
d Schwarze Wüstenkobra; Wüstenkobra
f Cobra noir du désert

8744 ***Wetmoreana***
Squamata - Anguidae
e Haitian Lizards

8745 ***Wetmoreana haetiana***
Squamata - Anguidae
e Cochran's Haitian Lizard; Earless Galliwasp

X

8746 ***Xantusia***
Squamata - Xantusiidae
e Night Lizards
d Nachtechsen; Wüstennachtechsen; Granitechsen
f Xantusies

8747 ***Xantusia bolsonae***
Squamata - Xantusiidae
e Bolson Night Lizard

8748 ***Xantusia henshawi***
Squamata - Xantusiidae
e Granite Night Lizard; Henshaw's Night Lizard
d Granitechse; Granitnachtechse
f Lézard nocturne de Henshaw; Xantusie de Henshaw

8749 ***Xantusia henshawi gracilis***
Squamata - Xantusiidae
e Sandstone Night Lizard

8750 ***Xantusia henshawi henshawi***
Squamata - Xantusiidae
e Henshaw Night Lizard; Granite Night Lizard

8751 ***Xantusia vigilis***
Squamata - Xantusiidae
e Desert Night Lizard
d Kleine Nachtechse; Wüstennachtechse; Yucca-Nachtechse
f Lézard nocturne du désert; Xantusie du désert

8752 ***Xantusia vigilis arizonoae***
Squamata - Xantusiidae
e Arizona Night Lizard

8753 ***Xantusia vigilis gilberti***
Squamata - Xantusiidae
e Baja California Night Lizard

8754 ***Xantusia vigilis sierrae***
Squamata - Xantusiidae
e Sierra Night Lizard

8755 ***Xantusia vigilis utahensis***
Squamata - Xantusiidae
e Utah Night Lizard

8756 ***Xantusia vigilis vigilis***
Squamata - Xantusiidae
e Common Night Lizard; Yucca Night Lizard; Desert Night Lizard

8757 **Xantusiidae**
Squamata
e Night Lizards
d Nachtechsen
f Xantusiidés; Lézards nocturnes

8758 ***Xenagama***
Squamata - Agamidae
e Strange Agamas

8759 ***Xenagama batillifera***
Squamata - Agamidae
e Somali Strange Agama

8760 ***Xenagama taylori***
Squamata - Agamidae
e Taylor's Strange Agama; Taylor's Agama
f Agame de Taylor

8761 ***Xenelaphis***
Squamata - Colubridae
e Malaysian Brown Snakes;

Brown Snakes

8762 ***Xenelaphis ellipsifer***
Squamata - Colubridae
e Ocellated Brown Snake

8763 ***Xenelaphis hexagonotus***
Squamata - Colubridae
e Malaysian Brown Snake

8764 ***Xenoboa***
Squamata - Boidae
e Cropan's Boas; Cropan's Tree Boas

8765 ***Xenoboa cropanii***
Squamata - Boidae
e Cropan's Boa; Cropan's Tree Boa

8766 ***Xenocalamus***
Squamata - Colubridae
e Quill-snouted Snakes; Slender Quill-snouted Snakes; Quill-nosed Snakes; Slender Quill-nosed Snakes
d Spatelnasennattern

8767 ***Xenocalamus bicolor***
Squamata - Colubridae
e Slender Quill-snouted Snake; Bi-coloured Quill-snouted Snake (CSA); Northern Quill-nosed Snake; Northern Quill-snouted Snake

8768 ***Xenocalamus bicolor australis***
Squamata - Colubridae
e Waterberg Quill-snouted Snake; Bi-coloured Quill-snouted Snake (CSA)

8769 ***Xenocalamus bicolor bicolor***
Squamata - Colubridae
e Bi-coloured Quill-snouted Snake

8770 ***Xenocalamus bicolor lineatus***
Squamata - Colubridae
e Lined Quill-snouted Snake; Bi-coloured Quill-snouted Snake (CSA)

8771 ***Xenocalamus mechowii***
Squamata - Colubridae
e Elongate Quill-snouted Snake; Western Quill-nosed Snake; Western Quill-snouted Snake; Mechow's Quill-snouted Snake (CSA)

8772 ***Xenocalamus michellii***
Squamata - Colubridae
e Michell's Quill-snouted Snake

8773 ***Xenocalamus sabiensis***
Squamata - Colubridae
e Sabi Quill-snouted Snake

8774 ***Xenocalamus transvaalensis***
Squamata - Colubridae
e Transvaal Quill-snouted Snake (CSA); Transvaal Quill-nosed Snake

8775 ***Xenochrophis***
Squamata - Colubridae
e Painted Keelbacks; Keelbacks
d Fischernattern

8776 ***Xenochrophis asperrimus***
Squamata - Colubridae
e Boulenger's Keelback

8777 ***Xenochrophis cerasogaster***
Squamata - Colubridae
e Painted Keelback; Dark-bellied Marsh Snake

8778 ***Xenochrophis flavipunctatus***
Squamata - Colubridae
e Gold-dotted Keelback

8779 ***Xenochrophis piscator***
Squamata - Colubridae
e Chequered Keelback; Asiatic Water Snake; Checkered Keelback (ISC); Chequered Keelback Water Snake
d Fischernatter; Fischnatter

8780 ***Xenochrophis punctulata***
Squamata - Colubridae
e Dotted Keelback

8781 ***Xenochrophis vittata***
Squamata - Colubridae
e Banded Keelback
d Indonesische Wassernatter

8782 ***Xenodermus***
Squamata - Colubridae
e Xenodermine Snakes; Xenodermon Snakes
d Javanische Höckernattern

8783 ***Xenodermus javanicus***
Squamata - Colubridae
e Xenodermine Snake; Xenodermon Snake
d Javanische Höckernatter

8784 ***Xenodon***
Squamata - Colubridae
e False Fer-de-Lances; False Lanceheads; False Vipers
d Haubennattern; Ungleichzahnige Nattern; Falsche Lanzenottern

8785 ***Xenodon bertholdi***
Squamata - Colubridae
e Berthold's False Fer-de-Lance; Berthold's False Lancehead

8786 ***Xenodon guentheri***
Squamata - Colubridae
e Günther's False Fer-de-Lance; Günther's False Lancehead

8787 ***Xenodon neuwiedii***
Squamata - Colubridae
e Neuwied's False Fer-de-Lance; Neuwied's False Lancehead

8788 ***Xenodon rabdocephalus***
Squamata - Colubridae
e False Fer-de-Lance; False Lancehead; Common False Viper
d Krötennatter

8789 ***Xenodon severus***
Squamata - Colubridae
e Amazon False Fer-de-Lance; Amazon False Lancehead; Giant False Viper

8790 **Xenopeltidae**
Squamata
e Sunbeam Snakes
d Erdschlangen; Wühlschlangen; Regenbogenschlangen
f Xenopeltidés

8791 ***Xenopeltis***
Squamata - Xenopeltidae
e Sunbeam Snakes; Iridescent Earth Snakes
d Regenbogenerdschlangen
f Serpents arc-en-ciel
i Serpenti ragio di sole

8792 ***Xenopeltis hainanensis***
Squamata - Xenopeltidae
e Chinese Sunbeam Snake

8793 ***Xenopeltis unicolor***
Squamata - Xenopeltidae
e Sunbeam Snake; Asian Sunbeam Snake; Iridescent Earth Snake
d Regenbogenerdschlange; Sonnenglanzschlange; Regenbogenschlange; Flachkopfschlange
f Serpent arc-en-ciel
i Xenopeltide unicolore

8794 ***Xenopholis***
Squamata - Colubridae
e Scaled Ground Snakes; Flat-headed Snakes

8795 ***Xenopholis scalaris***
Squamata - Colubridae

e Wucherer's Ground Snake; Flat-headed Snake

8796 ***Xenopholis undulatus***
Squamata - Colubridae
e Jensen's Ground Snake

8797 **Xenosauridae**
Squamata
e Knob-scaled Lizards; Guest Lizards; Strange Lizards
d Höckerechsen
f Xénosauridés

8798 ***Xenosaurus***
Squamata - Xenosauridae
e American Knob-scaled Lizard; Xenosaures
d Eigentliche Höckerechsen

8799 ***Xenosaurus arboreus***
Squamata - Xenosauridae
e King's Knob-scaled Lizard

8800 ***Xenosaurus grandis***
Squamata - Xenosauridae
e Knob-scaled Lizard
d Gemeine Höckerechse

8801 ***Xenosaurus newmanorum***
Squamata - Xenosauridae
e Newman's Knob-scaled Lizard

8802 ***Xenosaurus platyceps***
Squamata - Xenosauridae
e Flathead Knob-scaled Lizard

8803 ***Xenosaurus rackhami***
Squamata - Xenosauridae
e Rackham's Knob-scaled Lizard
d Rackhams Höckerechse

8804 ***Xenosaurus rectocollaris***
Squamata - Xenosauridae
e Pallid Knob-scaled Lizard

8805 ***Xylophis***
Squamata - Colubridae
e Mountain Snakes

8806 ***Xylophis perroteti***
Squamata - Colubridae
e Perrotet's Mountain Snake

8807 ***Xylophis stenorhynchus***
Squamata - Colubridae
e Günther's Mountain Snake

Z

8808 ***Zaocys***
Squamata - Colubridae
e Keeled Rat Snakes; Rat Snakes
d Orientalische Rattennattern; Chinesische Rattennattern
f Coryphodons

8809 ***Zaocys carinatus***
Squamata - Colubridae
e Keeled Rat Snake; Carinated Rat Snake
d Gekielte Rattennatter
f Coryphodon cariné

8810 ***Zaocys dhumnades***
Squamata - Colubridae
e Cantor's Rat Snake

8811 ***Zaocys fuscus***
Squamata - Colubridae
e Whitebelly Rat Snake; White-bellied Rat Snake

8812 ***Zaocys nigromarginatus***
Squamata - Colubridae
e Black-bordered Rat Snake; Blyth's Rat Snake

8813 ***Zonosaurus***
Squamata - Cordylidae
e Girdled Lizards
d Ringelschildechsen
f Zonosaures

8814 ***Zonosaurus aeneus***
Squamata - Cordylidae
e Bronze Girdled Lizard
d Erzringelschildechse

8815 ***Zonosaurus boettgeri***
Squamata - Cordylidae
e Boettger's Girdled Lizard

8816 ***Zonosaurus brygooi***
Squamata - Cordylidae
e Brygoo's Girdled Lizard

8817 ***Zonosaurus karsteni***
Squamata - Cordylidae
e Karsten's Girdled Lizard
d Karsten-Ringelschildechse

8818 ***Zonosaurus laticaudatus***
Squamata - Cordylidae
e Western Girdled Lizard; Girdled Lizard
d Breitschwanzringelschildechse; Bunte Felsenechse

8819 ***Zonosaurus madagascariensis***
Squamata - Cordylidae
e Madagascar Girdled Lizard; Girdled Lizard
d Madagassische Gürtelschildechse; Madagassische Ringelschildechse

8820 ***Zonosaurus maximus***
Squamata - Cordylidae
e Southeastern Girdled Lizard
d Große Ringelschildechse

8821 ***Zonosaurus ornatus***
Squamata - Cordylidae
e Ornate Girdled Lizard
d Schmuckringelschildechse

8822 ***Zonosaurus quadrilineatus***
Squamata - Cordylidae
e Four-lined Girdled Lizard
d Vierstreifenringelschildechse

8823 ***Zonosaurus rufipes***
Squamata - Cordylidae
e Red-legged Girdled Lizard

8824 ***Zonosaurus trilineatus***
Squamata - Cordylidae
e Three-lined Girdled Lizard

8825 ***Zygaspis***
Squamata - Amphisbaenidae
e Kalahari Round-snouted Worm Lizards; Purple Round-headed Worm Lizards

8826 ***Zygaspis quadrifrons***
Squamata - Amphisbaenidae
e Kalahari Round-snouted Worm Lizard; Kalahari Round-headed Worm Lizard; Round-snouted Amphisbaenian; Kalahari Round-snouted Amphisbaenian (CSA); Eastern Worm Lizard (CSA)

English Index

E

Deutsches Register

D

Index Français

F

Indice Italiano